Differential Equations: Concepts and Applications

Differential Equations: Concepts and Applications

Editor: Patrick McCann

NY RESEARCH PRESS
New York

Published by NY Research Press
118-35 Queens Blvd., Suite 400,
Forest Hills, NY 11375, USA
www.nyresearchpress.com

Differential Equations: Concepts and Applications
Edited by Patrick McCann

International Standard Book Number: 978-1-63238-762-2 (Hardback)

Cataloging-in-Publication Data

Differential equations : concepts and applications / edited by Patrick McCann.
p. cm.
Includes bibliographical references and index.
ISBN 978-1-63238-762-2
1. Differential equations. 2. Calculus. 3. Bessel functions. I. McCann, Patrick.
QA371 .D54 2020
515.35--dc23

Contents

Preface

The purpose of the book is to provide a glimpse into the dynamics and to present opinions and studies of some of the scientists engaged in the development of new ideas in the field from very different standpoints. This book will prove useful to students and researchers owing to its high content quality.

Differential Equation is a mathematical equation which denotes the relationship between a function and its derivatives. The functions usually represent physical quantities in the applications and the derivatives are used to show the rate of change. The study of differential equations involves pure and applied mathematics, physics and engineering. There are various types of differential equations such as ordinary differential equations, partial differential equations, linear differential equations and non - linear differential equations. Ordinary differential equation is an equation that contains unknown function of one real or complex variable. Partial differential equations are the differential equations that involve unknown multivariable functions and their partial derivatives. Linear differential equations are linear in the unknown function and its derivatives, whereas non - linear differential equations are formed by the products of the unknown function. This book unravels the recent studies in the field of differential equations. It elucidates new concepts and their applications in a multidisciplinary manner. It will serve as a valuable source of reference for graduate and post graduate students.

At the end, I would like to appreciate all the efforts made by the authors in completing their chapters professionally. I express my deepest gratitude to all of them for contributing to this book by sharing their valuable works. A special thanks to my family and friends for their constant support in this journey.

Editor

1

A Fractional Order Model for Viral Infection with Cure of Infected Cells and Humoral Immunity

Adnane Boukhouima,[1] Khalid Hattaf ,[1,2] and Noura Yousfi [1]

[1] *Laboratory of Analysis, Modeling and Simulation (LAMS), Faculty of Sciences Ben M'sik, Hassan II University, P.O Box 7955 Sidi Othman, Casablanca, Morocco*
[2] *Centre Régional des Métiers de l'Education et de la Formation (CRMEF), 20340 Derb Ghalef, Casablanca, Morocco*

Correspondence should be addressed to Khalid Hattaf; k.hattaf@yahoo.fr

Guest Editor: Nurcan B. Savaşaneril

In this paper, we study the dynamics of a viral infection model formulated by five fractional differential equations (FDEs) to describe the interactions between host cells, virus, and humoral immunity presented by antibodies. The infection transmission process is modeled by Hattaf-Yousfi functional response which covers several forms of incidence rate existing in the literature. We first show that the model is mathematically and biologically well-posed. By constructing suitable Lyapunov functionals, the global stability of equilibria is established and characterized by two threshold parameters. Finally, some numerical simulations are presented to illustrate our theoretical analysis.

1. Introduction

The immune response plays an important role to control the dynamics of viral infections such as human immunodeficiency virus (HIV), hepatitis B virus (HBV), hepatitis C virus (HCV), and human T-cell leukemia virus (HTLV). Therefore, many mathematical models have been developed to incorporate the role of immune response in viral infections. Some of these models considered the cellular immune response mediated by cytotoxic T lymphocytes (CTL) cells that attack and kill the infected cells [1–5] and the others considered the humoral immune response based on the antibodies which are produced by the B-cells and are programmed to neutralize the viruses [6–11]. However, all these models have been formulated by using ordinary differential equations (ODEs) in which the memory effect is neglected while the immune response involves memory [12, 13].

Fractional derivative is a generalization of integer derivative and it is a suitable tool to model real phenomena with memory which exists in most biological systems [14–16]. The fractional derivative is a nonlocal operator in contrast to integer derivative. This means that if we want to compute the fractional derivative at some point $t = t_1$, it is necessary to take into account the entire history from the starting point $t = t_0$ up to the point $t = t_1$. For these reasons, modeling some real process by using fractional derivative has drawn attention of several authors in various fields [17–22]. In biology, it has been shown that the fractional derivative is useful to analyse the rheological proprieties of cells [23]. Furthermore, it has been deduced that the membranes of cells of biological organism have fractional order electrical conductance [24]. Recently, much works have been done on modeling the dynamics of viral infections with FDEs [25–31]. These works ignored the impact of the immune response and the majority of them deal only with the local stability.

In some viral infections, the humoral immune response is more effective than cellular immune response [32]. For this reason, we improve the above ODE and FDE models by proposing a new fractional order model that describes the interactions between susceptible host cells, viral particles, and the humoral immune response mediated by the antibodies; that is,

$$D^{\alpha}x(t) = \lambda - dx - f(x, v)v + \rho l,$$

$$D^{\alpha}l(t) = f(x, v)v - (m + \rho + \gamma)l,$$

$$D^{\alpha}y(t) = \gamma l - ay,$$
$$D^{\alpha}v(t) = ky - \mu v - qvw,$$
$$D^{\alpha}w(t) = gvw - hw, \tag{1}$$

where $x(t)$, $l(t)$, $y(t)$, $v(t)$, and $w(t)$ are the concentrations of susceptible host cells, latently infected cells (infected cells which are not yet able to produce virions), productive infected cells, free virus particles, and antibodies at time t, respectively. Susceptible host cells are assumed to be produced at a constant rate λ, die at the rate dx, and become infected by virus at the rate $f(x, v)v$. Latently infected cells die at the rate ml and return to the uninfected state by loss of all covalently closed circular DNA (cccDNA) from their nucleus at the rate ρl. Productive infected cells are produced from latently infected cells at the rate γl and die at the rate ay. Free virus particles are produced from productive infected cells at the rate ky, cleared at the rate μv, and are neutralized by antibodies at the rate qvw. Antibodies are activated against virus at the rate gvw and die at the rate hw.

In system (1), D^{α} represents the Caputo fractional derivative of order α defined for an arbitrary function φ by

$$D^{\alpha}\varphi(t) = \frac{1}{\Gamma(1-\alpha)}\int_0^t \frac{\varphi'(u)}{(t-u)^{\alpha}}du, \tag{2}$$

with $0 < \alpha \leq 1$ [33]. Further, the infection transmission process in (1) is modeled by Hattaf-Yousfi functional response [34] which was recently used in [35, 36] and has the form $f(x, v) = \beta x/(\alpha_0 + \alpha_1 x + \alpha_2 v + \alpha_3 xv)$, where $\alpha_0, \alpha_1, \alpha_2, \alpha_3 \geq 0$ are the saturation factors measuring the psychological or inhibitory effect and $\beta > 0$ is the infection rate. In addition, this functional response generalizes many common types existing in the literature such as the specific functional response proposed by Hattaf et al. in [37] and used in [2, 31] when $\alpha_0 = 1$; the Crowley-Martin functional response introduced in [38] and used in [39] when $\alpha_0 = 1$ and $\alpha_3 = \alpha_1\alpha_2$; and the Beddington-DeAngelis functional response proposed in [40, 41] and used in [3, 4, 10] when $\alpha_0 = 1$ and $\alpha_3 = 0$. Also, the Hattaf-Yousfi functional response is reduced to the saturated incidence rate used in [9] when $\alpha_0 = 1$ and $\alpha_1 = \alpha_3 = 0$ and the standard incidence function used in [27] when $\alpha_0 = \alpha_3 = 0$ and $\alpha_1 = \alpha_2 = 1$, and it was simplified to the bilinear incidence rate used in [5, 6] when $\alpha_0 = 1$ and $\alpha_1 = \alpha_2 = \alpha_3 = 0$.

On the other hand, system (1) becomes a model with ODEs when $\alpha = 1$, which improves and generalizes the ODE model with bilinear incidence rate [42], the ODE model with saturated incidence rate [43], and the ODE model with specific functional response [44].

The rest of the paper is organized as follows. The next section deals with some basic proprieties of the solutions and the existence of equilibria. The global stability of equilibria is established in Section 3. To verify our theoretical results, we provide some numerical simulations in Section 4, and we conclude in Section 5.

2. Basic Properties and Equilibria

In this section, we will show that our model is well-posed and we discuss the existence of equilibria.

Since system (1) describes the evolution of cells, then we need to prove that the cell numbers should remain nonnegative and bounded. For biological considerations, we assume that the initial conditions of (1) satisfy

$$x(0) \geq 0,$$
$$l(0) \geq 0,$$
$$y(0) \geq 0, \tag{3}$$
$$v(0) \geq 0,$$
$$w(0) \geq 0.$$

Then we have the following result.

Theorem 1. *Assume that the initial conditions satisfy (3). Then there exists a unique solution of system (1) defined on $[0, +\infty)$. Moreover, this solution remains nonnegative and bounded for all $t \geq 0$.*

Proof. First, system (1) can be written as follows:

$$D^{\alpha}X(t) = F(X), \tag{4}$$

where

$$X(t) = \begin{pmatrix} x(t) \\ l(t) \\ y(t) \\ v(t) \\ w(t) \end{pmatrix}$$
$$\text{and } F(X) = \begin{pmatrix} \lambda - dx - f(x,v)v + \rho l \\ f(x,v)v - (m+\rho+\gamma)l \\ \gamma l - ay \\ ky - \mu v - qvw \\ gvw - hw \end{pmatrix}. \tag{5}$$

It is important to note that when $\alpha = 1$, (4) becomes a system with ODEs. In this case, we refer the reader to [45] for the existence of solutions and to the works [46–50] for the stability of equilibria. In the case of FDEs, we will use Lemma

2.4 in [31] to prove the existence and uniqueness of solutions. Hence, we put

$$\zeta = \begin{pmatrix} \lambda \\ 0 \\ 0 \\ 0 \\ 0 \end{pmatrix},$$

$$A = \begin{pmatrix} -d & \rho & 0 & 0 & 0 \\ 0 & -(m+\rho+\gamma) & 0 & 0 & 0 \\ 0 & \gamma & -a & 0 & 0 \\ 0 & 0 & k & -\mu & 0 \\ 0 & 0 & 0 & 0 & -h \end{pmatrix} \tag{6}$$

and $C = \begin{pmatrix} 0 & 0 & 0 & 0 & 0 \\ 0 & 0 & 0 & 0 & 0 \\ 0 & 0 & 0 & 0 & 0 \\ 0 & 0 & 0 & 0 & -q \\ 0 & 0 & 0 & 0 & g \end{pmatrix}$.

We discuss four cases:

(i) If $\alpha_0 \neq 0$, $F(X)$ can be formulated as follows:

$$F(X) = \zeta + AX + \frac{\alpha_0}{\alpha_0 + \alpha_1 x + \alpha_2 v + \alpha_3 x v} v B_0 X + vCX, \tag{7}$$

where

$$B_0 = \begin{pmatrix} -\frac{\beta}{\alpha_0} & 0 & 0 & 0 & 0 \\ \frac{\beta}{\alpha_0} & 0 & 0 & 0 & 0 \\ 0 & 0 & 0 & 0 & 0 \\ 0 & 0 & 0 & 0 & 0 \\ 0 & 0 & 0 & 0 & 0 \end{pmatrix}. \tag{8}$$

Hence,

$$\|F(X)\| \leq \|\zeta\| + (\|A\| + \|v\| (\|B_0\| + \|C\|)) \|X\|. \tag{9}$$

(ii) If $\alpha_1 \neq 0$, we can write $F(X)$ in the form

$$F(X) = \zeta + AX + \frac{\alpha_1 x}{\alpha_0 + \alpha_1 x + \alpha_2 v + \alpha_3 x v} B_1 X + vCX, \tag{10}$$

where

$$B_1 = \begin{pmatrix} 0 & 0 & 0 & -\frac{\beta}{\alpha_1} & 0 \\ 0 & 0 & 0 & \frac{\beta}{\alpha_1} & 0 \\ 0 & 0 & 0 & 0 & 0 \\ 0 & 0 & 0 & 0 & 0 \\ 0 & 0 & 0 & 0 & 0 \end{pmatrix}. \tag{11}$$

Moreover, we get

$$\|F(X)\| \leq \|\zeta\| + (\|A\| + \|B_1\| + \|v\| \|C\|) \|X\|. \tag{12}$$

(iii) If $\alpha_2 \neq 0$, we have

$$F(X) = \zeta + AX + \frac{\alpha_2 v}{\alpha_0 + \alpha_1 x + \alpha_2 v + \alpha_3 x v} B_2 X + vCX, \tag{13}$$

where

$$B_2 = \begin{pmatrix} -\frac{\beta}{\alpha_2} & 0 & 0 & 0 & 0 \\ \frac{\beta}{\alpha_2} & 0 & 0 & 0 & 0 \\ 0 & 0 & 0 & 0 & 0 \\ 0 & 0 & 0 & 0 & 0 \\ 0 & 0 & 0 & 0 & 0 \end{pmatrix}. \tag{14}$$

Further, we obtain

$$\|F(X)\| \leq \|\zeta\| + (\|A\| + \|B_2\| + \|v\| \|C\|) \|X\|. \tag{15}$$

(iv) If $\alpha_3 \neq 0$, we have

$$F(X) = \zeta + AX + \frac{\alpha_3 x v}{\alpha_0 + \alpha_1 x + \alpha_2 v + \alpha_3 x v} B_3 + vCX, \tag{16}$$

where

$$B_3 = \begin{pmatrix} -\frac{\beta}{\alpha_3} \\ \frac{\beta}{\alpha_3} \\ 0 \\ 0 \\ 0 \end{pmatrix}. \tag{17}$$

Then

$$\|F(X)\| \leq (\|\zeta\| + \|B_3\|) + (\|A\| + \|v\| \|C\|) \|X\|. \tag{18}$$

Hence, the conditions of Lemma 2.4 in [31] are verified. Then system (1) has a unique solution on $[0, +\infty)$. Now, we show the nonnegativity of solutions. By (1), we have

$$\begin{aligned} D^{\alpha} x(t)|_{x=0} &= \lambda + \rho l \geq 0, \\ D^{\alpha} l(t)|_{l=0} &= f(x, v) v \geq 0, \\ D^{\alpha} y(t)|_{y=0} &= \gamma l \geq 0, \\ D^{\alpha} v(t)|_{v=0} &= k y \geq 0, \\ D^{\alpha} w(t)|_{w=0} &= 0 \geq 0. \end{aligned} \tag{19}$$

As in [31, Theorem 2.7], we deduce that the solution of (1) is nonnegative.

Finally, we prove the boundedness of solutions. We define the function

$$T(t) = x(t) + l(t) + y(t) + \frac{a}{2k} v(t) + \frac{aq}{2kg} w(t). \tag{20}$$

Then, we have

$$\begin{aligned} D^{\alpha} T(t) &= D^{\alpha} x(t) + D^{\alpha} l(t) + D^{\alpha} y(t) + \frac{a}{2k} D^{\alpha} v(t) \\ &\quad + \frac{aq}{2kg} D^{\alpha} w(t) \\ &= \lambda - d x(t) - m l(t) - \frac{a}{2} y(t) - \frac{a\mu}{2k} v(t) \\ &\quad - \frac{aqh}{2kg} w(t) \leq \lambda - \delta T(t), \end{aligned} \tag{21}$$

where $\delta = \min\{d, m, a/2, \mu, h\}$. Thus, we obtain

$$T(t) \leq T(0) E_{\alpha}(-\delta t^{\alpha}) + \frac{\lambda}{\delta} [1 - E_{\alpha}(-\delta t^{\alpha})]. \tag{22}$$

Since $0 \leq E_{\alpha}(-\delta t^{\alpha}) \leq 1$, we get

$$T(t) \leq T(0) + \frac{\lambda}{\delta}. \tag{23}$$

This completes the proof. □

Now, we discuss the existence of equilibria. It is clear that system (1) has always an infection-free equilibrium $E_0(\lambda/d, 0, 0, 0, 0)$. Then the basic reproduction number of (1) is as follows:

$$R_0 = \frac{k\beta\lambda\gamma}{a\mu(m + \rho + \gamma)(d\alpha_0 + \lambda\alpha_1)}. \tag{24}$$

To find the other equilibria, we solve the following system:

$$\lambda - dx - f(x, v) v + \rho l = 0, \tag{25}$$

$$f(x, v) v - (m + \rho + \gamma) l = 0, \tag{26}$$

$$\gamma l - a y = 0, \tag{27}$$

$$k y - \mu v - q v w = 0, \tag{28}$$

$$g v w - h w = 0. \tag{29}$$

From (29), we get $w = 0$ or $v = h/g$. Then we discuss two cases.

If $w = 0$, by (25)-(28), we have $l = (\lambda - dx)/(m + \gamma)$, $y = \gamma(\lambda - dx)/a(m + \gamma)$, $v = k\gamma(\lambda - dx)/a\mu(m + \gamma)$, and

$$f\left(x, \frac{k\gamma(\lambda - dx)}{a\mu(m + \gamma)}\right) = \frac{a\mu(m + \rho + \gamma)}{k\gamma}. \tag{30}$$

Since $l \geq 0$, $y \geq 0$, and $v \geq 0$, then $x \leq \lambda/d$. Consequently, there is no equilibrium when $x > \lambda/d$.

We define the function h_1 on $[0, \lambda/d]$ by

$$h_1(x) = f\left(x, \frac{k\gamma(\lambda - dx)}{a\mu(m + \gamma)}\right) - \frac{a\mu(m + \rho + \gamma)}{k\gamma}. \tag{31}$$

We have $h_1(0) = -a\mu(m + \rho + \gamma)/k\gamma < 0$, $h_1'(x) = \partial f/\partial x - (k\gamma d/a\mu(m + \gamma))(\partial f/\partial v) > 0$, and $h_1(\lambda/d) = (a\mu(m + \rho + \gamma)/k\gamma)(R_0 - 1)$.

Hence if $R_0 > 1$, (30) has a unique root $x_1 \in (0, \lambda/d)$. As a result, when $R_0 > 1$ there exists an equilibrium $E_1(x_1, l_1, y_1, v_1, 0)$ satisfying $x_1 \in (0, \lambda/d)$, $l_1 = (\lambda - dx_1)/(m + \gamma)$, $y_1 = \gamma(\lambda - dx_1)/a(m + \gamma)$, and $v_1 = k\gamma(\lambda - dx_1)/a\mu(m + \gamma)$.

If $w \neq 0$, then $v = h/g$. By (25)-(27), we obtain $l = (\lambda - dx)/(m + \gamma)$, $y = \gamma(\lambda - dx)/a(m + \gamma)$, $w = k\gamma g(\lambda - dx)/aqh(m + \gamma) - \mu/q$, and

$$f\left(x, \frac{h}{g}\right) = \frac{g(m + \rho + \gamma)}{h(m + \gamma)}(\lambda - dx). \tag{32}$$

Since $l \geq 0$, $y \geq 0$, and $w \geq 0$, we have $x \leq \lambda/d - ah\mu(m + \gamma)/dkg\gamma$. Hence, there is no equilibrium if $x > \lambda/d - ah\mu(m + \gamma)/dkg\gamma$.

We define the function h_2 on $[0, \lambda/d - ah\mu(m + \gamma)/dkg\gamma]$ by

$$h_2(x) = f\left(x, \frac{h}{g}\right) - \frac{g(m + \rho + \gamma)}{h(m + \gamma)}(\lambda - dx). \tag{33}$$

We have $h_2(0) = -g\lambda(m + \rho + \gamma)/h(m + \gamma) < 0$, $h_2'(x) = \partial f/\partial x + gd(m + \rho + \gamma)/h(m + \gamma) > 0$, and $h_2(\lambda/d - ah\mu(m + \gamma)/dkg\gamma) = h_1(\lambda/d - ah\mu(m + \gamma)/dkg\gamma)$.

Let us introduce the reproduction number for humoral immunity as follows:

$$R_1 = \frac{g v_1}{h}, \tag{34}$$

which $1/h$ denotes the average life expectancy of antibodies and v_1 is the number of free viruses at E_1. For the biological significance, R_1 represents the average number of the antibodies activated by virus.

If $R_1 < 1$, we have $x_1 > \lambda/d - ah\mu(m + \gamma)/dkg\gamma$ and

$$h_2\left(\frac{\lambda}{d} - \frac{ah\mu(m + \gamma)}{dkg\gamma}\right) < h_1(x_1) = 0. \tag{35}$$

Therefore, there is no equilibrium when $R_1 < 1$.

If $R_1 > 1$, then $x_1 < \lambda/d - ah\mu(m + \gamma)/dkg\gamma$ and

$$h_2\left(\frac{\lambda}{d} - \frac{ah\mu(m + \gamma)}{dkg\gamma}\right) > h_1(x_1) = 0. \tag{36}$$

In this case, (32) has one root $x_2 \in (0, \lambda/d - ah\mu(m+\gamma)/dkg\gamma)$. Consequently, when $R_1 > 1$, there exists an equilibrium $E_2(x_2, l_2, y_2, v_2, w_2)$ satisfying $x_2 \in (0, \lambda/d - ah\mu(m+\gamma)/dkg\gamma)$, $l_2 = (\lambda - dx_2)/(m+\gamma)$, $y_2 = \gamma(\lambda - dx_2)/a(m+\gamma)$, $v_2 = h/g$, and $w_2 = k\gamma g(\lambda - dx_2)/aqh(m+\gamma) - \mu/q$. When $R_1 = 1$, $E_1 = E_2$.

We summarize the above discussions in the following theorem.

Theorem 2.

(i) *If $R_0 \le 1$, then system (1) has one infection-free equilibrium of the form $E_0(x_0, 0, 0, 0, 0)$, where $x_0 = \lambda/d$.*

(ii) *If $R_0 > 1$, then system (1) has an infection equilibrium without humoral immunity of the form $E_1(x_1, l_1, y_1, v_1, 0)$, where $x_1 \in (0, \lambda/d)$, $l_1 = (\lambda - dx_1)/(m+\gamma)$, $y_1 = \gamma(\lambda - dx_1)/a(m+\gamma)$, and $v_1 = k\gamma(\lambda - dx_1)/a\mu(m+\gamma)$.*

(iii) *If $R_1 > 1$, then system (1) has an infection equilibrium with humoral immunity of the form $E_2(x_2, l_2, y_2, v_2, w_2)$, where $x_2 \in (0, \lambda/d - ah\mu(m+\gamma)/dkg\gamma)$, $l_2 = (\lambda - dx_2)/(m+\gamma)$, $y_2 = \gamma(\lambda - dx_2)/a(m+\gamma)$, $v_2 = h/g$, and $w_2 = k\gamma g(\lambda - dx_1)/aqh(m+\gamma) - \mu/q$.*

3. Global Stability of Equilibria

In this section, we focus on the global stability of equilibria.

Theorem 3. *If $R_0 \le 1$, then the infection-free equilibrium E_0 is globally asymptotically stable and it becomes unstable if $R_0 > 1$.*

Proof. The proof of the first part of this theorem is based on the construction of a suitable Lyapunov functional that satisfies the conditions given in [51, Lemma 4.6]. Hence, we define a Lyapunov functional as follows:

$$\begin{aligned} L_0(t) &= \frac{\alpha_0}{\alpha_0 + \alpha_1 x_0} x_0 \Phi\left(\frac{x}{x_0}\right) + \frac{\rho\alpha_0}{2(d+m+\gamma)(\alpha_0+\alpha_1 x_0)x_0}(x - x_0 + l)^2 \\ &\quad + l + \frac{m+\rho+\gamma}{\gamma} y + \frac{a(m+\rho+\gamma)}{k\gamma} v + \frac{aq(m+\rho+\gamma)}{kg\gamma} w, \end{aligned} \tag{37}$$

where $\Phi(x) = x - 1 - \ln(x)$ for $x > 0$. It is not hard to show that the functional L_0 is nonnegative. In fact, the function Φ has a global minimum at $x = 1$. Consequently, $\Phi(x) \ge 0$ for all $x > 0$.

Calculating the fractional derivative of $L_0(t)$ along solutions of system (1) and using the results in [52], we get

$$\begin{aligned} D^\alpha L_0(t) &\le \frac{\alpha_0}{\alpha_0 + \alpha_1 x_0}\left(1 - \frac{x_0}{x}\right) D^\alpha x + \frac{\rho\alpha_0}{(d+m+\gamma)(\alpha_0+\alpha_1 x_0)x_0}(x - x_0 + l) \\ &\quad \cdot (D^\alpha x + D^\alpha l) + D^\alpha l + \frac{m+\rho+\gamma}{\gamma} D^\alpha y \\ &\quad + \frac{a(m+\rho+\gamma)}{k\gamma} D^\alpha v + \frac{aq(m+\rho+\gamma)}{kg\gamma} D^\alpha w. \end{aligned} \tag{38}$$

Using $\lambda = dx_0$, we obtain

$$\begin{aligned} D^\alpha L_0(t) &\le -\frac{d\alpha_0 (x - x_0)^2}{(\alpha_0 + \alpha_1 x_0)x} - \frac{\alpha_0}{\alpha_0+\alpha_1 x_0}\left(1 - \frac{x_0}{x}\right) \\ &\quad \cdot f(x,v)v + \frac{\rho\alpha_0}{\alpha_0+\alpha_1 x_0}\left(1 - \frac{x_0}{x}\right) l \\ &\quad \cdot \frac{\rho\alpha_0 (x - x_0 + l)}{(d+m+\gamma)(\alpha_0+\alpha_1 x_0)x_0} (d(x_0 - x) \\ &\quad - (m+\gamma) l) + f(x,v)v - \frac{a\mu(m+\rho+\gamma)}{k\gamma} v \\ &\quad - \frac{aqh(m+\rho+\gamma)}{kg\gamma} w \\ &\le -\left(\frac{1}{x} + \frac{\rho}{(d+m+\gamma)x_0}\right)\frac{d\alpha_0 (x-x_0)^2}{(\alpha_0+\alpha_1 x_0)} \\ &\quad - \frac{\rho\alpha_0 (m+\gamma) l^2}{(d+m+\gamma)(\alpha_0+\alpha_1 x_0)x_0} - \frac{\rho\alpha_0 (x-x_0)^2 l}{(\alpha_0+\alpha_1 x_0)xx_0} \\ &\quad + \frac{a\mu(m+\rho+\gamma)}{k\gamma}(R_0 - 1)v - \frac{aqh(m+\rho+\gamma)}{kg\gamma} w. \end{aligned} \tag{39}$$

Hence if $R_0 \le 1$, then $D^\alpha L_0(t) \le 0$. In addition, the equality holds if and only if $x = x_0$, $l = 0$, $y = 0$, $w = 0$, and $(R_0 - 1)v = 0$. If $R_0 < 1$, then $v = 0$. If $R_0 = 1$, from (1), we get $f(x_0, v)v = 0$ which implies that $v = 0$. Consequently, the largest invariant set of $\{(x, l, y, v, w) \in \mathbb{R}_+^5 : D^\alpha L_0(t) = 0\}$ is the singleton $\{E_0\}$. Therefore, by the LaSalle's invariance principle [51], E_0 is globally asymptotically stable.

The proof of the instability of E_0 is based on the computation of the Jacobean matrix of system (1) and the results presented in [53–55]. The Jacobean matrix of (1) at any equilibrium $E(x, l, y, v, w)$ is given by

$$\begin{pmatrix} -d - \frac{\partial f}{\partial x} v & \rho & 0 & -\frac{\partial f}{\partial v} v - f(x,v) & 0 \\ \frac{\partial f}{\partial x} v & -(m+\rho+\gamma) & 0 & \frac{\partial f}{\partial v} v + f(x,v) & 0 \\ 0 & \gamma & -a & 0 & 0 \\ 0 & 0 & k & -\mu - qw & -qv \\ 0 & 0 & 0 & gw & gv - h \end{pmatrix}. \tag{40}$$

We recall that E is locally asymptotically stable if the all eigenvalues ξ_i of (40) satisfy the following condition [53–55]:

$$|\arg(\xi_i)| > \frac{\alpha\pi}{2}. \quad (41)$$

From (40), the characteristic equation at E_0 is given as follows:

$$(d+\xi)(h+\xi)g_0(\xi) = 0, \quad (42)$$

where

$$g_0(\xi) = ((m+\rho+\gamma)+\xi)(a+\xi)(\mu+\xi) - \frac{k\gamma\beta\lambda}{d\alpha_0+\alpha_1\lambda}. \quad (43)$$

Obviously, (42) has the roots $\xi_1 = -d$ and $\xi_2 = -h$. If $R_0 > 1$, we have $g_0(0) = a\mu(m+\rho+\gamma)(1-R_0) < 0$ and $\lim_{\xi\to+\infty} g_0(\xi) = +\infty$. Then, there exists $\xi^* > 0$ satisfying $g_0(\xi^*) = 0$. In addition, we have $|\arg(\xi^*)| = 0 < \alpha\pi/2$. Consequently, when $R_0 > 1$, E_0 is unstable. □

Theorem 4.

(i) *The infection equilibrium without humoral immunity E_1 is globally asymptotically stable if $R_0 > 1$, $R_1 \le 1$, and*

$$R_0 \le 1 + \frac{(m+\rho+\gamma)[\alpha_0 ad\mu(m+\rho)+dk\lambda\gamma\alpha_2]+k\rho\gamma\alpha_3\lambda^2}{a\rho\mu(m+\rho+\gamma)(\alpha_0 d+\lambda\alpha_1)}. \quad (44)$$

(ii) *When $R_1 > 1$, E_1 is unstable.*

Proof. Define a Lyapunov functional as follows:

$$L_1(t) = \frac{\alpha_0+\alpha_2 v_1}{\alpha_0+\alpha_1x_1+\alpha_2v_1+\alpha_3x_1v_1}x_1\Phi\left(\frac{x}{x_1}\right) + l_1\Phi\left(\frac{l}{l_1}\right) + \frac{\rho(\alpha_0+\alpha_2v_1)}{2(d+m+\gamma)(\alpha_0+\alpha_1x_1+\alpha_2v_1+\alpha_3x_1v_1)x_1}(x - x_1 + l - l_1)^2 + \frac{m+\rho+\gamma}{\gamma}y_1\Phi\left(\frac{y}{y_1}\right) + \frac{a(m+\rho+\gamma)}{k\gamma}v_1\Phi\left(\frac{v}{v_1}\right) + \frac{aq(m+\rho+\gamma)}{kg\gamma}w. \quad (45)$$

Calculating the fractional derivative of $L_1(t)$, we get

$$D^\alpha L_1(t) = \frac{\alpha_0+\alpha_2v_1}{\alpha_0+\alpha_1x_1+\alpha_2v_1+\alpha_3x_1v_1}\left(1-\frac{x_1}{x}\right)D^\alpha x + \left(1-\frac{l_1}{l}\right)D^\alpha l + \frac{\rho(\alpha_0+\alpha_2v_1)(x-x_1+l-l_1)}{(d+m+\gamma)(\alpha_0+\alpha_1x_1+\alpha_2v_1+\alpha_3x_1v_1)x_1}(D^\alpha x + D^\alpha l) + \frac{m+\rho+\gamma}{\gamma}\left(1-\frac{y_1}{y}\right)D^\alpha y + \frac{a(m+\rho+\gamma)}{k\gamma}\left(1-\frac{v_1}{v}\right)D^\alpha v + \frac{aq(m+\rho+\gamma)}{kg\gamma}w. \quad (46)$$

Using $\lambda = dx_1+(m+\gamma)l_1$, $f(x_1,v_1)v_1 = (m+\rho+\gamma)l_1$, $\gamma l_1 = ay_1$, $ky_1 = \mu v_1$, and $1-f(x_i,v_i)/f(x,v_i) = ((\alpha_0+\alpha_2v_i)/(\alpha_0+\alpha_1x_i+\alpha_2v_i+\alpha_3x_iv_i))(1-x_i/x)\forall i \in \{1,2\}$, we obtain

$$\begin{aligned}
D^\alpha L_1(t) \le{}& d\left(1-\frac{f(x_1,v_1)}{f(x,v_1)}\right)(x_1-x) + (m+\rho+\gamma)l_1\left(1-\frac{f(x_1,v_1)}{f(x,v_1)}+\frac{v}{v_1}\frac{f(x,v)}{f(x,v_1)}\right) + (m+\rho+\gamma) \\
&\cdot l_1\left(1-\frac{l_1f(x,v)v}{lf(x_1,v_1)v_1}\right) + (m+\rho+\gamma)l_1\left(1-\frac{ly_1}{l_1y}\right) + (m+\rho+\gamma)l_1\left(1-\frac{v}{v_1}-\frac{yv_1}{y_1v}\right) + \rho(l-l_1) \\
&\cdot\left(1-\frac{f(x_1,v_1)}{f(x,v_1)}\right) - \frac{\rho(\alpha_0+\alpha_2v_1)\left[d(x-x_1)^2+(m+\gamma)(l-l_1)^2+(d+m+\gamma)(x-x_1)(l-l_1)\right]}{(d+m+\gamma)(\alpha_0+\alpha_1x_1+\alpha_2v_1+\alpha_3x_1v_1)x_1} \\
&+ \frac{aqh(m+\rho+\gamma)}{kg\gamma}\left(\frac{gv_1}{h}-1\right)w.
\end{aligned} \quad (47)$$

Hence,

$$\begin{aligned}
D^\alpha L_1(t) \le{}& -\frac{(\alpha_0+\alpha_2v_1)(x-x_1)^2}{xx_1(\alpha_0+\alpha_1x_1+\alpha_2v_1+\alpha_3x_1v_1)}\left((dx_1-\rho l_1)+\rho l + \frac{d\rho x}{d+m+\gamma}\right) \\
&- \frac{\rho(\alpha_0+\alpha_2v_1)(m+\gamma)(l-l_1)^2}{(m+\rho+\gamma)(\alpha_0+\alpha_1x_1+\alpha_2v_1+\alpha_3x_1v_1)x_1} + (m+\rho \\
&+\gamma)l_1\left(5-\frac{f(x_1,v_1)}{f(x,v_1)}-\frac{l_1f(x,v)v}{lf(x_1,v_1)v_1}-\frac{ly_1}{l_1y}-\frac{yv_1}{y_1v}-\frac{f(x,v_1)}{f(x,v)}\right) - (m+\rho+\gamma)l_1
\end{aligned}$$

$$\cdot \frac{(\alpha_0 + \alpha_1 x)(\alpha_2 + \alpha_3 x)(v - v_1)^2}{v_1 (\alpha_0 + \alpha_1 x + \alpha_2 v + \alpha_3 x v)(\alpha_0 + \alpha_1 x + \alpha_2 v_1 + \alpha_3 x v_1)} + \frac{aqh(m+\rho+\gamma)}{kg\gamma}(R_1 - 1) w. \tag{48}$$

Using the arithmetic-geometric inequality, we have

$$5 - \frac{f(x_i, v_i)}{f(x, v_i)} - \frac{l_i f(x, v) v}{l f(x_i, v_i) v_i} - \frac{l y_i}{l_i y} - \frac{y v_i}{y_i v} - \frac{f(x, v_i)}{f(x, v)} \le 0. \tag{49}$$

Since $R_1 \le 1$, we have $D^\alpha L_1(t) \le 0$ if $dx_1 \ge \rho l_1$. It is easy to see that this condition is equivalent to (44). Furthermore, $D^\alpha L_1(t) = 0$ if and only if $x = x_1, l = l_1, y = y_1, v = v_1$, and $(R_1 - 1)w = 0$. We discuss two cases: If $R_1 < 1$, then $w = 0$. If $R_1 = 1$, from (1), we get $D^\alpha v_1 = 0 = k y_1 - \mu v_1 - q v_1 w$, and then $w = 0$. Hence, the largest invariant set of $\{(x, l, y, v, w) \in \mathbb{R}_+^5 : D^\alpha L_1(t) = 0\}$ is the singleton $\{E_1\}$. By the LaSalle's invariance principle, E_1 is globally asymptotically stable.

At E_1, the characteristic equation of (40) is given as follows:

$$(g v_1 - h - \xi) g_1(\xi) = 0, \tag{50}$$

where

$$g_1(\xi) = \begin{vmatrix} -d - \frac{\partial f}{\partial x}(x_1, v_1) v_1 - \xi & \rho & 0 & -\frac{\partial f}{\partial v}(x_1, v_1) v_1 - f(x_1, v_1) \\ \frac{\partial f}{\partial x}(x_1, v_1) v_1 & -(m+\rho+\gamma) - \xi & 0 & \frac{\partial f}{\partial v}(x_1, v_1) v_1 + f(x_1, v_1) \\ 0 & \gamma & -a - \xi & 0 \\ 0 & 0 & k & -\mu - \xi \end{vmatrix}. \tag{51}$$

We can easily see that (50) has the root $\xi_1 = g v_1 - h$. Then, when $R_1 > 1$, we have $\xi_1 > 0$. In this case, E_1 is unstable. □

Theorem 5. *The infection equilibrium with humoral immunity E_2 is globally asymptotically stable if $R_1 > 1$ and*

$$\rho\beta h \le d(m+\rho+\gamma)(\alpha_0 g + \alpha_2 h) + \rho\lambda(\alpha_1 g + \alpha_3 h). \tag{52}$$

Proof. Consider the following Lyapunov functional:

$$L_2(t) = \frac{\alpha_0 + \alpha_2 v_2}{\alpha_0 + \alpha_1 x_2 + \alpha_2 v_2 + \alpha_3 x_2 v_2} x_2 \Phi\left(\frac{x}{x_2}\right) + l_2 \Phi\left(\frac{l}{l_2}\right) + \frac{\rho(\alpha_0 + \alpha_2 v_2)}{2(d+m+\gamma)(\alpha_0 + \alpha_1 x_2 + \alpha_2 v_2 + \alpha_3 x_2 v_2) x_2}(x - x_2 + l - l_2)^2 + \frac{m+\rho+\gamma}{\gamma} y_2 \Phi\left(\frac{y}{y_2}\right) + \frac{a(m+\rho+\gamma)}{k\gamma} v_2 \Phi\left(\frac{v}{v_2}\right) + \frac{aq(m+\rho+\gamma)}{kg\gamma} \cdot w_2 \Phi\left(\frac{w}{w_2}\right). \tag{53}$$

Computing the fractional derivative of $L_2(t)$ and using $\lambda = dx_2 + (m+\gamma) l_2$, $f(x_2, v_2) v_2 = (m+\rho+\gamma) l_2$, $\gamma l_2 = a y_2$, $k y_2 = (\mu + q w_2) v_2$, and $v_2 = h/g$, we get

$$\begin{aligned} D^\alpha L_2(t) \le{}& d\left(1 - \frac{f(x_2, v_2)}{f(x, v_2)}\right)(x_2 - x) + (m+\rho+\gamma) l_2 \left(1 - \frac{f(x_2, v_2)}{f(x, v_2)} + \frac{f(x, v) v}{f(x, v_2) v_2}\right) \\ &+ (m+\rho+\gamma) l_2 \left(1 - \frac{l_2 f(x, v) v}{l f(x_2, v_2) v_2}\right) + (m+\rho+\gamma) l_2 \left(1 - \frac{l y_2}{l_2 y}\right) + (m+\rho+\gamma) l_2 \left(1 - \frac{v}{v_2} - \frac{y v_2}{y_2 v}\right) \\ &+ \rho(l - l_2)\left(1 - \frac{f(x_2, v_2)}{f(x, v_2)}\right) \\ &- \frac{\rho(\alpha_0 + \alpha_2 v_2)\left[d(x - x_2)^2 + (m+\gamma)(l - l_2)^2 + (d+m+\gamma)(x - x_2)(l - l_2)\right]}{(d+m+\gamma)(\alpha_0 + \alpha_1 x_2 + \alpha_2 v_2 + \alpha_3 x_2 v_2) x_2} \end{aligned}$$

$$\leq -\frac{(\alpha_0+\alpha_2 v_2)(x-x_2)^2}{x x_2(\alpha_0+\alpha_1 x_2+\alpha_2 v_2+\alpha_3 x_2 v_2)}\left((dx_2-\rho l_2)+\rho l+\frac{d\rho x}{d+m+\gamma}\right)$$
$$-\frac{\rho(\alpha_0+\alpha_2 v_2)(m+\gamma)(l-l_2)^2}{(m+\rho+\gamma)(\alpha_0+\alpha_1 x_2+\alpha_2 v_2+\alpha_3 x_2 v_2)x_2}$$
$$+(m+\rho+\gamma)l_2\left(5-\frac{f(x_2,v_2)}{f(x,v_2)}-\frac{l_2 f(x,v)v}{l f(x_2,v_2)v_2}-\frac{l y_2}{l_2 y}-\frac{y v_2}{y_2 v}-\frac{f(x,v_2)}{f(x,v)}\right)$$
$$-(m+\rho+\gamma)l_2\frac{(\alpha_0+\alpha_1 x)(\alpha_2+\alpha_3 x)(v-v_2)^2}{v_2(\alpha_0+\alpha_1 x+\alpha_2 v+\alpha_3 x v)(\alpha_0+\alpha_1 x+\alpha_2 v_2+\alpha_3 x v_2)}. \tag{54}$$

From (49), we have $D^\alpha L_2(t)\leq 0$ when $dx_2\geq \rho l_2$. This condition is equivalent to (52). In addition, $D^\alpha L_2(t)=0$ if $x=x_2, l=l_2, y=y_2$, and $v=v_2$. Further, $D^\alpha v_2 = 0 = ky_2-\mu v_2-qv_2 w$; then $w=w_2$. Consequently, the largest invariant set of $\{(x,l,y,v,w)\in\mathbb{R}^5_+ : D^\alpha L_2(t)=0\}$ is the singleton $\{E_2\}$. By the LaSalle's invariance principle, E_2 is globally asymptotically stable. □

It is important to note that when ρ is sufficiently small or γ is sufficiently large, the two conditions (44) and (52) are satisfied. Then, we have the following corollary.

Corollary 6. *Assume that $R_0>1$. When ρ is sufficiently small or γ is sufficiently large, then we have the following:*

(i) *The infection equilibrium without humoral immunity E_1 is globally asymptotically stable if $R_1\leq 1$.*

(ii) *The infection equilibrium with humoral immunity E_2 is globally asymptotically stable if $R_1>1$.*

4. Numerical Simulations

In this section, we validate our theoretical results to HIV infection. Firstly, we take the parameter values as shown in Table 1.

By calculation, we have $R_0=0.4274\leq 1$. Then system (1) has an infection-free equilibrium $E_0(719.4245,0,0,0,0)$. By Theorem 3, the solution of (1) converges to E_0 (see Figure 1). Consequently, the virus is cleared and the infection dies out.

Now, we choose $\beta = 0.0012$ and we keep the other parameter values. Hence, we obtain $R_0=2.137$, $R_1=0.8334$, and

$$1+\frac{(m+\rho+\gamma)\left[\alpha_0 ad\mu(m+\rho)+dk\lambda\gamma\alpha_2\right]+k\rho\gamma\alpha_3\lambda^2}{a\rho\mu(m+\rho+\gamma)(\alpha_0 d+\lambda\alpha_1)} \tag{55}$$
$$=2.5934.$$

Consequently, condition (44) is satisfied. Therefore, the infection equilibrium without humoral immunity $E_1(176.6853, 168.7712, 6.2508, 1666.9, 0)$ is globally asymptotically stable. Figure 2 demonstrates this result. In this case, the infection becomes chronic.

Next, we take $g = 0.0004$ and do not change the other parameter values. In this case, we have $R_1 = 3.3338$, $\rho\beta h = 0.0000024$, and $d(m+\rho+\gamma)(\alpha_0 g+\alpha_2 h)+\rho\lambda(\alpha_1 g+\alpha_3 h) = 0.000006$. Hence, condition (52) is satisfied. Consequently, system (1) has an infection equilibrium with humoral immunity $E_2(423.4261, 92.0442, 3.4090, 500, 245.4473)$ which is globally asymptotically stable. Figure 3 illustrates this result. We can observe that the activation of the humoral immune response increases the healthy cells and decreases the productive infected cells and viral load to a lower levels but it is not able to eradicate the infection.

5. Conclusion

In the present paper, we have studied the dynamics of a viral infection model by taking into account the memory effect represented by the Caputo fractional derivative and the humoral immunity. We have proved that the solutions of the model are nonnegative and bounded which assure the well-posedness. We have shown that the proposed model has three infection equilibriums, namely, the infection-free equilibrium E_0, the infection equilibrium without humoral immunity E_1, and the infection equilibrium with humoral immunity E_2. By constructing suitable Lyapunov functionals, the global stability of these equilibria is fully determined by two threshold parameters R_0 and R_1. More precisely, when $R_0\leq 1$, E_0 is globally asymptotically stable, whereas if $R_0>1$, it becomes unstable and another equilibrium point appears, that is, E_1, which is globally asymptotically stable whenever $R_1\leq 1$ and condition (44) is satisfied. In the case that $R_1>1$, E_1 becomes unstable and there exists another equilibrium point E_2 which is globally asymptotically stable when condition (52) is satisfied. In addition, we remarked that when ρ is sufficiently small or γ is sufficiently large, conditions (44) and (52) are verified, and then the global stability of E_1 and E_2 is characterized only by R_0 and R_1.

From our theoretical and numerical results, we deduce that the order of the fractional derivative α has no effect on the dynamics of the model. However, when the value of α decreases (long memory), the solutions of our model converge rapidly to the steady states (see Figures 1–3). This behavior can be explained by the memory term $1/\Gamma(1-\alpha)(t-u)^\alpha$ included in the fractional derivative which represents

TABLE 1: Parameter values of system (1).

parameters	values	parameters	values	parameters	values
λ	10	a	0.27	h	0.2
d	0.0139	γ	0.01	g	0.0001
β	0.00024	k	800	α_0	1
ρ	0.01	μ	3	α_1	0.1
m	0.0347	q	0.01	α_2	0.01
		α_3	0.00001		

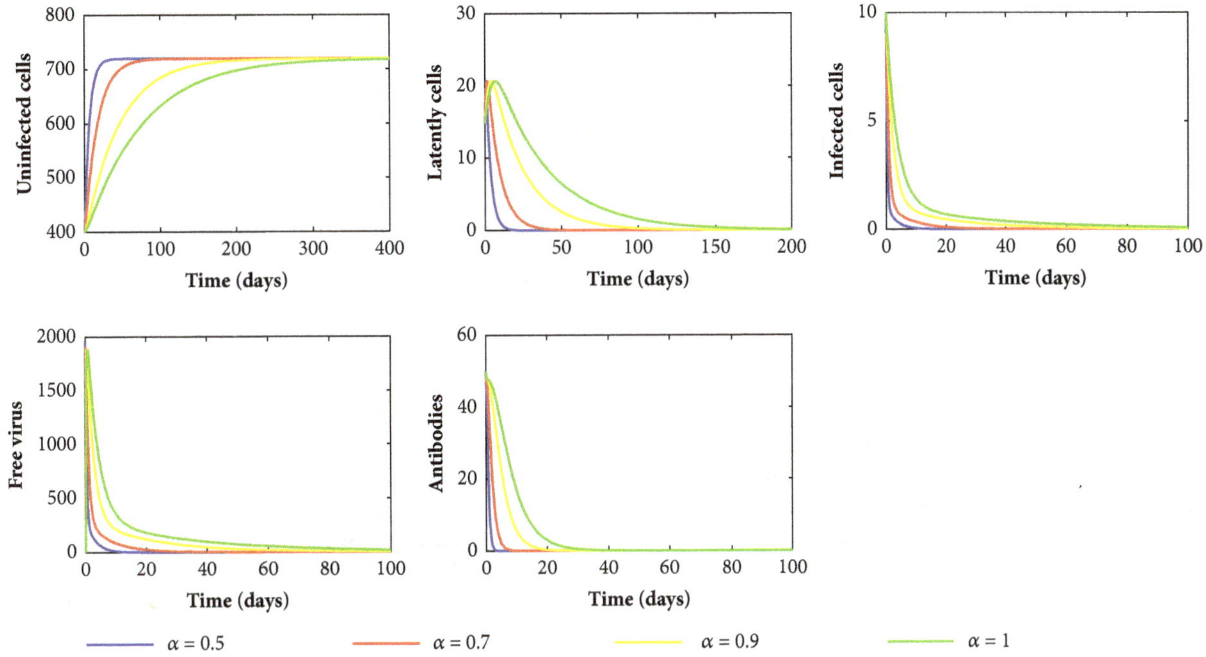

FIGURE 1: Stability of the infection-free equilibrium E_0.

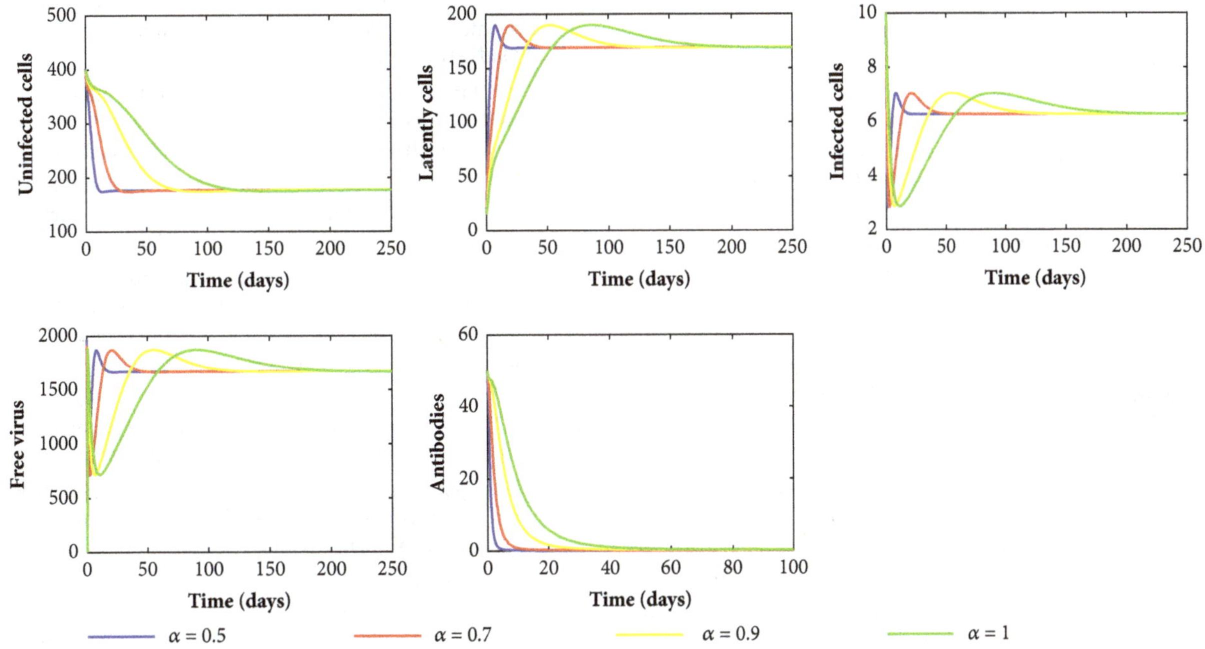

FIGURE 2: Stability of the infection equilibrium without humoral immunity E_1.

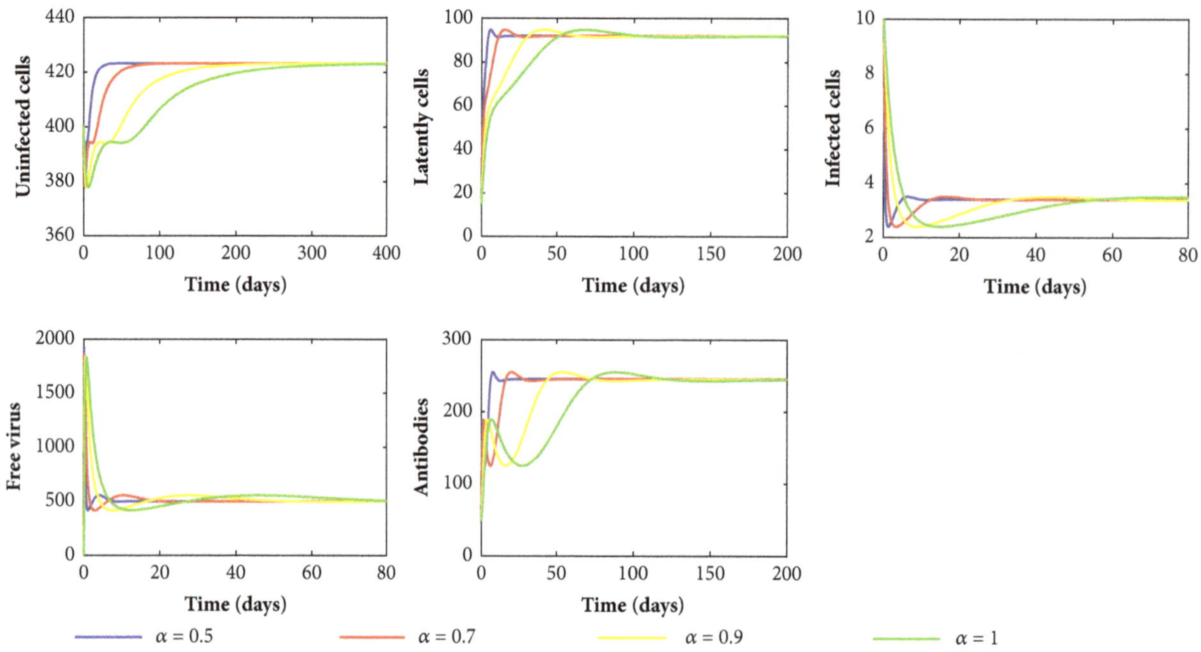

FIGURE 3: Stability of the infection equilibrium with humoral immunity E_2.

the time needed for the interaction between cells and viral particles and the time needed for the activation of humoral immune response. In fact, the knowledge about the infection and the activation of the humoral immune response in an early stage can help us to control the infection.

Conflicts of Interest

The authors declare that they have no conflicts of interest.

References

[1] K. Hattaf, N. Yousfi, and A. Tridane, "Global stability analysis of a generalized virus dynamics model with the immune response," *Canadian Applied Mathematics Quarterly*, vol. 20, no. 4, pp. 499–518, 2012.

[2] M. Maziane, K. Hattaf, and N. Yousfi, "Global stability for a class of HIV infection models with cure of infected cells in eclipse stage and CTL immune response," *International Journal of Dynamics and Control*, 2016.

[3] X. Wang, Y. Tao, and X. Song, "Global stability of a virus dynamics model with Beddington-DeAngelis incidence rate and CTL immune response," *Nonlinear Dynamics*, vol. 66, no. 4, pp. 825–830, 2011.

[4] C. Lv, L. Huang, and Z. Yuan, "Global stability for an HIV-1 infection model with Beddington-DeAngelis incidence rate and CTL immune response," *Communications in Nonlinear Science and Numerical Simulation*, vol. 19, no. 1, pp. 121–127, 2014.

[5] M. A. Nowak and C. R. M. Bangham, "Population dynamics of immune responses to persistent viruses," *Science*, vol. 272, no. 5258, pp. 74–79, 1996.

[6] A. Murase, T. Sasaki, and T. Kajiwara, "Stability analysis of pathogen-immune interaction dynamics," *Journal of Mathematical Biology*, vol. 51, no. 3, pp. 247–267, 2005.

[7] M. A. Obaid and A. M. Elaiw, "Stability of Virus Infection Models with Antibodies and Chronically Infected Cells," *Abstract and Applied Analysis*, vol. 2014, Article ID 650371, 12 pages, 2014.

[8] M. A. Obaid, "Dynamical behaviors of a nonlinear virus infection model with latently infected cells and immune response," *Journal of Computational Analysis and Applications*, vol. 21, no. 1, pp. 182–193, 2016.

[9] H. F. Huo, Y. L. Tang, and L. X. Feng, "A Virus Dynamics Model with Saturation Infection and Humoral Immunity," *Int. Journal of Math. Analysis*, vol. 6, no. 40, 2012.

[10] A. M. Elaiw, "Global stability analysis of humoral immunity virus dynamics model including latently infected cells," *Journal of Biological Dynamics*, vol. 9, no. 1, pp. 215–228, 2015.

[11] A. M. Elaiw and N. H. AlShamrani, "Global properties of nonlinear humoral immunity viral infection models," *International Journal of Biomathematics*, vol. 8, no. 5, 1550058, 53 pages, 2015.

[12] J. X. Velasco-Herna'ndez, J. A. Garci'a, and D. E. Kirschner, "Remarks on modeling host-viral dynamics and treatment, Mathematical Approaches for Emerging and Reemerging Infectious Diseases," *An Introduction to Models, Methods, and Theory*, vol. 125, pp. 287–308, 2002.

[13] A. S. Perelson, "Modelling viral and immune system dynamics," *Nature Reviews Immunology*, vol. 2, no. 1, pp. 28–36, 2002.

[14] R. L. Magin, "Fractional calculus models of complex dynamics in biological tissues," *Computers & Mathematics with Applications*, vol. 59, no. 5, pp. 1586–1593, 2010.

[15] A. A. Stanislavsky, "Memory effects and macroscopic manifestation of randomness," *Physical Review E: Statistical Physics, Plasmas, Fluids, and Related Interdisciplinary Topics*, vol. 61, no. 5, pp. 4752–4759, 2000.

[16] M. Saeedian, M. Khalighi, N. Azimi-Tafreshi, G. R. Jafari, and M. Ausloos, "Memory effects on epidemic evolution: The susceptible-infected-recovered epidemic model," *Physical Review E: Statistical, Nonlinear, and Soft Matter Physics*, vol. 95, no. 2, Article ID 022409, 2017.

[17] Y. A. Rossikhin and M. V. Shitikova, "Applications of fractional calculus to dynamic problems of linear and nonlinear hereditary mechanics of solids," *Applied Mechanics Reviews*, vol. 50, no. 1, pp. 15–67, 1997.

[18] R. J. Marks and M. W. Hall, "Differintegral Interpolation from a Bandlimited Signal's Samples," *IEEE Transactions on Signal Processing*, vol. 29, no. 4, pp. 872–877, 1981.

[19] G. L. Jia and Y. X. Ming, "Study on the viscoelasticity of cancellous bone based on higher-order fractional models," in *Proceedings of the 2nd International Conference on Bioinformatics and Biomedical Engineering (ICBBE '08)*, pp. 1733–1736, Shanghai, China, May 2008.

[20] R. Magin, "Fractional calculus in bioengineering," *Cretical reviews in biomedical engineering*, vol. 32, pp. 13–77, 2004.

[21] E. Scalas, R. Gorenflo, and F. Mainardi, "Fractional calculus and continuous-time finance," *Physica A: Statistical Mechanics and its Applications*, vol. 284, no. 1–4, pp. 376–384, 2000.

[22] R. Capponetto, G. Dongola, L. Fortuna, and I. Petras, "Fractional order systems: Modelling and control applications," *World Scientific Series in Nonlinear Science, Series A*, vol. 72, 2010.

[23] V. D. Djordjević, J. Jarić, B. Fabry, J. J. Fredberg, and D. Stamenović, "Fractional derivatives embody essential features of cell rheological behavior," *Annals of Biomedical Engineering*, vol. 31, no. 6, pp. 692–699, 2003.

[24] K. S. Cole, "Electric conductance of biological systems," *Cold Spring Harbor Symposium on Quantitative Biology*, vol. 1, pp. 107–116, 1933.

[25] F. A. Rihan, M. Sheek-Hussein, A. Tridane, and R. Yafia, "Dynamics of hepatitis C virus infection: mathematical modeling and parameter estimation," *Mathematical Modelling of Natural Phenomena*, vol. 12, no. 5, pp. 33–47, 2017.

[26] N. Khodabakhshi, S. M. Vaezpour, and D. Baleanu, "On Dynamics Of a Fractional-order Model Of HCV Infection," *Journal of Mathematical Analysis*, vol. 8, no. 1, pp. 16–27, 2017.

[27] X. Zhou and Q. Sun, "Stability analysis of a fractional-order HBV infection model," *International Journal of Advances in Applied Mathematics and Mechanics*, vol. 2, no. 2, pp. 1–6, 2014.

[28] S. S. M. and Y. A. M., "On a fractional-order model for HBV infection with cure of infected cells," *Journal of the Egyptian Mathematical Society*, vol. 25, no. 4, pp. 445–451, 2017.

[29] Y. Ding and H. Ye, "A fractional-order differential equation model of HIV infection of CD4+ T-cells," *Mathematical and Computer Modelling*, vol. 50, no. 3-4, pp. 386–392, 2009.

[30] C. M. Pinto and A. R. Carvalho, "The role of synaptic transmission in a HIV model with memory," *Applied Mathematics and Computation*, vol. 292, pp. 76–95, 2017.

[31] A. Boukhouima, K. Hattaf, and N. Yousfi, "Dynamics of a Fractional Order HIV Infection Model with Specific Functional Response and Cure Rate," *International Journal of Differential Equations*, vol. 2017, Article ID 8372140, 8 pages, 2017.

[32] J. A. Deans and S. Cohen, "Immunology of malaria," *Annual Review of Microbiology*, vol. 37, pp. 25–49, 1983.

[33] I. Podlubny, *Fractional Differential Equations: An Introduction to Fractional Derivatives, Fractional Differential Equations, to Methods of their Solution and Some of their Applications*, vol. 198, Academic press, 1998.

[34] K. Hattaf and N. Yousfi, "A class of delayed viral infection models with general incidence rate and adaptive immune response," *International Journal of Dynamics and Control*, vol. 4, no. 3, pp. 254–265, 2016.

[35] D. Riad, K. Hattaf, and N. Yousfi, "Dynamics of Capital-labour Model with Hattaf-Yousfi Functional Response," *British Journal of Mathematics & Computer Science*, vol. 18, no. 5, pp. 1–7, 2016.

[36] M. Mahrouf, K. Hattaf, and N. Yousfi, "Dynamics of a stochastic viral infection model with immune response," *Mathematical Modelling of Natural Phenomena*, vol. 12, no. 5, pp. 15–32, 2017.

[37] K. Hattaf, N. Yousfi, and A. Tridane, "Stability analysis of a virus dynamics model with general incidence rate and two delays," *Applied Mathematics and Computation*, vol. 221, pp. 514–521, 2013.

[38] P. Crowley and E. Martin, "Functional responses and interference within and between year classes of a dragonfly population," *Journal of the North American Benthological Society*, vol. 8, pp. 211–221, 1989.

[39] S. Xu, "Global stability of the virus dynamics model with Crowley-Martin functional response," *Electronic Journal of Qualitative Theory of Differential Equations*, No. 9, 10 pages, 2012.

[40] J. R. Beddington, "Mutual interference between parasites or predat ors and its effect on searching efficiency," *Journal of Animal Ecology*, vol. 44, pp. 331–340, 1975.

[41] D. L. DeAngelis, R. A. Goldstein, and R. V. ONeill, "A model for trophic interaction," *Ecology*, pp. 881–892, 1975.

[42] L. Rong, M. A. Gilchrist, Z. Feng, and A. S. Perelson, "Modeling within-host HIV-1 dynamics and the evolution of drug resistance: trade-offs between viral enzyme function and drug susceptibility," *Journal of Theoretical Biology*, vol. 247, no. 4, pp. 804–818, 2007.

[43] Z. Hu, W. Pang, F. Liao, and W. Ma, "Analysis of a cd4+ t cell viral infection model with a class of saturated infection rate," *Discrete and Continuous Dynamical Systems - Series B*, vol. 19, no. 3, pp. 735–745, 2014.

[44] M. Maziane, E. M. Lotfi, K. Hattaf, and N. Yousfi, "Dynamics of a Class of HIV Infection Models with Cure of Infected Cells in Eclipse Stage," *Acta Biotheoretica*, vol. 63, no. 4, pp. 363–380, 2015.

[45] J. K. Hale and S. M. Verduyn Lunel, *Introduction to Functional-Differential Equations*, Springer, Berlin, Germany, 1993.

[46] J. K. Hale, "Sufficient conditions for stability and instability of autonomous functional-differential equations," *Journal of Differential Equations*, vol. 1, pp. 452–482, 1965.

[47] T. A. Burton, *Stability and Periodic sSolutions of Ordinary and Functional Differential Equations*, Academic Press, Orlando, Fla, USA, 1985.

[48] B. S. Ogundare, "Stability and boundedness properties of solutions to certain fifth order nonlinear differential equations," *Matematicki Vesnik*, vol. 61, no. 4, pp. 257–268, 2009.

[49] C. Tunç, "A study of the stability and boundedness of the solutions of nonlinear differential equations of the fifth order," *Indian Journal of Pure and Applied Mathematics*, vol. 33, no. 4, pp. 519–529, 2002.

[50] C. Tunç, "New results on the stability and boundedness of nonlinear differential equations of fifth order with multiple deviating arguments," *Bulletin of the Malaysian Mathematical Sciences Society*, vol. 36, no. 3, pp. 671–682, 2013.

[51] J. Huo, H. Zhao, and L. Zhu, "The effect of vaccines on backward bifurcation in a fractional order HIV model," *Nonlinear Analysis: Real World Applications*, vol. 26, pp. 289–305, 2015.

[52] C. Vargas-De-Leon, "Volterra-type Lyapunov functions for fractional-order epidemic systems," *Communications in Nonlinear Science and Numerical Simulation*, vol. 24, no. 1–3, pp. 75–85, 2015.

[53] D. Matignon, "Stability results for fractional differential equations with applications to control processing," in *Computational Eng. in Sys. Appl*, vol. 2, p. 963, Lille, France, 1996.

[54] E. Ahmed, A. M. A. El-Sayed, and H. A. A. El-Saka, "On some Routh-Hurwitz conditions for fractional order differential equations and their applications in Lorenz, Rössler, Chua and Chen systems," *Physics Letters A*, vol. 358, no. 1, pp. 1–4, 2006.

[55] E. Ahmed, A. M. A. El-Sayed, and H. A. A. El-Saka, "Equilibrium points, stability and numerical solutions of fractional-order predator-prey and rabies models," *Journal of Mathematical Analysis and Applications*, vol. 325, no. 1, pp. 542–553, 2007.

2

An Asymptotic-Numerical Hybrid Method for Solving Singularly Perturbed Linear Delay Differential Equations

Süleyman Cengizci

Institute of Applied Mathematics, Middle East Technical University, 06800 Ankara, Turkey

Correspondence should be addressed to Süleyman Cengizci; cengizci.suleyman@metu.edu.tr

Academic Editor: Patricia J. Y. Wong

In this work, approximations to the solutions of singularly perturbed second-order linear delay differential equations are studied. We firstly use two-term Taylor series expansion for the delayed convection term and obtain a singularly perturbed ordinary differential equation (ODE). Later, an efficient and simple asymptotic method so called Successive Complementary Expansion Method (SCEM) is employed to obtain a uniformly valid approximation to this corresponding singularly perturbed ODE. As the final step, we employ a numerical procedure to solve the resulting equations that come from SCEM procedure. In order to show efficiency of this numerical-asymptotic hybrid method, we compare the results with exact solutions if possible; if not we compare with the results that are obtained by other reported methods.

1. Introduction

Almost all physical phenomena in nature are modelled using differential equations, and singularly perturbed problems are vital class of these kinds of problems. In general, a singular perturbation problem is defined as a differential equation that is controlled by a positive small parameter $0 < \varepsilon \ll 1$ that exists as multiplier to the highest derivative term in the differential equation. As ε tends to zero, the solution of problem exhibits interesting behaviors (rapid changes) since the order of the equation reduces. The region where these rapid changes occur is called *inner region* and the region in which the solution changes mildly is called *outer region*. As mentioned in [1, 2], these kinds of problems arise in almost all applied natural sciences. Some of these can be given as mechanical and electrical systems, celestial mechanics, fluid and solid mechanics, electromagnetics, particle and quantum physics, chemical and biochemical reactions, and economics and financial mathematics. Various methods are employed to solve singular perturbation problems analytically, numerically, or asymptotically such as the method of matched asymptotic expansions (MMAE), the method of multiple scales, the method of WKB approximation, Poincaré-Lindstedt method and periodic averaging method. Rigorous analysis and applications of these methods can be found in [3–8].

Modelling automatic systems often involve the idea of control because feedback is necessary in order to maintain a stable state. But much of this feedback require a finite time to sense information and react to it. A general way for describing this process is to formulate a delay differential equation (difference-differential equation). Delay differential equations (DDE) are widely used for modelling problems in population dynamics, nonlinear optics, fluid mechanics, mechanical engineering, evolutionary biology, and even modelling of HIV infection and human pupil-light reflex. One can refer to [10–14] for general theory and applications of DDEs.

In this paper, we study an important class of delay differential equations: singularly perturbed linear delay differential equations. A singularly perturbed delay differential equation is a differential equation in which the highest-order derivative is multiplied by a positive small ε parameter and involving at least one delay term. We restrict our attention to singularly perturbed second-order ordinary delay differential equations that contains the delay in convection term. Various methods have been used to solve singularly perturbed DDEs such as finite difference methods [9, 15, 16], finite element methods

[17, 18], homotopy perturbation method [19, 20], reproducing kernel method [21, 22], spline collocation methods [23, 24], and asymptotic approaches [25, 26]. We use an asymptotic-numerical hybrid method in order to find uniformly valid approximations to singularly perturbed ODEs. At the first step, two-term Taylor series expansion is used to vanish delayed term. Secondly, to obtain a uniformly valid approximation an efficient and easily applicable asymptotic method so called Successive Complementary Expansion Method (SCEM) that was introduced in [27] is employed. Finally, a numerical approach is used to solve resulting equations that come from SCEM process.

2. Description of the Method

In this section, we first give a short overview of asymptotic approximations and then explain Successive Complementary Expansion Method by which we obtain highly accurate approximations to solutions of singularly perturbed linear DDEs.

Consider two continuous functions of real numbers $g(\varepsilon)$ and $h(\varepsilon)$ that depend on a positive small parameter ε; then $g(\varepsilon) = O(h(\varepsilon))$ for $\varepsilon \to 0$ if there exist positive constants K and ε_0 such that $K \in (0, \varepsilon_0]$ with $|g(\varepsilon)| \leqslant K|h(\varepsilon)|$ for $\varepsilon \to 0$, and $g(\varepsilon) = o(h(\varepsilon))$ for $\varepsilon \to 0$ if $\lim_{\varepsilon\to 0}(g(\varepsilon)/h(\varepsilon)) = 0$. Let E be a set of real functions that depend on ε, strictly positive and continuous in $(0, \varepsilon_0]$, such that $\lim_{\varepsilon\to 0}\delta(\varepsilon)$ exists and $\delta_1(\varepsilon)\delta_2(\varepsilon) \in E$ for each $\delta_1(\varepsilon)$, $\delta_2(\varepsilon) \in E$. A function $\delta_i(\varepsilon)$ that satisfies these conditions is called *order function*. Given two functions $\phi(x,\varepsilon)$ and $\phi_a(x,\varepsilon)$ defined in a domain Ω are asymptotically identical to order $\delta(\varepsilon)$ if their difference is asymptotically smaller than $\delta(\varepsilon)$, where $\delta(\varepsilon)$ is an order function; that is,

$$\phi(x,\varepsilon) - \phi_a(x,\varepsilon) = o(\delta(\varepsilon)), \quad (1)$$

where ε is a positive small parameter arising from the physical problem under consideration. The function $\phi_a(x,\varepsilon)$ is named as *asymptotic approximation* of the function $\phi(x,\varepsilon)$. Asymptotic approximations in general form are defined by

$$\phi_a(x,\varepsilon) = \sum_{i=1}^{n} \delta_i(\varepsilon)\varphi_i(x,\varepsilon), \quad (2)$$

where $\delta_{i+1}(\varepsilon) = o(\delta_i(\varepsilon))$, as $\varepsilon \to 0$. Under these conditions, the approximation (2) is named as *generalized asymptotic expansion*. If the expansion (2) is written in the form of

$$\phi_a(x,\varepsilon) = E_0\phi = \sum_{i=1}^{n} \delta_i^{(0)}(\varepsilon)\varphi_i^{(0)}(x), \quad (3)$$

then it is called *regular asymptotic expansion* where the special operator E_0 is *outer expansion operator* of a given order $\delta(\varepsilon)$. Thus, $\phi - E_0\phi = o(\delta(\varepsilon))$. For more detailed information about the asymptotic approximations, we refer the interested reader to [3–8, 28, 29].

Interesting behaviors occur when the function $\phi(x,\varepsilon)$ is not regular in Ω so (2) or (3) is valid only in a restricted region $\Omega_0 \in \Omega$, called the outer region. We introduce an inner domain which can be formally denoted as $\Omega_1 = \Omega - \Omega_0$ and corresponding inner layer variable, located near the point $x = x_0$, as $\overline{x} = (x - x_0)/\eta(\varepsilon)$, with $\eta(\varepsilon)$ being the order of thickness of the boundary layer (the region in which the rapid changes-behaviors occur). If a regular expansion can be constructed in Ω_1, one can write down the approximation as

$$\phi_a(x,\varepsilon) = E_1\phi = \sum_{i=1}^{n} \delta_i^{(1)}(\varepsilon)\varphi_i^{(1)}(\overline{x}), \quad (4)$$

where the *inner expansion operator* E_1, defined in Ω_1, is of the same order $\delta(\varepsilon)$ as the outer expansion operator E_0; that is, $\phi - E_1\phi = o(\delta(\varepsilon))$. Thus,

$$\phi_a = E_0\phi + E_1\phi - E_1E_0\phi \quad (5)$$

is clearly uniformly valid approximation (UVA) [28–30].

Now let us consider second-order singularly perturbed DDE in its general form (delay in the convection term):

$$\varepsilon y''(x) + p(x)y'(x-\delta) + q(x)y(x) = r(x), \quad (6)$$

where $0 < \varepsilon \ll 1$ small parameter and $0 < x < 1$. Boundary and interval conditions are given as

$$y(x) = \phi(x), \quad -\delta \le x \le 0, \quad y(1) = \gamma, \quad (7)$$

where $p(x)$, $q(x)$, $r(x)$, and $\phi(x)$ are smooth functions, $\gamma \in \mathbb{R}$, and δ is delay term.

As ε tends to zero, the order of the differential equation reduces and so a layer occurs in the solution. The sign of $p(x)$ on the interval $[0,1]$ determines the type of the layer. If the sign changes on the interval, interior layer behavior occurs in the solution. If the sign of $p(x)$ does not change, there are two possibilities: if $p(x) < 0$ on $[0,1]$, then a boundary layer occurs at the right end (near the point $x = 1$) and if $p(x) > 0$ on $[0,1]$, then a boundary layer occurs at the left end (near the point $x = 0$).

Using Taylor series expansion we linearize the convection term; that is, $y'(x-\delta) = y'(x) - \delta y''(x)$ and substituting it into (6) one can reach

$$(\varepsilon - \delta p(x))y''(x) + p(x)y'(x) + q(x)y(x) = r(x). \quad (8)$$

Letting $\delta = \kappa\varepsilon$, where $\kappa \in \mathbb{R}$

$$\varepsilon(1-\kappa p(x))y''(x) + p(x)y'(x) + q(x)y(x) = r(x) \quad (9)$$

is found and it is clear that (9) is a singularly perturbed ordinary differential equation for $\varepsilon \to 0$ with the same boundary and interval conditions as given by (7). SCEM procedure is applicable at this stage.

The uniformly valid SCEM approximation is in the regular form given by

$$y_n^{\text{scem}}(x,\overline{x},\varepsilon) = \sum_{i=1}^{n} \delta_i(\varepsilon)\left[y_i(x) + \Psi_i(\overline{x})\right], \quad (10)$$

where $\{\delta_i(\varepsilon)\}$ is an asymptotic sequence and functions $\Psi_i(\overline{x})$ are the complementary functions that depend on $\overline{x}$.

If the functions $y_i(x)$ and $\Psi_i(\overline{x})$ depend also on ε, the uniformly valid SCEM approximation is called *generalized SCEM approximation* and given by

$$y_{ng}^{\text{scem}}(x,\overline{x},\varepsilon) = \sum_{i=1}^{n}\delta_i(\varepsilon)\left[y_i(x,\varepsilon) + \Psi_i(\overline{x},\varepsilon)\right]. \quad (11)$$

If only one-term SCEM approximation is desired, then one seeks a uniformly valid SCEM approximation in the form of

$$y_1^{\text{scem}}(x,\overline{x},\varepsilon) = y_1(x,\varepsilon) + \Psi_1(\overline{x},\varepsilon). \quad (12)$$

To improve the accuracy of approximation, (12) can be iterated using (11). It means that successive complementary terms will be added to the approximation. To this end, second SCEM approximation will be in the form of

$$\begin{aligned} y_2^{\text{scem}}(x,\overline{x},\varepsilon) &= y_1(x,\varepsilon) + \Psi_1(\overline{x},\varepsilon) \\ &\quad + \varepsilon\left(y_2(x,\varepsilon) + \Psi_2(\overline{x},\varepsilon)\right). \end{aligned} \quad (13)$$

In this work, we seek an approximation in our calculations in the form of (13).

Now, let us assume that problem (9) has a left boundary layer (near the point $x = 0$) and let $y_{\text{out}}(x)$ be asymptotic approximation to the outer solution and let $\Psi(\overline{x})$ be the complementary solution, where $\overline{x} = x/\varepsilon$ is boundary layer (stretching) variable. If approximations

$$\begin{aligned} y_{\text{out}}(x,\varepsilon) &= y_1(x) + \varepsilon y_2(x) + \varepsilon^2 y_3(x) + \cdots, \\ \Psi(\overline{x},\varepsilon) &= \Psi_1(\overline{x},\varepsilon) + \varepsilon\Psi_2(\overline{x},\varepsilon) + \varepsilon^2\Psi_3(\overline{x},\varepsilon) + \cdots \end{aligned} \quad (14)$$

are substituted into (9) and if each term is balanced with respect to the powers of ε (we balance just the terms $O(1)$ and $O(\varepsilon)$),

$$\begin{aligned} p(x)y_1'(x,\varepsilon) + q(x)y_1(x,\varepsilon) &= r(x), \\ y_1(1,\varepsilon) &= \gamma, \\ y_1''(x,\varepsilon) + p(x)y_2'(x,\varepsilon) + q(x)y_2(x,\varepsilon) &= 0, \\ y_2(1,\varepsilon) &= 0 \end{aligned} \quad (15)$$

are found. If the same procedure is applied for equations that involve complementary functions

$$\Psi_1''(\overline{x},\varepsilon) + p(x)\Psi_1'(\overline{x},\varepsilon) = r(x) \quad (16)$$

with the boundary conditions

$$\begin{aligned} &\Psi_1(0,\varepsilon) = \phi(0) - \gamma, \\ &\Psi_1\left(\frac{1}{\varepsilon},\varepsilon\right) = 0, \\ &\Psi_2''(\overline{x},\varepsilon) - \frac{\kappa}{\varepsilon}p(x)\Psi_1''(\overline{x},\varepsilon) + p(x)\Psi_2'(\overline{x},\varepsilon) \\ &\quad + q(x)\Psi_1(\overline{x},\varepsilon) = 0 \end{aligned} \quad (17)$$

with the boundary conditions

$$\begin{aligned} &\Psi_1(0,\varepsilon) = -y_2(0,\varepsilon), \\ &\Psi_1\left(\frac{1}{\varepsilon},\varepsilon\right) = 0 \end{aligned} \quad (18)$$

being obtained and so (13) gives uniformly valid second SCEM approximation.

3. Illustrative Examples

3.1. Left Boundary Layer Problem. Consider singularly perturbed DDE that exhibits a boundary layer at the left end of the interval:

$$\varepsilon y''(x) + y'(x-\delta) - y(x) = 0, \quad 0 \le x \le 1, \quad (19)$$

with the boundary conditions $y(0) = 1$ and $y(1) = 1$. The exact solution of this problem is given by $y(x) = ((1 - e^{m_2})e^{m_1 x} + (e^{m_1} - 1)e^{m_2 x})/(e^{m_1} - e^{m_2})$, where $m_{1,2} = (-1 \pm \sqrt{1 + 4(\varepsilon - \delta)})/2(\varepsilon - \delta)$. As the first step, we use two-term Taylor expansion for $y'(x - \delta) = y'(x) - \delta y''(x)$. If we substitute it into (19), the problem turns into

$$(\varepsilon - \delta)y''(x) + y'(x) - y(x) = 0, \quad 0 \le x \le 1. \quad (20)$$

In order to obtain a uniformly valid approximation (UVA), we first seek an outer solution which is valid far from the boundary layer (the boundary layer is near the point $x = 0$ for this problem) and then using SCEM we add complementary solution to it. Later, using the same idea, we will get more accurate approximations.

Outer Region Solutions. Let us take $\theta = \varepsilon - \delta$, assuming that δ depends on ε, and adopt a solution for the outer region in the form of $y_{\text{out}}(x) = y_1(x,\theta) + \theta y_2(x,\theta)$. Equation (20) turns into

$$\begin{aligned} &\theta\left(y_1''(x,\theta) + \theta y_2''(x,\theta)\right) + \left(y_1'(x,\theta) + \theta y_2'(x,\theta)\right) \\ &\quad - \left(y(x,\theta) + \theta y_2(x,\theta)\right) = 0 \end{aligned} \quad (21)$$

and balancing the terms of the order $O(1)$ and $O(\theta)$, we reach the equations

$$\begin{aligned} y_1'(x,\theta) - y_1(x,\theta) &= 0, \quad y_1(1,\theta) = 1, \\ y_1''(x,\theta) + y_2'(x,\theta) - y_2(x,\theta) &= 0, \quad y_2(1,\theta) = 0. \end{aligned} \quad (22)$$

One can easily find the exact solutions of these equations as

$$\begin{aligned} y_1(x,\theta) &= e^{x-1}, \\ y_2(x,\theta) &= e^{x-1}(1-x). \end{aligned} \quad (23)$$

It means that the outer solution is of the form

$$y_{\text{out}}(x,\theta) = e^{x-1} + \theta e^{x-1}(x-1). \quad (24)$$

Complementary Solutions. Now applying the stretching transformation $\overline{x} = x/\theta$ and adopting the complementary solution as $\Psi(\overline{x},\theta) = \Psi_1(\overline{x},\theta) + \theta\Psi_2(\overline{x},\theta)$, one can reach

$$\Psi_1''(\overline{x}) + \theta\Psi_2''(\overline{x},\theta) + \Psi_1'(\overline{x},\theta) + \theta\Psi_2'(\overline{x},\theta) - \theta\Psi_1(\overline{x},\theta) - \theta^2\Psi_2(\overline{x},\theta) = 0. \quad (25)$$

Balancing the terms of the order $O(1)$ and $O(\theta)$, we obtain

$$\Psi_1''(\overline{x},\theta) + \Psi_1'(\overline{x},\theta) = 0, \quad \Psi_1(0,\theta) = 1, \; \Psi_1\left(\frac{1}{\theta},\theta\right) = 0, \quad (26)$$

$$\Psi_2''(\overline{x},\theta) + \Psi_2'(\overline{x},\theta) - \Psi_1(\overline{x},\theta) = 0, \quad \Psi_2(0,\theta) = -e^{-1}, \; \Psi_2\left(\frac{1}{\theta},\theta\right) = 0. \quad (27)$$

Here we are able to solve $\Psi_1(\overline{x},\theta)$ and $\Psi_2(\overline{x},\theta)$ exactly, but in many cases to obtain analytical solution to $\Psi_1(\overline{x},\theta)$ and $\Psi_2(\overline{x},\theta)$ is really tedious process, even for many problems it is impossible. The solutions may be given as

$$\Psi_1(\overline{x},\theta) = \frac{e-1}{e(1-e^{-1/\theta})}\left(e^{-\overline{x}} - e^{-1/\theta}\right) \quad (28)$$

and $\Psi_2(\overline{x},\theta)$ is given as the solution of (27). Since we solve (27) numerically (*MATLAB bvp4c* routine) and the others using an asymptotic scheme, our present method is a hybrid method. As a result, we obtain first two SCEM approximations to problem (19) as follows:

$$y_1^{\text{scem}}(x,\overline{x},\theta) = e^{x-1} + \frac{e-1}{e(1-e^{-1/\theta})}\left(e^{-\overline{x}} - e^{-1/\theta}\right),$$
$$y_2^{\text{scem}}(x,\overline{x},\theta) = y_1^{\text{scem}}(x,\overline{x},\theta) + \theta\left[e^{x-1}(1-x) + \Psi_2(\overline{x},\theta)\right]. \quad (29)$$

3.2. Right Boundary Layer Problem. Consider singularly perturbed DDE that exhibits a boundary layer at the right end of the interval

$$\varepsilon y''(x) - y'(x-\delta) + y(x) = 0, \quad (30)$$

with the boundary and interval conditions

$$y(x) = 1, \quad -\delta \le x \le 0, \qquad y(1) = -1. \quad (31)$$

Using two-term Taylor expansion for the convection term, we reach $y'(x-\delta) = y'(x) - \delta y''(x)$ and applying it to (30) one can obtain

$$(\varepsilon+\delta)y''(x) - y'(x) + y(x) = 0, \quad 0 \le x \le 1 \quad (32)$$

with the boundary conditions $y(0) = 1$ and $y(1) = -1$.

In order to obtain a uniformly valid approximation, we first seek an outer solution which is valid for far from the boundary layer (the boundary layer is near the point $x = 1$ for this problem) and then using SCEM we add complementary solution to it. Later, using the same idea, we will get more accurate approximations.

Outer Region Solutions. Let us take $\theta = \varepsilon + \delta$ assuming that δ depends on ε and adopt an approximation for the outer region in the form of $y_{\text{out}}(x) = y_1(x,\theta) + \theta y_2(x,\theta)$. Equation (32) turns into

$$\theta\left(y_1''(x,\theta) + \theta y_2''(x,\theta)\right) - \left(y_1'(x,\theta) + \theta y_2'(x,\theta)\right) + \left(y(x,\theta) + \theta y_2(x,\theta)\right) = 0; \quad (33)$$

balancing the terms of the order $O(1)$ and $O(\theta)$, we reach the equations

$$y_1'(x,\theta) - y_1(x,\theta) = 0, \quad y_1(0,\theta) = 1,$$
$$y_1''(x,\theta) - y_2'(x,\theta) + y_2(x,\theta) = 0, \quad y_2(0,\theta) = 0. \quad (34)$$

One can easily find the exact solutions of these equations as

$$y_1(x,\theta) = e^x, \qquad y_2(x,\theta) = e^x x. \quad (35)$$

It means that the outer solution is of the form

$$y_{\text{out}}(x,\theta) = e^x + \theta e^x x. \quad (36)$$

Complementary Solutions. Now applying the stretching transformation $\overline{x} = (x-1)/\theta$ and adopting the complementary solution as $\Psi(\overline{x},\theta) = \Psi_1(\overline{x},\theta) + \theta\Psi_2(\overline{x},\theta)$ one can reach

$$\Psi_1''(\overline{x}) + \theta\Psi_2''(\overline{x},\theta) - \Psi_1'(\overline{x},\theta) - \theta\Psi_2'(\overline{x},\theta) + \theta\Psi_1(\overline{x},\theta) - \theta^2\Psi_2(\overline{x},\theta) = 0. \quad (37)$$

Balancing the terms of the order $O(1)$ and $O(\theta)$ we obtain

$$\Psi_1''(\overline{x},\theta) - \Psi_1'(\overline{x},\theta) = 0, \quad \Psi_1\left(-\frac{1}{\theta},\theta\right) = 0, \; \Psi_1(0,\theta) = -1-e, \quad (38)$$

$$\Psi_2''(\overline{x},\theta) - \Psi_2'(\overline{x},\theta) + \Psi_1(\overline{x},\theta) = 0, \quad \Psi_2\left(-\frac{1}{\theta},\theta\right) = 0, \; \Psi_2(0,\theta) = -1-e. \quad (39)$$

The solutions are given as

$$\Psi_1(\overline{x},\theta) = \frac{e+1}{e^{-1/\theta}}\left(e^{\overline{x}} - 1\right) - e - 1 \quad (40)$$

and $\Psi_2(\overline{x},\theta)$ is given as the solution of (39). Thus, we reach uniformly valid SCEM approximations as

$$y_1^{\text{scem}}(x,\overline{x},\theta) = e^x + \frac{e+1}{e^{-1/\theta}}\left(e^{\overline{x}} - 1\right) - e - 1,$$
$$y_2^{\text{scem}}(x,\overline{x},\theta) = y_1^{\text{scem}}(x,\overline{x},\theta) + \theta\left[e^x x + \Psi_2(\overline{x},\theta)\right]. \quad (41)$$

TABLE 1: Results of left layer problem for $\varepsilon = 10^{-3}$ and $\delta = 0.5\varepsilon$.

x	y_{exact}	y_1^{scem}	$\lvert y_{\text{exact}} - y_1^{\text{scem}} \rvert$	y_2^{scem}	$\lvert y_{\text{exact}} - y_2^{\text{scem}} \rvert$	Method [9]
0.0000	1.0000000	1.0000000	0.0000000	0.0000000	0.0000000	1.0000000
0.0010	0.4538692	0.45379572	$7.3488e-05$	0.4538692	$9.2048e-09$	0.3171426
0.0020	0.3803509	0.38019363	$1.5732e-04$	0.3803510	$9.4728e-08$	0.3603879
0.0030	0.3707303	0.37055161	$1.7865e-04$	0.3707303	$1.2683e-07$	0.3666117
0.0040	0.3697488	0.36956596	$1.8289e-04$	0.3697489	$1.3552e-07$	0.3678324
0.0050	0.3699358	0.36975214	$1.8364e-04$	0.3699359	$1.3753e-07$	0.3683816
0.0100	0.3717605	0.37157669	$1.8379e-04$	0.3717606	$1.3821e-07$	0.3708603
0.0150	0.3736230	0.37343923	$1.8378e-04$	0.3736231	$1.3848e-07$	0.3733556
0.0200	0.3754949	0.37531110	$1.8376e-04$	0.3754950	$1.3869e-07$	0.3758674
0.1000	0.4067525	0.40656966	$1.8281e-04$	0.4067526	$1.4163e-07$	0.4071563
0.2000	0.4495085	0.44932896	$1.7958e-04$	0.4495086	$1.4362e-07$	0.4499552
0.4000	0.5489761	0.54881164	$1.6450e-04$	0.5489762	$1.3978e-07$	0.5495225
0.6000	0.6704540	0.67032005	$1.3394e-04$	0.6704541	$1.2051e-07$	0.6711221
0.8000	0.8188125	0.81873075	$8.1795e-05$	0.8188126	$7.7685e-08$	0.8196300
0.9000	0.9048826	0.90483742	$4.5197e-05$	0.9048826	$4.4056e-08$	0.9057869
1.0000	1.0000000	1.0000000	0.0000000	0.0000000	0.0000000	1.0000000

TABLE 2: Results of left layer problem for $\varepsilon = 10^{-4}$ and $\delta = 0.5\varepsilon$.

x	Exact	y_1^{scem}	$\lvert y_{\text{exact}} - y_1^{\text{scem}} \rvert$	y_2^{scem}	$\lvert y_{\text{exact}} - y_2^{\text{scem}} \rvert$	Method [9]
0.0000	1.0000000	1.0000000	0.0000000	1.0000000	0.0000000	1.0000000
0.0001	0.4534718	0.4534644	$7.3497e-06$	0.4534717	$1.7565e-10$	0.3181581
0.0002	0.3795465	0.3795307	$1.5740e-05$	0.3795464	$1.1979e-09$	0.3612116
0.0003	0.3695746	0.3695566	$1.7877e-05$	0.3695745	$1.2808e-09$	0.3670360
0.0004	0.3682570	0.3682386	$1.8301e-05$	0.3682569	$1.3520e-09$	0.3678237
0.0005	0.3681105	0.3680921	$1.8377e-05$	0.3681105	$1.3718e-09$	0.3679338
0.0010	0.3682659	0.3682475	$1.8392e-05$	0.3682658	$1.3794e-09$	0.3682020
0.0015	0.3684501	0.3684316	$1.8392e-05$	0.3684500	$1.3806e-09$	0.3684702
0.0020	0.3686343	0.3686159	$1.8392e-05$	0.3686343	$1.3794e-09$	0.3687384
0.1000	0.4065880	0.4065696	$1.8294e-05$	0.4065879	$1.4170e-09$	0.4066914
0.2000	0.4493469	0.4493289	$1.7971e-05$	0.4493469	$1.4373e-09$	0.4494485
0.4000	0.5488281	0.5488116	$1.6462e-05$	0.5488281	$1.3990e-09$	0.5489215
0.6000	0.6703334	0.6703200	$1.3405e-05$	0.6703334	$1.2061e-09$	0.6704094
0.8000	0.8187389	0.8187307	$8.1865e-06$	0.8187389	$7.7755e-10$	0.8187855
0.9000	0.9048420	0.9048374	$4.5237e-06$	0.9048419	$4.4095e-10$	0.9048674
1.0000	1.0000000	1.0000000	0.0000000	1.0000000	0.0000000	1.0000000

4. Conclusion

In this paper, singularly perturbed second-order linear delay differential equations that have a delay in the convection term are considered. Firstly, the delayed terms are linearized using two-term Taylor series expansion. Later, an efficient asymptotic method so called Successive Complementary Expansion Method (SCEM) is employed so as to obtain a uniformly valid approximation scheme. At the last stage, the equations that come from the SCEM process are solved by a numerical procedure and so the present method is an asymptotic-numerical hybrid method. The method is easily applicable since it does not require any matching principle in contrast to the well-known method matched asymptotic expansions (MMAE). Highly accurate approximations are obtained in only few iterations and moreover boundary conditions are not satisfied asymptotically, but exactly. In Tables 1 and 2, exact solution, present method approximations, and approximations that are obtained by the method given in [9] are compared and to show the efficiency of present method, results are supported by Figures 1, 3, and 4. In Figures 2 and 5, the delay effects are compared and since the right

TABLE 3: Results of right layer problem for $\varepsilon = 10^{-3}$ and $\delta = 0.5\varepsilon$.

x	y_1^{scem}	y_2^{scem}	$\lvert y_2^{\text{scem}} - y_1^{\text{scem}} \rvert$	Numerical method
0.0000	1.0000000	1.0000000	0.0000000	1.0000000
0.1000	1.1051709	1.1052261	5.5258e − 05	1.1052262
0.2000	1.2214027	1.2215248	1.2214e − 04	1.2215250
0.4000	1.4918246	1.4921230	2.9836e − 04	1.4921233
0.6000	1.8221188	1.8226654	5.4663e − 04	1.8226660
0.8000	2.2255409	2.2264311	8.9021e − 04	2.2264322
0.9000	2.4596031	2.4607099	0.0011068	2.4607112
0.9980	2.6447479	2.6460768	0.0013288	2.6458021
0.9985	2.5290851	2.5303725	0.0012873	2.5299480
0.9990	2.2123501	2.2135226	0.00117247	2.2128733
0.9995	1.3490435	1.3499013	8.5777e − 04	1.3488799
0.9996	1.0464630	1.0472103	7.4736e − 04	1.0456402
0.9997	0.6768301	0.6774425	6.1241e − 04	0.6780147
0.9998	0.2252993	0.2257469	4.4754e − 04	0.2279200
0.9999	−0.3262616	−0.3260155	2.4612e − 04	−0.3247246
1.0000	−1.0000000	−1.0000000	0.0000000	−1.0000000

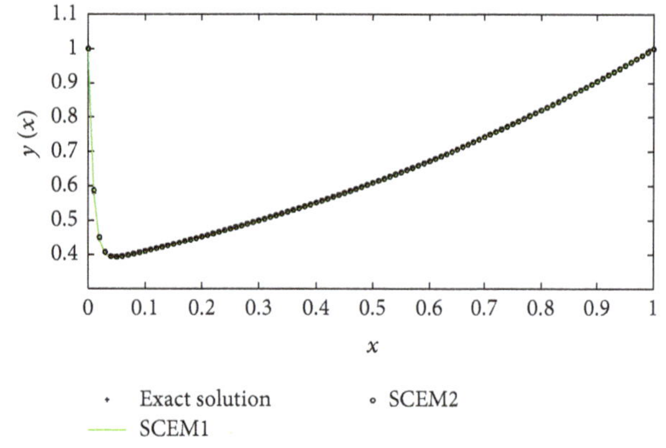

FIGURE 1: Left layer problem for $\varepsilon = 0.01$ and $\delta = 0.1\varepsilon$.

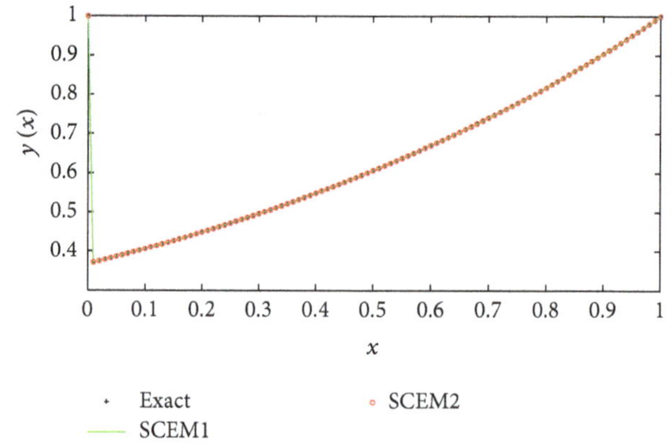

FIGURE 3: Left layer problem for $\varepsilon = 0.001$ and $\delta = 0.9\varepsilon$.

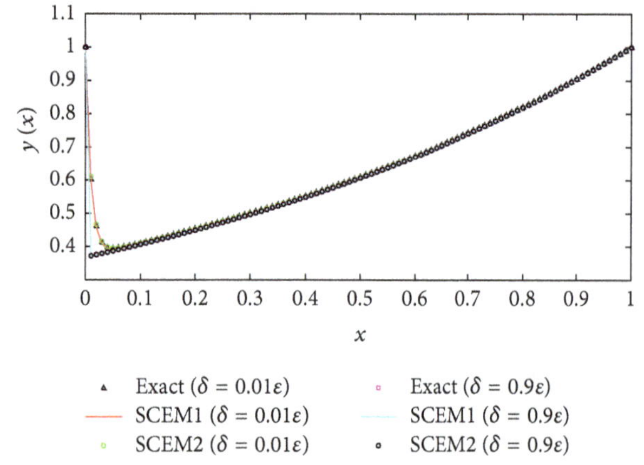

FIGURE 2: Delay effect on left layer problem for $\varepsilon = 0.01$, $\delta = 0.01\varepsilon$, and $\delta = 0.9\varepsilon$.

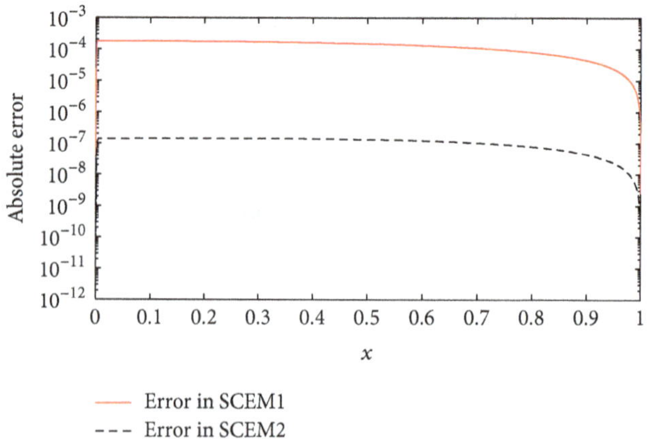

FIGURE 4: Absolute errors in SCEM approximations for left layer problem.

layer problem does not have an exact solution, the first two SCEM approximations are compared in Table 3. As a result, the present method is a simple and very efficient technique for solving singularly perturbed linear DDEs.

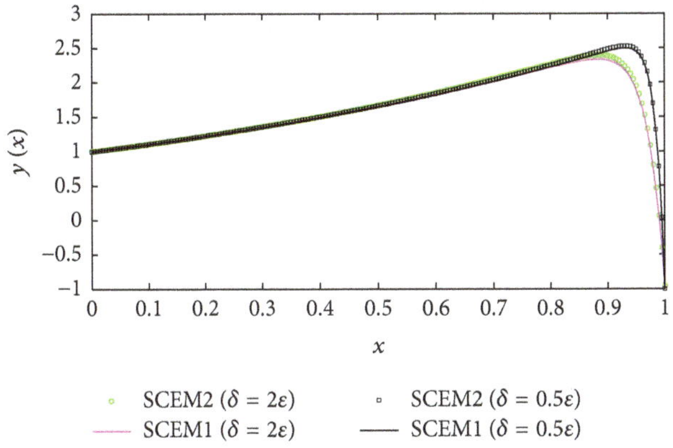

FIGURE 5: Delay effect on right layer problem for $\varepsilon = 0.01$, $\delta = 0.5\varepsilon$, and $\delta = 2\varepsilon$.

Competing Interests

The author declares that there is no conflict of interests.

Acknowledgments

The author would like to express his sincere thanks to his master thesis advisors Dr. A. Eryılmaz and Dr. M. T. Atay for their valuable comments and suggestions.

References

[1] M. Kumar, "Methods for solving singular perturbation problem sarising in science and engineering," *Mathematical and Computer Modelling*, vol. 54, no. 1, pp. 556–575, 2011.

[2] K. K. Sharma, P. Rai, and K. C. Patidar, "A review on singularly perturbed differential equations with turning points and interior layers," *Applied Mathematics and Computation*, vol. 219, no. 22, pp. 10575–10609, 2013.

[3] A. H. Nayfeh, *Perturbation Methods*, John Wiley & Sons, 2008.

[4] E. J. Hinch, *Perturbation Methods*, Cambridge Texts in Applied Mathematics, Cambridge University Press, Cambridge, UK, 1991.

[5] J. A. Murdock, *Perturbations: Theory and Methods*, vol. 27 of *Classics in Applied Mathematics*, SIAM, Philadelphia, Pa, USA, 1999.

[6] E. M. de Jager and J. F. Furu, *The Theory of Singular Perturbations*, vol. 42, Elsevier, 1996.

[7] P. A. Lagerstrom, *Matched asymptotic expansions*, vol. 76 of *Applied Mathematical Sciences*, Springer New York, 1988.

[8] J. K. Kevorkian and J. D. Cole, *Multiple Scale and Singular Perturbation Methods*, vol. 114, Springer Science & Business Media, Berlin, Germany, 2012.

[9] F. Gemechis and Y. N. Reddy, "Terminal boundary-value technique for solving singularly perturbed delay differential equations," *Journal of Taibah University for Sciences*, vol. 8, no. 3, pp. 289–300, 2014.

[10] T. Erneux, *Applied Delay Differential Equations*, vol. 3 of *Surveys and Tutorials in the Applied Mathematical Sciences*, Springer, New York, NY, USA, 2009.

[11] E. Pinney, *Ordinary Difference-Differential Equations*, University of California Press, 1958.

[12] R. D. Driver, *Ordinary and Delay Differential Equations*, vol. 20 of *Applied Mathematics*, Springer, Berlin, Germany, 1977.

[13] R. E. Bellman and K. L. Cooke, *Differential-Difference Equations*, Academic Press, New York, NY, USA, 1963.

[14] J. K. Hale, *Functional Differential Equations*, Springer, New York, NY, USA, 1971.

[15] M. K. Kadalbajoo and K. K. Sharma, "A numerical method based on finite difference for boundary value problems for singularly perturbed delay differential equations," *Applied Mathematics and Computation*, vol. 197, no. 2, pp. 692–707, 2008.

[16] R. Nageshwar Rao and P. Pramod Chakravarthy, "A finite difference method for singularly perturbed differential-difference equations with layer and oscillatory behavior," *Applied Mathematical Modelling*, vol. 37, no. 8, pp. 5743–5755, 2013.

[17] S. Nicaise and C. Xenophontos, "Robust approximation of singularly perturbed delay differential equations by the hp finite element method," *Computational Methods in Applied Mathematics*, vol. 13, no. 1, pp. 21–37, 2013.

[18] H. Zarin, "On discontinuous Galerkin finite element method for singularly perturbed delay differential equations," *Applied Mathematics Letters. An International Journal of Rapid Publication*, vol. 38, pp. 27–32, 2014.

[19] F. Shakeri and M. Dehghan, "Solution of delay differential equations via a homotopy perturbation method," *Mathematical and Computer Modelling*, vol. 48, no. 3-4, pp. 486–498, 2008.

[20] Q. Wang and F. Fu, "Solving delay differential equations with homotopy analysis method," *Communications in Computer and Information Science*, vol. 97, no. 1, pp. 144–153, 2010.

[21] A. Akgul and A. Kiliman, "Solving delay differential equations by an accurate method with interpolation," *Abstract and Applied Analysis*, vol. 2015, Article ID 676939, 7 pages, 2015.

[22] F. Z. Geng, S. P. Qian, and M. G. Cui, "Improved reproducing kernel method for singularly perturbed differential-difference equations with boundary layer behavior," *Applied Mathematics and Computation*, vol. 252, pp. 58–63, 2015.

[23] M. K. Kadalbajoo and D. Kumar, "Fitted mesh B-spline collocation method for singularly perturbed differential–difference equations with small delay," *Applied Mathematics and Computation*, vol. 204, no. 1, pp. 90–98, 2008.

[24] H. M. El-Hawary and S. M. Mahmoud, "Spline collocation methods for solving delay-differential equations," *Applied Mathematics and Computation*, vol. 146, no. 2-3, pp. 359–372, 2003.

[25] C. G. Lange and R. M. Miura, "Singular perturbation analysis of boundary value problems for differential-difference equations. v. small shifts with layer behavior," *SIAM Journal on Applied Mathematics*, vol. 54, no. 54, pp. 249–272, 1994.

[26] C. G. Lange and R. M. Miura, "Singular perturbation analysis of boundary-value problems for differential-difference equations. VI. Small shifts with rapid oscillations," *SIAM Journal on Applied Mathematics*, vol. 54, no. 1, pp. 273–283, 1994.

[27] J. Mauss and J. Cousteix, "Uniformly valid approximation for singular perturbation problems and matching principle," *Comptes Rendus - Mecanique*, vol. 330, no. 10, pp. 697–702, 2002.

[28] W. Eckhaus, *Asymptotic Analysis of Singular Perturbations*, vol. 9 of *Studies in Mathematics and Its Applications*, North-Holland, Amsterdam, The Netherlands, 1979.

[29] J. Cousteix and J. Mauss, *Asymptotic Analysis and Boundary Layers*, Scientific Computation, Springer, Berlin, Germany, 2007.

[30] S. Cengizci, M. T. Atay, and A. Eryılmaz, "A uniformly valid approximation algorithm for nonlinear ordinary singular perturbation problems with boundary layer solutions," *SpringerPlus*, vol. 5, no. 1, article no. 280, 2016.

3

A Family of Boundary Value Methods for Systems of Second-Order Boundary Value Problems

T. A. Biala[1] and S. N. Jator[2]

[1]*Department of Mathematics and Computer Science, Sule Lamido University, PMB 048, Kafin Hausa, Nigeria*
[2]*Department of Mathematics and Statistics, Austin Peay State University, Clarksville, TN 37044, USA*

Correspondence should be addressed to T. A. Biala; bialatoheeb@yahoo.com

Academic Editor: Elena Braverman

A family of boundary value methods (BVMs) with continuous coefficients is derived and used to obtain methods which are applied via the block unification approach. The methods obtained from these continuous BVMs are weighted the same and are used to simultaneously generate approximations to the exact solution of systems of second-order boundary value problems (BVPs) on the entire interval of integration. The convergence of the methods is analyzed. Numerical experiments were performed to show efficiency and accuracy advantages.

1. Introduction

In what follows, we consider the general system of second-order boundary value problems:

$$\begin{aligned} y'' &= f\left(x, y, y'\right), \quad x \in [a,b], \\ y(a) &= y_0, \\ y(b) &= y_N, \end{aligned} \tag{1}$$

where $f : \mathbb{R} \times \mathbb{R}^{2m} \rightarrow \mathbb{R}^m$ are continuous functions, y, y', and $y'' \in \mathbb{R}^m$, and m is the dimension of the system. These second-order boundary value problems are encountered in several areas of engineering and applied sciences such as celestial mechanics, circuit theory, astrophysics, chemical kinetics, and biology. Most of these problems cannot be solved analytically, thus the need for a numerical approach. In practice, (1) is solved by the multiple shooting technique and the finite difference methods. The construction and implementation of higher order methods for the latter approach are difficult while the former approach suffers from numerical instability if the BVP is stiff [1–3] and singularly perturbed.

In the past few decades, the boundary value methods (BVMs) have been used to solve first-order initial and boundary value problems [4–8]. Their stability and convergence properties have been fully discussed in [5]. These BVMs are also used to solve higher order initial and boundary value problems by first reducing the higher order differential equations into an equivalent first-order system. This approach increases the computational costs and time and also does not utilize additional information associated with specific differential equations such as the oscillatory nature of some solutions [9, 10].

Lambert and Watson [11] have derived symmetric schemes for periodic initial value problems of the special second-order $y'' = f(x, y)$. Brugnano and Trigiante [4–6] have also derived BVMs for the first-order initial and boundary value problems. Amodio and Iavernaro [12] used BVMs to solve the special second-order problem $y'' = f(x, y)$. Biala, Biala and Jator, Jator and Li [13–15] applied the BVMs to solve the general second-order problem $y'' = f(x, y, y')$ and Aceto et al. [16] constructed symmetric linear multistep methods (LMMs) which were used as BVMs for the special second-order problem $y'' = f(x, y)$. In this paper, we have derived a class of BVMs and given a general framework via the block unification approach on how to use the BVMs on systems of BVPs for the general second-order differential equations (ODEs).

The boundary value technique simultaneously generates approximate solution $(y_1, y_2, \ldots, y_{N-1})^T$ to the exact solution $(y(x_1), y(x_2), \ldots, y(x_{N-1}))^T$ of (1) on the entire interval of integration. The BVMs can only be successfully implemented if used together with appropriate additional methods [5]. In this regard, we have proposed methods which are obtained from the same continuous scheme and are derived via the interpolation and collocation approach [15, 17–19].

The paper is organised as follows. In Section 2, we derive a continuous approximation $U(x)$ of the exact solution $y(x)$. Section 3 gives the specification of the methods. The convergence of the methods is discussed in Section 4. The use and implementation of the methods on ODEs and partial differential equations (PDEs) are detailed in Section 5. Numerical tests and concluding remarks are given in Sections 6 and 7, respectively.

2. Derivation of Methods

In this section, we shall use the interpolation and collocation approach [17] to construct a 2ν-step continuous LMM (CLMM) which will be used to produce the main and additional formulas for solving (1).

Our starting point is to construct the CLMM which has the form

$$U(x) = \alpha_\nu(x) y_{n+\nu} + \alpha_0(x) y_n + h^2 \sum_{i=0}^{2\nu} \beta_i(x) f_{n+i}, \quad (2)$$

where $\alpha_0(x)$, $\alpha_\nu(x)$, and $\beta_i(x)$ are continuous coefficients and ν is chosen to be half the step number so that each formula, derived from (2), satisfies the root condition. The main and additional methods are then obtained by evaluating (2) at x_{n+j} $(j = 1(1)2\nu,\ j \neq \nu)$ to obtain the formulas of the form

$$y_{n+j} + 2y_{n+\nu} - y_n = h^2 \sum_{i=0}^{2\nu} \beta_i f_{n+i}, \qquad j = 1, \ldots, \nu-1, \nu+1, \ldots, 2\nu, \quad (3)$$

$$h y'_{n+k} + \alpha'_\nu y_{n+\nu} + \alpha'_0 y_n = h^2 \sum_{i=0}^{2\nu} \beta'_i f_{n+i}, \quad k = 0(1)(2\nu) \quad (4)$$

obtained from the first derivative of (2).

Next, we discuss the construction of (2) in the theorem that follows.

Theorem 1. *Let (2) satisfy the following equations:*

$$U(x_{n+r}) = y_{n+r} \quad r = 0, \nu,$$
$$U''(x_{n+i}) = f_{n+i} \quad i = 0(1)(2\nu). \quad (5)$$

Then, the continuous representation (2) is equivalent to

$$U(x) = \sum_{j=0}^{2\nu+2} \frac{\det(V_j)}{\det(V)} P_j(x), \quad (6)$$

where one defines the matrix V as

$$V = \begin{pmatrix} P_0(x_n) & P_1(x_n) & \cdots & P_{2\nu+2}(x_n) \\ P_0(x_{n+\nu}) & P_1(x_{n+\nu}) & \cdots & P_{2\nu+2}(x_{n+\nu}) \\ P''_0(x_n) & P''_1(x_n) & \cdots & P''_{2\nu+2}(x_n) \\ P''_0(x_{n+1}) & P''_1(x_{n+1}) & \cdots & P''_{2\nu+2}(x_{n+1}) \\ \vdots & \vdots & \vdots & \vdots \\ P''_0(x_{n+2\nu}) & P''_1(x_{n+2\nu}) & \cdots & P''_{2\nu+2}(x_{n+2\nu}) \end{pmatrix}, \quad (7)$$

V_j is obtained by replacing the jth column of V by

$$W = (y_n, y_{n+\nu}, f_n, f_{n+1}, \ldots, f_{n+2\nu})^T, \quad (8)$$

and $P_j(x) = x^j$, $j = 0(1)(2\nu+2)$ are basis functions.

Proof. We require that method (2) be defined by the assumed polynomial basis functions

$$\alpha_j(x) = \sum_{i=0}^{2\nu+2} \alpha_{i+1,j} P_i(x), \quad j = 0, \nu$$
$$h^2 \beta_j(x) = \sum_{i=0}^{2\nu+2} h^2 \beta_{i+1,j} P_i(x), \quad j = 0(1)(2\nu), \quad (9)$$

where $\alpha_{i+1,j}$ and $h^2\beta_{i+1,j}$ are coefficients to be determined.

Substituting (9) into (2), we have

$$U(x) = \sum_{i=0}^{2\nu+2} \alpha_{i+1,0} P_i(x) y_n + \sum_{i=0}^{2\nu+2} \alpha_{i+1,\nu} P_i(x) y_{n+\nu} + \sum_{j=0}^{2\nu} \sum_{i=0}^{2\nu+2} h^2 \beta_{i+1,j} P_i(x) f_{n+j} \quad (10)$$

which is simplified to

$$U(x) = \sum_{i=0}^{2\nu+2} \left\{ \alpha_{i+1,0} y_n + \alpha_{i+1,\nu} y_{n+\nu} + \sum_{j=0}^{2\nu} h^2 \beta_{i+1,j} f_{n+j} \right\} P_i(x) \quad (11)$$

and expressed in the form

$$U(x) = \sum_{i=0}^{2\nu+2} \tau_i P_i(x), \quad (12)$$

where

$$\tau_i = \alpha_{i+1,0} y_n + \alpha_{i+1,\nu} y_{n+\nu} + \sum_{j=0}^{2\nu} h^2 \beta_{i+1,j} f_{n+j}. \quad (13)$$

Imposing conditions (5) on (12), we obtain a system of $(2\nu+3)$ equations which can be expressed as $VL = W$, where

$L = (\tau_0, \tau_1, \ldots, \tau_{2\nu+2})^T$ is a vector of $(2\nu + 3)$ undetermined coefficients.

Using Crammer's rule, the elements of L are determined and given as

$$\tau_i = \frac{\det(V_j)}{\det(V)}, \quad j = 0\,(1)\,(2\nu + 2), \tag{14}$$

where V_j is obtained by replacing the jth column of V by W. We rewrite (12) using the newly found elements of L as in (6); that is,

$$U(x) = \sum_{j=0}^{2\nu+2} \frac{\det(V_j)}{\det(V)} P_j(x). \tag{15}$$

□

3. Specification of Methods

In this section, we specify the family of methods by evaluating the CLMM (2) at x_{n+i}, $i = i, \nu - 1, \nu + 1, \ldots, 2\nu$, which is also used to obtain the derivative formula given by

$$U'(x) = \frac{1}{h}\left(\alpha'_\nu(x)\, y_{n+\nu} + \alpha'_0(x)\, y_n + h^2 \sum_{i=0}^{2\nu} \beta'_i(x)\, f_{n+i}\right), \tag{16}$$

which is effectively applied by imposing that

$$U'(a) = y'_0, \qquad U'(b) = y'_N, \tag{17}$$

to produce derivative formulas of the form (4).

3.1. BVM of Orders 4, 6, and 8. For $\nu = 1$, the BVM of order 4 is given as follows (where we have denoted a BVM with k step number as BVMk):

BVM2

$$y_{n+2} - 2y_{n+1} + y_n = \frac{h^2}{12}(f_n + 10f_{n+1} + f_{n+2}), \tag{18}$$

with the derivative formulas

$$\begin{aligned} hy'_n - y_{n+1} + y_n &= \frac{h^2}{24}(-7f_n - 6f_{n+1} + f_{n+2}), \\ hy'_{n+1} - y_{n+1} + y_n &= \frac{h^2}{24}(3f_n + 10f_{n+1} - f_{n+2}), \\ hy'_{n+2} - y_{n+1} + y_n &= \frac{h^2}{24}(f_n + 26f_{n+1} + 9f_{n+2}). \end{aligned} \tag{19}$$

For $\nu = 2$, we obtain the BVM of order 6 given as follows:

BVM4

$$\begin{aligned} y_{n+1} - \frac{1}{2}y_{n+2} - \frac{1}{2}y_n &= \frac{h^2}{480}(-19f_n - 204f_{n+1} - 14f_{n+2} - 4f_{n+3} + f_{n+4}), \\ y_{n+3} - \frac{3}{2}y_{n+2} + \frac{1}{2}y_n &= \frac{h^2}{480}(17f_n + 252f_{n+1} + 402f_{n+2} + 52f_{n+3} - 3f_{n+4}), \\ y_{n+4} - 2y_{n+2} + y_n &= \frac{h^2}{15}(f_n + 16f_{n+1} + 26f_{n+2} + 16f_{n+3} + f_{n+4}), \\ & n = 0\,(4)\,(N-4), \end{aligned} \tag{20}$$

with the derivative formulas

$$\begin{aligned} hy'_n - \frac{1}{2}y_{n+2} + \frac{1}{2}y_n &= \frac{h^2}{180}(-53f_n - 144f_{n+1} + 30f_{n+2} - 16f_{n+3} + 3f_{n+4}), \\ hy'_{n+1} - \frac{1}{2}y_{n+2} + \frac{1}{2}y_n &= \frac{h^2}{720}(39f_n + 70f_{n+1} - 144f_{n+2} + 42f_{n+3} - 7f_{n+4}), \\ hy'_{n+2} - \frac{1}{2}y_{n+2} + \frac{1}{2}y_n &= \frac{h^2}{180}(5f_n + 104f_{n+1} + 78f_{n+2} - 8f_{n+3} + f_{n+4}), \\ hy'_{n+3} - \frac{1}{2}y_{n+2} + \frac{1}{2}y_n &= \frac{h^2}{720}(31f_n + 342f_{n+1} + 768f_{n+2} + 314f_{n+3} - 15f_{n+4}), \\ hy'_{n+4} - \frac{1}{2}y_{n+2} + \frac{1}{2}y_n &= \frac{h^2}{180}(3f_n + 112f_{n+1} + 56f_{n+2} + 240f_{n+3} + 59f_{n+4}). \end{aligned} \tag{21}$$

For $\nu = 3$, we obtain the BVM of order 8 given as follows:

BVM6

$$y_{n+1} - \frac{1}{3}y_{n+3} - \frac{2}{3}y_n = \frac{h^2}{60480}(-2803f_n - 37950f_{n+1} - 14913f_{n+2} - 7108f_{n+3} + 3147f_{n+4} - 990f_{n+5} + 137f_{n+6}),$$

$$y_{n+2} - \frac{2}{3}y_{n+3} - \frac{1}{3}y_n = \frac{h^2}{60480}(-1291f_n - 21906f_{n+1} - 32133f_{n+2} - 6288f_{n+3} + 1467f_{n+4} - 402f_{n+5} + 53f_{n+6}),$$

$$y_{n+4} - \frac{4}{3}y_{n+3} + \frac{1}{3}y_n = \frac{h^2}{30240}(661f_n + 10734f_{n+1} + 19323f_{n+2} + 27268f_{n+3} + 2523f_{n+4} - 18f_{n+5} - 11f_{n+6}),$$

$$y_{n+5} - \frac{5}{3}y_{n+3} + \frac{2}{3}y_n = \frac{h^2}{12096}(535f_n + 8550f_{n+1} + 15501f_{n+2} + 22900f_{n+3} + 11889f_{n+4} + 1158f_{n+5} - 53f_{n+6}),$$

$$y_{n+6} - 2y_{n+3} + y_n = \frac{h^2}{2240}(141f_n + 2430f_{n+1} + 4131f_{n+2} + 6756f_{n+3} + 4131f_{n+4} + 2430f_{n+5} + 141f_{n+6}),$$

$$n = 0(6)(N-6), \quad (22)$$

with the derivatives

$$hy'_n - \frac{1}{3}y_{n+3} + \frac{1}{3}y_n = \frac{h^2}{13440}(-3795f_n - 14850f_{n+1} + 2403f_{n+2} - 6300f_{n+3} + 3267f_{n+4} - 1026f_{n+5} + 141f_{n+6}),$$

$$hy'_{n+1} - \frac{1}{3}y_{n+3} + \frac{1}{3}y_n = \frac{h^2}{120960}(4019f_n - 3426f_{n+1} - 7125f_{n+2} + 18308f_{n+3} - 11019f_{n+4} + 3390f_{n+5} - 457f_{n+6}),$$

$$hy'_{n+2} - \frac{1}{3}y_{n+3} + \frac{1}{3}y_n = \frac{h^2}{120960}(2293f_n + 46830f_{n+1} + 22683f_{n+2} - 14204f_{n+3} + 3579f_{n+4} - 786f_{n+5} + 85f_{n+6}),$$

$$hy'_{n+3} - \frac{1}{3}y_{n+3} + \frac{1}{3}y_n = \frac{h^2}{13440}(315f_n + 4590f_{n+1} + 9369f_{n+2} + 6576f_{n+3} - 1107f_{n+4} + 270f_{n+5} - 33f_{n+6}),$$

$$hy'_{n+4} - \frac{1}{3}y_{n+3} + \frac{1}{3}y_n = \frac{h^2}{120960}(2453f_n + 44526f_{n+1} + 70779f_{n+2} + 135812f_{n+3} + 51675f_{n+4} - 3090f_{n+5} + 245f_{n+6}),$$

$$hy'_{n+5} - \frac{1}{3}y_{n+3} + \frac{1}{3}y_n = \frac{h^2}{120960}(2995f_n + 40350f_{n+1} + 85377f_{n+2} + 103300f_{n+3} + 145653f_{n+4} + 47166f_{n+5} - 1481f_{n+6}),$$

$$hy'_{n+6} - \frac{1}{3}y_{n+3} + \frac{1}{3}y_n = \frac{h^2}{13440}(141f_n + 5886f_{n+1} + 4995f_{n+2} + 19812f_{n+3} + 5859f_{n+4} + 19710f_{n+5} + 4077f_{n+6}),$$

$$n = 0(6)(N-6). \quad (23)$$

4. Convergence of the Methods

In this section, we shall establish the convergence of the BVMs derived in the previous section. We emphasize that we evaluate (2) at $x_{n+1}, x_{n+2}, \ldots, x_{n+\nu-1}, x_{n+\nu+1}, \ldots, x_{n+2\nu}$ to obtain

$$y_{n+1} + \alpha_\nu^{(1)} y_{n+\nu} + \alpha_0^{(1)} y_n = h^2 \sum_{i=0}^{2\nu} \beta_i^{(1)} f_{n+i}$$

$$y_{n+2} + \alpha_\nu^{(2)} y_{n+\nu} + \alpha_0^{(2)} y_n = h^2 \sum_{i=0}^{2\nu} \beta_i^{(2)} f_{n+i}$$

$$\vdots$$

$$y_{n+\nu-1} + \alpha_\nu^{(\nu-1)} y_{n+\nu} + \alpha_0^{(\nu-1)} y_n = h^2 \sum_{i=0}^{2\nu} \beta_i^{(\nu-1)} f_{n+i} \quad (24)$$

$$y_{n+\nu+1} + \alpha_\nu^{(\nu+1)} y_{n+\nu} + \alpha_0^{(\nu+1)} y_n = h^2 \sum_{i=0}^{2\nu} \beta_i^{(\nu+1)} f_{n+i}$$

$$\vdots$$

$$y_{n+2\nu} + \alpha_\nu^{(2\nu)} y_{n+\nu} + \alpha_0^{(2\nu)} y_n = h^2 \sum_{i=0}^{2\nu} \beta_i^{(2\nu)} f_{n+i}$$

and also evaluate $U'(x)$ at x_{n+i}, $i = 0(1)(2\nu)$, to obtain

$$hy'_n + \alpha_\nu'^{(0)} y_{n+\nu} + \alpha_0'^{(0)} y_n = h^2 \sum_{i=0}^{2\nu} \beta_i'^{(0)} f_{n+i}$$

$$hy'_{n+1} + \alpha_\nu'^{(1)} y_{n+\nu} + \alpha_0'^{(1)} y_n = h^2 \sum_{i=0}^{2\nu} \beta_i'^{(1)} f_{n+i}$$

$$\vdots$$

$$hy'_{n+2\nu} + \alpha_{\nu}'^{(2\nu)} y_{n+\nu} + \alpha_0'^{(2\nu)} y_n = h^2 \sum_{i=0}^{2\nu} \beta_i'^{(2\nu)} f_{n+i}. \tag{25}$$

We note that the formulas in (24) and 4 are $O(h^{2\nu+4})$.

We introduce the matrices P and Q such that systems (24) and 4 are given by

$$P\overline{Y} - QF(\overline{Y}) + C = 0, \tag{26}$$

and the exact form of the system is

$$PY - QF(Y) + C + L(h) = 0, \tag{27}$$

where

$$P = \left[\begin{array}{c|c} P_{11} & \mathbf{O} \\ \hline P_{21} & P_{22} \end{array}\right], \quad Q = \left[\begin{array}{c|c} Q_{11} & \mathbf{O} \\ \hline Q_{21} & \mathbf{O} \end{array}\right]. \tag{28}$$

P_{ij} and Q_{ij} are $N \times N$ matrices, $\mathbf{O}$ is the zero matrix, $\overline{Y} = (hy'_0, y_1, \ldots, y_{N-1}, hy'_1, \ldots, hy'_N)^T$, $F = (hf'_0, f_1, \ldots, f_{N-1}, hf'_1, \ldots, hf'_N)^T$, C is a vector of constants, and $L(h)$ is the truncation error vector of the formulas in (24) and 4.

Lemma 2. *Let P be a 2×2 block lower triangular matrix given by*

$$P = \left[\begin{array}{c|c} P_{11} & \mathbf{O} \\ \hline P_{21} & P_{22} \end{array}\right], \tag{29}$$

where each submatrix is of order N and $\mathbf{O}$ is the zero matrix. Then, P is invertible if and only if P_{11} and P_{22} are invertible. Moreover,

$$P^{-1} = \left[\begin{array}{c|c} P_{11}^{-1} & \mathbf{O} \\ \hline -P_{22}^{-1}P_{21}P_{11}^{-1} & P_{22}^{-1} \end{array}\right]. \tag{30}$$

P_{22} is an identity matrix so that $\|P_{22}^{-1}\| = 1$. Thus, to obtain an estimate for $\|P^{-1}\|$, it suffices to show the existence of the inverse of P_{11}.

Now, we define

$$\overline{P}_{11} = D_{11} - P_{11}, \tag{31}$$

where $D_{11} = \operatorname{diag}(P_{11})$ so that

$$D_{11}^{-1}P_{11} = I - D_{11}^{-1}\overline{P}_{11} \tag{32}$$

and consequently P_{11} is nonsingular provided $\rho(D_{11}^{-1}\overline{P}_{11}) < 1$ ([21]).

Thus, P^{-1} exists provided $\rho(D_{11}^{-1}\overline{P}_{11}) < 1$.

Lemma 3. *If $H < 1/h^2\|P^{-1}\|\|Q\|$, then the matrix $A = P - h^2QJ$ is monotone, that is $A^{-1} > 0$, where J is also a 2×2 block matrix of first partial derivatives and $H = \max\{|\partial f_i/\partial y_i|, |\partial f_i/\partial y'_i|, i = 1(1)N\}$.*

Proof.

$$\begin{aligned} A &= P - h^2QJ \\ AP^{-1} &= I - h^2QJP^{-1} \\ PA^{-1} &= \left(I - h^2QJP^{-1}\right)^{-1} = I + \left(h^2QJP^{-1}\right) \\ &\quad + \left(h^2QJP^{-1}\right)^2 + \left(h^2QJP^{-1}\right)^3 + \cdots \\ &= \left[I + h^2QJP^{-1}\right] \\ &\quad \cdot \left[I + \left(h^2QJP^{-1}\right)^2 + \left(h^2QJP^{-1}\right)^4 + \cdots\right]. \end{aligned} \tag{33}$$

The two series converge provided the spectral radius $\rho(h^2QJP^{-1}) < 1$:

$$\begin{aligned} A^{-1} &= \left[P^{-1} + h^2P^{-1}QJP^{-1}\right] \\ &\quad \cdot \left[I + \left(h^2QJP^{-1}\right)^2 + \left(h^2QJP^{-1}\right)^4 + \cdots\right]. \end{aligned} \tag{34}$$

The infinite series is nonnegative. Thus, to show that A is monotone, it suffices to show that

$$\begin{aligned} P^{-1} + h^2P^{-1}QJP^{-1} &> 0 \\ P^{-1} &> h^2P^{-1}QJP^{-1} \\ I &> h^2P^{-1}QJ \\ \left\|h^2P^{-1}QJ\right\| &\le h^2\left\|P^{-1}\right\|\|Q\|\|J\| < 1 \end{aligned} \tag{35}$$

for $\|J\| = H < 1/h^2\|P^{-1}\|\|Q\|$. □

Theorem 4. *Let $\overline{Y}$ be an approximation of the solution vector Y for the system obtained on a partition $\pi_N := \{a = x_0 < x_1 < x_2 < \cdots < x_{m-1} < x_m = b\}$ from systems (24) and 4. If $e_i = |y_i - y(x_i)|$ and $e'_i = |y'_i - y'(x_i)|$, where the exact solution $y(x)$ is assumed to be several times differentiable on $[a, b]$, and if $\|E\| = \|Y - \overline{Y}\|$, then, for sufficiently small h, $\|E\| = O(h^{2\nu+2})$.*

Proof. Subtracting (27) from (26), we obtain

$$AE = L(h). \tag{36}$$

Under the conditions of Lemma 3, A^{-1} exists and is nonnegative. Therefore,

$$\begin{aligned} E &= \left(P - h^2QJ\right)^{-1}L(h) \\ &= \left(I - h^2P^{-1}QJ\right)^{-1}P^{-1}L(h) \\ \|E\| &\le \left\|\left(I - h^2P^{-1}QJ\right)^{-1}\right\|\left\|P^{-1}\right\|\|L(h)\| \\ &\le \frac{\left\|P^{-1}\right\|\|L(h)\|}{1 - h^2\left\|P^{-1}\right\|\|Q\|\|J\|} \end{aligned} \tag{37}$$

provided $h^2\|P^{-1}\|\|Q\|\|J\| < 1$. Hence,

$$\|E\| \leq \frac{\|P^{-1}\| O\left(h^{2\nu+4}\right)}{1 - h^2 H \|P^{-1}\| \|Q\|} = O\left(h^{2\nu+2}\right). \quad (38)$$

□

5. Use of Methods

In this section, we discuss the use of methods in (16) and (17) for $n = 0(2\nu)(N - 2\nu)$, where N is a multiple of 2ν. We emphasize that the methods in (16) and (17) are all main methods since they are weighted the same and their use leads to a single matrix equation which can be solved for the unknowns. For example, for BVM6, we make use of each of the methods above in steps of 6; that is, $n = 0, 6, \ldots, N - 6$. This results in a system of $2N$ equations in $2N$ unknowns which can be easily solved for the unknowns. Below is an algorithm for the use of the methods.

The methods are implemented as BVMs by efficiently using the following steps.

Step 1. Use the methods in (16) and (17) for $n = 0$ to obtain $\mathbf{Y}_1$ in the interval $[y_0, y_{2\nu}]$ and for $n = 1$ $\mathbf{Y}_2$ is obtained in the interval $[y_{2\nu}, y_{4\nu}]$. Similarly, for $n = 2, 3, \ldots, (\Gamma - 1)$, we obtain $\mathbf{Y}_3, \ldots, \mathbf{Y}_\Gamma$, where $N = 2\nu \times \Gamma$ in the intervals, $[y_{4\nu}, y_{6\nu}], [y_{6\nu}, y_{8\nu}], \ldots, [y_{N-2\nu}, y_N]$, respectively.

Step 2. The unified block given by the system $\mathbf{Y}_1 \bigcup \mathbf{Y}_2 \bigcup \cdots \bigcup \mathbf{Y}_{\Gamma-1} \bigcup \mathbf{Y}_\Gamma$ obtained in Step 1 results in a system of $2N$ equations in $2N$ unknowns which can be easily solved.

Step 3. The values of the solution and the first derivatives of (1) are generated by the sequence $\{y_n\}$, $\{y_n'\}$, $n = 0, \ldots, N$, obtained as the solution in Step 2.

We note that all computations were carried out in Mathematica 10.0 enhanced by the feature FindRoot[ ].

6. Numerical Examples

In this section, we consider seven numerical examples. Examples 1 to 5 were solved using the BVMs $k = 4$, $k = 6$, and $k = 8$ (derived in this paper) of orders 6, 8, and 10, respectively. Also, these examples were solved using the Extended Trapezoidal Methods of the second kind (ETRs) and the Top Order Methods (TOMs) given in [5] of orders 6 and 10, respectively. Comparisons are made between the BVM $k = 4$ and the ETRs [5] as well as between the BVM $k = 8$ and the TOMs [5] by obtaining the maximum errors Ey in the interval of integration. We also compared our methods with the Sinc-Collocation method [20]. Examples 6 and 7 were solved using the BVMs of order 6. We note that the number of function evaluations (NFEs) involved in implementing the BVMs is $N \times 2\nu$ in the entire range of integration. The code was based on Newton's method which uses the feature FindRoot[ ] or NSolve[ ] for linear problems in Mathematica. The efficiency curves show the plot of the logarithm of Ey against the number of function evaluations for each method.

Example 1. We consider the linear system of second-order boundary value problems given in [20]

$$\frac{d^2y_1}{dx^2} + (2x-1)\frac{dy_1}{dx} + \cos(\pi x)\frac{dy_2}{dx} = f_1(x), \quad 0 < x < 1$$
$$\frac{d^2y_2}{dx^2} + xy_1 = f_2(x) \quad (39)$$
$$y_1(0) = y_2(0) = y_1(1) = y_2(1) = 0,$$

where

$$f_1(x) = -\pi^2 \sin(\pi x) + (2x-1)(\pi+1)\cos(\pi x),$$
$$f_2(x) = 2 + x\sin(\pi x) \quad (40)$$
$$\text{Exact: } y_1(x) = \sin(\pi x),$$
$$y_2(x) = x^2 - x.$$

This problem was solved using the ETRs and BVM of order 6 as well as the TOMs and BVM of order 10. The maximum Euclidean norm of the absolute errors in y_1 and y_2 was obtained in the entire interval of integration. In Table 1, we compared the Sinc-Collocation method [20] with the BVM of order 8. Table 2 shows the comparison between the ETRs, BVM4, TOMs, and BVM8. While the results of these methods are of approximate accuracy, we emphasize that the TOMs and ETRs use 20 function evaluations per step while the BVM4 and BVM8 use $8N$ and $16N$ function evaluations for this system. Hence, the BVMs are quite accurate and efficient. We also calculated the Rate of Convergence (ROC) using the formula $\log_2(E^{2h}/E^h)$, where E^h is the error obtained using step size h. The ROC of the BVM4 and ETRs shows that these methods are consistent with the theoretical order (order 6) behavior of the methods. We omit the ROC of the TOMs and BVM8 because their errors are mainly due to round-off errors rather than to truncation errors. Figure 1 also shows the efficiency curves of these methods.

Example 2. Consider the nonlinear BVP given in [22]

$$\frac{d^2y_1}{dx^2} + xy_1 + 2xy_2 + xy_1^2 = f_1(x), \quad 0 < x < 1$$
$$\frac{d^2y_2}{dx^2} + y_2 + x^2y_1 + \sin(x)\,y_2^2 = f_2(x) \quad (41)$$
$$y_1(0) = y_2(0) = y_1(1) = y_2(1) = 0,$$

where

$$f_1(x) = -2 + x\left(x - x^2\right) + x\left(x - x^2\right)^2 - 2x\sin(\pi x)$$

TABLE 1: Maximum errors for Example 1.

	Sinc-Coll. [20]			BVM6	
N	Ey_1	Ey_2	N	Ey_1	Ey_2
20	$3.128e-05$	$1.175e-06$	18	$5.309e-09$	$2.853e-10$
40	$1.829e-07$	$5.095e-09$	36	$2.332e-11$	$1.154e-12$
60	$3.573e-09$	$7.696e-11$	54	$8.883e-13$	$4.535e-14$
80	$1.287e-10$	$2.267e-11$	78	$4.763e-14$	$1.915e-15$
100	$6.389e-12$	$1.026e-13$	96	$9.326e-15$	$1.388e-16$

TABLE 2: Maximum Errors for different stepsizes for Example 1.

N	ETRs [5]	ROC	BVM4	ROC	TOMs [5]	BVM8
20	$7.281e-8$		$1.391e-7$		$5.736e-11$	$6.952e-12$
40	$1.165e-9$	5.96	$2.267e-9$	5.94	$1.958e-14$	$4.514e-14$
80	$2.023e-11$	5.86	$3.530e-11$	6.01	$3.818e-16$	$3.856e-16$
160	$3.325e-13$	5.93	$5.476e-13$	6.01	$4.578e-16$	$6.621e-16$

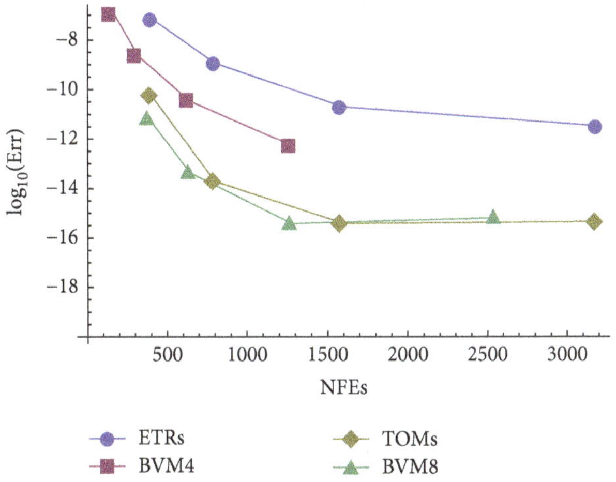

FIGURE 1: Efficiency curve for Example 1.

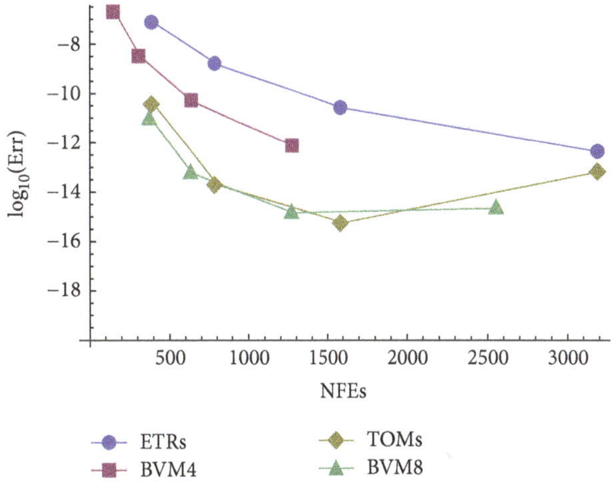

FIGURE 2: Efficiency curve for Example 2.

$$f_2(x) = x^3(1-x) + \sin(\pi x)(1+\sin(x)\sin(\pi x)) - \pi^2 \sin(\pi x)$$

$$\text{Exact: } y_1(x) = x - x^2, \quad y_2(x) = \sin(\pi x). \tag{42}$$

The maximum Euclidean norm of the absolute errors in y_1 and y_2 was obtained in the range of integration. Table 3 shows the comparison between the ETRs, BVM4, TOMs, and BVM8. While the results of these methods are of approximate accuracy, we emphasize that the TOMs and ETRs use 20 function evaluations per step while the BVM4 and BVM8 use $8N$ and $16N$ function evaluations for this system. We also calculated the ROC of the BVM4 and ETRs which shows that these methods are consistent with the theoretical order (order 6) behavior of the methods. We do not calculate the ROC of the TOMs and BVM8 because their errors are mainly due to round-off errors rather than to truncation errors. Figure 2 shows the efficiency curves of these methods.

Example 3. We consider the nonlinear BVP with mixed boundary conditions given in [23]

$$y'' = \frac{(y')^2 + y^2}{2e^x}, \quad 0 < x < 1$$

$$y(0) - y'(0) = 0, \quad y(1) + y'(1) = 2e \tag{43}$$

$$\text{Exact: } y(x) = e^x.$$

This problem was chosen to demonstrate the performance of the BVMs on nonlinear BVPs with mixed boundary

TABLE 3: Maximum Errors for different stepsizes for Example 2.

N	ETRs [5]	ROC	BVM4	ROC	TOMs [5]	BVM8
20	$7.448e-08$		$1.895e-07$		$3.189e-11$	$9.489e-12$
40	$1.480e-09$	5.65	$3.061e-09$	5.95	$1.972e-14$	$6.021e-14$
80	$2.576e-11$	5.84	$4.772e-11$	6.00	$5.722e-16$	$1.450e-15$
160	$4.234e-13$	5.93	$7.538e-13$	5.98	$6.732e-14$	$2.229e-15$

TABLE 4: Maximum errors for different step sizes for Example 3.

N	ETRs [5]	ROC	BVM4	ROC	TOMs [5]	BVM8
20	$2.476e-10$		$1.505e-10$		$8.882e-16$	$4.441e-16$
40	$6.019e-12$	5.36	$2.347e-12$	6.00	$4.441e-16$	$4.441e-16$
80	$6.402e-14$	6.55	$3.642e-14$	6.01	$4.441e-16$	$4.441e-16$
160	$1.010e-15$	5.99	$6.661e-16$	5.77	$4.441e-16$	$4.441e-16$

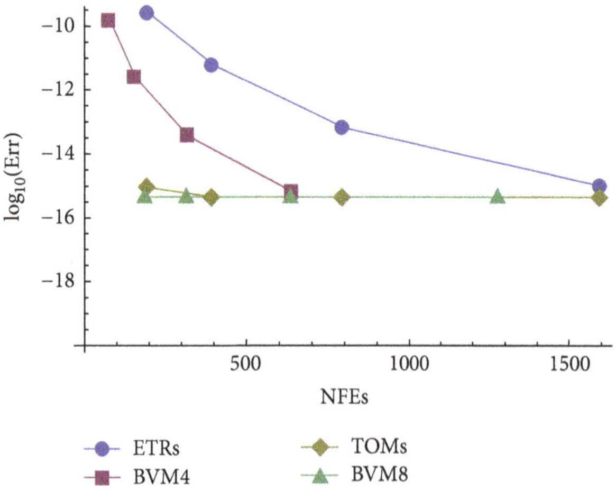

FIGURE 3: Efficiency curve for Example 3.

conditions. The maximum absolute errors were obtained in the range of integration. Table 4 shows the comparison between the ETRs, BVM4, TOMs, and BVM8. Figure 3 shows the efficiency curves of these methods.

Example 4. Consider the second-order BVP given in [24] (bvpT17)

$$\begin{aligned} \frac{d^2y}{dx^2} &= \frac{-3\lambda y}{(\lambda+x^2)^2}, \qquad -0.1 < x < 0.1 \\ y(-0.1) &= \frac{-0.1}{\sqrt{\lambda+0.001}}, \\ y(0.1) &= \frac{0.1}{\sqrt{\lambda+0.001}} \\ \text{Exact: } y(x) &= \frac{x}{\sqrt{\lambda+x^2}}. \end{aligned} \tag{44}$$

In order to assess the efficiency of our methods, we solve the boundary layer problem given in [24] (bvpT17). The maximum absolute errors were obtained in the range of integration. Tables 5 and 6 show the comparison between the ETRs, BVM4, TOMs, and BVM8 with $\lambda = 1$ and 0.1, respectively. Figure 4 shows the plot of the solution for values of $\lambda = 1, 0.1, 0.01, 0.001$ and the solution has a boundary layer at $x = 0$.

Example 5. Consider the second-order BVP given in [24] (bvpT20)

$$\begin{aligned} \lambda\frac{d^2y}{dx^2} &= -\left(\frac{dy}{dx}\right)^2 + 1, \quad 0 < x < 1 \\ y(0) &= 1 + \lambda\log\left(\cosh\left(-\frac{0.745}{\lambda}\right)\right), \\ y(1) &= 1 + \lambda\log\left(\cosh\left(\frac{0.255}{\lambda}\right)\right) \\ \text{Exact: } y(x) &= 1 + \lambda\log\left(\cosh\left(x - \frac{0.745}{\lambda}\right)\right). \end{aligned} \tag{45}$$

Also, the efficiency of the scheme is shown by solving the problem given in [24] (bvpT20). The maximum absolute errors were obtained in the range of integration. Tables 7 and 8 show the comparison between the ETRs, BVM4, TOMs, and BVM8 with $\lambda = 1$ and 0.1, respectively. Figure 5 shows the plot of the solution for values of $\lambda = 1, 0.1, 0.01$. From the figure, we see that the solution of the problem has a corner layer at $x = 0.745$.

Example 6. We consider the Poisson equation given in [25]

$$\begin{aligned} u_{xx}(x,y) + u_{yy}(x,y) &= g(x,y) \quad \text{on } R, \\ u(x,0) &= u(x,1), \\ u(x,2) &= e^{-2\pi}\sin(\pi x), \\ & \qquad 0 \le x \le 1 \\ u(0,y) &= \sin(\pi y), \\ u(1,y) &= e^{\pi}\sin(\pi y), \\ & \qquad 0 \le y \le 2, \end{aligned} \tag{46}$$

TABLE 5: Maximum Errors for different stepsizes for Example 4 for $\lambda = 1$.

N	ETRs [5]	ROC	BVM4	ROC	TOMs [5]	BVM8
20	$2.178e-12$		$8.635e-14$		$1.804e-16$	$4.163e-17$
40	$3.782e-14$	5.85	$1.200e-15$	6.17	$1.457e-16$	$4.857e-17$
80	$6.245e-16$	5.92	$1.110e-16$	3.43	$3.469e-17$	$1.749e-16$
160	$5.551e-17$	3.49	$3.747e-16$	1.76	$7.633e-17$	$4.163e-17$

TABLE 6: Maximum Errors for different stepsizes for Example 4 for $\lambda = 0.1$.

N	ETRs [5]	ROC	BVM4	ROC	TOMs [5]	BVM8
20	$3.559e-09$		$1.201e-09$		$5.038e-12$	$5.201e-14$
40	$5.279e-11$	6.07	$1.820e-11$	6.04	$2.498e-13$	$4.663e-15$
80	$7.931e-13$	6.06	$2.902e-13$	5.97	$1.110e-16$	$1.110e-16$
160	$1.221e-14$	6.02	$6.106-15$	5.57	$1.665e-16$	$4.163e-16$

TABLE 7: Maximum Errors for different stepsizes for Example 5 for $\lambda = 1$.

N	ETRs [5]	ROC	BVM4	ROC	TOMs [5]	BVM8
20	$2.235e-09$		$1.664e-09$		$1.998e-14$	$4.452e-14$
40	$3.570e-11$	5.96	$2.823e-11$	5.88	$1.776e-15$	$4.441e-15$
80	$5.607e-13$	5.99	$4.370e-13$	6.01	$6.661e-16$	$6.661e-16$
160	$9.104e-15$	5.94	$6.883e-15$	5.99	$6.661e-16$	$8.882e-16$

TABLE 8: Maximum Errors for different stepsizes for Example 5 for $\lambda = 0.1$.

N	ETRs [5]	ROC	BVM4	ROC	TOMs [5]	BVM8
20	$7.808e-05$		$2.000e-04$		$6.950e-08$	$8.527e-06$
40	$1.838e-06$	5.41	$4.090e-06$	5.61	$6.612e-09$	$4.707e-06$
80	$3.163e-08$	5.86	$5.784e-08$	6.14	$4.674e-12$	$8.212e-09$
160	$5.176e-10$	5.93	$7.066e-10$	6.36	$2.665e-15$	$6.771e-12$

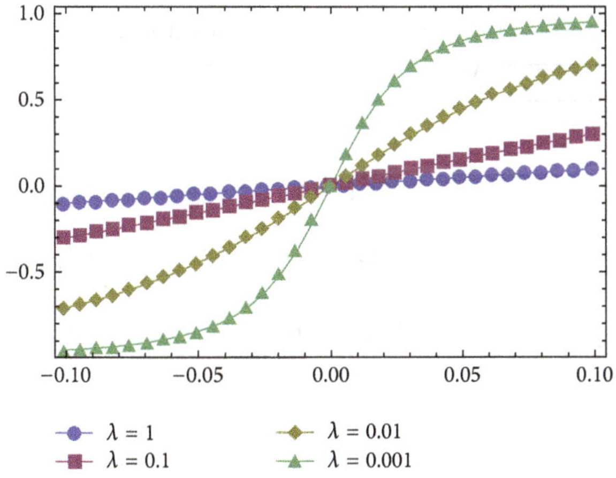

FIGURE 4: Plot of solution for Example 4.

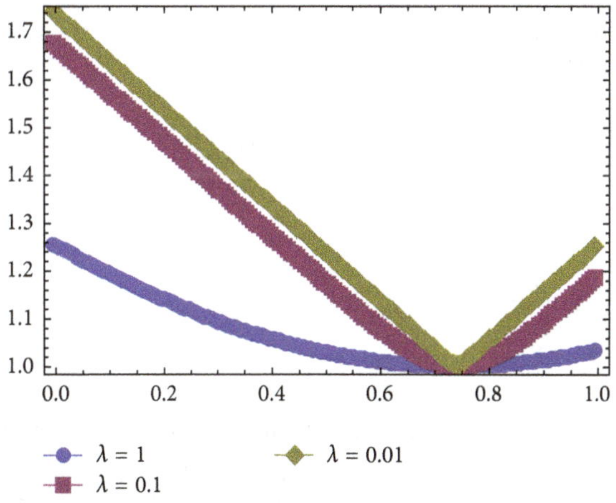

FIGURE 5: Plot of solution for Example 5.

where $R = \{(x, y) : 0 \le x \le 1,\ 0 \le y \le 2\}$ and $g(x, y) = 2\pi(2\pi y^2 - 2\pi y - 1)e^{\pi y(1-y)} \sin(\pi y)$:

$$\text{Exact: } u(x, y) = e^{\pi y} \sin(\pi y) + e^{\pi y(1-y)} \sin(\pi x). \quad (47)$$

This example shows the performance of the BVMs on the Poisson equation. In order to solve the equation using the BVMs, we carry out the semidiscretization of the spatial variable x using the second-order finite difference method

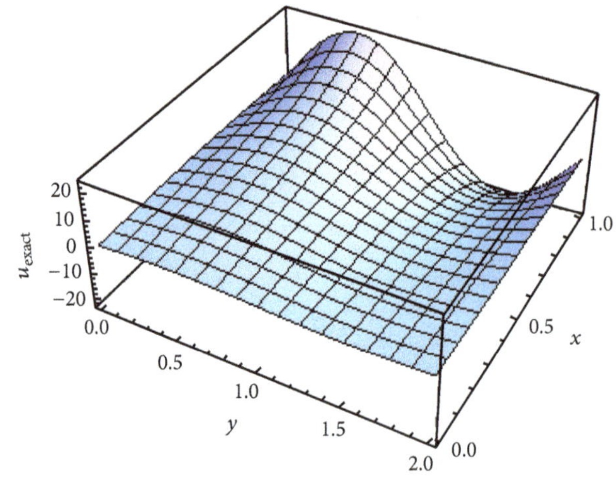

(a) Exact

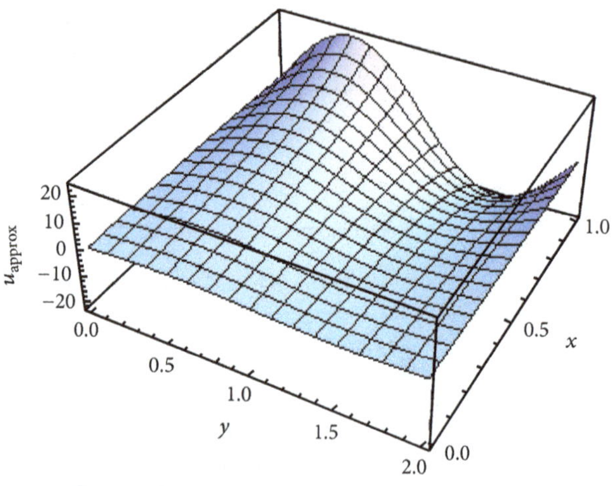

(b) Approximate

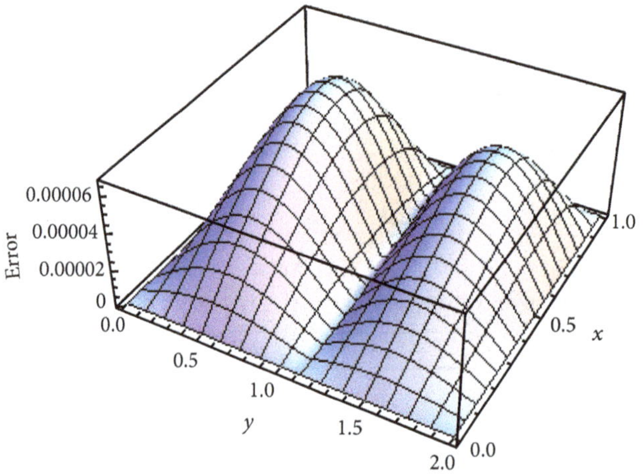

(c) Error function

FIGURE 6: Plot of solution for the Poisson equation.

to obtain the following second-order system in the second variable y:

$$\frac{\partial^2 u_m}{\partial y^2} + \frac{u_{m+1}(y) - 2u_m(y) + u_{m-1}(y)}{(\Delta x)^2} = g_m(y),$$

$$m = 1, \ldots, M-1, \quad (48)$$

$$u(x_m, 0) = u(x_m, 1),$$

$$u(x_m, 2) = e^{-2\pi} \sin(\pi x_m),$$

where $\Delta x = (b-a)/M$, $x_m = a + m\Delta x$, $m = 0, 1, \ldots, M$, $\mathbf{u} = [u_1(y), \ldots, u_M(y)]^T$, $\mathbf{g} = [g_1(y), \ldots, g_M(y)]^T$, $u_m(y) \approx u(x_m, y)$, and $g_m(y) \approx g(x_m, y)$ which can be written in the form

$$\mathbf{u}'' = \mathbf{f}(y, \mathbf{u}), \quad (49)$$

subject to the boundary conditions $\mathbf{u}(y_0) = \mathbf{u}(y_{M/2})$, $\mathbf{u}(y_M) = \mathbf{u}_M$, where $\mathbf{f}(y, \mathbf{u}) = \mathbf{A}\mathbf{u} + \mathbf{g}$ and $\mathbf{A}$ is an $M-1 \times M-1$ matrix arising from the semidiscretized system and $\mathbf{g}$ is a vector of constants. Table 9 shows the comparison between the BVM and the method in [25]. Figure 6 shows the plot of the exact, approximate, and error function of the problem.

TABLE 9: Maximum error for the Poisson equation on $y = 1$.

h	Method in [20]	BVM4
$\frac{1}{16}$	$3.266e-02$	$2.605e-03$
$\frac{1}{32}$	$8.210e-03$	$3.141e-04$
$\frac{1}{64}$	$2.053e-03$	$3.947e-05$
$\frac{1}{128}$	$5.128e-04$	$4.952e-06$
$\frac{1}{256}$	—	$6.199e-07$

Example 7. Lastly, we consider the Sine-Gordon nonlinear hyperbolic equation given in [26]

$$u_{yy}(x, y) = u_{xx}(x, y) + \sin(u) \quad \text{on } R, \quad (50)$$

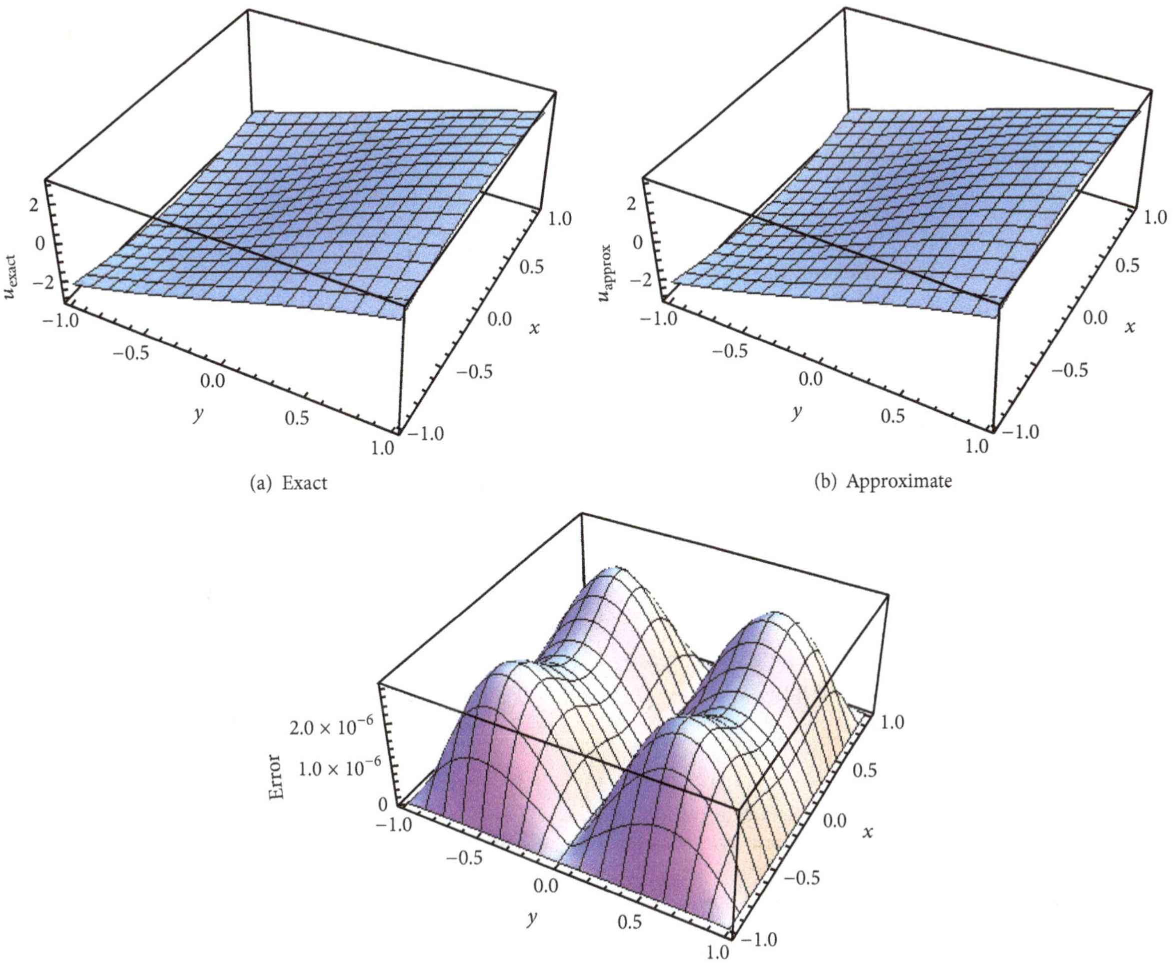

(a) Exact

(b) Approximate

(c) Error function

FIGURE 7: Plot of solution for the Sine-Gordon equation.

subject to the initial conditions

$$u(x,0) = 0,$$
$$u_y(x,0) = 4\,\text{sech}(\pi x), \qquad -1 \le x \le 1 \tag{51}$$

and the boundary conditions

$$u(0,y) = 4\tan^{-1}(y),$$
$$u(1,y) = 4\tan^{-1}(\text{sech}(1)\,y), \qquad -1 \le x \le 1, \tag{52}$$

where $R = \{(x,y) : -1 \le x \le 1,\ -1 \le y \le 1\}$. Hence,

$$\text{Exact: } u(x,y) = 4\tan^{-1}(\text{sech}(x)\,y). \tag{53}$$

This example shows the performance of the BVMs on the hyperbolic problem. We discretize the x variable using finite difference schemes to obtain the system

$$\frac{\partial^2 u_m}{\partial y^2} = \frac{u_{m+1}(y) - 2u_m(y) + u_{m-1}(y)}{(\Delta x)^2} + \sin(u_m(y)), \quad m = 1,\ldots,M-1, \tag{54}$$
$$u(x_m, 0) = 0,$$
$$u'(x_m, 0) = 4\,\text{sech}(\pi x_m),$$

where $\Delta x = (b-a)/M$, $x_m = a + m\Delta x$, $m = 0,1,\ldots,M$, $\mathbf{u} = [u_1(y),\ldots,u_M(y)]^T$, $\mathbf{g} = [g_1(y),\ldots,g_M(y)]^T$, $u_m(y) \approx u(x_m,y)$, and $g_m(y) \approx g(x_m,y)$ which can be written in the form

$$\mathbf{u}'' = \mathbf{f}(y,\mathbf{u}), \tag{55}$$

subject to the initial conditions $\mathbf{u}(y_0) = \mathbf{u}_0, \mathbf{u}'(y_0) = \mathbf{u}'_0$, where $\mathbf{f}(y,\mathbf{u}) = \mathbf{A}\mathbf{u} + \mathbf{g}$ and $\mathbf{A}$ is an $M-1 \times M-1$ matrix arising from the semidiscretized system and $\mathbf{g}$ is a vector of constants. Table 10 shows the computational results for this example using the BVM of order 6. Figure 7 shows the plot of the exact, approximate, and the error function of the problem.

TABLE 10: Maximum errors for the Sine-Gordon equation.

h	$\frac{1}{16}$	$\frac{1}{32}$	$\frac{1}{64}$	$\frac{1}{128}$	$\frac{1}{256}$
	$33.6e-04$	$4.7e-06$	$6.0e-07$	$7.6e-08$	$9.6e-09$

7. Conclusions

This paper is concerned with the solution of systems of second-order boundary value problems. This has been achieved by the construction and implementation of a family of BVMs. The methods are applied as a block unification method to obtain the solution on the entire interval of integration. We established the convergence of the methods. We have also shown that the methods are competitive with existing methods cited in the literature.

In the future, we would like to develop a variable step size version of the BVMs with an automatic error estimation.

Competing Interests

The authors declare that there are no competing interests regarding the publication of this manuscript.

References

[1] L. Brugnano and D. Trigiante, "On the characterization of stiffness for ODEs," *Dynamics of Continuous, Discrete and Impulsive Systems*, vol. 2, no. 3, pp. 317–335, 1996.

[2] L. Brugnano, F. Mazzia, and D. Trigiante, "Fifty years of stiffness," in *Recent Advances in Computational and Applied Mathematics*, T. E. Simos, Ed., pp. 1–21, Springer, 2011.

[3] F. Iavernaro, F. Mazzia, and D. Trigiante, "Stability and conditioning in numerical analysis," *Journal of Numerical Analysis, Industrial and Applied Mathematics*, vol. 1, no. 1, pp. 91–112, 2006.

[4] L. Brugnano and D. Trigiante, "High-order multistep methods for boundary value problems," *Applied Numerical Mathematics*, vol. 18, no. 1–3, pp. 79–94, 1995.

[5] L. Brugnano and D. Trigiante, *Solving Differential Problems by Multistep Initial and Boundary Value Methods*, vol. 6 of *Stability and Control: Theory, Methods and Applications*, Gordon and Breach Science, Amsterdam, The Netherlands, 1998.

[6] L. Brugnano and D. Trigiante, "Stability properties of some boundary value methods," *Applied Numerical Mathematics*, vol. 13, no. 4, pp. 291–304, 1993.

[7] P. Amodio and F. Mazzia, "A boundary value approach to the numerical solution of initial value problems by multistep methods," *Journal of Difference Equations and Applications*, vol. 1, no. 4, pp. 353–367, 1995.

[8] P. Amodio and L. Brugnano, "Parallel implementation of block boundary value methods for ODEs," *Journal of Computational and Applied Mathematics*, vol. 78, no. 2, pp. 197–211, 1997.

[9] D. O. Awoyemi, "A class of continuous methods for general second order initial value problems in ordinary differential equations," *International Journal of Computer Mathematics*, vol. 72, no. 1, pp. 29–37, 1999.

[10] J. Vigo-Aguilar and H. Ramos, "Variable stepsize implementation of multistep methods for $y'' = f(x, y, y')$," *Journal of Computational and Applied Mathematics*, vol. 192, no. 1, pp. 114–131, 2006.

[11] J. D. Lambert and I. A. Watson, "Symmetric multistep methods for periodic initial value problems," *Journal of the Institute of Mathematics and Its Applications*, vol. 18, no. 2, pp. 189–202, 1976.

[12] P. Amodio and F. Iavernaro, "Symmetric boundary value methods for second order initial and boundary value problems," *Mediterranean Journal of Mathematics*, vol. 3, no. 3-4, pp. 383–398, 2006.

[13] T. A. Biala, "A computational study of the boundary value methods and the block unification methods for $y'' = f(x, y, y')$," *Abstract and Applied Analysis*, vol. 2016, Article ID 8465103, 14 pages, 2016.

[14] T. A. Biala and S. N. Jator, "A boundary value approach for solving three-dimensional elliptic and hyperbolic partial differential equations," *SpringerPlus Journals*, vol. 4, no. 1, article no. 588, 2015.

[15] S. N. Jator and J. Li, "An algorithm for second order initial and boundary value problems with an automatic error estimate based on a third derivative method," *Numerical Algorithms*, vol. 59, no. 3, pp. 333–346, 2012.

[16] L. Aceto, P. Ghelardoni, and C. Magherini, "PGSCM: a family of P-stable Boundary Value Methods for second-order initial value problems," *Journal of Computational and Applied Mathematics*, vol. 236, no. 16, pp. 3857–3868, 2012.

[17] P. Onumanyi, U. W. Sirisena, and S. N. Jator, "Continuous finite difference approximations for solving differential equations," *International Journal of Computer Mathematics*, vol. 72, no. 1, pp. 15–27, 1999.

[18] T. A. Biala, S. N. Jator, and R. B. Adeniyi, "Numerical approximations of second order PDEs by boundary value methods and the method of lines," *Afrika Matematika*, 2016.

[19] P. L. Ndukum, T. A. Biala, S. N. Jator, and R. B. Adeniyi, "On a family of trigonometrically fitted extended backward differentiation formulas for stiff and oscillatory initial value problems," *Numerical Algorithms*, vol. 74, no. 1, pp. 267–287, 2017.

[20] M. El-Gamel, "Sinc-collocation method for solving linear and nonlinear system of second-order boundary value problems," *Applied Mathematics*, vol. 03, no. 11, pp. 1627–1633, 2012.

[21] P. A. Cook, "Estimates for the inverse of a matrix," *Linear Algebra and Its Applications*, vol. 10, pp. 41–53, 1975.

[22] F. Geng and M. Cui, "Homotopy perturbation-reproducing kernel method for nonlinear systems of second order boundary value problems," *Journal of Computational and Applied Mathematics*, vol. 235, no. 8, pp. 2405–2411, 2011.

[23] R. S. Stepleman, "Tridiagonal fourth order approximations to general two-point nonlinear boundary value problems with mixed boundary conditions," *Mathematics of Computation*, vol. 30, pp. 92–103, 1976.

[24] https://archimede.dm.uniba.it/~bvpsolvers/testsetbvpsolvers/?page_id=29.

[25] E. A. Volkov, A. A. Dosiyev, and S. C. Buranay, "On the solution of a nonlocal problem," *Computers & Mathematics with Applications*, vol. 66, no. 3, pp. 330–338, 2013.

[26] M. Dehghan and A. Shokri, "A numerical method for one-dimensional nonlinear sine-Gordon equation using collocation and radial basis functions," *Numerical Methods for Partial Differential Equations*, vol. 24, no. 2, pp. 687–698, 2008.

4

Affine Discontinuous Galerkin Method Approximation of Second-Order Linear Elliptic Equations in Divergence form with Right-Hand Side in L^1

Abdeluaab Lidouh and Rachid Messaoudi

Department of Mathematics and Computer Science, Laboratory LACSA, Faculty of Sciences, Mohammed I[st] University, BV Mohammed VI, P.O. Box 717, 60000 Oujda, Morocco

Correspondence should be addressed to Rachid Messaoudi; m_rachid_ens@yahoo.fr

Academic Editor: Patricia J. Y. Wong

We consider the standard **affine** discontinuous Galerkin method approximation of the second-order linear elliptic equation in divergence form with coefficients in $L^\infty(\Omega)$ and the right-hand side belongs to $L^1(\Omega)$; we extend the results where the case of **linear** finite elements approximation is considered. We prove that the unique solution of the discrete problem converges in $W_0^{1,q}(\Omega)$ for every q with $1 \le q < d/(d-1)$ ($d = 2$ or $d = 3$) to the unique renormalized solution of the problem. Statements and proofs remain valid in our case, which permits obtaining a weaker result when the right-hand side is a bounded Radon measure and, when the coefficients are smooth, an error estimate in $W_0^{1,q}(\Omega)$ when the right-hand side f belongs to $L^r(\Omega)$ verifying $T_k(f) \in H^1(\Omega)$ for every $k > 0$, for some $r > 1$.

1. Introduction

In this work we consider, in dimension $d = 2$ or 3, the $\mathbb{P}_1$ discontinuous Galerkin (dG) method approximation of the Dirichlet problem

$$-\operatorname{div}(A\nabla u) = f \quad \text{in } \Omega, \qquad u = 0 \quad \text{on } \partial\Omega, \tag{1}$$

where Ω is an open bounded set of $\mathbb{R}^d$, A is a coercive matrix with coefficients in $L^\infty(\Omega)$, and f belongs to $L^1(\Omega)$.

The solution of (1) does not belong to $H_0^1(\Omega)$ for a general right-hand side in $L^1(\Omega)$. Actually, in order to correctly define the solution of (1), one has to consider a specific framework, the concept of renormalized (or equivalently entropy) solution (see for example [1, 2]). These definitions allow one to prove that in this new sense problem (1) is well-posed in the terminology of Hadamard.

For this problem the standard $\mathbb{P}_1$-nonconforming finite elements approximation, related to a triangulation $\mathcal{T}_h$ of Ω, namely,

$$\text{Find } u_h \in V_h, \quad \forall v_h \in V_h, \quad a_h^{swip}(u_h, v_h) = \int_\Omega f v_h \, dx, \tag{2}$$

where

$$V_h = \left\{ v_h \in L^2(\Omega) : \forall T \in \mathcal{T}_h,\ v_h|_T \in \mathbb{P}_1[T],\ \forall F \in \mathcal{F}_h,\ \int_F [v_h] = 0 \right\} \tag{3}$$

with the discrete bilinear form a_h^{swip} yet to be designed, has a unique solution, since the right-hand side (2) $\int_\Omega f v_h dx$ is correctly defined for $f \in L^1(\Omega)$ and the bilinear form a_h^{swip} is consistent.

Using the ideas which are at the root of the SWIP (Symmetric Weighted Interior Penalty) method, in the case $f \in L^2(\Omega)$, D. A. Di Pietro and A. Ern have proved, in [3], that the unique solution u_h of (2) converges to the unique solution u of (1) in the following sense:

$$\begin{aligned} u_h &\longrightarrow u \quad \text{strongly in } L^2(\Omega), \\ \nabla_h u_h &\longrightarrow \nabla u \quad \text{strongly in } \left[L^2(\Omega)\right]^d, \\ |u_h|_J &\longrightarrow 0, \\ &\textit{when } h \longrightarrow 0, \end{aligned} \tag{4}$$

with the broken gradient ∇_h and the jump seminorm $|\cdot|_J$ yet to be designed.

To do that, the authors in [3] assume that the family of triangulations $\mathcal{T}_h$ belong to an admissible mesh sequence in the sense of 17 and is compatible with the partition $\mathcal{P}_\Omega$ (see Assumption 3).

The framework in this paper is the same as in [3]. The unique difference here is that $f \in L^1(\Omega)$ is considered instead of $f \in L^2(\Omega)$; and we ourselves focus on the two cases $d = 2$ and $d = 3$. The same convergence results are proved.

Notations. In the present work, Ω denotes an open bounded subset of $\mathbb{R}^d$ with $d = 2$ or $d = 3$. A particular case is the case where Ω is an open bounded polyhedron. We use the notation Avw for the scalar product of the vector Av by the vector w (which is often denoted by ${}^t w{\cdot}Av$). For a measurable set $S \subset \Omega$, we denote by $|S|$ the measure of S and by S^c the complement $\Omega \setminus S$ of S.

For $1 < p < +\infty$ and $m \geq 0$, we have

$$\begin{aligned} H^m(\mathcal{T}_h) &:= \left\{v \in L^2(\Omega) / \forall T \in \mathcal{T}_h,\ v_{|T} \in H^m(T)\right\}, \\ W^{m,p}(\mathcal{T}_h) & \\ &:= \left\{v \in L^p(\Omega) / \forall T \in \mathcal{T}_h,\ v_{|T} \in W^{m,p}(T)\right\}. \end{aligned} \tag{5}$$

We define also the following function spaces:

$$\begin{aligned} H(\operatorname{div}, \Omega) &:= \left\{\tau \in \left[L^2(\Omega)\right]^d / \nabla \cdot \tau \in L^2(\Omega)\right\}, \\ H(\operatorname{div}, \mathcal{T}_h) & \\ &:= \left\{\tau \in \left[L^2(\Omega)\right]^d / \forall T \in \mathcal{T}_h,\ \tau_{|T} \in H(\operatorname{div}, T)\right\}. \end{aligned} \tag{6}$$

For $k \geq 0$, we define the broken polynomial space

$$\mathbb{P}_d^k(\mathcal{T}_h) := \left\{v \in L^2(\Omega) / \forall T \in \mathcal{T}_h,\ v_{|T} \in \mathbb{P}_d^k(T)\right\}, \tag{7}$$

with polynomial degree $k \geq 1$.

In that case, $\mathbb{P}_d^1(\mathcal{T}_h) \subset H^1(\mathcal{T}_h)$, which leads us to define the broken gradient $\nabla_h : H^1(\mathcal{T}_h) \to [L^2(\Omega)]^d$ such that, $\forall v \in H^1(\mathcal{T}_h)$,

$$\forall T \in \mathcal{T}_h: \quad (\nabla_h v)|_T := \nabla\left(v_{|T}\right), \tag{8}$$

and the broken divergence operator $\nabla_h : H(\operatorname{div}, \mathcal{T}_h) \to [L^2(\Omega)]^d$ such that, $\forall v \in H(\operatorname{div}, \mathcal{T}_h)$,

$$\forall T \in \mathcal{T}_h: \quad (\nabla_h \cdot v)|_T := \nabla \cdot \left(v_{|T}\right). \tag{9}$$

Moreover, for any mesh element $T \in \mathcal{T}_h$, we denote

$$\begin{aligned} \forall T \in \mathcal{T}_h, \quad \mathcal{F}_T &:= \left\{F \in \mathcal{F}_h / F \in \partial T\right\}, \\ \forall F \in \mathcal{F}_h, \quad \mathcal{T}_F &:= \left\{T \in \mathcal{T}_h / F \in \partial T\right\}. \end{aligned} \tag{10}$$

And for a scalar-valued function v defined on Ω (which can admit two possible traces) the average of v is defined as

$$\begin{aligned} \{v\}_F(x) &:= \frac{1}{2}\left(v_{|T_1}(x) + v_{|T_2}(x)\right); \\ &\quad \textit{if } F \in \mathcal{F}_h^i \textit{ such that } F \subset \partial T_1 \cap \partial T_2, \\ \{v\}_F(x) &:= v_{|T}(x), \\ &\quad \textit{if } F \in \mathcal{F}_h^b \textit{ such that } F \subset \partial T \cap \partial\Omega, \end{aligned} \tag{11}$$

and the jump of v as

$$\begin{aligned} [v]_F(x) &:= v_{|T_1}(x) - v_{|T_2}(x), \\ &\quad \textit{if } F \in \mathcal{F}_h^i \textit{ such that } F \subset \partial T_1 \cap \partial T_2, \\ [v]_F(x) &:= v_{|T}(x), \\ &\quad \textit{if } F \in \mathcal{F}_h^b \textit{ such that } F \subset \partial T \cap \partial\Omega. \end{aligned} \tag{12}$$

For any face $F \in \mathcal{F}_h$ and for any integer $l \geq 0$, we define the (local) lifting operator $r_{F,A}^l : L^2(F) \to \mathbb{P}_d^l(\mathcal{T}_h)$ as follows. For all $\varphi \in L^2(F)$,

$$\int_\Omega A r_{F,A}^l(\varphi) \cdot \tau_h = \int_F \{A\tau_h\} \cdot n_F\, \varphi \qquad \forall \tau_h \in \left[\mathbb{P}_d^l(\mathcal{T}_h)\right]^d, \tag{13}$$

and for any function $v \in H^1(\mathcal{T}_h)$, we define the (global) lifting of its interface and boundary jumps as

$$R_{h,A}^l([v]) = \sum_{F \in \mathcal{F}_h} r_{F,A}^l([v]) \in \left[\mathbb{P}_d^l(\mathcal{T}_h)\right]^d. \tag{14}$$

We also introduce the normal diffusion coefficient to one face F as

$$\begin{aligned} &\forall F \in \mathcal{F}_h^i, \quad F = \partial T_1 \cap \partial T_2: \\ &A_i := \left(A_{|T_i} n_F\right) \cdot n_F, \quad i \in \{1, 2\}, \end{aligned} \tag{15}$$

the diffusion-dependent penalty parameter (harmonic average of normal diffusion) as

$$\gamma_{A,F} := \begin{cases} \dfrac{2A_1A_2}{A_1 + A_2} & \textit{if } F \subset \partial T_1 \cap \partial T_2, \\ \left(A_{|T} n\right) \cdot n & \textit{if } F \subset \partial T \cap \partial\Omega, \end{cases} \tag{16}$$

the weighted average operator for all $F \in \mathcal{F}_h^i$ such that $F \subset \partial T_1 \cap \partial T_2$ as

$$\omega_{T_1,F} := \frac{A_2}{A_1 + A_2}, \qquad \omega_{T_2,F} := \frac{A_1}{A_1 + A_2}, \tag{17}$$

the weighted average operator for all $F \in \mathcal{F}_h^i$ and for a.e. $x \in F$ as

$$\{v\}_{\omega,F}(x) := \omega_{T_1,F} v_{|T_1}(x) + \omega_{T_2,F} v_{|T_2}(x), \tag{18}$$

on boundary faces $F \in \mathcal{F}_h^b$ such that $F \subset \partial T \cap \partial \Omega$, we set

$$\{v\}_{\omega,F}(x) := v_{|T}(x), \tag{19}$$

and the skew-weighted average operator for all $F \in \mathcal{F}_h^i$ and for a.e. $x \in F$, as

$$\{v\}_{\overline{\omega},F}(x) := \omega_{T_2,F} v_{|T_1}(x) + \omega_{T_1,F} v_{|T_2}(x). \tag{20}$$

The SWIP bilinear form is defined by (see Lemma 4.47 in [3])

$$\begin{aligned} \mathrm{a}_h^{swip}(v,w) &:= \underbrace{\int_\Omega A\nabla_h v \nabla_h w \, dx}_{\text{consistency term}} + \underbrace{\sum_{F\in\mathcal{F}_h} \eta \frac{\gamma_{A,F}}{h_F} \int_F [v,w]}_{\text{penalty term}} \\ &\quad - \underbrace{\sum_{F\in\mathcal{F}_h} \int_F \left(\{A\nabla_h v\}_{\overline{\omega}} \cdot n_F [w] + \{A\nabla_h w\}_{\omega} \cdot n_F [v]\right)}_{\text{symmetry term}}, \end{aligned} \tag{21}$$

where the quantity $\eta > 0$ denotes a user-dependent penalty parameter which is independent of the diffusion coefficient.

And the SWIP norms are defined by

$$\begin{aligned} \|v\|_{swip} &:= \left(\left\| A^{1/2} \nabla_h v \right\|^2_{[L^2(\Omega)]^d} + |v|^2_{A,J} \right)^{1/2}, \\ \forall q \geq 1: \quad \|v\|_{swip,q} &:= \left(\left\| A^{1/2} \nabla_h v \right\|^q_{[L^q(\Omega)]^d} + |v|^q_{J,A,q} \right)^{1/q}, \end{aligned} \tag{22}$$

with the diffusion-dependent jump seminorms

$$\begin{aligned} |v|^2_{J,A} &:= \sum_{F\in\mathcal{F}_h} \frac{\gamma_{A,F}}{h_F} \|[v]\|^2_{L^2(F)}, \\ \forall q \geq 1: \quad |v|^q_{J,A,q} &:= \sum_{F\in\mathcal{F}_h} \frac{\gamma_{A,F}}{h_F^{q-1}} \|[v]\|^q_{L^q(F)}. \end{aligned} \tag{23}$$

The discrete Galerkin norm is defined by

$$\|v\|_{dG} := \left(\|\nabla_h v\|^2_{[L^2(\Omega)]^d} + |v|^2_J \right)^{1/2}, \tag{24}$$

with the jump seminorm

$$|v|^2_J := \sum_{F\in\mathcal{F}_h} \frac{1}{h_F} \|[v]\|^2_{L^2(F)}. \tag{25}$$

For every r with $1 < r < +\infty$, we denote by $L^{r,\infty}(\Omega)$ the Marcinkiewicz space whose norm is defined by

$$\|v\|_{L^{r,\infty}(\Omega)} = \sup_{\lambda>0} \lambda \left| \{x \in \Omega : |v(x)| \geq \lambda\} \right|^{1/r}. \tag{26}$$

For every real number $k > 0$ we define the truncation $T_k : \mathbb{R} \to \mathbb{R}$ by

$$T_k(s) = \max(-k, \min(k, s)) = \begin{cases} s & \text{if } |s| \leq k, \\ k\dfrac{s}{|s|} & \text{if } |s| \geq k. \end{cases} \tag{27}$$

For every d − simplex T in $\mathbb{R}^d$, we adopt the following notations:

(i) $a_{i,T}$, $i = 0, \ldots, d$, denote the vertices of T.

(ii) $m_{i,T}$, $i = 0, \ldots, d$, denote the centers of the faces $F_i \in T$.

(iii) $\lambda_{i,T}$, $i = 0, \ldots, d$, designate the barycentric coordinates with respect to the $a_{i,T}$'s.

(iv) for every $x \in \mathbb{R}^d$ we put

$$\varphi_{i,T}(x) := 1 - d\lambda_{i,T}(x) \quad \textit{for } i = 0, \ldots, d, \tag{28}$$

where $(\varphi_{i,T})_{0 \leq i \leq d}$ are the $\mathbb{P}_1$ shape functions related to T; it is known that

$$\begin{aligned} &\varphi_i \in L^2(\Omega_h), \\ &\varphi_{i|T} \in \mathbb{P}_1[T], \qquad \forall i \in \{0,1,\ldots,d\}, \ \forall T \in \mathcal{T}_h, \\ &\varphi_{i|F_i} = 1, \\ &\varphi_i(a_i) = 1 - d, \\ &\varphi_i(a_j) = 1 \ \textit{such that } j \neq i, \\ &\frac{1}{|F_j|} \int_{F_j} \varphi_i = \varphi_i(m_j) = \delta_{i,j}, \\ &\sum_{i=0}^d \varphi_{i,T}(x) = 1, \quad \textit{for every } \mathrm{x} \in \mathbb{R}^d, \end{aligned} \tag{29}$$

with $\Omega_h := \cup\{T, T \in \mathcal{T}_h\} \subset \overline{\Omega}$.

(v) If N designate the number of all interior centers m_i of faces F in $\mathcal{T}_h$ we define the interpolation operator Π_h and the truncated interpolation operator I_h^k by

$$\begin{aligned}
&\forall v \in L^2\left(\overline{\Omega}\right) \text{avec} \int_F [v_h] = 0,\\
&\Pi_h(v) \in V_h,\\
&\Pi_h(v) := \sum_{1\le i\le N} \alpha_i^v \varphi_i,\\
&I_h^k(v) := \sum_{1\le i\le N} T_k\left(\alpha_i^v\right)\varphi_i,
\end{aligned} \tag{30}$$

with $\alpha_i^v = \Pi_h(v)(m_i) = (1/|F_i|)\int_{F_i}\Pi_h(v) := (1/|F_i|)\int_{F_i} v$.

(vi) Finally, we define the $N\times N$ stiffness matrix $Q = (Q_{ij})$; namely,

$$Q_{i,j} = a_h^{swip}\left(\varphi_i, \varphi_j\right) \quad \text{for } i, j \text{ in } \{1, 2, \ldots, N\}. \tag{31}$$

As in [4], the main assumption of the present paper is that Q is a diagonally dominant matrix; namely,

$$\forall i \in \{1, 2, \ldots, N\}: \quad Q_{ii} - \sum_{\substack{1\le j\le N\\ j\ne i}} \left|Q_{ij}\right| \ge 0. \tag{32}$$

2. Statement of the Main Result

We consider a matrix A such that

$$A \in L^\infty(\Omega)^{d\times d}, \tag{33}$$

$$\text{a.e } x \in \Omega: \forall\xi \in \mathbb{R}^d: A(x)\xi\xi \ge \alpha|\xi|^2, \tag{34}$$

for some $\alpha > 0$, and a right-hand side f such that

$$f \in L^1(\Omega). \tag{35}$$

A function u is the renormalized solution of the problem (1) if u satisfies

$$u \in L^1(\Omega), \tag{36}$$

$$\forall k > 0, \quad T_k(u) \in H_0^1(\Omega), \tag{37}$$

$$\lim_{k\to\infty}\frac{1}{k}\int_\Omega |\nabla T_k(u)|^2\, dx = 0, \tag{38}$$

$$\begin{aligned}
&\forall k > 0,\ \forall S \in \mathcal{C}_c^1(\mathbb{R}) \text{ with } \operatorname{supp}(S) \subset [-k, k],\\
&\forall v \in H_0^1(\Omega) \cap L^\infty(\Omega),\\
&\int_\Omega A\nabla T_k(u)\nabla v S(u)\, dx\\
&\quad + \int_\Omega A\nabla T_k(u)\nabla T_k(u) S'(u)\, v\, dx\\
&\quad = \int_\Omega f S(u)\, v\, dx.
\end{aligned} \tag{39}$$

It is known (see [1, 5]) that when f belongs to $L^1(\Omega) \cap H^{-1}(\Omega)$, the usual weak solution of (1), namely,

$$\begin{aligned}
&u \in H_0^1(\Omega),\\
&\forall v \in H_0^1(\Omega),\\
&\int_\Omega A\nabla v\nabla u\, dx = \int_\Omega f v\, dx,
\end{aligned} \tag{40}$$

is a renormalized solution of (1) and conversely the main interest of definition of renormalized solution is the following existence, uniqueness, and continuity theorem (see [1, 4]).

Theorem 1. *Assume that A and f satisfy (33), (34), and (35). Then there exists a renormalized solution of (1). This solution is unique. Moreover this unique solution belongs to $W_0^{1,q}(\Omega)$ for every q with $1 \le q < d/(d-1)$. It depends continuously on the right-hand side f in the following sense: if f^ε is a sequence which satisfies*

$$f^\varepsilon \longrightarrow f \quad \text{strongly in } L^1(\Omega), \tag{41}$$

when ε tends to zero, then the sequence u^ε of the renormalized solutions of (1) for the right-hand sides f^ε satisfies for every $k > 0$ and for every q with $1 \le q < d/(d-1)$

$$\begin{aligned}
&T_k(u^\varepsilon) \longrightarrow T_k(u) \quad \text{strongly in } H_0^1(\Omega),\\
&u^\varepsilon \longrightarrow u \quad \text{strongly in } W_0^{1,q}(\Omega),
\end{aligned} \tag{42}$$

when ε tends to zero, where u is the renormalized solution of (1) for the right-hand side f. Finally, if f_1 and f_2 belong to $L^1(\Omega)$, and if u_1 and u_2 are the renormalized solutions of (1) for the right-hand sides f_1 and f_2, then, for every $k > 0$, the function $T_k(u_1 - u_2)$ belongs to $H_0^1(\Omega)$ and for every q with $1 \le q < d/(d-1)$ one has

$$\begin{aligned}
&\alpha\left\|T_k(u_1 - u_2)\right\|_{H_0^1(\Omega)}^2 \le k\left\|f_1 - f_2\right\|_{L^1(\Omega)},\\
&\alpha\left\|u_1 - u_2\right\|_{W_0^{1,q}(\Omega)} \le C(d, |\Omega|, q)\left\|f_1 - f_2\right\|_{L^1(\Omega)},
\end{aligned} \tag{43}$$

Where the constant $C(d, |\Omega|, q)$ only depends on d, $|\Omega|$, and q.

Remark 2. Throughout all this paper, we denote by $C(p_1, p_2, p_3, ..)$ any real constant which only depends on the parameters p_1, p_2, and p_3 .... We can use the same notation for different constants.

Now we consider a family of triangulations $\mathcal{T}_h$ satisfying for each $h > 0$ the following assumption:

$$\begin{aligned}&\text{the triangulation } \mathcal{T}_h \text{ is made of a finite number of closed}\\ &d\text{-simplices } T(\text{namely triangles when } d = 2 \text{ and tetrahedra}\\ &\quad\text{when } d = 3) \text{ such that :}\\ &\quad\text{(i) } \Omega_h = \bigcup \{T : T \in \mathcal{T}_h\} \subset \overline{\Omega},\\ &\text{(ii) for every compact set } K \text{ with } K \subset \Omega, \text{there exists}\\ &h_0(K) > 0 \text{ such that } K \subset \Omega_h \text{ for every } h \text{ with } h < h_0(K),\\ &\text{(iii) for } (T_1, T_2) \in \mathcal{T}_h^2 \text{ with } T_1 \neq T_2, \text{one has } |T_1 \cap T_2| = 0,\\ &\text{(iv) every face of every } T \text{ of } \mathcal{T}_h \text{ is either a subset of } \partial\Omega_h,\\ &\quad\text{or a face of another } T' \text{ of } \mathcal{T}_h.\end{aligned} \tag{44}$$

Note that because of (iv) the triangulations are conforming. A particular case is the case where Ω is a polyhedron of $\mathbb{R}^d$, and where Ω_h coincides with Ω for every h.

In practice, the diffusion coefficient (i.e., matrix A) has more regularity than just belonging to $L^\infty(\Omega)$. Henceforth, we make the following assumption (assumption 4.43 [3]):

$$\begin{aligned}&\text{there is a partition } P_\Omega\\ &\quad := \{\Omega_i\}_{1\le i\le N_\Omega} \text{ of } \Omega \text{ such that :}\\ &\text{(i) Each } \Omega_i,\ 1 \le i \le N_\Omega, \text{is apolyhedron;}\\ &\text{(ii) The restriction of A to each } \Omega_i,\ 1 \le i\\ &\quad \le N_\Omega, \text{is constant.}\end{aligned} \tag{45}$$

An important assumption on the mesh sequence $\mathcal{T}_h := (\mathcal{T}_{\mathcal{H}})_{h\in\mathcal{H}}$ is its compatibility with the partition P_Ω in the following sense (assumption 4.45 [3]).

Assumption 3 (mesh compatibility). We suppose that the admissible mesh sequence $\mathcal{T}_{\mathcal{H}}$ is such that, for each $h \in \mathcal{H}$, each $T \in \mathcal{T}_h$ is a subset of only one set Ω_i of the partition P_Ω. In this situation, the meshes are said to be compatible with the partition P_Ω.

For every $T \in \mathcal{T}_h$, we denote by h_T the diameter of T and by ρ_T the diameter of the ball inscribed in T. We set

$$h = \sup_{T\in\mathcal{T}_h} h_T \tag{46}$$

and we assume that h tends to zero.

We also assume that the family of triangulations $\mathcal{T}_h$ is regular in the sense of P. G. Ciarlet [6]; namely, there exists a constant σ such that

$$\forall h,\ \forall T \in \mathcal{T}_h,\quad \frac{h_T}{\rho_T} \le \sigma. \tag{47}$$

For every triangulation $\mathcal{T}_h$, we consider the discrete problem:

$$\begin{aligned}&\textit{Find } u_h \in V_h,\\ &\forall v_h \in V_h,\\ &a_h^{swip}(u_h, v_h) = \int_\Omega f v_h\,dx.\end{aligned} \tag{48}$$

Note that the right-hand side of (48) makes sense since $f \in L^1(\Omega)$ and $V_h \subset L^\infty(\Omega)$. The discrete bilinear forme a_h^{swip} is consistent and coercive (see (128)) on V_h, so a straightforward consequence of the Lax-Milgram Lemma is that the discrete problem (48) is well-posed. The solution u_h of (48) exists and is unique.

As in [4], the main result of this paper is the following.

Theorem 4. *Assume that A, f, and $\mathcal{T}_h$ satisfy (33), (34), (35), (44), (46), (47), and (32). Then the unique solution u_h of (48) satisfies for every $k > 0$ and for every q with $1 \le q < d/(d-1)$*

$$\begin{aligned}&u_h \longrightarrow u \quad \textit{strongly in } L^q(\Omega),\\ &\nabla_h u_h \longrightarrow \nabla u \quad \textit{strongly in } [L^q(\Omega)]^d,\\ &|u_h|_{J,A,q} \longrightarrow 0,\\ &I_h^k(u_h) \longrightarrow T_k(u) \quad \textit{strongly in } L^2(\Omega),\\ &\nabla_h I_h^k(u_h) \longrightarrow \nabla T_k(u) \quad \textit{strongly in } \left[L^2(\Omega)\right]^d,\\ &\left|I_h^k(u_h)\right|_{J,A} \longrightarrow 0,\end{aligned} \tag{49}$$

when h tends to zero, where u is the unique renormalized solution of (1).

This theorem will be proved in Section 4, using the tools that we will prepare in Section 3. In Section 5, we will explain why the results of [4] when f is a bounded Radon measure remain valid in our case. In Section 6 we also show that if we assume in addition that $T_k(f) \in H^1(\Omega)$ for every $k > 0$, we obtain for smooth solutions an $O(h^{4(1-1/r)})$ error estimate in $\|\ \|_{swip,q}$-norm (Section 6.1), and for Low-Regularity solutions an $O(h^{4\alpha_p(1-1/r)})$ error estimate in $\|\ \|_{swip,q}$-norm (Section 6.2). Finally, in Section 7 we show that in the case where A is the identity matrix, condition (32) remains satisfied when every inner angle of every d-simplex of $\mathcal{T}_h$ is acute.

3. Tools

We are going to prove Theorem 4 in several steps. We begin by proving the following result which is a piecewise $\mathbb{P}_1$ variant of a result of L. Boccardo & T. Gallouët [2, 5].

Theorem 5. *Assume that $v_h \in V_h$ satisfies*

$$\forall k > 0,\quad \int_\Omega \left|\nabla_h I_h^k(v_h)\right|^2 dx \le kM \tag{50}$$

for some $M > 0$. Then, for every q with $1 \leq q < d/(d-1)$,

$$\|v_h\|_{swip,q} \leq C_2\left(d, |\Omega|, q, \sigma, \|A\|_{L^\infty(\Omega)^{d\times d}}\right) M \quad (51)$$

where the constant $C_2(d, |\Omega|, q)$ only depends on d, $|\Omega|$, and q.

As in [4], to prove Theorem 5, we use the following lemmas.

Lemma 6. *Under assumption (47), for all $T \in \mathcal{T}_h$ and all $F \in \mathcal{F}_T$, one has*

$$h_F |F| \leq 2\sigma^d |T| \quad (52)$$

Proof. Indeed, let $T \in \mathcal{T}_h$ and $F \in \mathcal{F}_T$, so

$$h_F |F| \leq h_T^d \quad (53)$$

and by (47) one has

$$h_F |F| \leq \sigma^d \rho_T^d \quad (54)$$

which combined with the fact that

$$|T| \geq c\rho_T^d, \quad (55)$$

where $c = \Pi/4$ in 2D, and $c = \Pi/6$ in 3D, implies (52). □

Lemma 7. *Under assumption (47), for every q such that $1 \leq q$, the following bound holds for any $v_h \in V_h$:*

$$|v|^q_{J,A,q} \leq C\left(q, \sigma, d, \|A\|_{L^\infty(\Omega)^{d\times d}}\right) \|\nabla_h v\|_{L^q(\Omega)}. \quad (56)$$

Proof. For every $T \in \mathcal{T}_h$, we denote by $(\nabla_h v)_T$ the (constant) gradient of the restriction of v to T. With this notation, using the continuity of v across any F in $\mathcal{F}_T$ at the mass center x_F of any internal F, the fact that v vanishes at the mass center x_F of any external F, and the known inequality

$$\forall q \geq 1, \quad (|a| + |b|)^q \leq 2^{q-1}\left(|a|^q + |b|^q\right) \quad (57)$$

and using (52) we get

$$\begin{aligned}
|v|^q_{J,A,q} &= \sum_{F\in\mathcal{F}_h^i} \frac{\gamma_{A,F}}{h_F^{q-1}} \int_F [v]^q\,dx + \sum_{F\in\mathcal{F}_h^b} \frac{\gamma_{A,F}}{h_F^{q-1}} \int_F [v]^q\,dx \\
&\leq \sum_{\substack{F\in\mathcal{F}_h^i \\ F=T_1\cap T_2}} \frac{\gamma_{A,F}}{h_F^{q-1}} \\
&\quad \cdot \int_F \left(\left((\nabla_h v)_{T_1} - (\nabla_h v)_{T_2}\right)\cdot(x - x_F)\right)^q dx \\
&\quad + \sum_{\substack{F\in\mathcal{F}_h^b \\ F\in\mathcal{F}_T}} \int_F \frac{\gamma_{A,F}}{h_F^{q-1}} \left((\nabla_h v)_T \cdot (x - x_F)\right)^q dx \\
&\leq \sum_{\substack{F\in\mathcal{F}_h^i \\ F=T_1\cap T_2}} 2^q \min\left(\lambda_{T_1,F}, \lambda_{T_2,F}\right) h_F |F| \\
&\quad \cdot \left(\left|(\nabla_h v)_{T_1}\right|^q + \left|(\nabla_h v)_{T_2}\right|^q\right) + \sum_{\substack{F\in\mathcal{F}_h^b \\ F\in\mathcal{F}_T}} \lambda_{T,F} h_F |F| \\
&\quad \cdot \left|(\nabla_h v)_T\right|^q \leq 2\sigma^d \|A\|_{L^\infty(\Omega)^{d\times d}} \sum_{\substack{F\in\mathcal{F}_h^i \\ F=T_1\cap T_2}} 2^q \\
&\quad \cdot \left(|T_1| \left|(\nabla_h v)_{T_1}\right|^q + |T_2| \left|(\nabla_h v)_{T_2}\right|^q\right) \\
&\quad + 2\sigma^d \|A\|_{L^\infty(\Omega)^{d\times d}} \sum_{\substack{F\in\mathcal{F}_h^b \\ F\in\mathcal{F}_T}} |T| \left|(\nabla_h v)_T\right|^q \\
&\leq 2^{q+3}\sigma^d \|A\|_{L^\infty(\Omega)} \sum_{T\in\mathcal{T}_h} |T| \left|(\nabla_h v)_T\right|^q,
\end{aligned} \quad (58)$$

which is (56) with $C(q, \sigma, d, \|A\|_{L^\infty(\Omega)^{d\times d}}) = 2^{q+3}\sigma^d \|A\|_{L^\infty(\Omega)^{d\times d}}$. □

Lemma 8. *Let $v_h \in V_h$ and let $k > 0$. If for some $T \in \mathcal{T}_h$ there exists $y \in T$ with $|v_h(y)| \geq k$, then there exists a d-simplex $S \subset T$ with $|S| = c(d)|T|$ such that*

$$\forall x \in S: \quad \left|I_h^k(v_h(x))\right| \geq \frac{k}{2}, \quad (59)$$

and the strictly positive constant $c(d)$ only depends on d.

Proof. Let T be a d−simplex from the triangulation $\mathcal{T}_h$, $v_h \in V_h$, and $k > 0$, such that $\sup_T |v_h| \geq k$. Consider $w_h = I_h^k(v_h) \in \mathbb{P}_1(T)$.

(i) If $\max_{m_i\in T}|v_h| < k$, thus

$$\begin{aligned}
&\forall i \in \{0, 1, \ldots, d\}, \\
&T_k\left(\alpha_i^{v_h}\right) = T_k\left(v_h(m_i)\right) = v_h(m_i) = \alpha_i^{v_h}
\end{aligned} \quad (60)$$

so $w_h = v_h$, and $\exists y \in T$ such that $|w_h(y)| \geq k$.

(ii) If $\max_{m_i\in T}|v_h| \geq k$, thus $\exists m_i \in T$, such that $|v_h(m_i)| \geq k$

so $|T_k(v_h(m_i))| = k$, and $|w_h(m_i)| = T_k(\alpha_i^{v_h}) = k$.

In both cases, there exists an element y in T such that $|w_h(y)| \geq k$.

But $w_h \in \mathbb{P}_1(T)$, so

$$w_h(y) - w_h(x) = \nabla w_h(y - x), \quad (61)$$

In other words

$$\forall i \neq j \neq k \text{ dans } \{0, 1, \ldots, d\},$$
$$\nabla\lambda_i(a_i - a_k) = 1, \quad (62)$$
$$\nabla\lambda_i(a_j - a_k) = 0.$$

and since

$$y - x = \sum_{k=1}^{d} (\lambda_k(y) - \lambda_k(x))(a_k - a_0), \quad (63)$$

one obtains

$$\nabla w_h = -d\sum_{i=0}^{d} T_k(\alpha_i^{v_h})\nabla\lambda_i,$$
$$\nabla w_h(a_k - a_o) = -d\left(T_k(\alpha_k^{v_h}) - T_k(\alpha_0^{v_h})\right), \quad (64)$$

so

$$|\nabla w_h(a_k - a_o)| \leq 2kd, \quad (65)$$

and

$$|w_h(y) - w_h(x)| \leq 2d\sum_{k=1}^{d} |\lambda_k(y) - \lambda_k(x)| \leq \frac{k}{2}, \quad (66)$$

as soon as

$$\sum_{k=1}^{d} |\lambda_k(y) - \lambda_k(x)| \leq \frac{1}{4d}. \quad (67)$$

For this purpose we define the d-simplex $S = (y - a_0) + S_0$ such that

$$S_0 = \left\{x, \lambda_0 \geq 1 - \frac{1}{4d} = \frac{4d-1}{4d}\right\}, \quad (68)$$

so

$$|w_h(x)| \geq |w_h(y)| - |w_h(y) - w_h(x)| \geq k - \frac{k}{2} = \frac{k}{2}, \quad (69)$$

and to estimate the measure of S, it is clear to verify first that $|S| = |S_0|$. Let $\widehat{T}$ be the reference unit d-simplex with vertices $\widehat{a}_0 = 0, \widehat{a}_1 = e_1, \ldots, \widehat{a}_d = e_d$, where $\{e_i, i = 1, \ldots, d\}$ is the canonical basis of $\mathbb{R}^d$. Let F_T be the invertible affine mapping that maps $F_T(\widehat{T}) = T$ onto T and set $\widehat{S} = F_T^{-1}(S)$.

Since F_T is affine, it is easy to check that $\widehat{\lambda}_i = F_T^{-1} \circ \lambda_{i,T}$ for $i = 0, 1, \ldots, d$ are the barycentric coordinates with respect to the $\widehat{a}_i$'s and that

$$\widehat{S} = \left\{\widehat{x} \in \widehat{T} : \widehat{\lambda_0}(\widehat{x}) \geq \frac{4d-1}{4d}\right\},$$
$$|S| = \frac{|\widehat{S}|}{|\widehat{T}|}|T| = C(d)|T|, \quad (70)$$

where $c(d) = |\widehat{S}|/|\widehat{T}|$ is a constant that depends only on d. This proves the result. □

Lemma 9. *Assume that $v_h \in V_h$ satisfies (50), then*

$$|B_k(v_h)| \leq C(d, |\Omega|, 2^*)\left(\frac{M}{k}\right)^{2^*/2}, \quad (71)$$

for every $k > 0$, where $2^* = 2d/(d-2) = 6$ if $d = 3$ and 2^* is any real number with $2^* \geq 1$ if $d = 2$; $B_k(v_h)$ is defined by

$$B_k(v_h) = \bigcup\left\{T \in \mathcal{T}_h : \max_T |v_h| \geq k\right\}, \quad (72)$$

and $C(d, |\Omega|, 2^*)$ is a constant depending only on d, 2^*, θ_0, and $|\Omega|$.

Proof. Discrete Sobolev's theorem (see theorem 5.3 in [3]) asserts that

$$\forall v \in V_h, \quad \|v\|_{L^{2^*}(\Omega)} \leq \sigma_{2,2^*}(d, |\Omega|)\|v\|_{dG}, \quad (73)$$

and we will also need (56) with $q = 2$

$$\forall v \in V_h, \quad |v|_J^2 \leq C(\sigma, d)\|\nabla_h v\|_{L^2(\Omega)}. \quad (74)$$

Here $d = 2$ or $d = 3$ so $2^* = 2d/(d-2) = 6$ if $d = 3$ and 2^* can be any real number with $1 \leq 2^*$.

Fix $k > 0$. If $T \subset B_k(v_h)$, from Lemma 8, we know that there exists $S \subset T$, with $|S| = c(d)|T|$ and

$$\forall x \in S, \quad \left|I_h^k(v_h(x))\right| \geq \frac{k}{2}. \quad (75)$$

Therefore

$$\int_T \left|I_h^k(v_h(x))\right|^{2^*} dx \geq \int_S \left|I_h^k(v_h(x))\right|^{2^*} dx$$
$$\geq \left(\frac{k}{2}\right)^{2^*}|S| = c(d)|T|\left(\frac{k}{2}\right)^{2^*}. \quad (76)$$

Hence

$$|B_k(v_h)| = \sum_{T\subset B_k(v_h)} |T|$$
$$\leq \sum_{T\subset B_k(v_h)} \frac{1}{c(d)}\left(\frac{2}{k}\right)^{2^*}\int_T \left|I_h^k(v_h(x))\right|^{2^*} dx \quad (77)$$
$$\leq \frac{1}{c(d)}\left(\frac{2}{k}\right)^{2^*}\int_\Omega \left|I_h^k(v_h(x))\right|^{2^*} dx.$$

Combining with (73) and (74) one has

$$|B_k(v_h)| \le \frac{(2\sigma_{2,2^*}(d,|\Omega|))^{2^*}(1+C(\sigma,d))^{2^*/2}}{c(d)k^{2^*}}\left(\int_\Omega \left|\nabla_h I_h^k(v_h(x))\right|^2 dx\right)^{2^*/2}. \quad (78)$$

Finally using (50), we obtain

$$|B_k(v_h)| \le \frac{(2\sigma_{2,2^*}(d,|\Omega|))^{2^*}(1+C(\sigma,d))^{2^*/2}}{c(d)}\left(\frac{M}{k}\right)^{2^*/2}, \quad (79)$$

which is (71) with

$$C\left(d,|\Omega|,2^*,\sigma,\|A\|_{L^\infty(\Omega)^{d\times d}}\right) = \frac{(2\sigma_{2,2^*}(d,|\Omega|))^{2^*}}{C(d)}(1+C(\sigma,d))^{2^*/2}. \quad (80)$$

□

Proof (proof of Theorem 5; see [4]). Fix q with $1 \le q < d/(d-1)$. Take $r = (2\times 2^*)/(2+2^*) = 3/2$ in the case $d = 3$, and verify $(2\times 2^*)/(2+2^*) > q$, in the case $d = 2$.

The embedding inequality ($L^q(\Omega) \hookrightarrow L^{r,\infty}(\Omega)$) writes

$$\||\nabla_h v_h|\|_{L^q(\Omega)} \le C(q,r,|\Omega|)\||\nabla v_h|\|_{L^{r,\infty}(\Omega)}, \quad (81)$$

and by (56) one can have

$$\forall v \in W_0^{1,q}(\mathscr{T}_h),$$
$$|v|_{J,A,q}^q \le C\left(q,\sigma,d,\|A\|_{L^\infty(\Omega)^{d\times d}}\right)\|\nabla_h v\|_{L^q(\Omega)}. \quad (82)$$

So to prove Theorem 5, it suffices to estimate

$$\||\nabla v_h|\|_{L^{r,\infty}(\Omega)} = \sup_{\lambda>0}\left|\bigcup_{T\in T_h}\{x\in T : |\nabla_h v_h| \ge \lambda\}\right|^{1/r}. \quad (83)$$

So let $\lambda > 0$. For every $k > 0$, we can write

$$\begin{aligned}&\left|\bigcup_{T\in T_h}\{x\in T : |\nabla_h v_h| \ge \lambda\}\right| \\ &\le \left|\bigcup_{T\in T_h}\{x\in T : |\nabla_h v_h| \ge \lambda\}\cap B_k(v_h)\right| \\ &\quad + \left|\bigcup_{T\in T_h}\{x\in T : |\nabla_h v_h| \ge \lambda\}\cap [B_k(v_h)]^c\right| \\ &\le |B_k(v_h)| \\ &\quad + \left|\bigcup_{T\in T_h}\{x\in T : |\nabla_h v_h| \ge \lambda\}\cap [B_k(v_h)]^c\right|.\end{aligned} \quad (84)$$

But $\{x\in\Omega : |\nabla v_h(x)| \ge \lambda\}\cap[B_k(v_h)]^c$ coincides, up to a set of measure zero, with

$$\bigcup_{T\in T_h}\left(\{x\in T : |\nabla_h v_h| \ge \lambda\} \cap \left\{T\in\mathscr{T}_h : \max_T |v_h| < k\right\}\right). \quad (85)$$

On the other hand, if $\max_T|v_h| < k$, then $I_h^k(v_h)_{|_T} = v_{h|_T}$, so

$$\begin{aligned}&\left|\bigcup_{T\in T_h}\{x\in T : |\nabla_h v_h| \ge \lambda\}\cap [B_k(v_h)]^c\right| \\ &\le \left|\bigcup_{T\in T_h}\{x\in T : |\nabla\Pi_h T_k(v_h)| \ge \lambda\}\right| \\ &\le \frac{1}{\lambda^2}\int_\Omega |\nabla_h\Pi_h(T_k(v_h))|^2 dx.\end{aligned} \quad (86)$$

By the use of (50), one has

$$\left|\bigcup_{T\in T_h}\{x\in T : |\nabla_h v_h| \ge \lambda\}\cap [B_k(v_h)]^c\right| \le \frac{kM}{\lambda^2}. \quad (87)$$

We fix $k = \lambda^{2-r}M^{r-1}$ to have $kM/\lambda^2 = (M/\lambda)^r$; using (71) we obtain

$$\left|\bigcup_{T\in T_h}\{x\in T : |\nabla_h v_h| \ge \lambda\}\right| \le C(d,|\Omega|,2^*)\left(\frac{M}{k}\right)^{2^*/2} + \left(\frac{M}{\lambda}\right)^r, \quad (88)$$

so, for $r = (2\times 2^*)/(2+2^*)$ to have $(M/k)^{2^*/2} = (M/\lambda)^r$, in both cases $d = 2$ or $d = 3$, one has

$$\left|\bigcup_{T\in T_h}\{x\in T : |\nabla_h v_h| \ge \lambda\}\right| \le (C(d,|\Omega|,2^*)+1)\left(\frac{M}{\lambda}\right)^r, \quad (89)$$

and then

$$\lambda\left|\bigcup_{T\in T_h}\{x\in T : |\nabla_h v_h| \ge \lambda\}\right|^{1/r} \le C(d,|\Omega|,q)M, \quad (90)$$

for every $\lambda > 0$, which finishes the proof. □

Lemma 10. *Let $v \in V_h$. For every s, k with $0 \le s < k$, the set $B_{k,s}(v)$, defined by*

$$B_{k,s}(v) = \bigcup_{T \in T_h} \left\{ T \in \mathcal{T}_h : \min_T |v| \le s, \max_T |v| \ge k \right\}, \quad (91)$$

satisfies

$$|B_{k,s}(v)| \le \frac{h^2}{(k-s)^2} \int_\Omega |\nabla_h (v)|^2 \, dx. \quad (92)$$

Proof. If $T \in B_{k,s}(v)$, $\max_T |v| = |v(a)| \ge k$, and $\min_T |v| = |v(b)| \le s$, so

$$\begin{aligned} k - s &\le |v(a)| - |v(b)| \le |v(a) - v(b)| \\ &\le |\nabla (v)| \, |a - b| . \end{aligned} \quad (93)$$

Therefore

$$1 \le |\nabla (v)|^2 \frac{|a-b|^2}{(k-s)^2} \quad (94)$$

Hence

$$\begin{aligned} |B_{k,s}(v)| &= \sum_{T \in B_{k,s}(v)} |T| = \sum_{T \in B_{k,s}(v)} \int_T 1 dx \\ &\le \sum_{T \in B_{k,s}(v)} \int_T |\nabla (v)|^2 \frac{|a-b|^2}{(k-s)^2} dx \\ &\le \frac{h^2}{(k-s)^2} \int_\Omega |\nabla_h (v)|^2 \, dx. \end{aligned} \quad (95)$$

The estimate (92) follows. □

Lemma 11. *Let $v_h \in V_h$ and $0 \le s < k$, then the set $\widetilde{B}_{k,s}(v_h)$, defined by*

$$\begin{aligned} &\widetilde{B}_{k,s}(v_h) \\ &= \bigcup \left\{ T \in \mathcal{T}_h : \min_{m_i \in T} |v_h| \le s, \max_T |v_h| \ge k \right\}, \end{aligned} \quad (96)$$

satisfies

$$\left|\widetilde{B}_{k,s}(v_h)\right| \le \frac{h^2}{(k-s)^2} \int_\Omega \left|\nabla_h I_h^k (v_h)\right|^2 dx. \quad (97)$$

Proof. Indeed, if $T \subset \widetilde{B}_{k,s}(v_h)$ then there are two possibilities:

(i) If $\max_{m_i \in T} |v_h| \le k$ so $I_h^k(v_h)_{|T} = v_{h|T}$.

(ii) If $\max_{m_i \in T} |v_h| \ge k$ so $\max_T |I_h^k(v_h)| \ge k$ and obviously $\min_T |I_h^k(v_h)| \le s$.

In the two cases $T \subset B_{k,s}(I_h^k(v_h))$. The estimate (97) follows. □

Remark 12. It is then clear that, under hypothesis (50),

$$\begin{aligned} \left|\widetilde{B}_{k,s}(v_h)\right| &\underset{h \to 0}{\longrightarrow} 0, \\ \left|B_{k,s}\left(I_h^k(v_h)\right)\right| &\underset{h \to 0}{\longrightarrow} 0. \end{aligned} \quad (98)$$

In addition, one has the following.

Proposition 13. *Let $v_h \in V_h$ and $0 \le s < k$. If v_h satisfies (50), then*

$$\left|B_{k,s}(v_h)\right| \underset{h \to 0}{\longrightarrow} 0. \quad (99)$$

Proof. Fix $k > 0$ and $s > 0$, $s < k$. For $h > 0$ such that $1/h \ge k$, we can write

$$\begin{aligned} B_{k,s}(v_h) &= \left(B_{k,s}(v_h) \cap B_{1/h}(v_h)\right) \\ &\quad \cup \left(B_{k,s}(v_h) \cap \left[B_{1/h}(v_h)\right]^c\right). \end{aligned} \quad (100)$$

On the one hand, with $2^* = 6$ in (71), one has

$$\left|B_{k,s}(v_h) \cap B_{1/h}(v_h)\right| \le \left|B_{1/h}(v_h)\right| \le C(|\Omega|) h^3. \quad (101)$$

On the other hand,

$$B_{k,s}(v_h) \cap \left[B_{1/h}(v_h)\right]^c \subset B_{k,s}\left(I_h^{1/h}(v_h)\right). \quad (102)$$

Indeed, if $x \in B_{k,s}(v_h) \cap [B_{1/h}(v_h)]^c$ and $T \in \mathcal{T}_h$ such that $x \in T$, then $\max_T |v_h| \ge k$, $\min_T |v_h| \le s$, and, for every y in T, $|v_h(y)| \le 1/h$, which means $I_h^{1/h}(v_h)|_T = v_h|_T$, and $T \in B_{k,s}(I_h^{1/h}(v_h))$.

Therefore, with Lemma 10, and (50), one has

$$\begin{aligned} \left|B_{k,s}(v_h) \cap B_{1/h}^c(v_h)\right| &\le \left|B_{k,s}\left(I_h^{1/h}(v_h)\right)\right| \\ &\le \frac{h^2}{(k-s)^2} \int_\Omega \left|\nabla_h I_h^{1/h}(v_h)\right|^2 dx \\ &\le \frac{h}{(k-s)^2} M. \end{aligned} \quad (103)$$

The convergence (99) is then a consequence of (101) and (103). □

Lemma 14. *Let $v_h \in V_h$. For every s and every k with $0 < s < k$, one has*

$$\begin{aligned} &\left\{x \in \Omega : T_s\left(I_h^k(v_h)\right) \ne T_s(v_h(x))\right\} \\ &\quad \subset B_{k,s}(v_h) \cup B_{k,s}\left(I_h^k(v_h)\right). \end{aligned} \quad (104)$$

Proof. Let $x \in \Omega$ such that $T_s(I_h^k(v_h(x))) \ne T_s(v_h(x))$, and $T \in \mathcal{T}_h$ with $x \in T$. It is easily checked that $I_h^k(v_h)|_T \ne v_h|_T$, $\max_{\bar{T}} |v_h| \ge k$, and $\max_{\bar{T}} |I_h^k(v_h)| \ge k$. So there are three possibilities.

(i) $|I_h^k(v_h(x))| \ge s$ and $|v_h(x)| < s$, and then $T \subset B_{k,s}(v_h)$,

(ii) $|I_h^k(v_h(x))| < s$ and $|v_h(x)| \ge s$, and then $T \subset B_{k,s}(I_h^k(v_h))$,

(iii) $|I_h^k(v_h(x))| < s$ and $|v_h(x)| < s$, and then

$$T \subset B_{k,s}\left(I_h^k(v_h)\right) \cap B_{k,s}(v_h). \quad (105)$$

In all cases $x \in T \subset B_{k,s}(I_h^k(v_h)) \cup B_{k,s}(v_h)$, and (104) follows. □

Proposition 15. *Assume that $v_h \in V_h$ satisfies (50). Then, for every $k > 0$, one has*

$$I_h^k(v_h) - T_k(v_h) \longrightarrow 0 \quad \text{in measure,} \tag{106}$$

when h tends to zero.

Proof. Fix $k > 0$ and $\varepsilon > 0$ such that $\varepsilon < k$ and consider

$$I_\varepsilon = \left\{x \in \Omega : \left|I_h^k(v_h(x)) - T_k(v_h(x))\right| \geq \varepsilon\right\}. \tag{107}$$

Let $x \in I_\varepsilon$ and $T \in \mathcal{T}_h$ with $x \in T$. It is easily checked that

$$I_h^k(v_h)_{|T} \neq T_k(v_h)_{|T}, \tag{108}$$

which implies that $\max_T |v_h| > k$. So there are four possibilities.

(i) v_h changes sign in T; then, by continuity,

$$T \subset B_{k,s}(v_h) \quad \text{for every } s \in]0,k[. \tag{109}$$

(ii) $I_h^k(v_h)$ changes sign in T; then

$$T \subset B_{k,s}\left(I_h^k(v_h)\right) \quad \text{for every } s \in]0,k[. \tag{110}$$

(iii) $\max_{m_i \in T}|v_h| \leq k$ and $v_{h|_T} \geq 0$ (or $v_{h|_T} \leq 0$) so $I_h^k(v_h)_{|T} = v_{h|T}$.

(a) If $v_{h|_T} \geq 0$, then $v_h(x) \geq k + \varepsilon$ and $T \subset B_{k+\varepsilon,k}(I_h^k(v_h))$.

(b) If $v_{h|_T} \leq 0$, then $v_h(x) \leq -k - \varepsilon$ and $T \subset B_{k+\varepsilon,k}(I_h^k(v_h))$.

(iv) $\max_{m_i \in T}|v_h| > k$ and $v_{h|_T} \geq 0$ (or $v_{h|_T} \leq 0$). So

$$\begin{aligned} &\left|I_h^k(v_h(x)) - T_k(v_h(x))\right| \\ &\quad = \left|\left|I_h^k(v_h(x))\right| - \left|T_k(v_h(x))\right|\right|. \end{aligned} \tag{111}$$

(a) If $|I_h^k(v_h(x))| - |T_k(v_h(x))| \geq \varepsilon$. There are three possibilities:

case 1: $|v_h(x)| \geq k$, so $|I_h^k(v_h(x))| \geq k + \varepsilon$ and

$$T \subset B_{k+\varepsilon,k}\left(I_h^k(v_h)\right); \tag{112}$$

case 2: $|v_h(x)| < k - \varepsilon/2$, so $T \subset B_{k,k-\varepsilon/2}(v_h)$;

case 3: $k - \varepsilon/2 \leq |v_h(x)| < k$, so $|I_h^k(v_h(x))| \geq k + \varepsilon/2$ and

$$T \subset B_{k+\varepsilon/2,k}\left(I_h^k(v_h)\right). \tag{113}$$

(b) If $|T_k(v_h(x))| - |I_h^k(v_h(x))| \geq \varepsilon$, then

$$\left|I_h^k(v_h(x))\right| \leq \left|T_k(v_h(x))\right| - \varepsilon, \tag{114}$$

so

$$T \subset B_{k,k-\varepsilon}\left(I_h^k(v_h)\right). \tag{115}$$

We can then conclude that

$$\begin{aligned} I_\varepsilon \subset\ & B_{k+\varepsilon/2,k}\left(\Pi_h T_k(v_h)\right) \cup B_{k,k-\varepsilon}\left(\Pi_h T_k(v_h)\right) \\ & \cup B_{k,k-\varepsilon/2}(v_h). \end{aligned} \tag{116}$$

Convergence (106) is then consequence of (98) and (99). □

The result and the proof of the Proposition (2.7) in [4] can be conserved without changes.

Proposition 16. *Under assumption (32), one has for every $v_h \in V_h$ and every $k > 0$*

$$a_h^{swip}\left(v_h - I_h^k(v_h), I_h^k(v_h)\right) \geq 0. \tag{117}$$

Proof (proof of Proposition 2.7 in [4]). Since

$$\begin{aligned} v_h &= \sum_{i=0}^{d} \alpha_i^{v_h} \varphi_i, \\ I_h^k(v_h) &= \sum_{i=0}^{d} T_k\left(\alpha_i^{v_h}\right)\varphi_i, \end{aligned} \tag{118}$$

one has

$$\begin{aligned} a_h^{swip}&\left(v_h - I_h^k(v_h), I_h^k(v_h)\right) = \int_\Omega A\nabla_h\left(v_h - I_h^k(v_h)\right) \\ &\cdot \nabla_h I_h^k(v_h)\,dx \\ &- \sum_{F \in \mathcal{F}_h^i} \int_F \left\{A\nabla_h\left(v_h - I_h^k(v_h)\right)\right\}_{\overline{\omega}} \cdot n_F\left[I_h^k(v_h)\right]dx \\ &- \sum_{F \in \mathcal{F}_h^i} \int_F \left\{A\nabla_h I_h^k(v_h)\right\}_\omega \cdot n_F\left[v_h - I_h^k(v_h)\right]dx \\ &+ \sum_{F \in \mathcal{F}_h^i} \int_F \frac{\gamma_{A,F}}{h_F}\eta\left[v_h - I_h^k(v_h)\right]\left[I_h^k(v_h)\right]dx \\ &= \int_\Omega \left(\sum_{i=0}^{d}\left(\alpha_i^{v_h} + T_k\left(\alpha_i^{v_h}\right)\right)A\nabla_h\varphi_i\right)\left(\sum_{j=0}^{d} T_k\left(\alpha_j^{v_h}\right)\right. \\ &\left.\cdot A\nabla_h\varphi_j\right)dx \\ &- \sum_{F \in \mathcal{F}_h^i} \int_F \left(\sum_{i=0}^{d}\left(\alpha_i^{v_h} + T_k\left(\alpha_i^{v_h}\right)\right)\left\{A\nabla_h\varphi_i\right\}_{\overline{\omega}}\right) \\ &\cdot n_F\left(\sum_{j=0}^{d} T_k\left(\alpha_j^{v_h}\right)\left[\varphi_j\right]\right)dx \end{aligned}$$

$$
\begin{aligned}
&- \sum_{F\in\mathcal{F}_h^i} \int_F \left(\sum_{i=0}^{d} T_k\left(\alpha_i^{v_h}\right) \{\varphi_i\}_\omega \right) \\
&\cdot n_F \left(\sum_{j=0}^{d} \left(\alpha_j^{v_h} + T_k\left(\alpha_j^{v_h}\right)\right) \left[A\nabla_h \varphi_j\right] \right) dx \\
&+ \sum_{F\in\mathcal{F}_h^i} \int_F \left(\sum_{i=0}^{d} \left(\alpha_i^{v_h} + T_k\left(\alpha_i^{v_h}\right)\right) \left[A\nabla_h \varphi_i\right] \right) \\
&\cdot \left(\sum_{j=0}^{d} T_k\left(\alpha_j^{v_h}\right) [\varphi_j] \right) dx = \sum_{i=0}^{d} \left(\alpha_i^{v_h} - T_k\left(\alpha_i^{v_h}\right)\right) \\
&\cdot \sum_{j=0}^{d} T_k\left(\alpha_j^{v_h}\right) a_h^{swip}\left(\varphi_i, \varphi_j\right) = \sum_{i=0}^{d} \left(\alpha_i^{v_h} + T_k\left(\alpha_i^{v_h}\right)\right) \\
&\cdot \sum_{j=0}^{d} T_k\left(\alpha_j^{v_h}\right) Q_{i,j} = \sum_{i=0}^{d} S_i,
\end{aligned} \tag{119}
$$

where

$$
\begin{aligned}
S_i &= \left(\alpha_i^{v_h} - T_k\left(\alpha_i^{v_h}\right)\right) T_k\left(\alpha_i^{v_h}\right) Q_{i,i} \\
&+ \left(\alpha_i^{v_h} - T_k\left(\alpha_i^{v_h}\right)\right) \sum_{\substack{0\le j\le d \\ j\ne i}} \alpha_j^{T_k(v_h)} Q_{i,j}.
\end{aligned} \tag{120}
$$

Fix $\in \{0, 1, \ldots, d\}$;

(i) if $|T_k(v_h(m_i))| < k$, then $T_k(v_h(m_i)) = v_h(m_i)$ et $\alpha_i^{v_h} - T_k(\alpha_i^{v_h}) = 0$ and $S_i = 0$,

(ii) if $|T_k(v_h(m_i))| = k$,

then

$$
\begin{aligned}
&\left(\alpha_i^{v_h} - T_k\left(\alpha_i^{v_h}\right)\right) T_k\left(\alpha_i^{v_h}\right) \\
&= \left(v_h(m_i) - T_k\left(v_h(m_i)\right)\right) T_k\left(v_h(m_i)\right) \\
&= k\left|v_h(m_i) - T_k\left(v_h(m_i)\right)\right|,
\end{aligned} \tag{121}
$$

and therefore

$$
S_i \ge k\left|v_h(m_i) - T_k\left(v_h(m_i)\right)\right| \left(Q_{i,i} - \sum_{\substack{0\le j\le d \\ j\ne i}} \left|Q_{i,j}\right| \right) \tag{122}
$$

in the light of the assumption (32). This proves that $S_i \ge 0$ for every i in $\{1, 2, \ldots, N\}$; and so (117). □

4. Proof of the Main Theorem

We first show an a priori estimate (compared with (50)) on the solution u_h of (48).

Lemma 17. *Let $v_h \in V_h$, $T \in \mathcal{T}_h$, and $k > 0$ such that $\max_{m_i\in T}|v_h| \le k$. Then*

$$
\max_T |v_h| \le k\left(2d^2 - 1\right). \tag{123}
$$

Proof. Let $v_h \in V_h$, $T \in \mathcal{T}_h$, and $k > 0$ such that $\max_{m_i\in T}|v_h| \le k$. For every $x \in T$,

$$
\begin{aligned}
\left|v_h(x)\right| &= \left|\sum_{i=0}^{d} \alpha_i^{v_h} \varphi_i(x)\right| \le \sum_{i=0}^{d} \left|\alpha_i^{v_h} \varphi_i(x)\right| \\
&\le \sum_{i=0}^{d} \left|v_h(m_i)\right| \left|\varphi_i(x)\right| \\
&\le -k \left(\sum_{\substack{i=0 \\ \lambda_i<1/d}}^{d} \varphi_i(x) \right) + k \underbrace{\left(\sum_{\substack{i=0 \\ \lambda_i\ge 1/d}}^{d} \varphi_i(x) \right)}_{=1-\sum_{i=0,\ \lambda_i<1/d}^{d} \varphi_i(x)} \\
&\le k\left(1 + 2\sum_{i=0}^{d} (d-1)\right) \le k\left(2d^2 - 1\right),
\end{aligned} \tag{124}
$$

the inequality (123) is then proved. □

Proposition 18. *Under the assumption of Theorem 4, the solution u_h of (48) satisfies for every $k > 0$ and every $h > 0$*

$$
a_h^{swip}\left(I_h^k(u_h), I_h^k(u_h)\right) \le \int_\Omega f I_h^k(u_h)\, dx, \tag{125}
$$

and, in particular, u_h satisfies

$$
\int_\Omega \left|\nabla_h I_h^k(u_h)\right|^2 dx \le k \frac{2d^2 - 1}{\alpha C_\eta} \|f\|_{L^1(\Omega)}, \tag{126}
$$

where the constant C_η only depends on η, σ, and d.

The proof makes use of results appearing in [3] that we reproduce as follows.

Lemma 19 (Lemma 1.46 in [3], discrete trace inequality). *Under assumption (47), one has for all $v_h \in \mathbb{P}_d^1(\mathcal{T}_h)$, all $T \in \mathcal{T}_h$, and all $F \in \mathcal{F}_T$*

$$
h_T^{1/2} \|v_h\|_{L^2(F)} \le C_{tr} \|v_h\|_{L^2(T)} \tag{127}
$$

where C_{tr} only depends on σ and d.

Lemma 20 (Lemma 4.51 in [3], discrete coercivity). *For all $\eta > \underline{\eta} := (d+1)C_{tr}^2$, the SWIP bilinear form a_h^{swip} is coercive on V_h with respect to the $\|\cdot\|_{swip}$ – norm; i.e.,*

$$
\forall v_h \in V_h, \quad a_h^{swip}(v_h, v_h) \ge C_\eta \|v_h\|_{swip}^2 \tag{128}
$$

where $C_\eta = (\eta - (d+1)C_{tr}^2)/(1+\eta)$.

Proof (of Proposition 18). Using $I_h^k(u_h)$ as a test function in (48) one has

$$
a_h^{swip}\left(u_h, I_h^k(u_h)\right) = \int_\Omega f I_h^k(u_h)\, dx, \tag{129}
$$

and from (117) we obtain (125).

On the other hand, by combining (125) with coercivities (34) and (128) one obtains successively

$$\begin{aligned}
&\alpha C_\eta \int_\Omega \left|\nabla_h \left(I_h^k (u_h)\right)\right|^2 dx \\
&\leq C_\eta \int_\Omega \left|A\nabla_h \left(I_h^k (u_h)\right) \nabla_h \left(I_h^k (u_h)\right)\right| dx \\
&\leq C_\eta \left\|I_h^k (u_h)\right\|_{swip}^2 \leq a_h^{swip} \left(I_h^k (u_h), I_h^k (u_h)\right) \\
&\leq \int_\Omega f I_h^k (u_h)\, dx \leq \max_\Omega \left|I_h^k (u_h)\right| \int_\Omega |f|\, dx \\
&\leq k \left(2d^2 - 1\right) \|f\|_{L^1(\Omega)},
\end{aligned} \tag{130}$$

and by (123) the estimation (126) is then proved. □

Theorem 21. *Under the assumptions of Theorem 4, the solution u_h of (48) satisfies for every q with $1 \leq q < d/(d-1)$*

$$\begin{aligned}
u_h &\longrightarrow u \quad \text{strongly in } L^q(\Omega), \\
\nabla_h u_h &\longrightarrow \nabla u \quad \text{strongly in } \left[L^q(\Omega)\right]^d, \\
|u_h|_{J,A,q} &\longrightarrow 0,
\end{aligned} \tag{131}$$

when h tends to zero, where u is the unique renormalized solution of (1).

Proof (proof of Theorem 3.2 in [4]). Consider a sequence f^ε in $L^2(\Omega)$, converging strongly in $L^1(\Omega)$ to f (for example, $f^\varepsilon = T_{1/\varepsilon}(f)$). Let u_h^ε to be the unique solution of (48) for the right-hand side f^ε. Then $u_h - u_h^\varepsilon$ satisfies

$$\begin{aligned}
&u_h - u_h^\varepsilon \in V_h, \\
&\forall v_h \in V_h, \\
&a_h^{swip} \left(u_h - u_h^\varepsilon, v_h\right) = \int_\Omega \left(f - f^\varepsilon\right) v_h\, dx.
\end{aligned} \tag{132}$$

Applying estimate (126) to this problem, we obtain for every $k > 0$, every $h > 0$, and every $\varepsilon > 0$

$$\begin{aligned}
&\int_\Omega \left|\nabla_h \left(I_h^k \left(u_h - u_h^\varepsilon\right)\right)\right|^2 dx \\
&\leq k \frac{2d^2 - 1}{\alpha C_\eta} \left\|f - f^\varepsilon\right\|_{L^1(\Omega)},
\end{aligned} \tag{133}$$

Which implies by Theorem 5 that for every q with $1 \leq q < d/(d-1)$, every $h > 0$, and every $\varepsilon > 0$

$$\begin{aligned}
&\left\|u_h - u_h^\varepsilon\right\|_{swip,q} \\
&\leq \frac{C_2 \left(d, |\Omega|, q, \sigma, \|A\|_{L^\infty(\Omega)^{d\times d}}\right)}{\alpha C_\eta} \left\|f - f^\varepsilon\right\|_{L^1(\Omega)}.
\end{aligned} \tag{134}$$

On the other hand, since $f^\varepsilon \in L^2(\Omega)$ and $\mathcal{T}_h$ satisfies (44), (46), and (47), it is known (see [3]) that for every fixed ε

$$u_h^\varepsilon \longrightarrow u^\varepsilon \quad \text{strongly in } L^2(\Omega), \tag{135}$$

$$\nabla_h u_h^\varepsilon \longrightarrow \nabla u^\varepsilon \quad \text{strongly in } \left[L^2(\Omega)\right]^d, \tag{136}$$

$$\left|u_h^\varepsilon\right|_{J,A} \longrightarrow 0, \tag{137}$$

when h tends to zero, where u^ε is the unique solution of

$$\begin{aligned}
&u^\varepsilon \in H_0^1(\Omega), \\
&-\operatorname{div}\left(A\nabla u^\varepsilon\right) = f^\varepsilon \quad \text{in } \mathcal{D}'(\Omega).
\end{aligned} \tag{138}$$

Finally, the function u^ε is also the unique renormalized solution of the problem

$$\begin{aligned}
-\operatorname{div}\left(A\nabla u^\varepsilon\right) &= f^\varepsilon \quad \text{in } \Omega, \\
u^\varepsilon &= 0 \quad \text{on } \partial\Omega.
\end{aligned} \tag{139}$$

The estimate (43) combined with the inequality

$$\left\|u^\varepsilon - u\right\|_{swip,q} \leq \|A\|_{L^\infty(\Omega)^{d\times d}} \left\|u^\varepsilon - u\right\|_{W_0^{1,q}(\Omega)} \tag{140}$$

allows one to have

$$\begin{aligned}
&\left\|u^\varepsilon - u\right\|_{swip,q} \\
&\leq \frac{\|A\|_{L^\infty(\Omega)^{d\times d}} C_1 \left(d, |\Omega|, q\right)}{\alpha} \left\|f - f^\varepsilon\right\|_{L^1(\Omega)},
\end{aligned} \tag{141}$$

for every q with $1 \leq q < d/(d-1)$, where u is the unique renormalized solution of (1).

Writing now

$$\begin{aligned}
\left\|u_h - u\right\|_{swip,q} \leq{}& \left\|u_h - u_h^\varepsilon\right\|_{swip,q} + \left\|u^\varepsilon - u\right\|_{swip,q} \\
&+ \left\|u_h^\varepsilon - u^\varepsilon\right\|_{swip,q},
\end{aligned} \tag{142}$$

Using (134), (136), (137), and (141), one has for every $\varepsilon > 0$ and every q with $1 \leq q < d/(d-1)$

$$\begin{aligned}
&\limsup_{h\to 0} \left\|u_h - u\right\|_{swip,q} \\
&\leq \frac{C_3 \left(d, |\Omega|, q, \sigma, \|A\|_{L^\infty(\Omega)^{d\times d}}\right)}{\alpha C_\eta} \left\|f - f^\varepsilon\right\|_{L^1(\Omega)},
\end{aligned} \tag{143}$$

and passing to the limit when ε tends to zero proves Theorem 21. □

To complete the proof of Theorem 4, it remains to prove the following proposition.

Proposition 22. *Under the assumptions of Theorem 4, the solution u_h of (48) satisfies*

$$I_h^k (u_h) \underset{h\to 0}{\longrightarrow} T_k(u) \quad \text{strongly in } L^2(\Omega), \tag{144}$$

$$\nabla_h \left(I_h^k (u_h)\right) \underset{h\to 0}{\longrightarrow} \nabla T_k(u) \quad \text{strongly in } \left[L^2(\Omega)\right]^d, \tag{145}$$

$$\left|I_h^k (u_h)\right|_{J,A} \underset{h\to 0}{\longrightarrow} 0, \tag{146}$$

for every $k > 0$.

Proof (proof of Proposition 3.3 in [4]). First by result of Proposition 15 and the estimate (126) one can have (see theorem 5.7 in [3]) the two following convergences:

$$I_h^k(u_h) \underset{h\to 0}{\longrightarrow} T_k(u) \quad \text{strongly in } L^2(\Omega), \tag{147}$$

$$G_h^l\left(I_h^k(u_h)\right) \underset{h\to 0}{\rightharpoonup} \nabla T_k(u) \quad \text{weakly in } \left[L^2(\Omega)\right]^d, \tag{148}$$

for every $k > 0$.

On the other hand, using (123) one has

$$\left|f I_h^k(u_h)\right| \le k\left(2d^2 - 1\right)|f| \in L^1(\Omega), \tag{149}$$

so, by Lebesgue's dominated convergence theorem combined with discrete Rellich-Kondrachov's compactness theorem (see theorem 5.6 in [3]) one has

$$\int_\Omega f I_h^k(u_h)\,\mathrm{d}x \underset{h\to 0}{\longrightarrow} \int_\Omega f T_k(u)\,\mathrm{d}x. \tag{150}$$

Therefore passing to the limit with respect to h in (125) yields

$$\limsup_{h\to 0} a_h^{swip}\left(I_h^k(u_h), I_h^k(u_h)\right) \le \int_\Omega f T_k(u)\,dx. \tag{151}$$

Consequently

$$\begin{aligned}\limsup_{h\to 0} &\int_\Omega A G_h^l\left(I_h^k(u_h)\right) G_h^l\left(I_h^k(u_h)\right) \\ &\le \int_\Omega f T_k(u)\,dx.\end{aligned} \tag{152}$$

By the fact that

$$\begin{aligned}\int_\Omega & A G_h^l\left(I_h^k(u_h)\right) G_h^l\left(I_h^k(u_h)\right) \\ &\le a_h^{swip}\left(I_h^k(u_h), I_h^k(u_h)\right),\end{aligned} \tag{153}$$

and since u is the renormalized solution of (1), it is known that (see [4])

$$\int_\Omega A\nabla T_k(u)\nabla T_k(u) = \int_\Omega f T_k(u)\,dx. \tag{154}$$

Finally, from (152) and (154), we deduce that

$$\begin{aligned}\limsup_{h\to 0} &\int_\Omega A G_h^l\left(I_h^k(u_h)\right) G_h^l\left(I_h^k(u_h)\right) \\ &\le \int_\Omega A\nabla T_k(u)\nabla T_k(u),\end{aligned} \tag{155}$$

which combined with the weak convergence (148) implies

$$G_h^l\left(I_h^k(u_h)\right) \underset{h\to 0}{\longrightarrow} \nabla T_k(u) \quad \text{strongly in } \left[L^2(\Omega)\right]^d. \tag{156}$$

Owing to Proposition 4.36 in [3], for all $v_h \in V_h$ and all $\eta > (d+1)C_{tr}^2$, one can have

$$\begin{aligned}&\left|I_h^k(u_h)\right|_{J,A} \\ &\le \frac{a_h^{swip}\left(I_h^k(u_h), I_h^k(u_h)\right) - \left\|A^{1/2}G_h^l\left(I_h^k(u_h)\right)\right\|_{[L^2(\Omega)]^d}}{\eta - (d+1)C_{tr}^2}\end{aligned} \tag{157}$$

and, since the right-hand side tends to zero, the result (146) holds.

Finally, using the result of Proposition 4.34 in [3]

$$\left\|R_h^l\left(\left[I_h^k(u_h)\right]\right)\right\|_{[L^2(\Omega)]^d} \le \sqrt{d+1}C_{tr}\left|I_h^k(u_h)\right|_{J,A} \tag{158}$$

with the triangle inequality that yields

$$\begin{aligned}&\left\|\nabla_h I_h^k(u_h) - \nabla T_k(u)\right\|_{[L^2(\Omega)]^d} \\ &\quad\le \left\|G_h^l\left(I_h^k(u_h)\right) - \nabla T_k(u)\right\|_{[L^2(\Omega)]^d} \\ &\quad\quad + \left\|R_h^l\left(\left[I_h^k(u_h)\right]\right)\right\|_{[L^2(\Omega)]^d}\end{aligned} \tag{159}$$

concluding the proof of (145). □

5. The Case Where f Is a Bounded Radon Measure

The materials used in [4], to handle the case where f belongs to $\mathcal{M}_b(\Omega)$, are not specific to the case of $\mathbb{P}_1$ finite elements approximation; only the weak convergence (148) requires clarification; in our approach it is based on the result of Proposition 15 whose proof involves only properties of v_h not f. So we can also state the following convergence result.

Theorem 23 (Theorem 4.1 in [4]). *Assume that f belongs to $\mathcal{M}_b(\Omega)$ and A and $\mathcal{T}_h$ satisfy (46), (33), (35), (44), and (47) and (32). Then there exist a subsequence, still denoted by h, and a function u such that for every $k > 0$ and for every q with $1 \le q < d/(d-1)$ one has*

$$\begin{gathered}G_h^l\left(I_h^k(u_h)\right) \rightharpoonup \nabla T_k(u) \quad \text{weakly in } \left[L^2(\Omega)\right]^d, \\ \nabla_h u_h \rightharpoonup \nabla u \quad \text{weakly in } \left[L^q(\Omega)\right]^d,\end{gathered} \tag{160}$$

when h tends to zero along this subsequence, where u satisfies

$$\begin{gathered}\forall k > 0, \quad T_k(u) \in H_0^1(\Omega), \\ \forall q \text{ with } 1 \le q < \frac{d}{d-1}, \quad u \in W_0^{1,q}(\Omega), \\ \forall v \in \mathscr{C}_c^\infty(\Omega), \quad \int_\Omega A\nabla u\nabla v\,dx = \int_\Omega v\,df.\end{gathered} \tag{161}$$

6. Convergence Rate Estimation

6.1. Error Estimates for Smooth Solutions

Assumption 24 (regularity of exact solution and space V_*). As in [3], we assume that $\mathcal{T}_h$ is compatible with the partition P_Ω

in the sense of Assumption 3, and the unique solution v is such that

$$v \in V_* := H_0^1(\Omega) \cap H^2(\mathcal{T}_h). \tag{162}$$

And we set

$$V_{*h} := V_* + V_h. \tag{163}$$

The convergence analysis is performed in the spirit of (Theorem 1.35 [3]) by establishing discrete coercivity, consistency, and boundedness for a_h^{swip}. The discrete bilinear form a_h^{swip} is extended to $V_{*h} \times V_h$.

Without further knowledge on the exact solution v apart from the domain Ω and the datum $g \in L^2(\Omega)$, Assumption 24 can be asserted for instance if the domain Ω is convex; see Grisvard [7].

A straightforward consequence of the Lax-Milgram Lemma is that the discrete problem (48) is well-posed.

Theorem 25. *Under the assumptions of Theorem 4 ($d = 2$ or $d = 3$), if f belongs to $L^{r,\infty}(\Omega)$ for some r such that $1 < r < 2$, Ω a convex polyhedron ($\Omega_h = \Omega$), and $A \in [W^{1,\infty}(\Omega)]^{d\times d}$, then $\forall q$ with $1 \le q < d/(d-1)$*

$$\|u_h - u\|_{swip,q} \le C\left(d, |\Omega|, q, r, \alpha, \|A\|_{W^{1,\infty}(\Omega)^{d\times d}}, \sigma\right) \cdot h^{2(1-1/r)} \|f\|_{L^{r,\infty}(\Omega)}. \tag{164}$$

Proof (proof of Theorem 5.1 in [4]). From [3] the unique solution u^ϵ of (48) with right-hand side $f^\epsilon = T_{1/\epsilon}(f) \in L^\infty(\Omega) \subset L^2(\Omega)$ verifies

$$\|u_h^\epsilon - u^\epsilon\|_{swip} \le C\left(\|A\|_{W^{1,\infty}(\Omega)^{d\times d}}\right) h \|f^\epsilon\|_{L^2(\Omega)} \tag{165}$$

so

$$\|u_h^\epsilon - u^\epsilon\|_{swip,q} \le C\left(q, d, \eta, |\Omega|, \|A\|_{W^{1,\infty}(\Omega)^{d\times d}}\right) h \|f^\epsilon\|_{L^2(\Omega)} \tag{166}$$

Combined with (142), (134), and (141) allows one to have

$$\|u_h - u\|_{swip,q} \le C\left(q, d, \eta, |\Omega|, \|A\|_{W^{1,\infty}(\Omega)^{d\times d}}\right) \cdot \left(\|f - f^\epsilon\|_{L^1(\Omega)} + h\|f^\epsilon\|_{L^2(\Omega)}\right), \tag{167}$$

and by proceeding as in [4], we obtain (173). □

Remark 26. If $f \in L^{r,\infty}(\Omega)$ for some r with $1 < r < 2$ and

$$\forall k > 0: \quad T_k(f) \in H^1(\Omega), \tag{168}$$

then a small adaptation of the proof given in [4] provides an $O(h^{4(1-1/r)})$ error estimate in $\| \|_{swip,q}$-norm, with $1 \le q < d/(d-1)$, since, with (168), it is known that

$$\|u_h^\epsilon - u^\epsilon\|_{swip,q} \le C\left(d, |\Omega|, q, r, \alpha, \|A\|_{W^{1,\infty}(\Omega)^{d\times d}}, \sigma\right) h^2 \|f^\epsilon\|_{L^2(\Omega)}. \tag{169}$$

6.2. Error Estimates for Low-Regularity Solutions

Assumption 27 (regularity of exact solution and space V_*). As in [3], we assume that the mesh $\mathcal{T}_h$ is compatible with the partition P_Ω in the sense of Assumption 3, $d \ge 2$, and that there is p such that $2d/(d+2) < p \le 2$; the unique solution v is such that

$$v \in V_* := H_0^1(\Omega) \cap W^{2,p}(\mathcal{T}_h), \tag{170}$$

where $W^{2,p}(P_\Omega) = W^{2,p}(\mathcal{T})$ designate that the mesh $\mathcal{T}$ is compatible with the partition P_Ω, and we set

$$V_{*h} := V_* + V_h. \tag{171}$$

We also assume $p < 2$ since, in the case $p = 2$, Assumption 27 amounts to Assumption 24.

Assumption 27 requires $p > 1$ for $d = 2$ and $p > 6/5$ for $d = 3$. In particular, we observe that, in two space dimensions, $v \in W^{2,p}(P_\Omega)$ with $p > 1$ holds true in polygonal domains; see, e.g., Dauge [8]. Moreover, using Sobolev embeddings (see [Evans [9], Sect. 5.6] or [Brézis [10], Sect. IX.3]), Assumption 27 implies

$$v \in H^{1+\alpha_p}(\Omega), \quad \text{with } \alpha_p = \frac{d+2}{2} - \frac{d}{p} > 0. \tag{172}$$

Theorem 28. *Under the assumptions of Theorem 4 ($d = 2$ or $d = 3$), if f belongs to $L^{r,\infty}(\Omega)$ for some r such that $1 < r < 2$, Ω a convex polyhedron ($\Omega_h = \Omega$), and $A \in [W^{1,\infty}(\Omega)]^{d\times d}$, then $\forall q$ with $1 \le q < p$*

$$\|u_h - u\|_{swip,q} \le C\left(d, |\Omega|, q, r, \alpha, \|A\|_{W^{1,\infty}(\Omega)^{d\times d}}, \sigma\right) \cdot h^{2\alpha_p(1-1/r)} \|f\|_{L^{r,\infty}(\Omega)}. \tag{173}$$

Remark 29. Moreover, under the assumption in Remark 26 one can have an $O(h^{4\alpha_p(1-1/r)})$ error estimate in $\| \|_{swip,q}$-norm, with $1 \le q < p$.

7. The Case Where A Is the Identity Matrix

Returning to the definition (31) of mass matrix Q where $A = I$, one has for i, j in $\{1, 2, \ldots, N\}$

$$Q_{i,j} = a_h^{swip}\left(\varphi_i, \varphi_j\right)$$
$$= \underbrace{\int_{\Omega} \nabla_h \varphi_i \nabla_h \varphi_j dx}_{=I_1} + \underbrace{\sum_{F \in \mathcal{F}_h^i} \frac{1}{h_F} \eta \left[\varphi_i\right]\left[\varphi_j\right] dx}_{=I_2} - \underbrace{\left(\sum_{F \in \mathcal{F}_h^i} \int_F \{\nabla_h \varphi_i\}_\omega \cdot n_F \left[\varphi_j\right] dx + \sum_{F \in \mathcal{F}_h^i} \int_F \{\nabla_h \varphi_j\}_\omega \cdot n_F \left[\varphi_i\right] dx\right)}_{=I_3}. \quad (174)$$

If $i = j$, so

$$I_1 = \int_\Omega \nabla_h \varphi_i^2 dx \geq 0,$$
$$I_2 = \sum_{F \in \mathcal{F}_h^i} \int_F \frac{\eta}{h_F} \left[\varphi_i\right]^2 dx \geq 0,$$
$$I_3 = 2\left(\int_{F_i} \{\nabla_h \varphi_i\}_\omega \cdot n_F \left[\varphi_i\right] dx \right. \quad (175)$$
$$\left. + \sum_{\substack{F \in \mathcal{F}_h^i \\ F \neq F_i}} \{\nabla_h \varphi_i\}_\omega \cdot n_F \left[\varphi_i\right] dx \right) = 0,$$

since $[\varphi_i]_{F_i} = 0$, $\{\nabla_h \varphi_i\}_\omega \cdot n_F = -(d/h_{F_i}) n_i \cdot n_F$, and $\int_{F \neq F_i} \varphi_i dx = 0$.

If $i \neq j$, so

$$I_1 = \int_\Omega \nabla_h \varphi_i \nabla_h \varphi_j dx = \frac{1}{d^2} \left|F_i\right| \left|F_j\right| n_i \cdot n_j \leq 0 \quad (176)$$

(see Proposition 6.1 in [4]),

$$I_2 = \int_{F \in F_i \cup F_j} \frac{\eta}{h_F} \left[\varphi_i\right]\left[\varphi_j\right] dx + \sum_{\substack{F \in \mathcal{F}_h^i \\ F \neq F_i \neq F_j}} \frac{\eta}{h_F} \left[\varphi_i\right]\left[\varphi_j\right] dx$$
$$= -\frac{\eta}{d+1} \sum_{\substack{F \in \mathcal{F}_h^i \\ F \neq F_i \neq F_j}} \frac{|F|}{h_F} \leq 0, \quad (177)$$
$$I_3 = -d \sum_{\substack{F \in \mathcal{F}_h^i \\ F \neq F_i \neq F_j}} \left(\frac{1}{h_{F_i}} n_i \cdot n_F \varphi_j + \frac{1}{h_{F_j}} n_j \cdot n_F \varphi_i\right) dx = 0.$$

It is therefore concluded that, under the condition $n_l \cdot n_k \leq 0$ $(l \neq k)$, one can have $Q_{lk} \leq 0$. Thus, the matrix Q verifies (32).

Conflicts of Interest

The authors declare that they have no conflicts of interest.

References

[1] G. Dal Maso, F. Murat, L. Orsina, and A. Prignet, "Renormalized solutions of elliptic equations with general measure data," *Annali della Scuola Normale Superiore di Pisa. Classe di Scienze. Serie IV*, vol. 28, no. 4, pp. 741–808, 1999.

[2] P. Bénilan, L. Boccardo, T. Gallouët, R. Gariepy, M. Pierre, and J. L. Vázquez, "An L1-theory of existence and uniqueness of solutions of nonlinear elliptic equations," *Ann. Scuola Norm. Sup. Pisa*, vol. 22, no. 2, pp. 241-273, 1995.

[3] D. A. Di Pietro and A. Ern, "Mathematical aspects of discontinuous Galerkin methods," *Mathématiques et Applications*, vol. 69, 2011.

[4] J. Casado-Daz, T. Chacón Rebollo, V. Girault, M. Gómez Mármol, and F. Murat, "Finite elements approximation of second order linear elliptic equations in divergence form with right-hand side in L1," *Numerische Mathematik*, vol. 105, no. 3, pp. 337–374, 2007.

[5] L. Boccardo and T. Gallouët, "Nonlinear elliptic and parabolic equations involving measure data," *Journal of Functional Analysis*, vol. 87, no. 1, pp. 149–169, 1989.

[6] P. G. Ciarlet, *The Finite Element Method for Elliptic Problems*, North-Holland, Amsterdam, The Netherlands, 1978.

[7] P. Grisvard, *Singularities in Boundary Value Problems*, Masson, Paris, 1992.

[8] M. Dauge, *Elliptic Boundary Value Problems on Corner Domains*, vol. 1341 of *Lecture Notes in Mathematics*, Springer, Berlin, Germany, 1988.

[9] L. C. Evans, *Partial Differential Equations*, vol. 19 of *Graduate Studies in Mathematics*, American Mathematical Society, Providence, RI, USA, 1998.

[10] H. Brezis, *Analyse Fonctionnelle*, Masson, Paris, France, 1983.

5

Critical Oscillation Constant for Euler Type Half-Linear Differential Equation Having Multi-Different Periodic Coefficients

Adil Misir and Banu Mermerkaya

Department of Mathematics, Faculty of Science, Gazi University, Teknikokullar, 06500 Ankara, Turkey

Correspondence should be addressed to Adil Misir; adilm@gazi.edu.tr

Academic Editor: Said R. Grace

We compute explicitly the oscillation constant for Euler type half-linear second-order differential equation having multi-different periodic coefficients.

1. Introduction

In literature, half-linear second-order differential equations are given by

$$\left(r(t)\Phi\left(x'\right)\right)' + c(t)\Phi(x) = 0,\qquad \Phi(s) = |s|^{p-2}s,\quad p>1, \tag{1}$$

where r, c are continuous functions and $r(t) > 0$. It is well known that oscillation theory of (1) is very similar to that of the linear Sturm-Liouville differential equation, which is the special case of $p = 2$ in (1); see [1].

In particular, (1) with $\lambda c(t)$ instead of $c(t)$ is said to be conditionally oscillatory if there exists a constant λ_0 such that this equation is oscillatory for $\lambda > \lambda_0$ and nonoscillatory for $\lambda < \lambda_0$. λ_0 is called the critical oscillation constant of this equation; see [2].

The half-linear Euler differential equation

$$\left(\Phi\left(x'\right)\right)' + \frac{\gamma_p}{t^p}\Phi(x) = 0, \tag{2}$$

with the so-called critical oscillation constant $\gamma_p = ((p-1)/p)^p$, plays an important role in the conditionally oscillatory half-linear differential equation.

Equation (2) can be regarded as a good comparative equation in the sense that (2) with γ instead of γ_p is oscillatory if and only if $\gamma > \gamma_p$ (see [3]) and if $r(t) = 1$ in (1), then this equation is oscillatory provided

$$\liminf_{t\to\infty} t^p c(t) > \gamma_p \tag{3}$$

and nonoscillatory if

$$\limsup_{t\to\infty} t^p c(t) < \gamma_p; \tag{4}$$

see [4].

In [5], perturbations of (2) being of the form

$$\left(\Phi\left(x'\right)\right)' + \frac{1}{t^p}\left(\gamma_p + \sum_{l=1}^{n}\frac{\beta_j}{\mathrm{Log}_j^2 t}\right)\Phi(x) = 0 \tag{5}$$

are investigated when $\lim_{t\to\infty} t^p c(t) = \gamma_p$ for constant β_j $(j = 1, 2, \ldots, n)$. Here the notation

$$\begin{aligned}\mathrm{Log}_k t &= \prod_{j=1}^{k}\log_j t,\\ \mathrm{Log}_k t &= \log_{k-1}(\log t),\\ \mathrm{Log}_1 t &= \log t\end{aligned} \tag{6}$$

is used. It is shown that the constant $\mu_p := (1/2)((p-1)/p)^{p-1}$ plays a crucial role in (5). In particular, if $n = 1$ in (5) this equation reduces to the so-called Riemann-Weber half-linear differential equation, and this equation is oscillatory if $\beta_1 > \mu_p$ and nonoscillatory otherwise. In general, if $\beta_j = \mu_p$ for $j = 1, 2, \ldots, n-1$, then (5) is oscillatory if and only if $\beta_n > \mu_p$.

One of the typical problems in the qualitative theory of various differential equations is to study what happens when constants in an equation are replaced by periodic functions which have same periods and different periods. Our investigation follows this line and it is mainly motivated by the paper [6].

In [7], the half-linear differential equation being of the form

$$\left(\left(1+\sum_{j=1}^{n}\frac{\alpha_j}{\mathrm{Log}_j^2 t}\right)^{1-p}\Phi\left(x'\right)\right)' + \frac{1}{t^p}\left(\gamma_p+\sum_{j=1}^{n}\frac{\beta_j}{\mathrm{Log}_j^2 t}\right)\Phi(x)=0 \tag{7}$$

is investigated for α_j and β_j are constants and the following result is obtained.

Theorem 1. *Suppose that there exists $k \in \{2, \ldots, n\}$ such that*

$$\beta_j + (p-1)\gamma_p\alpha_j = \mu_p, \quad j = 1, \ldots, k-1 \tag{8}$$

and $\beta_k + (p-1)\gamma_p\alpha_k \neq \mu_p$. Then (7) is oscillatory if $\beta_k + (p-1)\gamma_p\alpha_k > \mu_p$ and nonoscillatory if $\beta_k + (p-1)\gamma_p\alpha_k < \mu_p$.

In [8], the half-linear differential equation being of the form

$$\left(r(t)\Phi\left(x'\right)\right)' + \frac{\gamma c(t)}{t^p}\Phi(x) = 0 \tag{9}$$

is considered for α-periodic positive functions r and c and it is shown that (9) is oscillatory if $\gamma > K$ and nonoscillatory if $\gamma < K$, where K is given by

$$K = q^{-p}\left(\frac{1}{\alpha}\int_0^\alpha\frac{d\tau}{r^{q-1}}\right)^{1-p}\left(\frac{1}{\alpha}\int_0^\alpha c(\tau)\,d\tau\right)^{-1} \tag{10}$$

for p and q are conjugate numbers; that is, $1/p + 1/q = 1$.

In [9], (9) and the half-linear differential equation being of the form

$$\left(r(t)\Phi\left(x'\right)\right)' + \frac{1}{t^p}\left(\gamma c(t) + \frac{\mu d(t)}{\log^2 t}\right)\Phi(x) = 0 \tag{11}$$

are considered for r, c, and d are α-periodic, positive functions defined on $[0, \infty)$ and it is shown that (9) is nonoscillatory if and only if $\gamma \le \gamma_{rc}$, where γ_{rc} is given by

$$\gamma_{rc} := \frac{\alpha^p\gamma_p}{\left(\int_0^\alpha r^{1-q}(t)\,dt\right)^{p-1}\int_0^\alpha c(t)\,dt}. \tag{12}$$

In the limiting case $\gamma = \gamma_{rc}$ (11) is nonoscillatory if $\mu < \mu_{rd}$ and it is oscillatory if $\mu > \mu_{rd}$, where μ_{rd} is given by

$$\mu_{rd} = \frac{\alpha^p\mu_p}{\left(\int_0^\alpha r^{1-q}(t)\,dt\right)^{p-1}\int_0^\alpha d(t)\,dt}. \tag{13}$$

In [10], the half-linear differential equation being of the form

$$\left(r(t)\Phi\left(x'\right)\right)' + \frac{c(t)}{t^p}\Phi(x) = 0 \tag{14}$$

is considered for $r : [a, \infty) \to \mathbb{R}$, $(a > 0)$, is a continuous function for which mean value $M(r^{1-q})$ exists and for which

$$0 < \inf_{t\in[a,\infty)} r(t) \le \sup_{t\in[a,\infty)} r(t) < \infty \tag{15}$$

holds and $c : [a, \infty) \to \mathbb{R}$, $(a > 0)$, is a continuous function having mean value $M(c)$ and it was shown that (14) is oscillatory if $M(c) > \Gamma$ and nonoscillatory if $M(c) < \Gamma$, where Γ is given by

$$\Gamma = q^{-p}\left[M\left(r^{1-q}\right)\right]^{1-p}. \tag{16}$$

In [6], the half-linear differential equation being of the form

$$\left(\left(r(t)+\sum_{j=1}^{n}\frac{\alpha_j(t)}{\mathrm{Log}_j^2 t}\right)^{1-p}\Phi\left(x'\right)\right)' + \frac{1}{t^p}\left(c(t)+\sum_{j=1}^{n}\frac{\beta_j(t)}{\mathrm{Log}_j^2 t}\right)\Phi(x)=0 \tag{17}$$

is considered for T-periodic functions r, c, α_j, and β_j, $j = 1, 2, \ldots, n$, and $r(t) > 0$ and the following result was obtained.

Theorem 2. *Let r, c, α_j, and β_j $(j = 1, 2, \ldots, n)$ be T-periodic continuous functions, $r(t) > 0$, and their mean values over the period T are denoted by $\tilde{r}, \tilde{c}, \tilde{\alpha}_j$, and $\tilde{\beta}_j$ $(j = 1, 2, \ldots, n)$.*

(i) *If $\tilde{c}\tilde{r}^{p-1} > \gamma_p$, then (17) is oscillatory and if $\tilde{c}\tilde{r}^{p-1} < \gamma_p$, then it is nonoscillatory.*

(ii) *Let $\tilde{c}\tilde{r}^{p-1} = \gamma_p$. If there exists $k \in \{1, \ldots, n\}$ such that*

$$\tilde{\beta}_j\tilde{r}^{p-1} + (p-1)\gamma_p\tilde{\alpha}_j\tilde{r}^{-1} = \mu_p, \quad j = 1, 2, \ldots, k-1 \tag{18}$$

(if $k \neq 1$), and $\tilde{\beta}_k\tilde{r}^{p-1} + (p-1)\gamma_p\alpha_k\tilde{r}^{-1} \neq \mu_p$, then (17) is oscillatory if

$$\tilde{\beta}_k\tilde{r}^{p-1} + (p-1)\gamma_p\tilde{\alpha}_k\tilde{r}^{-1} > \mu_p \tag{19}$$

and nonoscillatory if

$$\tilde{\beta}_k\tilde{r}^{p-1} + (p-1)\gamma_p\tilde{\alpha}_k\tilde{r}^{-1} < \mu_p. \tag{20}$$

Our research is motivated by the paper [6], where the oscillation constant is computed for (17) with the periodic coefficients having same T-period. However, if these periodic functions have different periods what would be the oscillation constant is not investigated. Thus, in this paper we investigate the oscillation constant for (17) with periodic coefficients having different periods. In this paper we consider two types of periodic coefficients which have different periods for (17). In the first type we consider these periodic coefficient functions having the least common multiple and in the second type, we consider these periodic coefficient functions which do not have least common multiple. We give some corollaries which illustrate the first type's cases that our results compile the known results in [6] but in the second type only our results can be applied.

In Section 2, we recall the concept of half-linear-trigonometric functions and their properties. In Section 3 we compute the oscillation constant for (17) with periodic coefficients which have different periods. Additionally we show that if the different periods coincide, then our results compile with the known results in [6]. Thus, our results extend and improve the results of [6].

2. Preliminaries

We start this section with recalling the concept of half-linear-trigonometric functions; see [1] or [4]. Consider the following special half-linear equation being of the form

$$\left(\Phi\left(x'\right)\right)' + (p-1)\Phi(x) = 0 \tag{21}$$

and denote its solution by $x = x(t)$ given by the initial conditions $x(0) = 0$, $x'(0) = 1$. We see that the behavior of this solution is very similar to that of the classical sine function. We denote this solution by $\sin_p t$ and its derivative by $(\sin_p t)' = \cos_p t$. These functions are $2\pi_p$-periodic, where $\pi_p := 2\pi/p\sin(\pi/p)$, and satisfy the half-linear Pythagorean identity

$$\left|\sin_p t\right|^p + \left|\cos_p t\right|^p = 1, \quad t \in \mathbb{R}. \tag{22}$$

Every solution of (21) is of the form $x(t) = C\sin_p(t+\varphi)$, where C and φ are real constants; that is, it is bounded together with its derivative and periodic with the period $2\pi_p$. The function $u = \Phi(\cos_p t)$ is a solution to the reciprocal equation of (21);

$$\begin{gathered}\left(\Phi^{-1}\left(u'\right)\right)' + (p-1)^{q-1}\Phi^{-1}(u) = 0,\\ \Phi^{-1}(u) = |u|^{q-2}u,\ q = \frac{p}{p-1},\end{gathered} \tag{23}$$

which is an equation of the form as in (21), so the functions u and u' are also bounded.

Let $x(t)$ be a nontrivial solution of (1) and we consider the half-linear Prüfer transformation which is introduced using the half-linear-trigonometric functions

$$\begin{aligned} x(t) &= \rho(t)\sin_p\varphi(t),\\ x'(t) &= r^{1-q}(t)\frac{\rho(t)}{t}\cos_p\varphi(t), \end{aligned} \tag{24}$$

where $\rho(t) = \sqrt{|x(t)|^p + r^q(t)|x'(t)|^p}$ and Prüfer angle $\varphi(t)$ is a continuous function defined at all points where $x(t) \neq 0$.

Then $\varphi(t)$ satisfies the following differential equation:

$$\begin{aligned} \varphi'(t) = \frac{1}{t}\Big[& r^{1-q}(t)\left|\cos_p\varphi(t)\right|^p \\ & - \Phi\left(\cos_p\varphi(t)\right)\sin_p\varphi(t) + \frac{t^p c(t)}{p-1}\left|\sin_p\varphi(t)\right|^p\Big]; \end{aligned} \tag{25}$$

see [9].

3. Main Results

We need the following lemma in order to prove our main Theorem 4.

Lemma 3. *Let $\varphi(t) = \varphi_1(t) + \sum_{j=1}^n \varphi_{2_j}(t) + \varphi_3(t) + \varphi_4(t) + \sum_{j=1}^n \varphi_{5_j}(t) + M$ (M is a suitable constant) be a solution of the equation*

$$\begin{aligned} \varphi'(t) = {} & \varphi_1'(t) + \sum_{j=1}^n \varphi_{2_j}'(t) + \varphi_3'(t) + \varphi_4'(t) \\ & + \sum_{j=1}^n \varphi_{5_j}'(t), \end{aligned} \tag{26}$$

where

$$\begin{aligned} \varphi_1'(t) &= \frac{1}{t}r(t)\left|\cos_p\varphi(t)\right|^p,\\ \varphi_{2_j}'(t) &= \frac{\alpha_j(t)}{t\mathrm{Log}_j^2 t}\left|\cos_p\varphi(t)\right|^p \quad (j = 1,\dots,n),\\ \varphi_3'(t) &= -\frac{1}{t}\Phi\left(\cos_p\varphi(t)\right)\sin_p\varphi(t),\\ \varphi_4'(t) &= \frac{c(t)}{(p-1)t}\left|\sin_p\varphi(t)\right|^p,\\ \varphi_{5_j}'(t) &= \frac{\beta_j(t)}{(p-1)t\mathrm{Log}_j^2 t}\left|\sin_p\varphi(t)\right|^p \quad (j = 1,\dots,n), \end{aligned} \tag{27}$$

with r, c, α_j, and β_j ($j = 1, 2, \dots, n$) are periodic functions having different T_1, T_2, P_j, and Q_j ($j = 1, 2, \dots, n$) periods, respectively, and $r(t) > 0$ and

$$\begin{aligned} \theta(t) = {} & \frac{1}{T_1}\int_t^{t+T_1}\varphi_1(s)\,ds + \sum_{j=1}^n\frac{1}{P_j}\int_t^{t+P_j}\varphi_{2_j}(s)\,ds \\ & + \frac{1}{\xi}\int_t^{t+\xi}\varphi_3(s)\,ds + \frac{1}{T_2}\int_t^{t+T_2}\varphi_4(s)\,ds \\ & + \sum_{j=1}^n\frac{1}{Q_j}\int_t^{t+Q_j}\varphi_{5_j}(s)\,ds, \end{aligned} \tag{28}$$

where ξ is one of the following T_1, T_2, P_j, and Q_j ($j = 1, 2, \ldots, n$) periods. Then $\theta(t)$ is a solution of

$$\theta'(t) = \frac{1}{t}\left[\overset{*}{r} + \sum_{j=1}^{n} \frac{\overset{*}{\alpha}_j}{\mathrm{Log}_j^2 t} + \frac{o(1)}{\mathrm{Log}_n^2 t}\right] \left|\cos_p \theta(t)\right|^p + \frac{1}{(p-1)t}\left[\overset{*}{c} + \sum_{j=1}^{n} \frac{\overset{*}{\beta}_j}{\mathrm{Log}_j^2 t} + \frac{o(1)}{\mathrm{Log}_n^2 t}\right] \left|\sin_p \theta(t)\right|^p - \frac{1}{t}\Phi\left(\cos_p \theta(t)\right) \sin_p \theta(t), \tag{29}$$

where

$$\overset{*}{r} = \frac{1}{T_1}\int_0^{T_1} r(s)\,ds, \quad \overset{*}{c} = \frac{1}{T_2}\int_0^{T_2} c(s)\,ds, \quad \overset{*}{\alpha}_j = \frac{1}{P_j}\int_0^{P_j} \alpha_j(s)\,ds, \quad \overset{*}{\beta}_j = \frac{1}{Q_j}\int_0^{Q_j} \beta_j(s)\,ds \quad \text{for } j = 1, 2, \ldots, n \tag{30}$$

and $\varphi(\tau) - \theta(t) = \circ(1)$ as $t \to \infty$.

Proof. Taking derivative of $\theta(t)$, we have

$$\begin{aligned}\theta'(t) &= \frac{1}{T_1}\int_t^{t+T_1} \varphi_1'(s)\,ds + \sum_{j=1}^{n}\frac{1}{P_j}\int_t^{t+P_j} \varphi_{2_j}'(s)\,ds + \frac{1}{\xi}\int_t^{t+\xi} \varphi_3'(s)\,ds + \frac{1}{T_2}\int_t^{t+T_2} \varphi_4'(s)\,ds + \sum_{j=1}^{n}\frac{1}{Q_j}\int_t^{t+Q_j} \varphi_{5_j}'(s)\,ds \\ &= \frac{1}{T_1}\int_t^{t+T_1} \frac{1}{s} r(s)\left|\cos_p \varphi(s)\right|^p ds + \sum_{j=1}^{n}\frac{1}{P_j}\int_t^{t+P_j} \frac{\alpha_j(s)}{s\mathrm{Log}_j^2 s}\left|\cos_p \varphi(s)\right|^p ds - \frac{1}{\xi}\int_t^{t+\xi} \frac{1}{s}\Phi\left(\cos_p \varphi(s)\right)\sin_p \varphi(s)\,ds \\ &\quad + \frac{1}{T_2}\int_t^{t+T_2} \frac{c(s)}{(p-1)s}\left|\sin_p \varphi(s)\right|^p ds + \sum_{j=1}^{n}\frac{1}{Q_j}\int_t^{t+Q_j} \frac{\beta_j(s)}{s(p-1)\mathrm{Log}_j^2 s}\left|\sin_p \varphi(s)\right|^p ds.\end{aligned} \tag{31}$$

Using integration by parts, we get

$$\begin{aligned}\theta'(t) &= \frac{1}{T_1 t}\int_t^{t+T_1} r(\tau)\left|\cos_p \varphi(\tau)\right|^p d\tau + \frac{1}{t}\sum_{j=1}^{n}\frac{1}{P_j} \cdot \int_t^{t+P_j} \frac{\alpha_j(\tau)}{\mathrm{Log}_j^2 \tau}\left|\cos_p \varphi(\tau)\right|^p d\tau - \frac{1}{\xi t} \cdot \int_t^{t+\xi} \Phi\left(\cos_p \varphi(\tau)\right)\sin_p \varphi(\tau)\,d\tau + \frac{1}{T_2 t} \\ &\quad \cdot \int_t^{t+T_2} \frac{c(\tau)}{(p-1)}\left|\sin_p \varphi(\tau)\right|^p d\tau + \frac{1}{t}\sum_{j=1}^{n}\frac{1}{Q_j} \cdot \int_t^{t+Q_j} \frac{\beta_j(\tau)}{(p-1)\mathrm{Log}_j^2 \tau}\left|\sin_p \varphi(\tau)\right|^p d\tau - \frac{1}{T_1} \\ &\quad \cdot \int_t^{t+T_1} \frac{1}{s^2}\int_s^{t+T_1} r(\tau)\left|\cos_p \varphi(\tau)\right|^p d\tau\,ds - \sum_{j=1}^{n}\frac{1}{P_j} \cdot \int_t^{t+P_j} \frac{1}{s^2}\int_t^{t+P_j} \frac{\alpha_j(\tau)}{\mathrm{Log}_j^2 \tau}\left|\cos_p \varphi(\tau)\right|^p d\tau\,ds + \frac{1}{\xi} \\ &\quad \cdot \int_t^{t+\xi} \frac{1}{s^2}\int_s^{t+\xi} \Phi\left(\cos_p \varphi(\tau)\right)\sin_p \varphi(\tau)\,d\tau\,ds - \frac{1}{T_2} \cdot \int_t^{t+T_2} \frac{1}{s^2}\int_s^{t+T_2} \frac{c(\tau)}{(p-1)}\left|\sin_p \varphi(\tau)\right|^p d\tau\,ds \\ &\quad - \sum_{j=1}^{n}\frac{1}{Q_j} \cdot \int_t^{t+Q_j} \frac{1}{s^2}\int_t^{t+Q_j} \frac{\beta_j(\tau)}{(p-1)\mathrm{Log}_j^2 \tau}\left|\sin_p \varphi(\tau)\right|^p d\tau\,ds.\end{aligned} \tag{32}$$

Let f be a continuous T-periodic function and $\overset{*}{f} = (1/T)\int_0^T f(s)ds$; then integration by parts yields

$$\frac{1}{T}\int_t^{t+T} \frac{f(s)}{\log_j^2 s}ds = \frac{\overset{*}{f}}{\log_j^2 t}\left[1 + O\left(\frac{1}{t\log t}\right)\right]. \tag{33}$$

By using (33) and $\int_t^{t+T} f(s)ds = \int_0^T f(s)ds$ for any T-periodic function and Pythagorean identity, the expressions

$$r^{1-q}(t)\left|\cos_p \varphi(t)\right|^p, \quad -\Phi\left(\cos_p \varphi(t)\right)\sin_p \varphi(t), \quad \frac{c(t)}{p-1}\left|\sin_p \varphi(t)\right|^p \tag{34}$$

are bounded. Thus we get

$$\theta'(t) = \frac{1}{T_1 t}\int_t^{t+T} r(\tau)\left|\cos_p \varphi(\tau)\right|^p d\tau$$

$$+\frac{1}{t}\sum_{j=1}^{n}\frac{1}{P_j}\int_t^{t+P_j}\frac{\alpha_j(\tau)}{\mathrm{Log}_j^2\tau}\left|\cos_p\varphi(\tau)\right|^p d\tau$$
$$-\frac{1}{\xi t}\int_t^{t+\xi}\Phi\left(\cos_p\varphi(\tau)\right)\sin_p\varphi(\tau)\,d\tau$$
$$+\frac{1}{T_2 t}\int_t^{t+T_2}\frac{c(\tau)}{(p-1)}\left|\sin_p\varphi(\tau)\right|^p d\tau$$
$$+\frac{1}{t}\sum_{j=1}^{n}\frac{1}{Q_j}\int_t^{t+Q_j}\frac{\beta_j(\tau)}{(p-1)\mathrm{Log}_j^2\tau}\left|\sin_p\varphi(\tau)\right|^p d\tau$$
$$+O\left(\frac{1}{t}\right). \tag{35}$$

If we add and subtract the below terms in the right side of this equation

$$\frac{1}{T_1 t}\int_t^{t+T_1}r(\tau)\left|\cos_p\theta(t)\right|^p d\tau$$
$$+\frac{1}{t}\sum_{j=1}^{n}\frac{1}{P_j}\int_t^{t+P_j}\frac{\alpha_j(\tau)}{\mathrm{Log}_j^2\tau}\left|\cos_p\theta(t)\right|^p d\tau$$
$$-\frac{1}{\xi t}\int_t^{t+\xi}\Phi\left(\cos_p\theta(t)\right)\sin_p\varphi(\tau)\,d\tau \tag{36}$$
$$+\frac{1}{T_2 t}\int_t^{t+T_2}\frac{c(\tau)}{(p-1)}\left|\sin_p\theta(t)\right|^p d\tau$$
$$+\frac{1}{t}\sum_{j=1}^{n}\frac{1}{Q_j}\int_t^{t+Q_j}\frac{\beta_j(\tau)}{(p-1)\mathrm{Log}_j^2\tau}\left|\sin_p\theta(t)\right|^p d\tau$$

we can rewrite this equation as

$$\theta'(t)=\frac{1}{T_1 t}\int_t^{t+T_1}r(\tau)\left|\cos_p\theta(t)\right|^p d\tau+\frac{1}{t}\sum_{j=1}^{n}\frac{1}{P_j}$$
$$\cdot\int_t^{t+P_j}\frac{\alpha_j(\tau)}{\mathrm{Log}_j^2\tau}\left|\cos_p\theta(t)\right|^p d\tau-\frac{1}{\xi t}$$
$$\cdot\int_t^{t+\xi}\Phi\left(\cos_p\theta(t)\right)\sin_p\varphi(\tau)\,d\tau+\frac{1}{T_2 t}$$
$$\cdot\int_t^{t+T}\frac{c(\tau)}{(p-1)}\left|\sin_p\theta(t)\right|^p d\tau+\frac{1}{t}\sum_{j=1}^{n}\frac{1}{Q_j}$$
$$\cdot\int_t^{t+Q_j}\frac{\beta_j(\tau)}{(p-1)\mathrm{Log}_j^2\tau}\left|\sin_p\theta(t)\right|^p d\tau+\frac{1}{T_1 t}$$
$$\cdot\int_t^{t+T_1}r(\tau)\left\{\left|\cos_p\varphi(\tau)\right|^p-\left|\cos_p\theta(t)\right|^p\right\}d\tau$$
$$+\frac{1}{t}\sum_{j=1}^{n}\frac{1}{P_j}\int_t^{t+P_j}\frac{\alpha_j(\tau)}{\mathrm{Log}_j^2\tau}\left\{\left|\cos_p\varphi(\tau)\right|^p\right.$$
$$\left.-\left|\cos_p\theta(t)\right|^p\right\}d\tau-\frac{1}{\xi t}$$
$$\cdot\int_t^{t+\xi}\left\{\Phi\left(\cos_p\varphi(\tau)\right)\sin_p\varphi(\tau)\right.$$
$$\left.-\Phi\left(\cos_p\theta(t)\right)\sin\theta(t)\right\}d\tau+\frac{1}{T_2 t}$$
$$\cdot\int_t^{t+T_2}\frac{c(\tau)}{(p-1)}\left\{\left|\sin_p\varphi(t)\right|^p-\left|\sin_p\theta(t)\right|^p\right\}d\tau$$
$$+\frac{1}{t}\sum_{j=1}^{n}\frac{1}{Q_j}\int_t^{t+Q_j}\frac{\beta_j(\tau)}{(p-1)\mathrm{Log}_j^2\tau}\left\{\left|\sin_p\varphi(t)\right|^p\right.$$
$$\left.-\left|\sin_p\theta(t)\right|^p\right\}d\tau+O\left(\frac{1}{t}\right). \tag{37}$$

And using the half-linear-trigonometric functions, we have

$$\left|\left|\cos_p\varphi(\tau)\right|^p-\left|\cos_p\theta(t)\right|^p\right|$$
$$\le p\left|\int_{\theta(t)}^{\varphi(\tau)}\left|\Phi\left(\cos_p s\right)\left(\cos_p s\right)'\right|ds\right|$$
$$\le \mathrm{const}\left|\varphi(\tau)-\theta(t)\right|,$$
$$\left|\Phi\left(\cos_p\varphi(\tau)\right)\sin_p\varphi(\tau)-\Phi\left(\cos_p\theta(t)\right)\sin_p\theta(t)\right| \tag{38}$$
$$\le\left|\int_{\theta(t)}^{\varphi(\tau)}\left|\left(\Phi\left(\cos_p s\right)\sin_p s\right)'\right|ds\right|$$
$$\le \mathrm{const}\left|\varphi(\tau)-\theta(t)\right|,$$
$$\left|\left|\sin_p\varphi(t)\right|^p-\left|\sin_p\theta(t)\right|^p\right|\le \mathrm{const}\left|\varphi(\tau)-\theta(t)\right|.$$

By the Mean Value Theorem we can write

$$\theta(t)=\varphi_1(t_1)+\sum_{j=1}^{n}\varphi_{2_j}\left(t_{2_j}\right)+\varphi_3(t_3)+\varphi_4(t_4)$$
$$+\sum_{j=1}^{n}\varphi_{5_j}\left(t_{5_j}\right) \tag{39}$$

for $t_1\in[t,t+T_1]$, $t_{2_j}\in[t,t+P_j]$, $j=1,2,\ldots,n$, $t_3\in[t,t+\xi]$, $t_4\in[t,t+T_2]$, and $t_{5_j}\in[t,t+Q_j]$, $j=1,2,\ldots,n$; thus

$$\left|\varphi(\tau)-\theta(t)\right|\le\left|\varphi_1(\tau)-\varphi_1(t_1)\right|$$
$$+\left|\sum_{j=1}^{n}\left(\varphi_{2_j}(\tau)-\varphi_{2j}\left(t_{2_j}\right)\right)\right|$$
$$+\left|\varphi_3(\tau)-\varphi_3(t_3)\right| \tag{40}$$
$$+\left|\varphi_4(\tau)-\varphi_4(t_4)\right|$$
$$+\left|\sum_{j=1}^{n}\left(\varphi_{5_j}(\tau)-\varphi_{5j}\left(t_{5_j}\right)\right)\right|.$$

This implies that

$$|\varphi(\tau) - \theta(t)| \le o\left(\frac{1}{t}\right), \quad \varphi(\tau) - \theta(t) = o(1) \quad \text{as } t \longrightarrow \infty. \tag{41}$$

And using $\mathring{r}$, $\mathring{c}$, $\mathring{\alpha}_j$, $\mathring{\beta}_j$, and (33), we get

$$\theta'(t) = \frac{1}{t}\left[\mathring{r} + \left\{1 + O\left(\frac{1}{t\log t}\right)\right\}\sum_{j=1}^{n}\frac{\mathring{\alpha}_j}{\text{Log}_j^2 t}\right] \cdot \left|\cos_p\theta(t)\right|^p + \frac{1}{(p-1)t}\left[\mathring{c} + \left\{1 + O\left(\frac{1}{t\log t}\right)\right\}\sum_{j=1}^{n}\frac{\mathring{\beta}_j}{\text{Log}_j^2 t}\right] \cdot \left|\sin_p\theta(t)\right|^p - \frac{1}{t}\Phi\left(\cos_p\theta(t)\right)\sin_p\theta(t) + O\left(\frac{1}{t}\right). \tag{42}$$

The term $O(1/t)$ can be written as $(|\cos_p\theta|^p + |\sin_p\theta|^p)O(1/t)$; hence we get

$$\theta'(t) = \frac{1}{t}\left[\mathring{r} + \left[1 + O\left(\frac{1}{t\log t}\right)\right]\sum_{j=1}^{n}\frac{\mathring{\alpha}_j}{\text{Log}_j^2 t} + O\left(\frac{1}{t}\right)\right]\left|\cos_p\theta(t)\right|^p + \frac{1}{(p-1)t}\left[\mathring{c} + \left\{1 + O\left(\frac{1}{t\log t}\right)\right\}\sum_{j=1}^{n}\frac{\mathring{\beta}_j}{\text{Log}_j^2 t} + O\left(\frac{1}{t}\right)\right] \cdot \left|\sin_p\theta(t)\right|^p - \frac{1}{t}\Phi\left(\cos_p\theta(t)\right)\sin_p\theta(t). \tag{43}$$

Now since all the terms of $O(1/t\log t)/\log_j^2 t$ are $O(1/t)$ as $t \to \infty$ for $j = 1, 2, \ldots, n$, then all these terms are asymptotically less than $o(1)/\log_n^2 t$. Hence we get

$$\theta'(t) = \frac{1}{t}\left[\mathring{r} + \sum_{j=1}^{n}\frac{\mathring{\alpha}_j}{\text{Log}_j^2 t} + \frac{o(1)}{\text{Log}_n^2 t}\right]\left|\cos_p\theta(t)\right|^p + \frac{1}{(p-1)t}\left[\mathring{c} + \sum_{j=1}^{n}\frac{\mathring{\beta}_j}{\text{Log}_j^2 t} + \frac{o(1)}{\text{Log}_n^2 t}\right]\left|\sin_p\theta(t)\right|^p - \frac{1}{t}\Phi\left(\cos_p\theta(t)\right)\sin_p\theta(t). \tag{44}$$

□

The main result of this paper is as follows.

Theorem 4. *Let r, c, α_j, and β_j, $j = 1, 2, \ldots, n$, are periodic functions which have different T_1, T_2, P_j, and Q_j, $j = 1, 2, \ldots, n$, periods, respectively, and $r(t) > 0$ in (17).*

(i) *(17) is oscillatory if $\mathring{c}\mathring{r}^{p-1} > \gamma_p$ and nonoscillatory if $\mathring{c}\mathring{r}^{p-1} < \gamma_p$, where $\mathring{r}$ and $\mathring{c}$ are defined in Lemma 3.*

(ii) *Let $\mathring{c}\mathring{r}^{p-1} = \gamma_p$. If there exists $k \in \{2, \ldots, n\}$ such that*

$$\mathring{\beta}_j\mathring{r}^{p-1} + (p-1)\gamma_p\mathring{\alpha}_j\mathring{r}^{-1} = \mu_p, \quad j = 1, \ldots, k-1 \tag{45}$$

and $\mathring{\beta}_k\mathring{r}^{p-1} + (p-1)\gamma_p\mathring{\alpha}_k\mathring{r}^{-1} \neq \mu_p$, then (17) is oscillatory if

$$\mathring{\beta}_k\mathring{r}^{p-1} + (p-1)\gamma_p\mathring{\alpha}_k\mathring{r}^{-1} > \mu_p \tag{46}$$

and nonoscillatory if

$$\mathring{\beta}_k\mathring{r}^{p-1} + (p-1)\gamma_p\mathring{\alpha}_k\mathring{r}^{-1} < \mu_p, \tag{47}$$

where $\mathring{\alpha}_j$ and $\mathring{\beta}_j$, $j = 1, 2, \ldots, n$, are defined in Lemma 3.

Proof. The statement (i) is proved in [10]. It remains to prove the statement (ii) in full generality.

We consider (17); let $x(t)$ be the nontrivial solution of (17) and $\varphi(t)$ is the Prüfer angle of (17) given in (24). Then

$$\varphi(t) = \varphi_1(t) + \sum_{j=1}^{n}\varphi_{2_j}(t) + \varphi_3(t) + \varphi_4(t) + \sum_{j=1}^{n}\varphi_{5_j}(t) + M \tag{48}$$

is a solution of

$$\varphi'(t) = \varphi_1'(t) + \sum_{j=1}^{n}\varphi_{2_j}'(t) + \varphi_3'(t) + \varphi_4'(t) + \sum_{j=1}^{n}\varphi_{5_j}'(t), \tag{49}$$

where

$$\begin{aligned}
\varphi_1'(t) &= \frac{1}{t}r(t)\left|\cos_p\varphi(t)\right|^p, \\
\varphi_{2_j}'(t) &= \frac{\alpha_j(t)}{t\text{Log}_j^2 t}\left|\cos_p\varphi(t)\right|^p \quad (j = 1, 2, \ldots, n), \\
\varphi_3'(t) &= -\frac{1}{t}\Phi\left(\cos_p\varphi(t)\right)\sin_p\varphi(t), \\
\varphi_4'(t) &= \frac{c(t)}{(p-1)t}\left|\sin_p\varphi(t)\right|^p, \\
\varphi_{5_j}'(t) &= \frac{\beta_j(t)}{(p-1)t\text{Log}_j^2 t}\left|\sin_p\varphi(t)\right|^p \quad (j = 1, 2, \ldots, n).
\end{aligned} \tag{50}$$

By the help of Lemma 3, $\theta(t)$ is a solution of

$$\theta'(t) = \frac{1}{t}\left[\overset{*}{r} + \sum_{j=1}^{n}\frac{\overset{*}{\alpha}_j}{\mathrm{Log}_j^2 t} + \frac{o(1)}{\mathrm{Log}_n^2 t}\right]\left|\cos_p\theta(t)\right|^p + \frac{1}{(p-1)t}\left[\overset{*}{c} + \sum_{j=1}^{n}\frac{\overset{*}{\beta}_j}{\mathrm{Log}_j^2 t} + \frac{o(1)}{\mathrm{Log}_n^2 t}\right]\left|\sin_p\theta(t)\right|^p - \frac{1}{t}\Phi\left(\cos_p\theta(t)\right)\sin_p\theta(t), \quad (51)$$

where $\overset{*}{r}, \overset{*}{c}, \overset{*}{\alpha}_j$, and $\overset{*}{\beta}_j$, $j = 1, 2, \ldots, n$, are given in Lemma 3.

This equation is a "Prüfer angle" equation for the following second-order half-linear differential equation

$$\left(\left(\overset{*}{r} + \sum_{j=1}^{n}\frac{\overset{*}{\alpha}_j}{\mathrm{Log}_j^2 t} + \frac{o(1)}{\mathrm{Log}_n^2 t}\right)^{1-p}\Phi(x')\right)' + \frac{1}{t^p}\left(\overset{*}{c} + \sum_{j=1}^{n}\frac{\overset{*}{\beta}_j}{\mathrm{Log}_j^2 t} + \frac{o(1)}{\mathrm{Log}_n^2 t}\right)\Phi(x) = 0, \quad (52)$$

which is the same as the following equation:

$$\left(R(t)\Phi(x')\right)' + \frac{1}{t^p}\left(\overset{*}{c}\overset{*}{r}^{p-1} + \sum_{j=1}^{n}\frac{\overset{*}{\beta}_j\overset{*}{r}^{p-1}}{\mathrm{Log}_j^2 t} + \frac{o(1)}{\mathrm{Log}_n^2 t}\right)\Phi(x) = 0. \quad (53)$$

Suppose that assumption (ii) of Theorem 4 is satisfied and that (46) holds for $k \in \{1, 2, \ldots, n-1\}$. Then (53) is oscillatory as a direct consequence of Theorem 1. If (46) holds for $k = n$, let $\varepsilon > 0$ be so small that still

$$\overset{*}{\beta}_n\overset{*}{r}^{p-1} - \varepsilon + (p-1)\gamma_p\left(\overset{*}{r}^{-1}\overset{*}{\alpha}_j - \varepsilon\right) > \mu_p \quad (54)$$

and consider the following equation:

$$\left(R_1(t)\Phi(x')\right)' + \frac{1}{t^p}\left(\overset{*}{c}\overset{*}{r}^{p-1} + \sum_{j=1}^{n}\frac{\overset{*}{\beta}_j\overset{*}{r}^{p-1}}{\mathrm{Log}_j^2 t} + \frac{\overset{*}{\beta}_n\overset{*}{r}^{p-1} - \varepsilon}{\mathrm{Log}_n^2 t}\right)\Phi(x) = 0, \quad (55)$$

where $R_1(t) = (1 + \sum_{j=1}^{n}((\overset{*}{\alpha}_j/\overset{*}{r})/\log_j^2 t) + (\overset{*}{\alpha}_n/\overset{*}{r} - \varepsilon)/\log_n^2 t)^{1-p}$. This equation is a Sturmian minorant for sufficiently large t in (53) and (54) and Theorem 1 implies that this minorant equation is oscillatory and hence (53) is oscillatory as well. This means that the Prüfer angle $\theta(t)$ of the solution of (52) is unbounded and by Lemma 3 the Prüfer angle $\varphi(t)$ of the solution of (17) is unbounded as well. Thus, (17) is oscillatory. A slightly modified argument implies that (17) is nonoscillatory provided that (47) holds. □

Corollary 5. *If the periods of the functions r, c, α_j, and β_j, $j = 1, 2, \ldots, n$, in (17) coincide with T-period, which is given in [6], then our oscillation constants overlap to their oscillation constants and our main result compiles with the result given in [6].*

Corollary 6. *If there exists a $\mathrm{lcm}(T_1, T_2, P_j, Q_j)$, $j = 1, 2, \ldots, n$, and the period T which is given in [6] is chosen as $\mathrm{lcm}(T_1, T_2, P_j, Q_j)$, $j = 1, 2, \ldots, n$, then our oscillation constants overlap to their oscillation constants and our main result compiles with the result given in [6].*

Remark 7. If for $j = 1, 2, \ldots, n$ $\mathrm{lcm}(T_1, T_2, P_j, Q_j)$ is not defined, then only our result can be applied whereas the result given in [6] can not.

Example 8. Consider the nonlinear equation (17) for $p = 3$, $r(t) = 2+\cos(ax+b)$, $(a, b \in \mathbb{R})$, $\alpha_1(t) = \cos 3t$, $\alpha_2(t) = \sin 8t$, $\beta_1(t) = \sin 4t$, $\beta_2(t) = \sin 2t$, and $c(t) = 2 + \sin 6t$. In this case $T_1 = 2\pi/|a|$, $P_1 = 2\pi/3$, $P_2 = \pi/4$, $Q_1 = \pi/2$, $Q_2 = \pi$, and $T_2 = \pi/3$ are periods of these functions, respectively. Because of these functions being periodic functions and $r(t)$ positive defined we can use Theorem 4 for all $a \neq 0$ and we obtain

$$\overset{*}{c}\overset{*}{r}^{p-1} = \left(\frac{|a|}{2\pi}\int_0^{2\pi/|a|}(2 + \cos(as + b))\,ds\right) \cdot \left(\frac{3}{\pi}\int_0^{\pi/3}(2 + \sin 6s)\,ds\right)^{3-1} = 8, \quad (56)$$

$$\gamma_3 = \left(\frac{3-1}{3}\right)^3 = \frac{8}{27}.$$

Thus we get $\overset{*}{c}\overset{*}{r}^{p-1} > \gamma_3$ for all $a \neq 0$ and considered equation is oscillatory. Here the important point to note is that while we cannot apply Theorem 2 which is given in [6] for this example if we choose $a = \sqrt{5}$, then $\mathrm{lcm}(2\pi/|a|, 2\pi/3, \pi/4, \pi/2, \pi, \pi/3)$ is not defined, we can apply our Theorem 4.

Competing Interests

The authors declare that there is no conflict of interests regarding the publication of this paper.

References

[1] A. Elbert, "A half-linear second order differential equation," in *Qualitative Theory of Differential Equations, Volume I, II*, vol. 30 of *Colloqula Mathematica Societatis Janos Bolyai*, pp. 153–180, North-Holland, Amsterdam, The Netherlands, 1981.

[2] T. Kusano, Y. Naito, and A. Ogata, "Strong oscillation and nonoscillation of quasilinear differential equations of second order," *Differential Equations and Dynamical Systems*, vol. 2, no. 1, pp. 1–10, 1994.

[3] A. Elbert, "Asymptotic behaviour of autonomous half-linear differential systems on the plane," *Studia Scientiarum Mathematicarum Hungarica*, vol. 19, pp. 447–464, 1984.

[4] O. Dosly and P. Rehak, *Half-Linear Differential Equations*, vol. 202 of *North-Holland Mathematics Studies*, Elsevier Science B.V., Amsterdam, The Netherlands, 2005.

[5] A. Elbert and A. Schneider, "Perturbations of the half-linear Euler differential equation," *Results in Mathematics*, vol. 37, no. 1-2, pp. 56–83, 2000.

[6] O. Došlý and H. Funková, "Euler type half-linear differential equation with periodic coefficients," *Abstract and Applied Analysis*, vol. 2013, Article ID 714263, 6 pages, 2013.

[7] O. Došlý and H. Funková, "Perturbations of half-linear Euler differential equation and transformations of modified Riccati equation," *Abstract and Applied Analysis*, vol. 2012, Article ID 738472, 19 pages, 2012.

[8] P. Hasil, "Conditional oscillation of half-linear differential equations with periodic coefficients," *Archiv der Mathematik*, vol. 44, pp. 119–131, 2008.

[9] O. r. Došlý and P. Hasil, "Critical oscillation constant for half-linear differential equations with periodic coefficients," *Annali di Matematica Pura ed Applicata*, vol. 190, no. 3, pp. 395–408, 2011.

[10] P. Hasil, R. Mařík, and M. Veselý, "Conditional oscillation of half-linear differential equations with coefficients having mean values," *Abstract and Applied Analysis*, vol. 2014, Article ID 258159, 14 pages, 2014.

6

Modeling and Analysis of Integrated Pest Control Strategies via Impulsive Differential Equations

Joseph Páez Chávez,[1,2] Dirk Jungmann,[3] and Stefan Siegmund[2]

[1] *Center for Applied Dynamical Systems and Computational Methods (CADSCOM), Faculty of Natural Sciences and Mathematics, Escuela Superior Politécnica del Litoral, P.O. Box 09-01-5863, Guayaquil, Ecuador*
[2] *Center for Dynamics, Department of Mathematics, TU Dresden, 01062 Dresden, Germany*
[3] *Institute of Hydrobiology, Faculty of Environmental Sciences, TU Dresden, 01062 Dresden, Germany*

Correspondence should be addressed to Joseph Paez Chavez; jpaez@espol.edu.ec

Academic Editor: Guodong Zhang

The paper is concerned with the development and numerical analysis of mathematical models used to describe complex biological systems in the framework of Integrated Pest Management (IPM). Established in the late 1950s, IPM is a pest management paradigm that involves the combination of different pest control methods in ways that complement one another, so as to reduce excessive use of pesticides and minimize environmental impact. Since the introduction of the IPM concept, a rich set of mathematical models has emerged, and the present work discusses the development in this area in recent years. Furthermore, a comprehensive parametric study of an IPM-based impulsive control scheme is carried out via path-following techniques. The analysis addresses practical questions, such as how to determine the parameter values of the system yielding an optimal pest control, in terms of operation costs and environmental damage. The numerical study concludes with an exploration of the dynamical features of the impulsive model, which reveals the presence of codimension-1 bifurcations of limit cycles, hysteretic effects, and period-doubling cascades, which is a precursor to the onset of chaos.

1. Introduction

Food losses due to pests and plant diseases are nowadays one of the major threats to food security, particularly in large parts of the developing world. As reported by the United Nations [1], the world's population in 2014 was estimated at 7.2 billion, with an approximate yearly growth of 82 million, a quarter of which occurs in the least developed countries. This unprecedented amount of people in the world poses serious challenges for food producers and policy-makers, especially regarding the minimization of crop losses due to pests and plant diseases, which have been estimated to be as high as 40% of the world production [2]. This issue has been a matter of active research for many decades, where the main challenge lies in the unavoidable trade-off between pest reduction, financial costs, effects on human health, and environmental impact. Therefore, the problem of pest control has necessarily to be addressed in an integrated manner, which has motivated the development of various integrated approaches, such as *Integrated Pest Management* (IPM) [2, 3]. IPM's basic principle consists in the judicious and coordinated use of multiple pest control mechanisms (e.g., biological control, cultural practices, and selected chemical methods) in ways that complement one another, maintaining pest damage below acceptable economic levels, while minimizing hazards to humans, animals, plants, and the environment. The literature on Integrated Pest Management is vast, and Section 2 will present a discussion of the historical development of the IPM concept over the past decades, which will serve as motivation for the mathematical description of the underlying pest control methods (see below).

One of the critical factors of success in the implementation of IPM programmes is the fundamental understanding of the interplay between the different elements of the associated agricultural ecosystems, such as crops, pests, natural enemies, and biopesticides, which quite often can only be achieved via a certain degree of mathematical abstraction. The mathematical description of ecological processes is one of the main subjects of study in the field of *Ecological Modeling*, which can be considered to a great extent as a nonlinear science, since almost all ecological interactions, both trophic and competitive, are nonlinear. In this respect, one of the main challenges in this area is the construction of mathematical models that provide reliable predictions and understanding of field observations in real ecosystems, in such a way that these models can be used as decision support tools. The possible approaches that can be employed to tackle this issue have been a matter of extensive debate among scientists for at least four decades [4, 5], and discussions on this topic are still going on [6–11]. In the particular case of agricultural pest control, the challenge consists in developing robust mathematical models able to at least qualitatively describe different pest control methods, so as to make the development of pest control strategies and policies less intuitive or empirical.

Since the origins of the IPM concept in the late 1950s [12, 13], a rich set of mathematical models has emerged in the literature focusing on different features of IPM-based applications, and in the present work we will briefly discuss several aspects of the recent development in this area (Section 3). One of the first systematic reviews of IPM-related mathematical models was published by Shoemaker [5], which was based upon her doctoral dissertation submitted in 1971. She considered suitable combinations of chemical and biological pest control applied to simplified ecosystems consisting of crops, pests, and parasites. The parasites are able to naturally control the pest population to a certain extent, but this effect is compromised by the pesticides applications, as Shoemaker assumed that the parasites are also killed by the chemical. As a practical example, she developed charts that the growers can use to determine whether pesticides should be sprayed, given in terms of the time until harvest and pest and parasite densities. Subsequent surveys of mathematical models for controlling pest populations have been carried out by Jaquette [4], Wickwire [14], Beddington et al. [15], and Barclay [16], and Section 3 will discuss some representative models that have been further introduced since then.

Despite the vast literature on modeling pest control strategies in agricultural ecosystems, comprehensive parametric studies of the mathematical models are rather scarce, with numerical investigations conducted mainly at the simulation level. This fact motivates the main part of the present work (Section 4), which will be devoted to the utilization of specialized numerical techniques in order to address practical questions relevant to IPM applications. Specifically, an impulsive pest control model will be chosen and reformulated as a *hybrid dynamical system* [17], thus allowing the parametric study of the periodic response of the system by means of numerical continuation (path-following) methods [18]. In this way, we will be able to tackle questions as to the optimal implementation of the impulsive pest control, in terms of minimizing operation costs and environmental damage. The paper will end with further numerical investigations of the dynamic response of the impulsive model, which will reveal the presence of codimension-1 bifurcations of limit cycles, hysteretic phenomena, and period-doubling cascades, which is a precursor to the onset of chaos.

2. A Brief Overview of Integrated Pest Management

An apparently promising approach to reduce crop losses due to pests appeared during the 1940s, with the discovery of synthetic pesticides such as DDT, which marked a new era in pest control. Pesticides became soon popular in the agricultural industry, as they were easy to apply and effectively killed a significant amount of the targeted pest, due to which their use spread rapidly worldwide. However, the overreliance on synthetic pesticides was proven to be unsustainable already by the end of the 1950s. Just a few years after the first use of DDT, resistance to the chemical was observed in a variety of insect pests [2], which meant more frequent applications and higher dosages of pesticides in order to keep acceptable pest population levels. Another negative effect was the reduction of beneficial species (such as natural enemies), which intensified the problem of pest resurgence and allowed nonpest species to increase in number and become pests themselves. In addition to these on-site crop problems produced by the overuse of pesticides, their negative impact extended beyond the agricultural framework, causing damage to water sources and further ecosystems, as well as posing serious human health hazards due to, for example, pesticide residuals in food and pesticide exposition [19, 20].

Recognition of the problems associated with the indiscriminate use of synthetic pesticides encouraged the development of alternative pest control paradigms, such as the concept of *Integrated Pest Management* (IPM) [2, 3, 12, 13]. Having its origins in the seminal work by Stern et al. [21], IPM is an interdisciplinary pest control approach that relies heavily upon natural mortality factors, such as natural predators and environmental conditions, combined with further control mechanisms. These include biological control, selected chemical methods, and cultural practices. The basic idea is that, instead of employing a single control method, efforts are directed to the judicious and coordinated use of multiple tactics in ways that complement one another, maintaining pest damage below acceptable levels, while minimizing hazards to humans, animals, plants, and the environment.

Integrated Pest Management (IPM) has been recognized as one of the most robust constructs in agricultural sciences to deal with the challenges related to the excessive use of pesticides [12], already outlined above (see also [22–25]). The key concept for the implementation of a pest control programme in an IPM framework is that of *economic injury level*. This term was introduced for the first time by Stern et al. [21] (see also [3]) and means the lowest pest population density that will cause *economic damage*. The latter term can be defined as the amount of injury that justifies the application of control measures, and it can vary depending on the area, season, and other economic or ecological factors.

In general, it is assumed that a number of pest control mechanisms are available, for instance, biological methods, cultural practices, natural enemies, habitat management, and synthetic pesticides. The basic decision rules rely on a predefined economic injury level and an *economic threshold*, which gives the pest population density above which control actions must be taken so as to prevent the pest population from reaching the economic injury level. An IPM-based pest control scheme, in its simplest form, will then require that whenever the amount of pests is less than the economic threshold only ecologically benign control measures are applied, that is, those that enhance natural control. If natural control is not capable of preventing the pest population from reaching the economic injury level, then synthetic pesticides come into play, nevertheless, in adequate combination with environmentally friendly control measures so as to minimize the amount of pesticides released into the underlying ecosystem. In practice, however, the task of developing and implementing an IPM-based pest control programme sustainable in both ecological and economic terms is by no means a trivial one.

As has been recognized in the past [2, 12, 16], the implementation of IPM-based control strategies requires a profound knowledge of the interactions between the different components of the underlying agricultural ecosystem, for instance, crops, pests, natural enemies, and habitat conditions. Given the remarkable complexity of such ecosystems, the required fundamental understanding of the involved biological interactions can quite often only be obtained via a certain degree of mathematical abstraction. Ideally, the final result should be a mathematical model tailored for a specific application, which provides reliable predictions and understanding of field observations, in such a way that the model can be used as a decision support tool for devising effective pest control schemes [26]. Nevertheless, the main challenge lies in the sheer complexity of the involved biological processes, which often hinders the search for appropriate mathematical representations of the laws governing the ecosystem. In the next section we will discuss in detail some representative mathematical models used for describing various pest control methods, including synthetic (pesticides), biological (natural enemy predation, biopesticides), and cultural (roguing, replanting), which are precisely some of the most common control mechanisms used in combination in IPM-based control programmes.

3. Mathematical Models for Pest Control

In this section we will present a short overview on the available mathematical models used to describe the ecological interactions in pest control. The discussion will be mainly guided by two criteria: model type, in a mathematical sense, and the underlying ecological phenomena in the framework of agricultural pest control. Special attention will be paid to those models related to Integrated Pest Management (IPM), where the main purpose is to minimize damage to nontarget organisms and harmful environmental effects, by combining classical chemical strategies with alternative control methods, such as biological control, host-plant resistance breeding, crop rotation, harvest management, and cultural techniques (see Section 2).

Before starting our discussion, let us give some remarks regarding notation. As is well known, biological pest control is characterized by the reduction of pest population as a result of the introduction of a natural enemy [15]. The interaction between a pest and its natural enemies can be understood in terms of the dynamics observed in prey-predator models. Therefore, most of the models presented in our discussion will have in their core a certain type of prey-predator system. In this context, state variables related to pest (prey) and natural enemies (predator) populations will be denoted by scalar, nonnegative variables x and y, respectively. Furthermore, all system parameters are assumed to be positive numbers, unless otherwise stated, and they will be used throughout the manuscript in a consistent manner, in such a way that, whenever possible, they will have the same meaning in different models. Finally, the prime symbol will denote time differentiation.

3.1. Pest Control as a Time-Continuous Process

Pest Diseases and Natural Enemies as Control Measures. In the context of Integrated Pest Management, one of the most representative control strategies is that of an artificial spread of an infection among a pest population combined with a different control method, such as continuous pesticide spraying or natural enemy predation. Typically, the pest population is divided into two classes: *susceptible* and *infective*. The infective population is used to spread a certain disease or virus created in a laboratory. At the beginning, a small amount of infected pest is introduced into the ecosystem with the purpose of generating an epidemic. The susceptible population becomes infected through direct contact with the infective pest, thereby causing a significant reduction of the pest population as a direct consequence of the disease, or due to a decrease in its reproductive ability. A model describing this control mechanism has been recently proposed by Jana and Kar [27], based on the classical susceptible-infective (SI) paradigm [28]:

$$\begin{aligned}
x_S'(t) &= r x_S(t)\left(1 - \frac{x_S(t) + \eta x_I(t)}{K}\right) - \alpha x_S(t) x_I(t) \\
&\quad - \frac{\beta x_S(t) y(t)}{a + x_S(t)}, \\
x_I'(t) &= \alpha x_S(t) x_I(t) - \gamma x_I(t) y(t) - \sigma x_I(t), \\
y'(t) &= \frac{C\beta x_S(t) y(t)}{a + x_S(t)} + \gamma(p - q) x_I(t) y(t) \\
&\quad + d y(t)\left(1 - \frac{x_S(t) + \eta x_I(t)}{K}\right) - \mu y(t) \\
&\quad - \delta y(t)^2.
\end{aligned} \tag{1}$$

The subscripts S and I denote the susceptible and infective pest population, respectively, while y represents the natural enemy (predator) population. From the two classes of pest, only the susceptible population is able to reproduce, according to a logistic law with intrinsic growth rate r and environmental carrying capacity K. The infective pest is assumed to also contribute towards the carrying capacity of the ecosystem; however, the significance of this environmental impact can differ from that of the susceptible population, which is reflected by the parameter η. The disease spreads within the susceptible pest through direct contact with the infected population, whose effect can be seen in the second term of the first equation of system (1), where the coefficient α represents the force of infection. A second control action in this ecosystem is given by the presence of a natural enemy (predator). The predator population feeds on the pest according to a Holling type II trophic function [29–31], where β is the maximum predator's capturing rate and a stands for the susceptible pest population density at which the capturing rate is half the maximum value.

As already mentioned above, the infective population is not able to reproduce; hence its survival relies heavily upon the ability of the disease to turn the susceptible population into infected pest, which depends on the force of infection α. On the other hand, the infective pest is assumed to have a natural death rate σ. This produces an exponential decay of the infective population in the absence of natural enemies and susceptible pest. The infective population is attacked by the natural enemy following a Holling type I functional response with capturing rate γ; see the second equation of the model (1).

The third equation in (1) describes the evolution of the natural enemy population. Here, C and p represent the conversion factors from susceptible and infective pest, respectively, into predators. A negative effect of the infection on the predator population is accounted for by the parameter q. In addition, μ denotes the mortality rate and δ measures the intensity of competition among the natural enemies for space, food, and so on. At low pest population densities, the natural enemies are able to find alternative food sources, leading to an additional growth rate d. As the pest population grows, the predators make less use of the alternative food, and when the pest approaches the environmental carrying capacity, the natural enemy feeds almost on the pest only.

A systematic study of the dynamics of system (1) has been carried out by Jana and Kar [27]. They analyze in great detail the effect of the relevant parameters on the proposed ecosystem, and the theoretical results are illustrated by representative numerical simulations. As has been already explained above, the pest population is controlled by two methods: infection release and natural enemy predation, which fall into the class of biological pest control. From a practical point of view, the role of the alternative food source for the natural enemy has been shown to be crucial. If the predator growth rate due to the alternative food is greater than its mortality rate (i.e., $d > \mu$), a nontrivial pest-free equilibrium is feasible, where the natural enemy is the only species present in the ecosystem. Furthermore, if d exceeds a certain threshold, the equilibrium is asymptotically stable, which means that small (pest) perturbations in the ecosystem will be controlled entirely by the natural enemies. Jana and Kar conclude their investigation by considering a third control measure in the system consisting in chemical methods (pesticides), and they discuss how to combine these three control techniques so as to minimize the deadly effects on the natural enemies and environmental damage, while effectively reducing the pest population.

Plant Disease Control via Cultural Methods. Another non-chemical mechanism to deal with plant diseases can be implemented via cultural practices, which fall into the category of biological control. They are carried out by means of human actions only, such as roguing (removing) infected plants, replantation of disease-free plants, crop rotation, intercropping, and strip farming [20]. The main purpose is to reduce the negative consequences of a disease to levels that are acceptable in economic terms, while causing minimal damage to the environment. However, there are certain limitations for the application of such control methods, since they typically involve high labor costs and a complete eradication of plant diseases through cultural practices is generally not possible.

A low-dimensional model considering cultural practices as the only disease control action is proposed by van den Bosch et al. [32]:

$$\begin{aligned} z_S'(t) &= r\phi + r(1-\phi)\frac{q(1-p)z_I(t) + z_S(t)}{(1-p)z_I(t) + z_S(t)} \\ &\quad - \mu z_S(t) - \alpha z_S(t) z_I(t), \\ z_I'(t) &= r(1-\phi)\frac{(1-q)(1-p)z_I(t)}{(1-p)z_I(t) + z_S(t)} - (\mu+\sigma) z_I(t) \\ &\quad + \alpha z_S(t) z_I(t), \end{aligned} \tag{2}$$

with $0 < \phi, p, q < 1$. Here, z_S and z_I denote the population of susceptible and infected plants, respectively, in an agricultural crop field contaminated with a certain viral disease. New plants are introduced in the ecosystem at a constant rate r, from which $r\phi$ corresponds to *in vitro*-germinated healthy plants and $r(1-\phi)$ to cuttings taken from the crop (susceptible and infected). By visual inspection or other diagnostic methods, infected plants from the cuttings are discarded with probability p. Furthermore, it is assumed that the infected plants are able to recover due to reversion with probability q. Consequently, the introduction of new plants contribute to both the susceptible and infected populations (see, e.g., the first term in the second equation of (2), which gives the proportion of infected plants that enter the crop as a result of the introduction of $r(1-\phi)$ cuttings). In addition to the usual death rate μ, the infected plants are removed from the system (roguing) at a rate σ. As can be seen, only cultural actions are considered as a measure of disease control. On the other hand, another parameter considered in the ecosystem is the within-plant virus titre (denoted hereafter by w). As pointed out in [32], this parameter deeply influences both the viral transmission in the crops and the corresponding disease symptoms, which in turn affect the roguing rates and recovery and detection probabilities. Therefore, van den Bosch et al. assume the parameters α, σ, p, and q to be w-dependent, and they suggest certain functional relations to quantify these interactions, motivated by some previous field studies.

A remarkable feature of model (2) is the fact that it illustrates two well-known disease transmission modes: horizontal and vertical. The first one typically occurs through herbivorous insects that ingest spores on host-plant leaves and carry the disease from one plant to another. The intensity of this transmission mode is quantified by the transmission coefficient α. On the other hand, the vertical transmission is caused by the use of cuttings to establish new crops. In this respect, van den Bosch et al. focused special attention on the effects of a trade-off between horizontal and vertical transmission modes operating in plant pathogens. Specifically, the authors analyze in detail the underlying cultural practices and how they are reflected in the proposed ecosystem (2). Their results then suggest that the considered control mechanisms should be carefully adapted and combined in order to minimize the risks of failure, which to a great extent depends on how effectively the vertical and horizontal transmission modes are adequately dealt with. Analogous models considering disease control based on cultural practices have been proposed and investigated in the past and can be found in, for example, [33–38].

3.2. Pest Control Strategies via Impulsive Perturbations. As already explained before, pest control strategies based on an Integrated Pest Management approach involve a suitable combination of biological, cultural, and chemical control techniques, including pesticide spraying, pest harvesting or trapping, plant roguing, and acceleration of the pest mortality through the introduction of natural enemies or pest diseases. Some of these control methods were mathematically described in the models presented in Section 3.1, in terms of smooth ordinary differential equations. In doing so, the pest control strategies were conceived as a time-continuous process of autonomous type, where the system parameters have to be chosen in such a way that the underlying control mechanisms effectively reduce the negative pest effects on agricultural crop fields, without any external perturbation. However, this approach neglects the fact that pest control methods are in reality implemented in a discontinuous manner, which is a consequence of the discrete nature of human activities. Furthermore, there can be exogenous factors in the ecosystems (e.g., temperature, air composition, and further human actions) that may lead to pest population densities changing very rapidly in a short period of time, which can be modeled via impulsive perturbations.

In our case, the above-outlined sudden changes in the ecosystems will be described in the framework of impulsive differential equations. This class of models is particularly suited for the representation of dynamical phenomena subject to short-term perturbations whose duration is negligible in comparison to the duration of the system evolution. Therefore, these perturbations can be assumed to act instantaneously in the form of impulses, which generally leads to jumps in the state space (discontinuous evolution). Processes of this nature can be found in numerous applications, for instance, in mechanics, population dynamics, ecology, biology, and economy, and the theoretical foundations have been developed to a great extent [39–42].

Susceptible-Infective Control Scheme with Impulsive Roguing and Replanting. A straightforward mathematical description of a pest control strategy based on impulsive perturbations is given by Tang et al. [43]:

$$\begin{aligned}
z_S'(t) &= -\alpha z_S(t)\, z_I(t) - \mu z_S(t), \\
z_I'(t) &= \alpha z_S(t)\, z_I(t) - \mu z_I(t), \\
&\qquad\qquad t \neq nT, \\
z_S(t^+) &= z_S(t^-) + r\phi, \\
z_I(t^+) &= (1-\sigma)\, z_I(t^-) + r(1-\phi), \\
&\qquad t = nT,\ n \in \mathbb{N}\ \text{(replanting and roguing)}.
\end{aligned} \tag{3}$$

Here, cultural practices (roguing and replanting) are considered for pest control, and the state variables and parameters are of the same nature as those of system (2), with $0 \leq \sigma < 1$. The actions of planting and roguing are assumed to be carried out in a periodic and impulsive manner, with period $T > 0$, which differs from the approach employed in the model (2), where those actions are continuously executed. As can be seen, the key parameters in this control scheme are the roguing coefficient σ and the impulse period T, that is, the parameters that are directly influenced by human decisions. Consequently, the main question here is how to choose those parameters so as to keep the number of infected plants below a certain predefined critical level. In this respect, Tang et al. identify a lower bound for the impulse period that guarantees the extinction of the infected plants, for the special case $\phi = 1$ (i.e., replanted plants contribute only to the healthy class). Specifically, they prove the existence of a periodic solution of system (3) of the form $(\tilde{z}_S(t), 0)$, $t > 0$, which is asymptotically stable provided T is large enough.

Periodic Impulsive Control at Different Fixed Times. When different control measures are combined, for example, natural enemy release and pesticide spraying, it can be beneficial to carry out those actions at different moments within a period of control. This is particularly convenient when the chemicals used to control the pest population also have the collateral effect of killing the natural enemies. Therefore, a suitable amount of natural enemies has to be regularly introduced into the ecosystem at an appropriate time, in order to compensate for the undesired pesticide-induced predator mortality. Another reason for which control measures applied at different times can be advantageous is that some natural enemies are effective only at certain life stages of the pest population. For instance, some predators are able to attack the adult population only, hence leaving larval or other previous stages of the pests unaffected. In such cases, the natural enemy releases and pesticide spraying have to be carried out according to the life cycle of the pest population at different times, so as to effectively cover all possible life stages of the pest.

Zhang and coworkers have proposed several pest control models based on the scheme described above, for example [44],

$$\begin{aligned}
x'(t) &= x(t)\left(1 - \frac{x(t)}{K}\right) - \frac{\beta x(t) y(t)}{a + k_x x(t) + k_y y(t)},\\
y'(t) &= \frac{C\beta x(t) y(t)}{a + k_x x(t) + k_y y(t)} - \mu y(t),\\
&\qquad t \neq (n + l - 1) T,\ t \neq nT,\\
x(t^+) &= (1 - \sigma_x) x(t^-),\\
y(t^+) &= (1 - \sigma_y) y(t^-),\\
&\qquad t = (n + l - 1) T\ \ \text{(pesticide spraying)},\\
x(t^+) &= x(t^-),\\
y(t^+) &= y(t^-) + d,\\
&\qquad t = nT,\ n \in \mathbb{N}\ \ \text{(natural enemy release)},
\end{aligned} \tag{4}$$

where x and y stand for the densities of the pest and natural enemy population, respectively, and the parameters have the usual meaning as in the previous models, with $0 \leq \sigma_{x,y} \leq 1$ and $0 < l < 1$. As can be seen, two control actions applied at different times are considered in model (4). The first one consists in spraying pesticides into the ecosystem, which has not only the effect of killing pest individuals but also of reducing the predator population, by proportions σ_x and σ_y, respectively. In order to compensate for the undesired effect of the applied pesticide, a fixed amount of d natural enemies is periodically introduced into the ecosystem, which corresponds to the second control action in system (4).

The trophic interaction between the pest and its natural enemy is characterized by a Beddington–DeAngelis functional [31, 45, 46] (see the second term in the first equation of model (4)), and it has some similarities with the Holling type II trophic function described before. The coefficient k_x is a weighting factor that determines how fast the predator's capturing rate approaches its saturation value as the pest population increases. In addition, the Beddington–DeAngelis functional considers mutual interference between the natural enemies, and the intensity of this effect is regulated by the parameter k_y. Furthermore, the constant a gives a measure of the abundance of pests and natural enemies relative to the ecosystem in which they interact, and it can be interpreted as a protection provided to the pest by the environment.

In the study presented in [44], the authors established a critical value of the impulse period that separates the system behavior into two cases. In the first one, model (4) admits a stable periodic solution of the form $(0, \tilde{y}(t))$, $t > 0$; that is, the pest population can be completely eradicated by means of the control measures. To prove this, the authors give an explicit construction of the fundamental solution matrix of the linear variational equation around the pest-free periodic orbit, along with the corresponding transition (also referred to as saltation [47]) matrix. Based on the resulting explicit forms, an integral condition is derived so as to guarantee that the Floquet multipliers of the periodic orbit lie within the unit circle, thus ensuring the local stability of the pest-free solution. If the so obtained integral stability condition is not satisfied, then a second type of system response takes place in which the pest-free trajectory is not stable anymore. Instead, the system possesses a periodic solution corresponding to the case when the pest and its natural enemies coexist, and the stability of this system response is shown to strongly depend on the value of the impulse period T and the proportion of pest population killed by the pesticide σ_x.

Apart from the preliminary results mentioned above, there is little information on how the remaining system parameters actually affect the behavior of model (4), although various studies related to this class of pest control are available [48–54]. For instance, a crucial feature of this model is that the control actions of natural enemy release and pesticide spraying are carried out at different times, and this effect is controlled by the parameter l, whose influence on the underlying pest control strategy has not been systematically discussed in the past. Therefore, in Section 4 we will use system (4) as a toy model to show how specialized numerical methods (based on path-following algorithms for nonsmooth dynamical systems) can be used to study in detail the behavior of such models under parameter variations.

4. Numerical Analysis of a Pest Control Scheme with Impulsive Effects

In the previous section, various existing mathematical models were presented and discussed in detail, with special attention given to those describing agricultural pest control methods in the framework of Integrated Pest Management. The model types ranged from smooth ordinary differential equations to differential equations with impulsive perturbations. As was pointed out, several authors have contributed to the theoretical analysis of those systems, with particular emphasis on the existence and stability of pest-free solutions, as well as finding explicit thresholds for control parameters at which stability is lost (bifurcations). While this is generally a straightforward task for models belonging to the class of smooth ordinary differential equations, the situation can be more involved for other types of systems, for example, impulsive differential equations; see [43, 44, 48, 55, 56]. Although impulsive systems describing pest control methods have received a good deal of attention in the past, numerical investigations of such models are rather scarce in the literature and are mostly carried out at the simulation level. This is the main motivation of the present section, in which we will present a comprehensive numerical analysis of one of the impulsive models introduced before, namely, system (4), with particular emphasis on how specialized numerical techniques can be employed to study the model behavior and gain insight into the underlying pest control methods.

In order to investigate the dynamics of the impulsive system (4), we will employ two different kinds of numerical approaches, namely, direct numerical integration and path-following methods. As is well known [57–59], impulsive models of the type of (4) can be formulated in the framework of hybrid dynamical systems, which are characterized by a continuous-time behavior interrupted by discrete-time events [17]. In our case, these interruptions are defined by the impulse times at which the pest control actions are carried out. In order to get reliable numerical simulations of the model behavior, the impulse times need to be accurately detected, which can be achieved by means of the standard MATLAB ODE solvers together with their built-in event location routines [60, 61], as suggested in [62]. In this way, direct numerical integration will be implemented in the present work.

As will be seen later, our investigation will primarily focus on the periodic behavior of system (4), with special attention given to the pest population and its response to the control actions. Since the impulsive model is parameter-dependent, a family of periodic solutions can generically be tracked by varying one (control) parameter, which can be numerically realized via path-following (continuation) methods. These are well-established techniques in applied mathematics [18] that enable a systematic study of system solutions subject to parameter variations, without having to make recourse to direct numerical integration, which sometimes can be time-demanding and inefficient. For the analysis of periodic solutions of hybrid dynamical systems, various specialized computational tools are available, such as COCO [63], SlideCont [64], and TC-HAT [17], and the latter will be employed in the current work for the numerical study of the impulsive model. In the next section we will formulate system (4) as a hybrid dynamical system, which will allow the implementation of the model in TC-HAT.

4.1. Impulsive Pest Control Model as a Hybrid Dynamical System. Before starting the numerical investigation of model (4), it is convenient to introduce a rescaling of the system as follows:

$$\begin{aligned}\tilde{x} &= \frac{x}{K},\\ \tilde{y} &= \frac{y}{K},\\ \tilde{\beta} &= \frac{\beta K}{a},\\ \tilde{k}_x &= \frac{k_x K}{a},\\ \tilde{k}_y &= \frac{k_y K}{a},\\ \tilde{d} &= \frac{d}{K}.\end{aligned} \tag{5}$$

According to these transformations, we obtain the following scaled version of the impulsive system (4):

$$\begin{aligned}x'(t) &= x(t)(1-x(t)) - \frac{\beta x(t)y(t)}{1+k_x x(t)+k_y y(t)},\\ y'(t) &= \frac{C\beta x(t)y(t)}{1+k_x x(t)+k_y y(t)} - \mu y(t),\\ &\qquad t \neq (n+l-1)T,\ t\neq nT,\\ x(t^+) &= (1-\sigma_x)x(t^-),\\ y(t^+) &= (1-\sigma_y)y(t^-),\\ &\qquad t=(n+l-1)T\ \text{(pesticide spraying)},\\ x(t^+) &= x(t^-),\\ y(t^+) &= y(t^-)+d,\\ &\qquad t=nT,\ n\in\mathbb{N}\ \text{(natural enemy release)},\end{aligned} \tag{6}$$

where the tildes have been dropped for the sake of simplicity.

As mentioned earlier, this impulsive model can be formulated as a hybrid dynamical system. For this purpose, the trajectories are divided into smooth *segments* consisting of the following components: a smooth vector field that governs the system behavior during the segment; a smooth *event function* whose zeroes define the terminal point of the segment; and a smooth *jump function*, which maps the terminal point of the current segment to the initial point of the next one. Each segment is labeled with an *index* I_i, $i\in\mathbb{N}$, so that any solution of the hybrid dynamical system is fully characterized by its *solution signature* $\{I_i\}_{i=1}^{M}$, where $M\in\mathbb{N}$ defines the length of the signature. This mathematical framework enables the application of path-following algorithms by means of the software package TC-HAT [17], a driver of AUTO 97 [65] for numerical continuation and bifurcation detection of periodic solutions of hybrid dynamical systems. Recent applications of TC-HAT can be found in [66–70], where the continuation package is employed to study the bifurcation scenario of various engineering applications.

Denote by $\alpha := (\beta, k_x, k_y, C, \mu, d, T, \sigma_x, \sigma_y, l) \in (\mathbb{R}^+)^7 \times [0,1]\times[0,1]\times(0,1)$ and $u := (x, y, s)^T \in (\mathbb{R}_0^+)^3$ the parameters and state variables of the system, respectively, with $\mathbb{R}_0^+$ being the set of nonnegative numbers. The auxiliary variable s will be used to embed the time into the state space, in such a way that each impulsive period $[(n-1)T, nT]$, $n\in\mathbb{N}$, will be mapped to the interval $[0,T]$. In this setting, a solution of the impulsive model (6) will be divided into smooth segments, as defined as follows.

Pesticide Spraying (I_1, P-Spr). This segment occurs for $0\le s\le lT$, and the dynamics of the system during this regime is governed by (cf. (6))

$$u'(t) = f(u(t),\alpha) := \begin{pmatrix} x(t)(1-x(t)) - \dfrac{\beta x(t)y(t)}{1+k_x x(t)+k_y y(t)} \\ \dfrac{C\beta x(t)y(t)}{1+k_x x(t)+k_y y(t)} - \mu y(t) \\ 1 \end{pmatrix}. \tag{7}$$

This segment terminates when a crossing with the discontinuity boundary

$$\Sigma_{\text{P-Spr}} := \left\{(x, y, s) \in (\mathbb{R}_0^+)^3 : h_{\text{P-Spr}}(u, \alpha) := s - lT = 0\right\} \quad (8)$$

is detected. According to the pest control scheme, pesticide is sprayed at this terminal point, and this is implemented in the hybrid dynamical system via the jump function

$$g_{\text{P-Spr}}(u, \alpha) := \begin{pmatrix} (1 - \sigma_x) x \\ (1 - \sigma_y) y \\ s \end{pmatrix}, \quad (9)$$

which gives the initial point for the next segment.

Predator Release (I_2, Pr-Re). In this segment we have that $lT < s \leq T$, and the system behavior is determined by the ODE (7). The segment ends when the solution hits the discontinuity boundary

$$\Sigma_{\text{Pr-Re}} := \left\{(x, y, s) \in (\mathbb{R}_0^+)^3 : h_{\text{Pr-Re}}(u, \alpha) := s - T = 0\right\}, \quad (10)$$

with the initial point for the next segment given by the jump function

$$g_{\text{Pr-Re}}(u, \alpha) := \begin{pmatrix} x \\ y + d \\ s - T \end{pmatrix}. \quad (11)$$

The second component of this function represents the control action of introducing natural enemies into the ecosystem (cf. (6)), while the third one has the purpose of resetting the variable s, so that it is always kept within the interval $[0, T]$.

Moreover, throughout our numerical investigations the following solution measures will be used:

$$M_P := \max_{0 \leq t \leq T} x(t), \qquad M_E := \max_{0 \leq t \leq T} y(t), \quad (12)$$

where $(x(t), y(t))$ is assumed to be a T-periodic solution of the impulsive model (6). The quantities M_P and M_E give the maximum amount of pest and natural enemy populations attained in a period, respectively, and can be used to investigate the impact of the system parameters on the pest control scheme from a practical point of view. For instance, if we consider the auxiliary boundary condition

$$h(u, \alpha) := x - x_{\text{ET}} = 0, \quad (13)$$

it is possible to trace a curve in a two-parameter space (see Section 4.2.2) for which the pest population achieves a maximum, fixed critical value $M_P = x_{\text{ET}} > 0$ corresponding to, for example, a predefined economic threshold (see Section 2).

According to the mathematical framework presented above, the pest control model (6) can be written in compact form as follows:

$$\begin{aligned} u'(t) &= f(u(t), \alpha), \quad s \neq lT,\ s \neq T, \\ u(t^+) &= g_{\text{P-Spr}}(u(t^-), \alpha), \\ &\qquad s = lT \ \text{(pesticide spraying)}, \\ u(t^+) &= g_{\text{Pr-Re}}(u(t^-), \alpha), \\ &\qquad s = T \ \text{(natural enemy release)}. \end{aligned} \quad (14)$$

In Figure 1 we present a periodic orbit of this system illustrating the solution segmentation introduced above. With this mathematical framework we are now ready to carry out specialized numerical investigations of this type of periodic response, via the numerical package TC-HAT.

4.2. Numerical Results. This section will be devoted to a detailed numerical study of the impulsive pest control scheme introduced in Section 3.2 and modeled by system (14). The focus will be on the effect of the system parameters on M_P (see (12)), which can be used to monitor the amount of pest population in the ecosystem. One of the main questions here will be how the control parameters should be chosen so as to keep the pest population below certain admissible levels. In addition, various dynamical phenomena will be investigated, such as fold and period-doubling bifurcations, as well as chaotic responses. Unless otherwise indicated, the parameter values used in the numerical results reported here are given in Table 1.

4.2.1. Behavior of the Pest Control Method under One-Parameter Perturbations. As was already mentioned in Section 3.2, Zhang et al. [44] studied the stability of pest-free (also referred to as pest-eradication) periodic solutions of system (14), as the impulse period T is varied. They determined a threshold T_0, depending on some of the remaining system parameters, so that for $T < T_0$ the pest-free solution is asymptotically stable, while for $T > T_0$ the system response is dominated by periodic solutions for which the pest and its natural enemies coexist. Specifically, Zhang et al. showed that the pest-free solution undergoes a change of stability (bifurcation) at $T = T_0$; however, they did not determine the actual type of bifurcation that produces this qualitative change, and this is precisely the first question that will be addressed numerically in this section.

Let us then begin our study with the numerical continuation of a pest-free solution with respect to T, as shown in Figure 2(a). In this picture, the solid black line denotes stable pest-free solutions as displayed in Figure 2(d). If the impulse period is increased, a critical value $T_0 \approx 2.2736$ is found, at which the pest-free response loses stability. As was confirmed numerically, this bifurcation corresponds to a branching point [71], wherefrom two branches of periodic solutions emanate (black dashed and solid green lines). The dashed curve represents unstable pest-free trajectories, while the green one corresponds to stable periodic solutions with

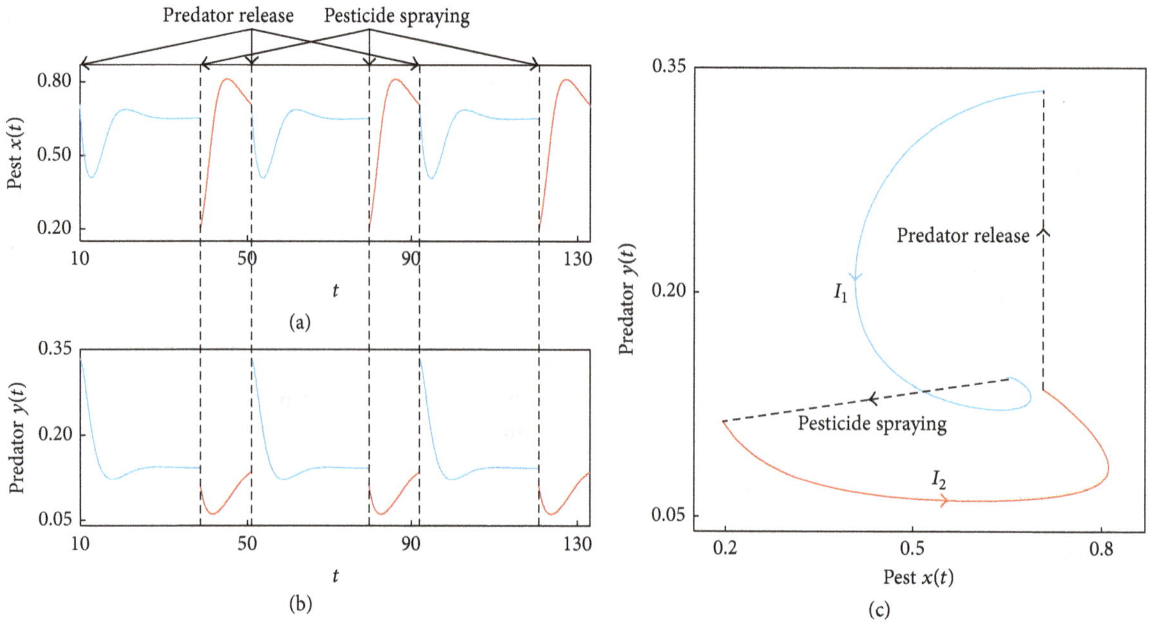

FIGURE 1: Periodic solution of the impulsive system (14) computed for the parameter values given in Table 1. (a) and (b) show the time histories of the pest and predator populations, respectively, while (c) presents the corresponding phase plot with the solution segments I_1 (pesticide spraying, blue) and I_2 (predator release, red). Hence, the displayed periodic trajectory has a cyclic solution signature $\{I_1, I_2\}$. In what follows, this color code will be used to distinguish the solution segments.

TABLE 1: Parameter values used for the numerical investigation of the impulsive model (14).

Parameter	Symbol	Value
Maximum predator's capturing rate	β	5.25
Weighting factor for saturation of predator's capturing rate	k_x	1.5
Weighting factor for interference among predators	k_y	1.125
Pest-to-predator conversion coefficient	C	0.5
Predator's mortality rate	μ	0.8
Impulse period	T	41
Tuning coefficient for pesticide spraying time	l	0.7
Amount of predators introduced per impulse period	d	0.2
Proportion of pest population killed by pesticides	σ_x	0.7
Proportion of predator population killed by pesticides	σ_y	0.2

pests and predators coexisting in the ecosystem. The green branch has a critical point $T_c \approx 10.4465$ where the curve loses smoothness. This singularity is produced by a change in the position of the peak value of the pest population. For $T < T_c$, the maximum amount of pest is attained exactly at the end of the segment I_1 (Figure 2(b)), while for $T > T_c$ the maximum value occurs at the end of the segment I_2 (Figure 2(c)). Another feature of this curve is that the amount of pest population, measured by M_P, increases as T grows. From a practical point of view, this means that, in order to keep low levels of pest population in the ecosystem, the impulse period should be chosen as small as possible, ideally below T_0. Nevertheless, having small impulse periods may amount to high operation costs, since the control actions are carried out more frequently if T is reduced, and therefore a compromise should be made.

The next step in our numerical investigation is to study the model response when further system parameters are varied, one at a time. The result can be seen in Figure 3. Figures 3(a) and 3(b) correspond to the numerical continuation of the periodic response of the impulsive model (14) with respect to β and k_y, respectively. In both diagrams, it can be observed that the predator population is not significantly affected by those parameters. On the other hand, the pest population shows a decreasing tendency when β is increased, whereas the effect of k_y on the pest population is exactly the opposite. This is consistent with the biological meaning of those parameters: an increment in β means that the natural enemy enhances its ability to catch pest individuals, while a larger k_y implies more competition between predators, which has a detrimental effect on their capturing rate. Although the numerical results indicate that both β and k_y can be

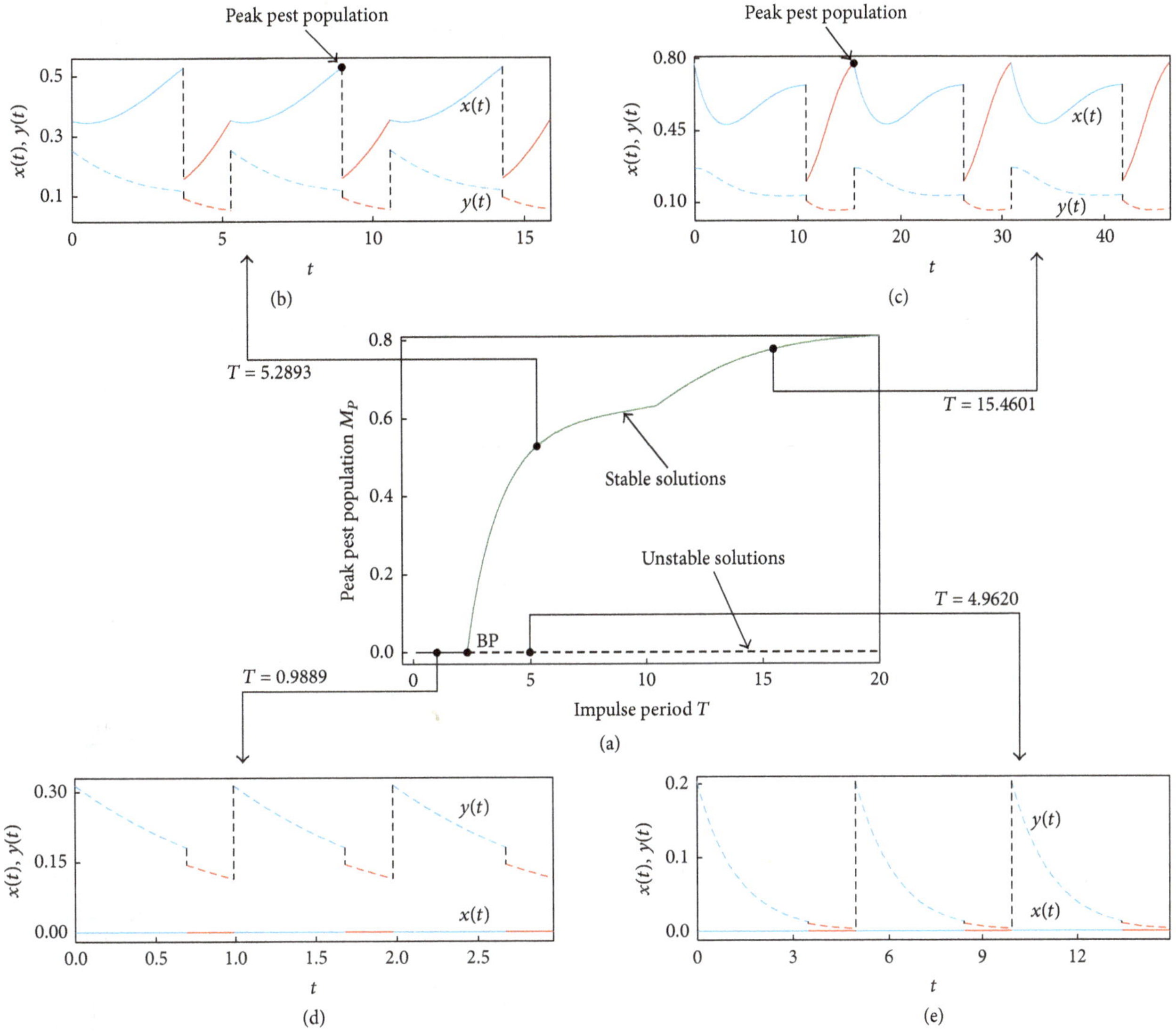

FIGURE 2: (a) One-parameter continuation of the periodic response of system (14), with respect to the impulse period T. The solid and dashed lines stand for stable and unstable solutions, respectively. In what follows, this convention will be used to denote stability in a bifurcation diagram. (b)–(e) present time histories for different values of impulse period. Here, the pest and natural enemy populations are displayed with solid and dashed lines, respectively. Throughout the paper, pest and natural enemies will be distinguished in a time history in this way.

effectively used to reduce the pest population, this may be in practice difficult or too expensive to implement, as it would require a certain mechanism to modify biological attributes of the natural enemy (via, e.g., genetic engineering or selective breeding).

From a practical perspective, variations in the control parameters σ_x, σ_y, d, and l may be more accessible for the users. As can be seen in Figure 3(d), the influence of σ_y (proportion of natural enemies killed by pesticides) on the system response is rather marginal. Even if one compares the extreme cases $\sigma_y \approx 0$ and $\sigma_y \approx 1$, no significant difference can be observed. This is due to the fact that the pest control scheme introduces periodically a certain amount of natural enemies into the ecosystem, which compensates for the mortality of the predators due to the pesticide. On the other hand, the proportion of pest individuals killed by pesticides σ_x affects significantly the presence of pests in the ecosystem; see Figure 3(c). As this parameter approaches the upper boundary $\sigma_x = 1$, the pest population suffers a significant reduction, which suggests that the pesticide mortality should be maximized in order to eradicate the pest. However, this can have well-known negative consequences for the environment; hence the effectiveness of the pest control method cannot be based upon such a strategy. Alternatively, one can try to find an environmentally acceptable, yet optimal, pest mortality σ_x. For instance, the local minimum $\sigma_x \approx 0.4345$ shown in Figure 3(c).

The next parameter to be discussed is the number of predators introduced periodically into the ecosystem d, whose impact on the model response is presented in Figure 3(e). The result is biologically consistent in that the larger d, the larger the amount of predators in the ecosystem and

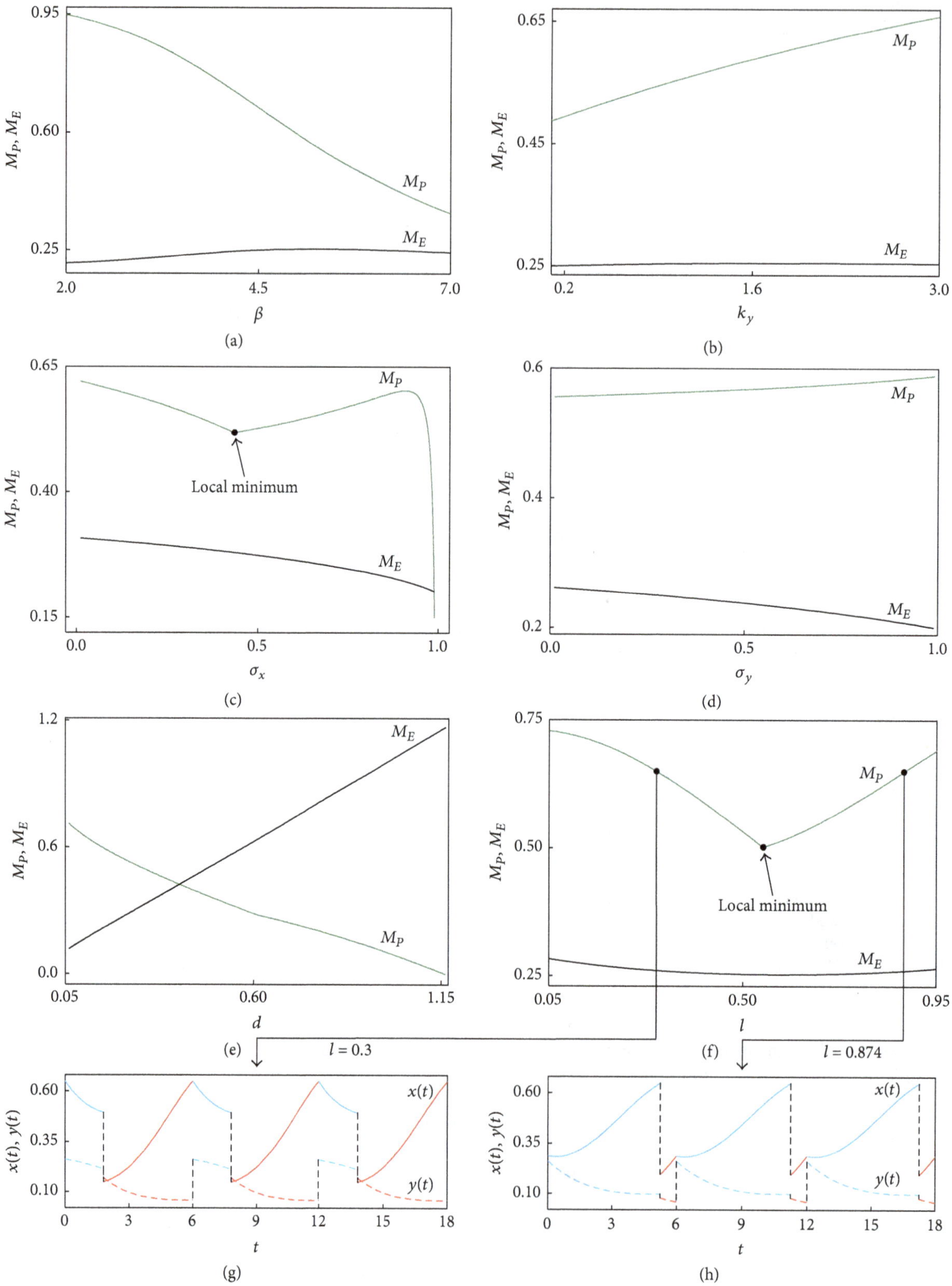

FIGURE 3: Numerical continuation of periodic solutions of system (14) with pests and natural enemies coexisting in the ecosystem, computed for impulse period $T = 6$. (a)–(f) show the continuation with respect to β, k_y, σ_x, σ_y, d, and l, respectively. In each diagram, two curves are shown, corresponding to the peak values of pest (M_P, green) and natural enemy (M_E, black) populations. (g) and (h) present time histories for which $M_P = 0.65$.

the smaller the size of the pest population. This provides us with another effective way to control the pest; however, an increment in d may imply higher operation costs or negative consequences for other species present in the ecosystem, and therefore this parameter must also be chosen with certain precaution.

From the system parameters considered in Figure 3, l is probably the one that can be varied without causing major ecological damage. This parameter determines the instant at which pesticide is sprayed in each impulsive period, and its effect on the pest and predator populations is shown in Figure 3(f). The picture reveals an optimal point $l \approx 0.5464$ where the peak size of pest population in the ecosystem achieves a minimum value $M_P \approx 0.5021$. Practically speaking, this indicates that the performance of the pest control method can be improved by just changing the time at which pesticide is sprayed, keeping all remaining parameters fixed. This represents another avenue that can be explored in order to apply the impulsive control scheme in the most effective way, with acceptable levels of environmental impact.

4.2.2. Optimal Implementation of the Impulsive Control Method. In the previous section, we presented a systematic parametric study of the pest control scheme modeled by system (14). Special emphasis was put on trying to find optimal operation points when one parameter is changed at a time. This process can be further refined by defining objective functions and constraints and letting two or more parameters vary simultaneously. Let us assume that, for a certain application, we have to minimize the utilization of pesticides and introduction of natural enemies in the framework of the pest control scheme studied in the preceding section. This can be motivated by the associated operation costs or environmental damage, as discussed earlier. Nevertheless, a reduction in the usage of pesticides and natural enemies evidently increases the risk of high levels of pest population in the ecosystem. Hence, we need to introduce a restriction for the optimization problem, which can be defined in terms of the size of the pest population not exceeding a predefined economic threshold $x_{\mathrm{ET}} > 0$. Mathematically speaking, we will consider the following optimization problem:

$$\begin{aligned} &\text{Minimize} \quad F(d, \sigma_x) := \sqrt{d^2 + \sigma_x^2}, \\ &\qquad\qquad \text{under the constraint } M_P = x_{\mathrm{ET}}. \end{aligned} \tag{15}$$

Here, the objective function F is nothing but the Euclidean norm of the vector (d, σ_x), which gives a measure of the amount of pesticides and natural enemies used in the control scheme (control effort). Depending on the specific application, this functional can be further refined by, for example, introducing weighting coefficients, but for the sake of clarity we will carry out the numerical implementation with the simple objective function defined above.

The optimization problem (15) will be solved numerically via path-following techniques, as shown in Figure 4. Figure 4(a) presents a curve in the d-σ_x plane corresponding to the combination of pesticide and introduced predators yielding a constant peak pest population $M_P = x_{\mathrm{ET}} = 0.6$. A trajectory along this curve is presented in Figure 4(c), where it can be verified that the pest population indeed does not exceed the predefined economic threshold x_{ET}. Figures 4(e) and 4(f) display solutions corresponding to parameter values above and below the point considered in Figure 4(c), respectively. These three panels demonstrate how M_P moves away from the imposed economic threshold when the operation point is perturbed around the computed curve.

As was confirmed numerically, the curve shown in Figure 4(a) divides the d-σ_x plane locally into two parts. The one to the left corresponds to parameter values for which the peak pest population M_P exceeds the economic threshold x_{ET}, while in the region to the right we have that $M_P < x_{\mathrm{ET}}$. Therefore, the optimal operation point must lie on the right part of the d-σ_x plane, including the computed boundary curve. This point can be located numerically by monitoring the values of the objective function on the boundary and is found to be $(d, \sigma_x) \approx (0.1807, 0.1411)$; see Figure 4(b). The system response corresponding to this operation point is displayed in Figure 4(d). In comparison to the solution shown in Figure 4(c), it can be observed that the reduction of the pest population due to pesticide spraying is significantly smaller, meaning that less pesticide is being used in the optimal case. This is nonetheless compensated with a slight increment of the amount of predators introduced periodically into the ecosystem (from $d = 0.15$ to $d = 0.1807$). In both cases the condition $M_P = x_{\mathrm{ET}} = 0.6$ is satisfied; however, the environmental damage, measured in terms of the objective function defined above, is minimized for the optimal parameter values found.

4.2.3. Further Dynamical Analysis of the Pest Control Scheme. So far the main tool in our numerical study has been path-following algorithms, which enabled a detailed parametric study of the periodic response of the pest control scheme described by the impulsive model (14). In this way, we were able to tackle practical questions such as how to find the most suitable operation conditions in terms of pest reduction and minimization of harmful environmental effects. Nevertheless, the dynamic behavior of impulsive systems is a subject of scientific interest in its own right, and the present section will be devoted to this matter. Specifically, we will employ both path-following methods and direct numerical integration in order to gain a deeper understanding of the dynamics of the impulsive pest control scheme considered in our investigation.

The theoretical foundations for a numerical study of the impulsive system (14) have been established by Zhang et al. [44]. They addressed the question of existence and uniqueness of nontrivial periodic solutions in terms of a fixed point problem of a suitably defined operator. Following this approach, they also determined a threshold for the impulse period after which pest-free solutions lose stability. Moreover, in the conclusion part of [44] the authors raised the question whether chaotic behavior may be present in the system, and this will be precisely one of the main motivations for the numerical study presented in this section. Unless otherwise

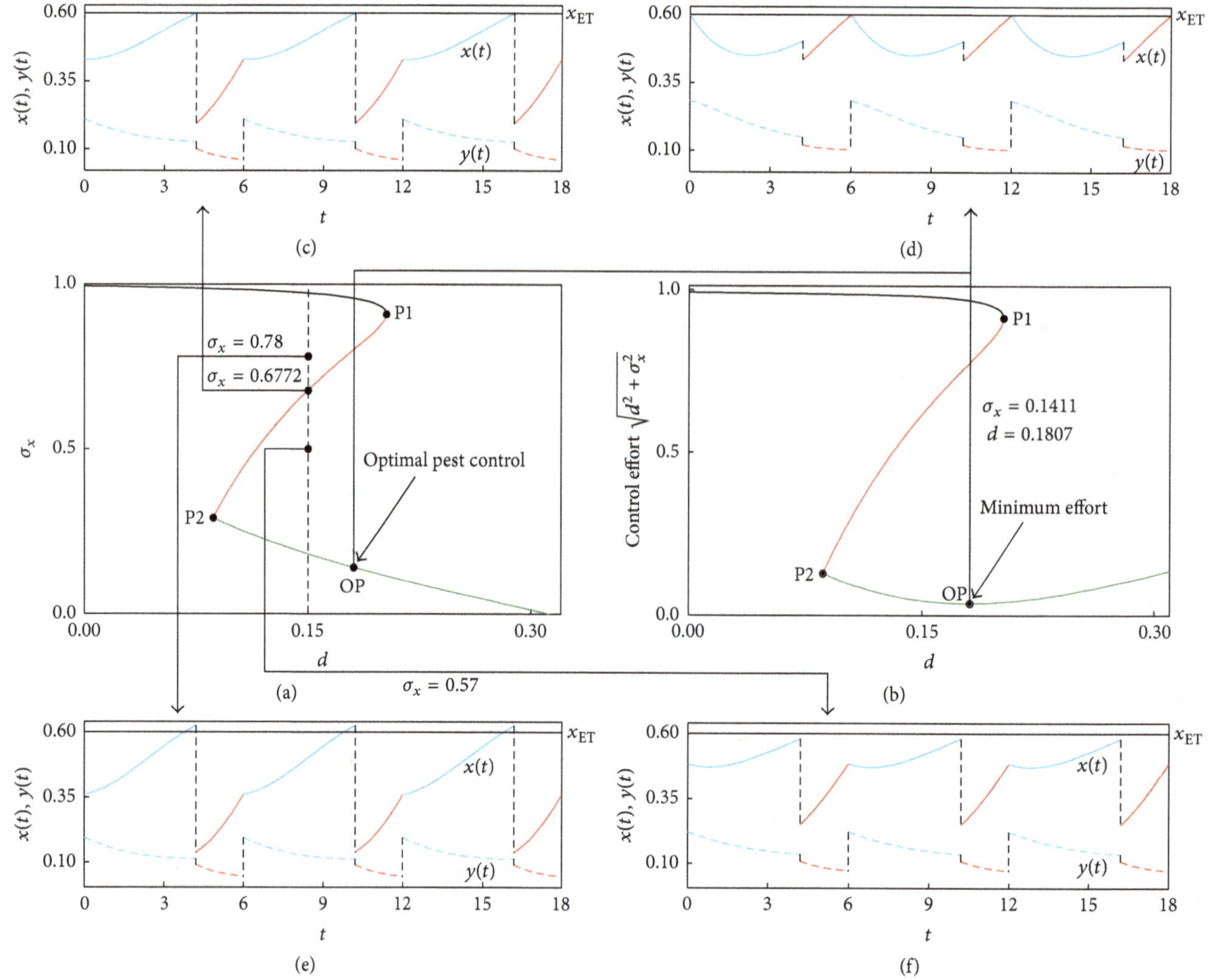

FIGURE 4: Numerical continuation of the periodic response of system (14) with respect to d and σ_x, computed for impulse period $T = 6$. (a) Along this curve, the auxiliary boundary condition (13) is used to keep $M_P = x_{ET} = 0.6$ constant (economic threshold). (b) depicts the same result, but in this case the control effort, computed as the modulus of the vector (d, σ_x), is monitored. In this picture, the optimal operation point is defined as the parameter values for which the control effort is minimum. (c)–(f) present solutions of the system for different combinations of d and σ_x.

indicated, the parameter values used in the discussion below are those given in Table 1.

We begin our numerical study with the continuation of the periodic solution shown in Figure 1 with respect to the predator's mortality rate μ, see Figure 5(a). As the parameter is varied from larger to lower values, a first qualitative change is observed at $\mu \approx 0.1323$ (PD1). Here, one real Floquet multiplier of the periodic orbit crosses the complex unit circle from the inside, passing through -1. This phenomenon is referred to as a period-doubling (flip) bifurcation of limit cycles [71] and is characterized, in the supercritical case, by the birth of a stable periodic solution with twice the period of the original limit cycle, which in turn loses stability (schematically represented by the dashed line in Figure 5(a)). This unstable solution regains stability via another flip bifurcation at $\mu \approx 0.1001$ (PD2), where the critical Floquet multiplier comes back inside the unit circle and the stable periodic solution with double period disappears. If μ is further decreased, a turning point (also known as fold bifurcation) is found at $\mu \approx 0.0722$ (F1), in which case a pair of stable and unstable periodic orbits collide and then disappear for lower parameter values. From this point a branch (dashed segment) of unstable solutions is born, which finishes at F2 ($\mu \approx 0.0951$), where the system undergoes another fold bifurcation of limit cycles, and hence stability is regained. The last stability change is found at $\mu \approx 0.0792$ (PD3), corresponding to a supercritical flip bifurcation, from which a branch of unstable periodic solutions emanates. Another feature of the bifurcation diagram shown in Figure 5(a) is the presence of a parameter window $0.0792 < \mu < 0.0951$ for which the impulsive system possesses two stable periodic solutions that coexist (coexisting attractors [72]). This is produced by the interplay between the fold bifurcations F1

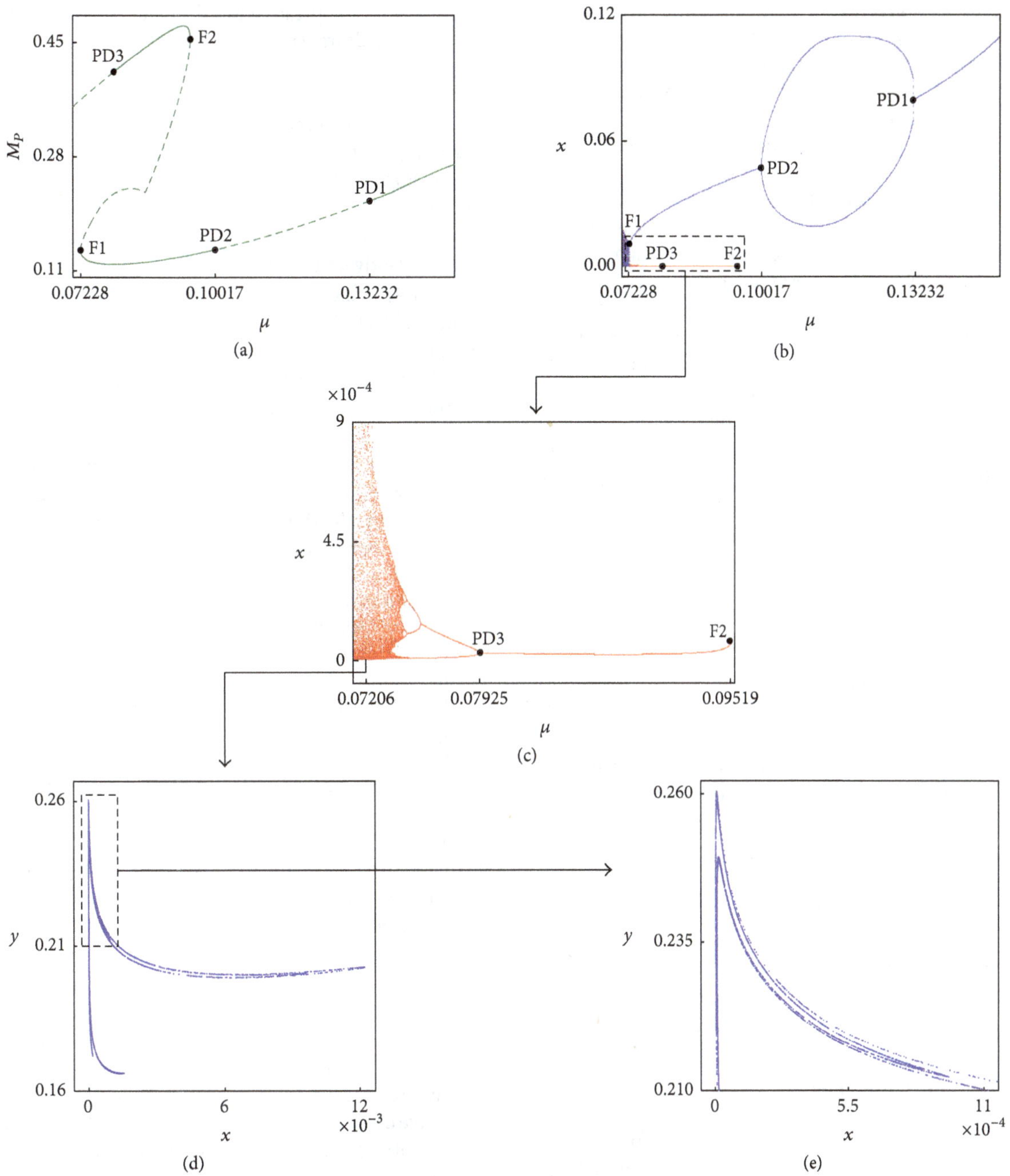

FIGURE 5: Dynamical behavior of the pest control method modeled by system (14), computed for the parameter values shown in Table 1. (a) Numerical continuation of the periodic response of the model with respect to the predator's mortality rate μ. The points labeled PDi and Fi denote period-doubling and fold bifurcations of periodic orbits, respectively. (b) Bifurcation diagram computed via direct numerical integration of system (14). The red and blue colors mark the parameter sweeps in the increasing and decreasing directions, respectively. (c) Blow-up of a part of the bifurcation diagram shown in (b). (d) Poincaré map associated with a chaotic attractor computed for $\mu = 0.07206$. (e) Enlargement of a portion of the attractor displayed in (d).

and F2, typically giving rise to hysteretic effects, as will be discussed below.

In order to gain further insight into the long-time dynamics of the pest control model, we will carry out a parametric study of the impulsive system (14) via direct numerical integration, when the predator's mortality rate μ is varied. For this purpose, we fix a starting value for μ and then integrate the system over 300 impulse periods to allow for the decay of transients. After this, we extend the numerical integration for another interval of 100 periods and store samples of the extended solution at the times $t = (i + \epsilon - 1)T$, $i = 1, 2, \dots, 100$, with $0 < \epsilon < l$ being a fixed shift coefficient. This parameter is introduced so as to avoid sampling the solution at the impulse times $t = (n+l-1)T$ and $t = nT, n \in \mathbb{N}$.

Next, μ is increased (or decreased, depending on the sweep direction) by a small amount and then the same procedure is repeated, now using the last sample of the previous step as initial value. This process terminates when a predefined final value for μ is reached.

The result of the numerical procedure described above is shown in Figure 5(b). The picture confirms the qualitative changes (bifurcations) predicted in Figure 5(a), labeled PD*i* and F*i*. In particular, the bifurcation diagram shown in Figure 5(b) allows us to visualize the creation and disappearance of orbits with double period, observed at the flip bifurcations $\mu \approx 0.1001$ and $\mu \approx 0.1323$ detected before. The increasing (red) and decreasing (blue) parameter sweeps reveal the presence of parameter hysteresis in the system, produced by the coexistence of periodic attractors, as predicted above. The blow-up of the red part of the bifurcation diagram, shown in Figure 5(c), contains the bifurcation points PD3 and F2 found in Figure 5(a). These two points define a parameter window in which a stable orbit of period T survives. When μ decreases through PD3, the period-T solution becomes unstable and a stable periodic orbit of twice the period appears. If μ is further reduced, the period-$2T$ trajectory loses stability via another flip bifurcation ($\mu \approx 0.0755$), giving rise to a stable periodic solution of period $4T$. This phenomenon repeats again and again as the parameter continues to decrease, until a critical value is reached where the flip bifurcations accumulate, leading to chaotic behavior. This infinite sequence of period-doubling bifurcations caused by the variation of a parameter over a finite interval is referred to as a period-doubling cascade and is one of the classical routes to chaos in dynamical systems [72]. Figure 5(d) shows the intersection of a chaotic attractor of the impulsive system (14) with a Poincaré section, for $\mu = 0.07206$. This numerical study gives a positive answer to one of the open questions outlined by Zhang et al. [44] related to chaotic behavior, since we have identified a parameter set and a mechanism through which chaos can appear in the pest control model.

5. Concluding Remarks

The intrinsic premise in the present work is that pest control is a dynamical process. As such, mathematical models are essential for understanding and providing useful abstractions of the underlying biological phenomena and ecological interactions taking place in pest control applications. Since the introduction of the notion of Integrated Pest Management (IPM) in the late 1950s [20, 21], a rich set of mathematical models has emerged focusing on various aspects of IPM-based applications, and Section 3 was devoted to providing the reader with an overview of the development in this area. The discussion presented here was guided by two criteria: model class (in mathematical sense) and type of pest control method. As a result, the models considered in this section ranged from classical smooth differential equations to differential equations with impulses. Moreover, the types of pest control methods considered in our discussion can be briefly grouped in the following categories: chemical (pesticides), biological (natural enemy predation, biopesticides), and cultural (roguing, replanting), which are suitably combined in the framework of IPM in ways that complement one another.

Although the literature on modeling pest control strategies in agricultural ecosystems is vast, comprehensive parametric studies of the underlying mathematical models have received rather little attention in the past, with numerical investigations carried out primarily at the simulation level. This fact motivated the main contribution of the present work (Section 4), which concerned the application of specialized numerical techniques in order to address practical questions relevant to IPM. For this purpose, an impulsive pest control model was chosen (namely system (4)) and reformulated in the framework of hybrid dynamical systems, thus enabling the employment of path-following (continuation) methods for the numerical study of the impulsive system under parameter variations via the software package TC-HAT [17]. With the introduction of appropriate numerical indicators, a comprehensive parametric study was carried out in Sections 4.2.1 and 4.2.2, where the main question was how to determine the parameter values of the system so as to control the pest population in an optimal way, in terms of minimizing operation costs and environmental damage. Further numerical investigations of the model (4) were conducted in Section 4.2.3, with special attention focused on the dynamical features of the system. This study revealed the presence of fold and flip bifurcations of limit cycles, period-doubling cascades leading to chaotic behavior, and hysteretic effects.

As has been pointed out in the past [2, 12, 16], a decisive factor of success in the application of IPM programmes is the fundamental understanding of the interactions between the different components of the underlying agricultural ecosystem, such as crops, pests, natural enemies, and biopesticides. Much of the challenge lies in the sheer complexity of the involved biological processes, which often hinders the search for appropriate mathematical representations of the laws governing the ecosystem. The resulting models are usually the product of a trade-off between model simplicity, predictive capability and accuracy, biological consistency, and amenability for experimental validation and calibration. It is then reasonable to assume that the validity of a model correlates with how well it satisfies the aforementioned criteria, which makes model development by no means a trivial task. Once a valid model has been obtained, it is crucial to tackle formal questions regarding the well-posedness of the model in a mathematical sense, for instance, those related to existence and uniqueness of solutions, degree of smoothness, and dependence on initial values, which enable a confident application of numerical methods in order to explore the model behavior. Future progress in this area will therefore require a multidisciplinary collaborative work between agronomists, ecological modelers, mathematical analysts, and numerical specialists, aimed at constructing mathematical models that provide reliable predictions and understanding of field observations in real ecosystems, in such a way that these models can be used as support tools for the development of new pest control strategies and paradigms.

Conflicts of Interest

The authors declare that there are no conflicts of interest regarding the publication of this paper.

Acknowledgments

The first author wishes to express his gratitude to the Alexander von Humboldt Foundation (Georg Forster Research Fellowship Programme) for the financial support to carry out the present research.

References

[1] United Nations, *Concise Report on the World Population Situation*, New York, NY, USA, 2014.

[2] R. Peshin and A. K. Dhawan, *Integrated Pest Management: Innovation-Development Process*, Springer Netherlands, Dordrecht, 2009.

[3] S. E. Naranjo and P. C. Ellsworth, "Fifty years of the integrated control concept: Moving the model and implementation forward in Arizona," *Pest Management Science*, vol. 65, no. 12, pp. 1267–1286, 2009.

[4] D. L. Jaquette, "Mathematical models for controlling growing biological populations: a survey," *Operations Research*, vol. 20, no. 6, pp. 1142–1151, 1972.

[5] C. Shoemaker, "Optimization of agricultural pest management II: formulation of a control model," *Mathematical Biosciences*, vol. 17, no. 3-4, pp. 357–365, 1973.

[6] Y. M. Svirezhev, "Nonlinearities in mathematical ecology: Phenomena and models. Would we live in Volterra's world?" *Ecological Modelling*, vol. 216, no. 2, pp. 89–101, 2008.

[7] J. M. Baker, "Use and Abuse of Crop Simulation Models," *Agronomy Journal*, vol. 88, no. 5, p. 689, 1996.

[8] D. R. McCullough and R. H. Barrett, *Wildlife 2001: Populations*, Springer Netherlands, Dordrecht, 1992.

[9] F. Affholder, P. Tittonell, M. Corbeels et al., "Ad hoc modeling in agronomy: What have we learned in the last 15 years?" *Agronomy Journal*, vol. 104, no. 3, pp. 735–748, 2012.

[10] L. Prost, M. Cerf, and M.-H. Jeuffroy, "Lack of consideration for end-users during the design of agronomic models. A review," *Agronomy for Sustainable Development*, vol. 32, no. 2, pp. 581–594, 2012.

[11] N. J. Cunniffe, B. Koskella, C. J. E. Metcalf, S. Parnell, T. R. Gottwald, and C. A. Gilligan, "Thirteen challenges in modelling plant diseases," *Epidemics*, vol. 10, pp. 6–10, 2015.

[12] M. Kogan, "Integrated pest management: Historical perspectives and contemporary developments," *Annual Review of Entomology*, vol. 43, pp. 243–270, 1998.

[13] W. I. Bajwa and M. Kogan, "Compendium of IPM Definitions (CID). What is IPM and how is it defined in the Worldwide Literature?" *IPPC Publication No. 998*, 2002.

[14] K. Wickwire, "Mathematical models for the control of pests and infectious diseases: a survey," *Theoretical Population Biology. An International Journal*, vol. 11, no. 2, pp. 182–238, 1977.

[15] J. R. Beddington, C. A. Free, and J. H. Lawton, "Characteristics of successful natural enemies in models of biological control of insect pests," *Nature*, vol. 273, no. 5663, pp. 513–519, 1978.

[16] H. J. Barclay, "Models for pest control using predator release, habitat management and pesticide release in combination," *Journal of Applied Ecology*, vol. 19, no. 2, pp. 337–348, 1982.

[17] P. Thota and H. Dankowicz, "TC-HAT: A Novel Toolbox for the Continuation of Periodic Trajectories in Hybrid Dynamical Systems," *SIAM Journal on Applied Dynamical Systems*, vol. 7, no. 4, pp. 1283–1322, 2008.

[18] B. Krauskopf, H. M. Osinga, and J. Galán-Vioque, "A continuing influence in dynamics," *Numerical Continuation Methods for Dynamical Systems: Path following and boundary value problems*, pp. V–VII, 2007.

[19] F. González-Andrade, R. López-Pulles, and E. Estévez, "Acute pesticide poisoning in Ecuador: A short epidemiological report," *Journal of Public Health*, vol. 18, no. 5, pp. 437–442, 2010.

[20] J. A. Lucas, "Plant Pathogens and Plant Diseases.," *Plant Pathology*, vol. 47, no. 4, pp. 542-542, 1998.

[21] V. M. Stern, R. F. Smith, R. van den Bosch, and K. S. Hagen, "The Integrated Control Concept," *Hilgardia*, vol. 29, no. 2, pp. 81–101, 1959.

[22] M. Rybicki, C. Winkelmann, C. Hellmann, P. Bartels, and D. Jungmann, "Herbicide indirectly reduces physiological condition of a benthic grazer," *Aquatic Biology*, vol. 17, no. 2, pp. 153–166, 2012.

[23] O. Licht, D. Jungmann, and R. Nagel, "Method development on aufwuchs and mayfly larvae to determine the effects of chemicals in artificial indoor streams," *Fresenius Environmental Bulletin*, vol. 12, no. 6, pp. 594–600, 2003.

[24] K. Brust, O. Licht, V. Hultsch, D. Jungmann, and R. Nagel, "Effects of terbutryn on aufwuchs and Lumbriculus variegatus in artificial indoor streams," *Environmental Toxicology and Chemistry*, vol. 20, no. 9, pp. 2000–2007, 2001.

[25] O. Licht, D. Jungmann, K.-U. Ludwichowski, and R. Nagel, "Long-term effects of fenoxycarb on two mayfly species in artificial indoor streams," *Ecotoxicology and Environmental Safety*, vol. 58, no. 2, pp. 246–255, 2004.

[26] E. Böckmann and R. Meyhöfer, "AEP – Eine automatische Entscheidungshilfe-Software für den integrierten Pflanzenschutz," *Gesunde Pflanzen*, vol. 67, no. 1, pp. 1–10, 2015.

[27] S. Jana and T. K. Kar, "A mathematical study of a prey-predator model in relevance to pest control," *Nonlinear Dynamics*, vol. 74, no. 3, pp. 667–683, 2013.

[28] H. W. Hethcote, "Qualitative analyses of communicable disease models," *Mathematical Biosciences*, vol. 28, no. 3/4, pp. 335–356, 1976.

[29] C. S. Holling, "The functional response of predators to prey density and its role in mimicry and population regulation," *Memoirs of the Entomological Society of Canada*, vol. 97, supplement 45, pp. 5–60, 1965.

[30] A. D. Bazykin, *Nonlinear Dynamics of Interacting Populations*, vol. 11 of *World Scientific Series on Nonlinear Science, Series A*, World Scientific, River Edge, NJ, USA, 1998.

[31] R. K. Upadhyay, "Observability of chaos and cycles in ecological systems: lessons from predator-prey models," *International Journal of Bifurcation and Chaos*, vol. 19, no. 10, pp. 3169–3234, 2009.

[32] F. van den Bosch, M. J. Jeger, and C. A. Gilligan, "Disease control and its selection for damaging plant virus strains in vegetatively propagated staple food crops; a theoretical assessment," *Proceedings of the Royal Society B Biological Science*, vol. 274, no. 1606, pp. 11–18, 2007.

[33] C. A. Gilligan and F. Van Den Bosch, "Epidemiological models for invasion and persistence of pathogens," *Annual Review of Phytopathology*, vol. 46, pp. 385–418, 2008.

[34] X. Meng and Z. Li, "The dynamics of plant disease models with continuous and impulsive cultural control strategies," *Journal of Theoretical Biology*, vol. 266, no. 1, pp. 29–40, 2010.

[35] M. J. Jeger, "An analytical model of plant virus disease dynamics with roguing and replanting," *Journal of Applied Ecology*, vol. 31, no. 3, pp. 413–427, 1994.

[36] M. J. Jeger, L. V. Madden, and F. van den Bosch, "The effect of transmission route on plant virus epidemic development and disease control," *Journal of Theoretical Biology*, vol. 258, no. 2, pp. 198–207, 2009.

[37] L. Xia, S. Gao, Q. Zou, and J. Wang, "Analysis of a nonautonomous plant disease model with latent period," *Applied Mathematics and Computation*, vol. 223, pp. 147–159, 2013.

[38] Z. H. Zhang and Y. H. Suo, "Stability and sensitivity analysis of a plant disease model with continuous cultural control strategy," *Journal of Applied Mathematics*, vol. 2014, Article ID 207959, 15 pages, 2014.

[39] V. Lakshmikantham, D. D. Bainov, and P. S. Simeonov, *Theory of Impulsive Differential Equations*, World Scientific, Singapore, 1989.

[40] D. D. Bainov and P. S. Simeonov, *Impulsive Differential Equations*, vol. 28 of *Series on Advances in Mathematics for Applied Sciences*, World Scientific, Singapore, 1995.

[41] A. M. Samoilenko and N. A. Perestyuk, *Impulsive differential equations*, vol. 14 of *Series A: Monographs and Treatises*, World Scientific, New Jersey, 1995.

[42] N. A. Perestyuk, V. A. Plotnikov, A. M. Samoilenko, and N. V. Skripnik, *Differential equations with impulse effects. Multivalued right-hand sides with discontinuities*, vol. 40 of *De Gruyter Studies in Mathematics*, de Gruyter, Berlin, Germany, 2011.

[43] S. Tang, Y. Xiao, and R. A. Cheke, "Dynamical analysis of plant disease models with cultural control strategies and economic thresholds," *Mathematics and Computers in Simulation*, vol. 80, no. 5, pp. 894–921, 2010.

[44] H. Zhang, P. Georgescu, and L. S. Chen, "On the impulsive controllability and bifurcation of a predator-pest model of IPM," *BioSystems*, vol. 93, no. 3, pp. 151–171, 2008.

[45] D. L. DeAngelis, R. A. Goldstein, and R. V. O'Neill, "A model for trophic interaction," *Ecology*, vol. 56, pp. 881–892, 1975.

[46] J. R. Beddington, "Mutual interference between parasites or predators and its effect on searching efficiency," *Journal of Animal Ecology*, vol. 44, pp. 331–340, 1975.

[47] M. di Bernardo, C. J. Budd, and A. R. Champneys, *Piecewise-Smooth Dynamical Systems: Theory and Applications*, vol. 163 of *Applied Mathematical Sciences*, Springer, London, UK, 2008.

[48] B. Liu, Y. Zhang, and L. Chen, "Dynamic complexities of a Holling I predator-prey model concerning periodic biological and chemical control," *Chaos, Solitons & Fractals*, vol. 22, no. 1, pp. 123–134, 2004.

[49] Z. Xiang, X. Song, and F. Zhang, "Bifurcation and complex dynamics of a two-prey two-predator system concerning periodic biological and chemical control," *Chaos, Solitons and Fractals*, vol. 37, no. 2, pp. 424–437, 2008.

[50] H. Zhang, L. S. Chen, and P. Georgescu, "Impulsive control strategies for pest management," *Journal of Biological Systems*, vol. 15, no. 2, pp. 235–260, 2007.

[51] Y. Pei, S. Liu, and C. Li, "Complex dynamics of an impulsive control system in which predator species share a common prey," *Journal of Nonlinear Science*, vol. 19, no. 3, pp. 249–266, 2009.

[52] B. Liu, Y. Wang, and B. Kang, "Dynamics on a pest management SI model with control strategies of different frequencies," *Nonlinear Analysis: Hybrid Systems*, vol. 12, pp. 66–78, 2014.

[53] H. Su, B. Dai, Y. Chen, and K. Li, "Dynamic complexities of a predator-prey model with generalized Holling type III functional response and impulsive effects," *Computers & Mathematics with Applications. An International Journal*, vol. 56, no. 7, pp. 1715–1725, 2008.

[54] G. Pang and L. Chen, "Dynamic analysis of a pest-epidemic model with impulsive control," *Mathematics and Computers in Simulation*, vol. 79, no. 1, pp. 72–84, 2008.

[55] S. Tang, Y. Xiao, L. Chen, and R. A. Cheke, "Integrated pest management models and their dynamical behaviour," *Bulletin of Mathematical Biology*, vol. 67, no. 1, pp. 115–135, 2005.

[56] K. S. Jatav and J. Dhar, "Hybrid approach for pest control with impulsive releasing of natural enemies and chemical pesticides: a plant-pest-natural enemy model," *Nonlinear Analysis: Hybrid Systems*, vol. 12, pp. 79–92, 2014.

[57] A. Michel, K. Wang, and B. Hu, *Qualitative theory of dynamical systems*, vol. 239 of *Monographs and Textbooks in Pure and Applied Mathematics*, Basel: Marcel Dekker AG Publishers, Second edition, 2001.

[58] R. Goebel, R. G. Sanfelice, and A. R. Teel, *Hybrid Dynamical Systems: Modeling, Stability, and Robustness*, Princeton University Press, 2012.

[59] M. Akhmet, "Principles of discontinuous dynamical systems," *Principles of Discontinuous Dynamical Systems*, pp. 1–176, 2010.

[60] L. F. Shampine, I. Gladwell, and R. W. Brankin, "Reliable solution of special event location problems for ODEs," *ACM Transactions on Mathematical Software*, vol. 17, no. 1, pp. 11–25, 1991.

[61] L. F. Shampine and S. Thompson, "Event location for ordinary differential equations," *Computers & Mathematics with Applications*, vol. 39, no. 5-6, pp. 43–54, 2000.

[62] P. T. Piiroinen and Y. A. Kuznetsov, "An event-driven method to simulate Filippov systems with accurate computing of sliding motions," *ACM Transactions on Mathematical Software*, vol. 34, no. 3, article no. 13, 2008.

[63] H. Dankowicz and F. Schilder, *Recipes for continuation*, SIAM, Computational Science and Engineering, Philadelphia, Pennsylvania, 2013.

[64] F. Dercole and Y. A. Kuznetsov, "SlideCont: an Auto97 driver for bifurcation analysis of Filippov systems," *ACM Transactions on Mathematical Software*, vol. 31, no. 1, pp. 95–119, 2005.

[65] E. J. Doedel, Champneys, T. F. A. R.Fairgrieve, Y. A. Kuznetsov, and B. Sandstede, *Auto97: Continuation and bifurcation software for ordinary differential equations (with* HomCont*)*, Computer Science, Concordia University, Montreal, Canada, 1997.

[66] J. Páez Chávez and M. Wiercigroch, "Bifurcation analysis of periodic orbits of a non-smooth Jeffcott rotor model," *Communications in Nonlinear Science and Numerical Simulation*, vol. 18, no. 9, pp. 2571–2580, 2013.

[67] J. Páez Chávez, E. Pavlovskaia, and M. Wiercigroch, "Bifurcation analysis of a piecewise-linear impact oscillator with drift," *Nonlinear Dynamics*, vol. 77, no. 1-2, pp. 213–227, 2014.

[68] M. Liao, J. Ing, J. Páez Chávez, and M. Wiercigroch, "Bifurcation techniques for stiffness identification of an impact oscillator," *Communications in Nonlinear Science and Numerical Simulation*, vol. 41, pp. 19–31, 2016.

[69] J. Páez Chávez, Y. Liu, E. Pavlovskaia, and M. Wiercigroch, "Path-following analysis of the dynamical response of a

piecewise-linear capsule system," *Communications in Nonlinear Science and Numerical Simulation*, vol. 37, pp. 102–114, 2016.

[70] J. Páez Chávez, A. Voigt, J. Schreiter, U. Marschner, S. Siegmund, and A. Richter, "A new self-excited chemo-fluidic oscillator based on stimuli-responsive hydrogels: Mathematical modeling and dynamic behavior," *Applied Mathematical Modelling*, vol. 40, no. 23-24, pp. 1339–1351, 2016.

[71] Y. A. Kuznetsov, *Elements of Applied Bifurcation Theory*, vol. 112 of *Applied Mathematical Sciences*, Springer, New York, NY, USA, 3rd edition, 2004.

[72] T. Alligood, T. D. Sauer, and J. A. Yorke, *Chaos: An Introduction to Dynamical Systems*, Springer, New York, NY, USA, 1996.

7

Application of Residual Power Series Method to Fractional Coupled Physical Equations Arising in Fluids Flow

Anas Arafa[1] **and Ghada Elmahdy**[2]

[1]*Department of Mathematics and Computer Science, Faculty of Science, Port Said University, Port Said, Egypt*
[2]*Department of Basic science, Canal High Institute of Engineering and Technology, Suez, Egypt*

Correspondence should be addressed to Ghada Elmahdy; ghada.elmahdy91@gmail.com

Academic Editor: Julio D. Rossi

The approximate analytical solution of the fractional Cahn-Hilliard and Gardner equations has been acquired successfully via residual power series method (RPSM). The approximate solutions obtained by RPSM are compared with the exact solutions as well as the solutions obtained by homotopy perturbation method (HPM) and q-homotopy analysis method (q-HAM). Numerical results are known through different graphs and tables. The fractional derivatives are described in the Caputo sense. The results light the power, efficiency, simplicity, and reliability of the proposed method.

1. Introduction

Fractional differential equations (FDEs) have found applications in many problems in physics and engineering [1, 2]. Since most of the nonlinear FDEs cannot be solved exactly, approximate and numerical methods must be used. Some of the recent analytical methods for solving nonlinear problems include the Adomian decomposition method [3, 4], variational iteration method [5], homotopy perturbation method [6, 7], homotopy analysis method [8, 9], spectral collocation method [10], the tanh-coth method [11], exp-function method [12], Mittag-Leffler function method [13], differential quadrature method [14], and reproducing kernel Hilbert space method [15, 16].

The Gardner equation [17] (combined KdV-mKdV equation) is a useful model for the description of internal solitary waves in shallow water,

$$u_t + 6uu_x \pm 6u^2u_x + u_{xxx} = 0. \quad (1)$$

Those two models will be classified as positive Gardner equation and negative Gardner equation depending on the sign of the cubic nonlinear term [18, 19]. Gardner equation is widely used in various branches of physics, such as plasma physics, fluid physics, and quantum field theory [20, 21]. It also describes a variety of wave phenomena in plasma and solid state [22, 23].

The Cahn-Hilliard equation [24] is one type of partial differential equations (PDEs) and was first introduced in 1958 as a model for process of phase separation of a binary alloy under the critical temperature [25],

$$u_t = \gamma u_x + 6uu_x^2 + \left(3u^2 - 1\right)u_{xx} - u_{xxxx}, \quad \gamma \geq 0. \quad (2)$$

This equation is related to a number of interesting physical phenomena like the spinodal decomposition, phase separation, and phase ordering dynamics. On the other hand it becomes important in material sciences [26, 27].

The aim of this paper is to study the time-fractional Gardner equation [28–30] and time-fractional Cahn-Hilliard equation [31–37] of this form,

$$D_t^{\alpha}u(x,t) + 6\left(u - \varepsilon^2u^2\right)u_x + u_{xxx} = 0, \quad (3)$$

$$D_t^{\alpha}u(x,t) - u_x - 6uu_x^2 - \left(3u^2 - 1\right)u_{xx} + u_{xxxx} = 0, \quad (4)$$

where $0 < \alpha \leq 1$, $-\infty < x < \infty$, and $0 \leq t < R$. Numerous methods have been used to solve this equations, for example, q-Homotopy analysis method [28], the new version of F-expansion method [29], reduced differential transform method [30], the generalized tanh-coth method [38], the generalized Kudryashov method [39], extended fractional Riccati expansion method [31], subequation method [32], homotopy analysis method [33], the Adomian decomposition method [34], improved $(\grave{G}/G)$–expansion method [35], homotopy perturbation method [36], and variational iteration method [37]. We solve Cahn-Hilliard equation and Gardner equation by RPSM.

The RPSM was first devised in 2013 by the Jordanian mathematician Omar Abu Arqub as an efficient method for determining values of coefficients of the power series solution for first and the second-order fuzzy differential equations [40]. The RPSM is an effective and easy to construct power series solution for strongly linear and nonlinear equations without linearization, perturbation, or discretization. In the last few years, the RPSM has been applied to solve a growing number of nonlinear ordinary and PDEs of different types, classifications, and orders. It has been successfully applied in the numerical solution of the generalized Lane-Emden equation [41], which is a highly nonlinear singular differential equation, in the numerical solution of higher-order regular differential equations [42], in approximate solution of the nonlinear fractional KdV-Burgers equation [43], in construct and predict the solitary pattern solutions for nonlinear time-fractional dispersive PDEs [44], and in predicting and representing the multiplicity of solutions to boundary value problems of fractional order [45]. The RPSM distinguishes itself from various other analytical and numerical methods in several important aspects [46]. Firstly, the RPSM does not need to compare the coefficients of the corresponding terms and a recursion relation is not required. Secondly, the RPSM provides a simple way to ensure the convergence of the series solution by minimizing the related residual error. Thirdly, the RPSM is not affected by computational rounding errors and does not require large computer memory and time. Fourthly, the RPSM does not require any converting while switching from the low-order to the higher-order and from simple linearity to complex nonlinearity; as a result, the method can be applied directly to the given problem by choosing an appropriate initial guess approximation.

2. Fundamental Concepts

Definition 1 (see [43]). The Caputo time-fractional derivatives of order $\alpha > 0$ of $u(x,t)$ is defined as

$$D_t^\alpha u(x,t) = \begin{cases} \dfrac{1}{\Gamma(n-\alpha)} \displaystyle\int_0^t (t-\tau)^{n-\alpha-1} \frac{\partial^n u(x,\tau)}{\partial \tau^n} d\tau, & n-1<\alpha<n, \\ \dfrac{\partial^n u(x,t)}{\partial t^n}, & \alpha = n \in N. \end{cases} \quad (5)$$

Definition 2 (see [47, 48]). A power series representation of the form

$$\sum_{n=0}^{\infty} C_n (t-t_0)^{n\alpha} = C_0 + C_1 (t-t_0)^\alpha + C_2 (t-t_0)^{2\alpha} + \dots \quad (6)$$

where $0 \le n-1 < \alpha \le n, n \in N$ *and* $t \ge t_0$ is called fractional power series about t_0.

Theorem 3 (see [47, 48]). *Suppose that f has a fractional power series representation at t_0 of the form*

$$f(t) = \sum_{n=0}^{\infty} C_n (t-t_0)^{n\alpha}, \quad (7)$$

where $0 \le n-1 < \alpha \le n$ *and* $t_0 \le t < t_0 + R$.

If $D^{n\alpha} f(t)$ are continuous on $(t_0, t_0+R), n = 0,1,2,3,\dots$, then coefficients C_n will take the form

$$C_n = \frac{D^{n\alpha} f(t_0)}{\Gamma(n\alpha+1)}. \quad (8)$$

Definition 4 (see [43]). A power series representation of the form

$$\sum_{n=0}^{\infty} f_n(x)(t-t_0)^{n\alpha} = f_0(x) + f_1(x)(t-t_0)^\alpha + f_2(x)(t-t_0)^{2\alpha} + \dots \quad (9)$$

is called a multiple fractional power series about $t = t_0$.

Theorem 5 (see [43, 44]). *Suppose that $u(x,t)$ has a multiple fractional Power series representation at t_0 of the form*

$$u(x,t) = \sum_{n=0}^{\infty} f_n(x)(t-t_0)^{n\alpha}, \quad (10)$$

where $x \in I$, $0 \le n-1 < \alpha \le n$ *and* $t_0 \le t < t_0 + R$.

If $D_t^{n\alpha} u(x,t)$ are continuous on $I \times (t_0, t_0+R), n = 0,1,2,3,\dots$, then coefficients $f_n(x)$ will take the form

$$f_n(x) = \frac{D_t^{n\alpha} u(x,t_0)}{\Gamma(n\alpha+1)}. \quad (11)$$

Corollary 6 (see [44]). *Suppose that $u(x,y,t)$ has a multiple fractional Power series representation at t_0 of the form*

$$u(x,y,t) = \sum_{n=0}^{\infty} f_n(x,y)(t-t_0)^{n\alpha}, \quad (x,y) \in I_1 \times I_2,\ t_0 \le t < t_0 + R. \quad (12)$$

If $D_t^{n\alpha} u(x,y,t)$ are continuous on $I_1 \times I_2 \times (t_0, t_0 + R), n = 0,1,2,3,\dots$, then $f_n(x,y)$ will take the form

$$f_n(x,y) = \frac{D_t^{n\alpha} u(x,y,t_0)}{\Gamma(n\alpha+1)}. \quad (13)$$

3. Basic Idea of RPSM

To give the approximate solution of nonlinear fractional order differential equations by means of the RPSM, we consider a general nonlinear fractional differential equation:

$$D^\alpha u(x,t) = N(u) + R(u) \quad (14)$$

where $N(u)$ is nonlinear term and $R(u)$ is a linear term. Subject to the initial condition

$$u(x,0) = f(x). \tag{15}$$

The RPSM proposes the solution for (14) as a fractional power series about the initial point $t = 0$,

$$u(x,t) = \sum_{n=0}^{\infty} f_n(x) \frac{t^{n\alpha}}{\Gamma(1+n\alpha)}, \quad 0 < \alpha \le 1,\ -\infty < x < \infty,\ 0 \le t < R. \tag{16}$$

Next we let $u_k(x,t)$ denote the kth truncated series of $u(x,t)$,

$$u_k(x,t) = \sum_{n=0}^{k} f_n(x) \frac{t^{n\alpha}}{\Gamma(1+n\alpha)}. \tag{17}$$

The 0th RPS approximate solution of $u(x,t)$ is

$$u_0(x,t) = u(x,0) = f(x). \tag{18}$$

Equation (17) can be written as

$$u_k(x,t) = f(x) + \sum_{n=1}^{k} f_n(x) \frac{t^{n\alpha}}{\Gamma(1+n\alpha)}, \quad k = 1,2,3,\ldots. \tag{19}$$

We define the residual function for (14)

$$Res_u(x,t) = D_t^{\alpha} u(x,t) - N(u) - R(u). \tag{20}$$

Therefore, the kth residual function $Res_{u,k}$ is

$$Res_{u,k}(x,t) = D_t^{\alpha} u_k(x,t) - N(u_k) - R(u_k). \tag{21}$$

As in [40, 41], $Resu(x,t) = 0$ *and* $\lim_{k\to\infty} Res_k(x,t) = Res(x,t)$.*Therefore*, $D_t^{n\alpha} Res(x,t) = 0$ since the fractional derivative of a constant in the Caputo sense is zero and the fractional derivatives $D_t^{n\alpha}$ of $Res(x,t)$ and $Res_k(x,t)$ are matching at $t = 0$ for each $n = 0,1,2,\ldots,k$.; that is, $D_t^{n\alpha} Res(x,0) = D_t^{n\alpha} Res_k(x,0) = 0$, $n = 0,1,2,\ldots,k$.

To determine $f_1(x), f_2(x), f_3(x), \ldots$ we consider $k = 1,2,3,..$ in (19) and substitute it into (21), applying the fractional derivative $D_t^{(k-1)\alpha}$ in both sides, $k = 1,2,3,\ldots$, and finally we solve

$$D_t^{(k-1)\alpha} Res_{u,k}(x,0) = 0, \quad k = 1,2,3,\ldots. \tag{22}$$

4. Applications

To illustrate the basic idea of RPSM, we consider the following two time-fractional Gardner and Cahn-Hilliard equations.

4.1. Time-Fractional Gardner Equation. Consider the time-fractional homogeneous Gardner equation

$$D_t^{\alpha} u(x,t) + 6\left(u - \varepsilon^2 u^2\right) u_x + u_{xxx} = 0. \tag{23}$$

Subject to the initial Condition

$$u(x,0) = \frac{1}{2} + \frac{1}{2}\tanh\left[\frac{x}{2}\right]. \tag{24}$$

The exact solution when $\varepsilon = 1, \alpha = 1$ is

$$u(x,t) = \frac{1}{2} + \frac{1}{2}\tanh\left[\frac{x-t}{2}\right]. \tag{25}$$

We define the residual function for (23) as

$$Res_u(x,t) = D_t^{\alpha} u(x,t) + 6\left(u - \varepsilon^2 u^2\right) u_x + u_{xxx}, \tag{26}$$

therefore, for the kth residual function $Res_{u,k}(x,t)$,

$$Res_{u,k}(x,t) = D_t^{\alpha} u_k + 6\left(u_k - \varepsilon^2 u_k^2\right) u_{kx} + u_{kxxx}. \tag{27}$$

To determine $f_1(x)$, we consider $(k = 1)$ in (27)

$$Res_{u,1}(x,t) = D_t^{\alpha} u_1 + 6u_1 u_{1x} - 6\varepsilon^2 u_1^2 u_{1x} + u_{1xxx}. \tag{28}$$

But from (19) at $k = 1$,

$$u_1(x,t) = f(x) + f_1(x) \frac{t^{\alpha}}{\Gamma(1+\alpha)}, \tag{29}$$

$$\begin{aligned} Res_{u,1}(x,t) &= f_1 + 6ff_x - 6\varepsilon^2 f_x f^2 + f_{xxx} + \left[6ff_{1x} + 6f_1 f_x - 12\varepsilon^2 f_x f f_1 - 6\varepsilon^2 f_{1x} f^2 + f_{1xxx}\right] \\ &\cdot \frac{t^{\alpha}}{\Gamma(1+\alpha)} + \left[6f_1 f_{1x} - 6\varepsilon^2 f_x f_1^2 - 12\varepsilon^2 f_{1x} f f_1\right] \\ &\cdot \frac{t^{2\alpha}}{\Gamma(1+\alpha)^2} - 6\varepsilon^2 f_{1x} f_1^2 \frac{t^{3\alpha}}{\Gamma(1+\alpha)^3}. \end{aligned} \tag{30}$$

Now depending on the result of (22) In the case of k=1, we have $Res_{u_1}(x,0) = 0$,

$$f_1 = -6ff_x + 6\varepsilon^2 f_x f^2 - f_{xxx}, \tag{31}$$

$$f_1(x) = \frac{1}{8}\operatorname{sech}\left[\frac{x}{2}\right]^4 \left(-1 + \left(-4 + 3\varepsilon^2\right)\cosh[x] + 3\left(-1 + \varepsilon^2\right)\sinh[x]\right). \tag{32}$$

To determine $f_2(x)$, we consider $(k = 2)$ in (27)

$$Res_{u,2}(x,t) = D_t^{\alpha} u_2 + 6u_2 u_{2x} - 6\varepsilon^2 u_2^2 u_{2x} + u_{2xxx}. \tag{33}$$

But from (19) at $k = 2$,

$$u_2(x,t) = f(x) + f_1(x)\frac{t^{\alpha}}{\Gamma(1+\alpha)} + f_2(x) \cdot \frac{t^{2\alpha}}{\Gamma(1+2\alpha)}, \quad (34)$$

$$\begin{aligned} Res_{u,2}(x,t) &= f_1 + 6ff_x - 6\varepsilon^2 f_x f^2 + f_{xxx} + \left[f_2 + 6ff_{1x} + 6f_1 f_x - 12\varepsilon^2 f_x f f_1 - 6\varepsilon^2 f_{1x} f^2 + f_{1xxx}\right] \\ &\cdot \frac{t^{\alpha}}{\Gamma(1+\alpha)} + \left[6ff_{2x} + 6f_2 f_x - 12\varepsilon^2 f_x f f_2 - 6\varepsilon^2 f_{2x} f^2 + f_{2xxx}\right]\frac{t^{2\alpha}}{\Gamma(1+2\alpha)} + \left[6f_1 f_{1x} - 6\varepsilon^2 f_x f_1^2 - 12\varepsilon^2 f_{1x} f f_1\right]\frac{t^{2\alpha}}{\Gamma(1+\alpha)^2} \\ &+ \left[6f_1 f_{2x} + 6f_2 f_{1x} - 12\varepsilon^2 f_x f_1 f_2 - 12\varepsilon^2 f_{1x} f f_2 - 12\varepsilon^2 f_{2x} f f_1\right]\frac{t^{3\alpha}}{\Gamma(1+\alpha)\Gamma(1+2\alpha)} - 6\varepsilon^2 f_{1x} f_1^2 \\ &\cdot \frac{t^{3\alpha}}{\Gamma(1+\alpha)^3} + \left[6f_2 f_{2x} - 6\varepsilon^2 f_x f_2^2 - 12\varepsilon^2 f_{2x} f f_2\right] \cdot \frac{t^{4\alpha}}{\Gamma(1+2\alpha)^2} + \left[-12\varepsilon^2 f_{1x} f_1 f_2 - 6\varepsilon^2 f_{2x} f_1^2\right] \\ &\cdot \frac{t^{4\alpha}}{\Gamma(1+\alpha)^2\Gamma(1+2\alpha)} + \left[-6\varepsilon^2 f_{1x} f_2^2 - 12\varepsilon^2 f_{2x} f_1 f_2\right]\frac{t^{5\alpha}}{\Gamma(1+\alpha)\Gamma(1+2\alpha)^2} - 6\varepsilon^2 f_{2x} f_2^2 \cdot \frac{t^{6\alpha}}{\Gamma(1+2\alpha)^3}. \end{aligned} \quad (35)$$

Applying D_t^{α} on both sides and solving the equation $D_t^{\alpha} Res_{u,2}(x,0) = 0$, then we get

$$f_2 = -6ff_{1x} - 6f_1 f_x + 12\varepsilon^2 f_x f f_1 + 6\varepsilon^2 f_{1x} f^2 - f_{1xxx}, \quad (36)$$

$$\begin{aligned} f_2(x) &= \frac{-1}{64}\operatorname{sech}\left[\frac{x}{2}\right]^7 \left(-24\left(-1+\varepsilon^2\right)\cosh\left[\frac{x}{2}\right] - 6\left(22 - 37\varepsilon^2 + 15\varepsilon^4\right)\cosh\left[\frac{3x}{2}\right] + 24\cosh\left[\frac{5x}{2}\right] \right. \\ &- 42\varepsilon^2\cosh\left[\frac{5x}{2}\right] + 18\varepsilon^4\cosh\left[\frac{5x}{2}\right] + 206\sinh\left[\frac{x}{2}\right] - 204\varepsilon^2\sinh\left[\frac{x}{2}\right] - 129\sinh\left[\frac{3x}{2}\right] + 222\varepsilon^2\sinh\left[\frac{3x}{2}\right] \\ &\left. - 90\varepsilon^4\sinh\left[\frac{3x}{2}\right] + 25\sinh\left[\frac{5x}{2}\right] - 42\varepsilon^2\sinh\left[\frac{5x}{2}\right] + 18\varepsilon^4\sinh\left[\frac{5x}{2}\right]\right). \end{aligned} \quad (37)$$

The solution in series form is given by

$$u(x,t) = f(x) + f_1(x)\frac{t^{\alpha}}{\Gamma(1+\alpha)} + f_2(x)\frac{t^{2\alpha}}{\Gamma(1+2\alpha)} + \dots \quad (38)$$

4.2. Time-Fractional Cahn-Hilliard Equation. Consider the time-fractional Cahn-Hilliard equation

$$D_t^{\alpha} u(x,t) - u_x - 6uu_x^2 - \left(3u^2 - 1\right)u_{xx} + u_{xxxx} = 0. \quad (39)$$

Subject to the initial condition

$$u(x,0) = \tanh\left[\frac{\sqrt{2}}{2}x\right]. \quad (40)$$

The exact solution when $\alpha = 1$ is

$$u(x,t) = \tanh\left[\frac{\sqrt{2}}{2}(x+t)\right]. \quad (41)$$

We define the residual function for (39) as

$$Res_u(x,t) = D_t^{\alpha} u(x,t) - u_x - 6uu_x^2 - \left(3u^2 - 1\right)u_{xx} + u_{xxxx}, \quad (42)$$

therefore, for the kth residual function $Res_{u,k}(x,t)$,

$$Res_{u,k}(x,t) = D_t^{\alpha} u_k - u_{kx} - 6u_k u_{kx}^2 - \left(3u_k^2 - 1\right)u_{kxx} + u_{kxxxx}. \quad (43)$$

To determine $f_1(x)$, we consider $(k = 1)$ in (43)

$$Res_{u,1}(x,t) = D_t^{\alpha} u_1 - u_{1x} - 6u_1 u_{1x}^2 - \left(3u_1^2 - 1\right)u_{1xx} + u_{1xxxx}. \quad (44)$$

From (19) at $k = 1$,

$$u_1(x,t) = f(x) + f_1(x)\frac{t^{\alpha}}{\Gamma(1+\alpha)}, \quad (45)$$

$$\begin{aligned} Res_{u,1}(x,t) &= f_1 - f_x - 6ff_x^2 - 3f^2 f_{xx} + f_{xx} + f_{xxxx} + \left[-f_{1x} - 6f_1 f_x^2 - 12ff_x f_{1x} - 6ff_1 f_{xx} - 3f^2 f_{1xx} + f_{1xx} + f_{1xxxx}\right]\frac{t^{\alpha}}{\Gamma(1+\alpha)} \\ &+ \left[-6ff_{1x}^2 - 12f_1 f_x f_{1x} - 3f_1^2 f_{xx} - 6ff_1 f_{1xx}\right]\frac{t^{2\alpha}}{\Gamma(1+\alpha)^2} + \left[-6f_1 f_{1x}^2 - 3f_1^2 f_{1xx}\right]\frac{t^{3\alpha}}{\Gamma(1+\alpha)^3}. \end{aligned} \quad (46)$$

If we put $Res_{u,1}(x,0) = 0$, then

$$f_1(x) = f_x + 6f(x)f_x^2 + 3f^2 f_{xx} - f_{xx} - f_{xxxx}, \quad (47)$$

$$f_1(x) = \frac{\operatorname{sech}\left[x/\sqrt{2}\right]^2}{\sqrt{2}}. \quad (48)$$

Similarity, to determine $f_2(x)$, we substitute

$$u_2(x,t) = f(x) + f_1(x)\frac{t^\alpha}{\Gamma(1+\alpha)} + f_2(x)\frac{t^{2\alpha}}{\Gamma(1+2\alpha)}, \quad (49)$$

into (43) where $k = 2$,

$$\begin{aligned} Res_{u,2}(x,t) &= \left[f_1 - f_x - 6ff_x^2 - 3f^2 f_{xx} + f_{xx} + f_{xxxx}\right] + \left[f_2 - f_{1x} - 6f_1 f_x^2 - 12ff_x f_{1x} - 6ff_1 f_{xx} - 3f^2 f_{1xx} + f_{1xx} + f_{1xxxx}\right]\frac{t^\alpha}{\Gamma(1+\alpha)} \\ &+ \left[-f_{2x} - 12ff_x f_{2x} - 6f_2 f_x^2 - 6ff_2 f_{xx} - 3f_{2xx}f^2 + f_{2xx} + f_{2xxxx}\right]\frac{t^{2\alpha}}{\Gamma(1+2\alpha)} + \left[-6ff_{1x}^2 - 12f_1 f_x f_{1x} - 3f_1^2 f_{xx} - 6ff_1 f_{1xx}\right]\frac{t^{2\alpha}}{\Gamma(1+\alpha)^2} \\ &+ [-12ff_{1x}f_{2x} - 12f_2 f_x f_{1x} - 6f_1 f_2 f_{xx} - 6ff_2 f_{1xx} - 12f_1 f_x f_{2x} - 6f_{2xx}ff_1] \cdot \frac{t^{3\alpha}}{\Gamma(1+\alpha)\Gamma(1+2\alpha)} + \left[-6f_1 f_{1x}^2 - 3f_1^2 f_{1xx}\right] \cdot \frac{t^{3\alpha}}{\Gamma(1+\alpha)^3} \\ &+ \left[-6ff_{2x}^2 - 12f_2 f_x f_{2x} - 3f_2^2 f_{xx} - 6f_{2xx}ff_2\right]\frac{t^{4\alpha}}{\Gamma(1+2\alpha)^2} + \left[-12f_1 f_{1x}f_{2x} - 6f_2 f_{1x}^2 - 6f_1 f_2 f_{1xx} - 3f_{2xx}f_1^2\right]\frac{t^{4\alpha}}{\Gamma(1+\alpha)^2\Gamma(1+2\alpha)} \\ &+ \left[-6f_1 f_{2x}^2 - 12f_2 f_{1x}f_{2x} - 3f_2^2 f_{1xx} - 6f_{2xx}f_1 f_2\right] \cdot \frac{t^{5\alpha}}{\Gamma(1+\alpha)\Gamma(1+2\alpha)^2} + \left[-6f_2 f_{2x}^2 - 3f_{2xx}f_2^2\right] \cdot \frac{t^{6\alpha}}{\Gamma(1+2\alpha)^3}. \end{aligned} \quad (50)$$

Solving the equation $D_t^\alpha Res_{u,2}(x,0) = 0$, we find that

$$f_2(x) = f_{1x} + 6f_1 f_x^2 + 12ff_x f_{1x} + 6ff_1 f_{xx} + 3f^2 f_{1xx} - f_{1xx} - f_{1xxxx}, \quad (51)$$

$$f_2(x) = -\operatorname{sech}\left[\frac{x}{\sqrt{2}}\right]^2 \tanh\left[\frac{x}{\sqrt{2}}\right]. \quad (52)$$

To determine $f_3(x)$, we substitute

$$u_3(x,t) = f(x) + f_1(x)\frac{t^\alpha}{\Gamma(1+\alpha)} + f_2(x)\frac{t^{2\alpha}}{\Gamma(1+2\alpha)} + f_3(x)\frac{t^{3\alpha}}{\Gamma(1+3\alpha)}, \quad (53)$$

into (43) where k=3,

$$\begin{aligned} Res_{u,3}(x,t) &= f_1 - f_x - 6ff_x^2 - 3f^2 f_{xx} + f_{xx} + f_{xxxx} + \left[f_2 - f_{1x} - 12ff_x f_{1x} - 6f_1 f_x^2 - 6ff_1 f_{xx} - 3f^2 f_{1xx} + f_{1xx} + f_{1xxxx}\right]\frac{t^\alpha}{\Gamma(1+\alpha)} \\ &+ \left[f_3 - f_{2x} - 12ff_x f_{2x} - 6f_2 f_x^2 - 6ff_2 f_{xx} - 3f_{2xx}f^2 + f_{2xx} + f_{2xxxx}\right]\frac{t^{2\alpha}}{\Gamma(1+2\alpha)} + \left[-6ff_{1x}^2 - 12f_1 f_x f_{1x} - 3f_1^2 f_{xx} - 6ff_1 f_{1xx}\right]\frac{t^{2\alpha}}{\Gamma(1+\alpha)^2} \\ &+ \left[-12ff_x f_{3x} - f_{3x} - 6f_3 f_x^2 - 6ff_3 f_{xx} - 3f^2 f_{3xx} + f_{3xx} + f_{3xxxx}\right]\frac{t^{3\alpha}}{\Gamma(1+3\alpha)} \\ &+ [-12ff_{1x}f_{2x} - 12f_1 f_x f_{2x} - 12f_2 f_x f_{1x} - 6f_1 f_2 f_{xx} - 6ff_2 f_{1xx} - 6f_{2xx}ff_1] \cdot \frac{t^{3\alpha}}{\Gamma(1+\alpha)\Gamma(1+2\alpha)} + \left[-6f_1 f_{1x}^2 - 3f_1^2 f_{1xx}\right] \cdot \frac{t^{3\alpha}}{\Gamma(1+\alpha)^3} \\ &+ \left[-12f_1 f_{1x}f_{2x} - 6f_2 f_{1x}^2 - 6f_1 f_2 f_{1xx} - 3f_{2xx}f_1^2\right]\frac{t^{4\alpha}}{\Gamma(1+\alpha)^2\Gamma(1+2\alpha)} + \left[-6ff_{2x}^2 - 12f_2 f_x f_{2x} - 3f_2^2 f_{xx} - 6f_{2xx}ff_2\right] \cdot \frac{t^{4\alpha}}{\Gamma(1+2\alpha)^2} \\ &+ [-12ff_{1x}f_{3x} - 12f_1 f_x f_{3x} - 12f_3 f_x f_{1x} - 6f_1 f_3 f_{xx} - 6ff_3 f_{1xx} - 6ff_1 f_{3xx}] \cdot \frac{t^{4\alpha}}{\Gamma(1+\alpha)\Gamma(1+3\alpha)} + \left[-6f_1 f_{2x}^2 - 12f_2 f_{1x}f_{2x}\right. \end{aligned}$$

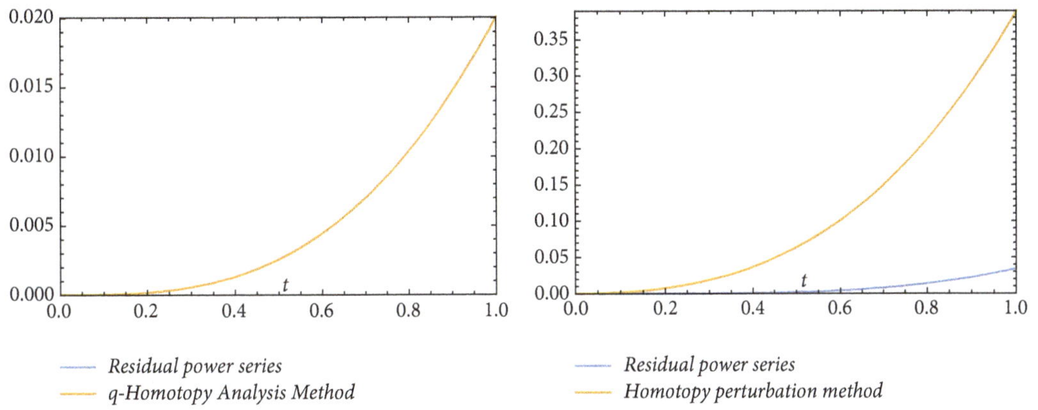

(a) (b)

FIGURE 1: (a) Absolute errors of fractional Gardner equation, at $x = 2, \varepsilon = 1$. (b) Absolute errors of fractional Chan-Hilliard equation at $x = 2$.

$$\begin{aligned}
&- 3f_2^2 f_{1xx} - 6f_{2xx} f_1 f_2\Big] \frac{t^{5\alpha}}{\Gamma(1+\alpha)\Gamma(1+2\alpha)^2} \\
&+ \Big[-12 f f_{2x} f_{3x} - 12 f_2 f_x f_{3x} - 12 f_3 f_x f_{2x} \\
&- 6 f_2 f_3 f_{xx} - 6 f f_3 f_{2xx} - 6 f f_2 f_{3xx}\Big] \\
&\cdot \frac{t^{5\alpha}}{\Gamma(1+2\alpha)\Gamma(1+3\alpha)} + \Big[-12 f_1 f_{1x} f_{3x} - 6 f_3 f_{1x}^2 \\
&- 3 f_1^2 f_{3xx} - 6 f_1 f_3 f_{1xx}\Big] \frac{t^{5\alpha}}{\Gamma(1+\alpha)^2\Gamma(1+3\alpha)} \\
&+ \Big[-12 f_1 f_{2x} f_{3x} - 12 f_2 f_{1x} f_{3x} - 12 f_3 f_{1x} f_{2x} \\
&- 6 f_2 f_3 f_{1xx} - 6 f_1 f_3 f_{2xx} - 6 f_1 f_2 f_{3xx}\Big] \\
&\cdot \frac{t^{6\alpha}}{\Gamma(1+\alpha)\Gamma(1+2\alpha)\Gamma(1+3\alpha)} + \Big[-6 f_2 f_{2x}^2 \\
&- 3 f_{2xx} f_2^2\Big] \frac{t^{6\alpha}}{\Gamma(1+2\alpha)^3} + \Big[-6 f f_{3x}^2 - 12 f_3 f_x f_{3x} \\
&- 3 f_3^2 f_{xx} - 6 f f_3 f_{3xx}\Big] \frac{t^{6\alpha}}{\Gamma(1+3\alpha)^2} + \Big[-6 f_1 f_{3x}^2 \\
&- 12 f_3 f_{1x} f_{3x} - 3 f_3^2 f_{1xx} - 6 f_1 f_3 f_{3xx}\Big] \\
&\cdot \frac{t^{7\alpha}}{\Gamma(1+\alpha)\Gamma(1+3\alpha)^2} + \Big[-12 f_2 f_{2x} f_{3x} - 6 f_3 f_{2x}^2 \\
&- 6 f_2 f_3 f_{2xx} - 3 f_2^2 f_{3xx}\Big] \frac{t^{7\alpha}}{\Gamma(1+3\alpha)\Gamma(1+2\alpha)^2} \\
&+ \Big[-6 f_2 f_{3x}^2 - 12 f_3 f_{2x} f_{3x} - 3 f_3^2 f_{2xx} - 6 f_{3xx} f_3 f_2\Big] \\
&\cdot \frac{t^{8\alpha}}{\Gamma(1+2\alpha)\Gamma(1+3\alpha)^2} + \Big[-6 f_3 f_{3x}^2 - 3 f_3^2 f_{3xx}\Big] \\
&\cdot \frac{t^{9\alpha}}{\Gamma(1+3\alpha)^3}.
\end{aligned} \tag{54}$$

Applying $D_t^{2\alpha}$ on both sides and then solving the equation $D_t^{2\alpha} Res_{u,3}(x, 0) = 0$, we get

$$\begin{aligned}
f_3(x) &= \Big[f_{2x} + 12 f f_x f_{2x} + 6 f_2 f_x^2 + 6 f f_2 f_{xx} \\
&+ 3 f_{2xx} f^2 - f_{2xx} - f_{2xxxx}\Big] + \Big[6 f f_{1x}^2 + 12 f_1 f_x f_{1x} \\
&+ 3 f_1^2 f_{xx} + 6 f f_1 f_{1xx}\Big] \frac{\Gamma(1+2\alpha)}{\Gamma(1+\alpha)^2},
\end{aligned} \tag{55}$$

$$\begin{aligned}
f_3(x) &= \frac{1}{8} \operatorname{sech}\left[\frac{x}{\sqrt{2}}\right]^6 \Big(-4\sqrt{2} \\
&+ \left(264 - 96\cosh\left[\sqrt{2}x\right] + \sqrt{2}\sinh\left[2\sqrt{2}x\right]\right) \\
&\cdot \tanh\left[\frac{x}{\sqrt{2}}\right]\Big) + \left(\frac{-21}{2}\operatorname{sech}\left[\frac{x}{\sqrt{2}}\right]^6 \tanh\left[\frac{x}{\sqrt{2}}\right]\right. \\
&\left. + 12\operatorname{sech}\left[\frac{x}{\sqrt{2}}\right]^4 \tanh\left[\frac{x}{\sqrt{2}}\right]^3\right) \frac{\Gamma(1+2\alpha)}{\Gamma(1+\alpha)^2}.
\end{aligned} \tag{56}$$

The solution in series form is given by

$$\begin{aligned}
u(x,t) &= f(x) + f_1(x)\frac{t^{\alpha}}{\Gamma(1+\alpha)} + f_2(x)\frac{t^{2\alpha}}{\Gamma(1+2\alpha)} \\
&+ f_3(x)\frac{t^{3\alpha}}{\Gamma(1+3\alpha)} + \ldots
\end{aligned} \tag{57}$$

5. Numerical Results

This section deals with the approximate analytical solutions obtained by RPSM for Gardner and Cahn-Hilliard equations. In classical case($\alpha \longrightarrow 1$), Figure 1 and Tables 1 and 2 describe the comparison between RPSM with q-HAM [28] and HPM [36]. In fractional case, Figures 2, 3, and 4 describe the geometrical behavior of the solutions obtained by RPSM for different fractional value α of the two equations.

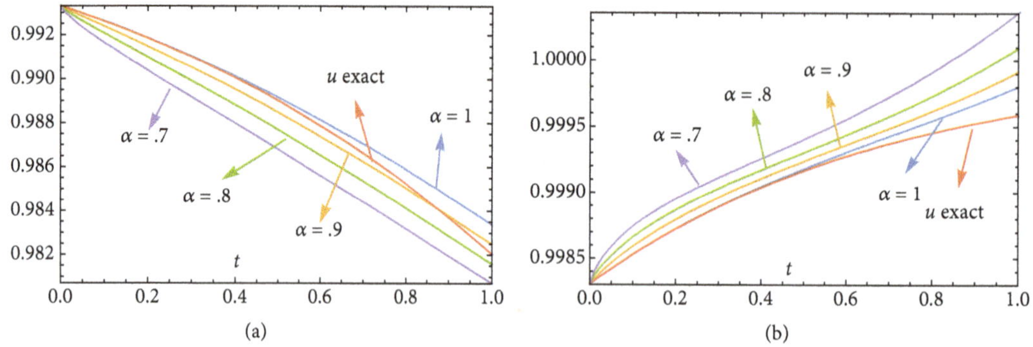

FIGURE 2: (a) Fractional Gardner equation at $x = 5, \varepsilon = 1$. (b) Fractional Chan-Hilliard equation at $x = 5$.

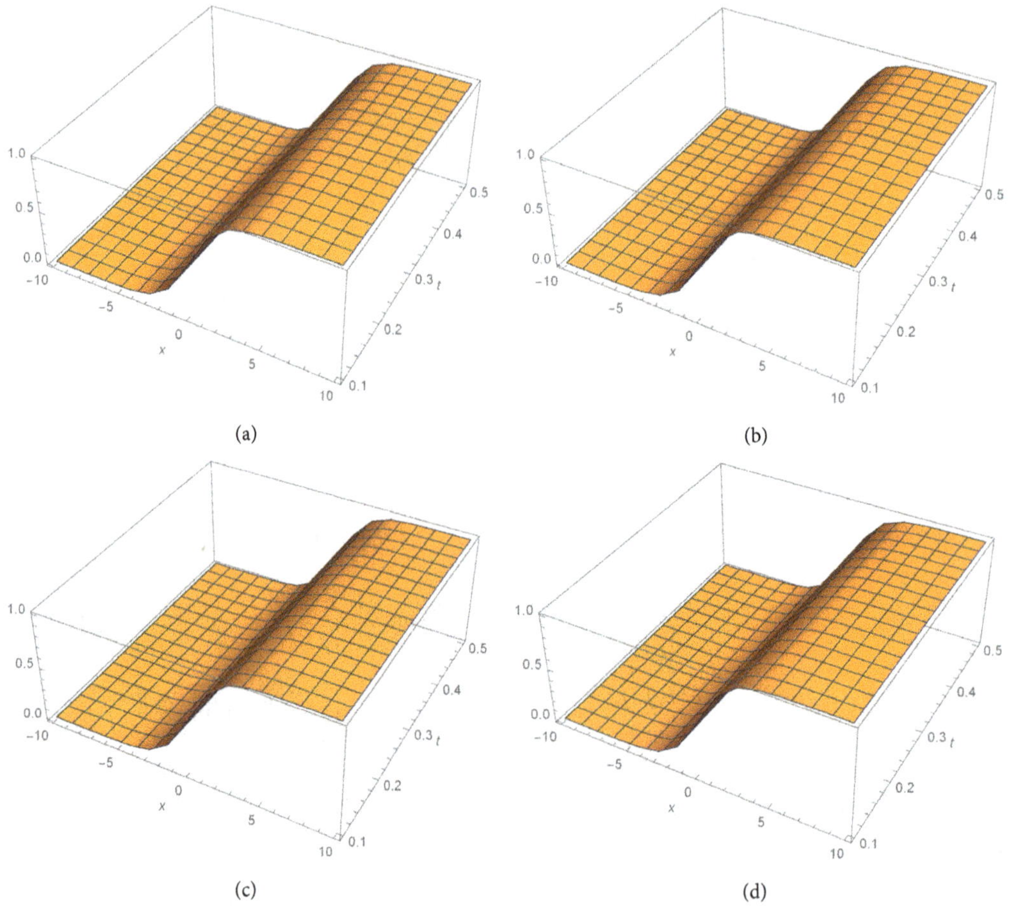

FIGURE 3: The approximate solution for fractional Gardner equation at $\varepsilon = 1$: (a) $\alpha = 1$, (b) Exact solution, (c) $\alpha = .99$, and (d) $\alpha = .95$.

6. Conclusions

This work has used the RPSM for finding the solution of the time-fractional Gardner and Cahn-Hilliard equations. A very good agreement between the results obtained by the RPSM and q-HAM [28] was observed in Figure 1(a) and Table 1. Figure 1(b) and Table 2 indicate that the mentioned method achieves a higher level of accuracy than HPM [36]. Consequently, the work emphasized that the method introduces a significant improvement in this field over existing techniques.

Data Availability

[1] The [approximate solution obtained by q-homotopy analysis method] data used to support the findings of this study have been deposited in the [article] repository ([doi.org/10.1016/j.asej.2014.03.014]) [28]. [2] The [approximate solution obtained by homotopy perturbation method] data used to support the findings of this study have been deposited in the [article] repository ([doi.org/10.1080/10288457.2013.867627]) [36].

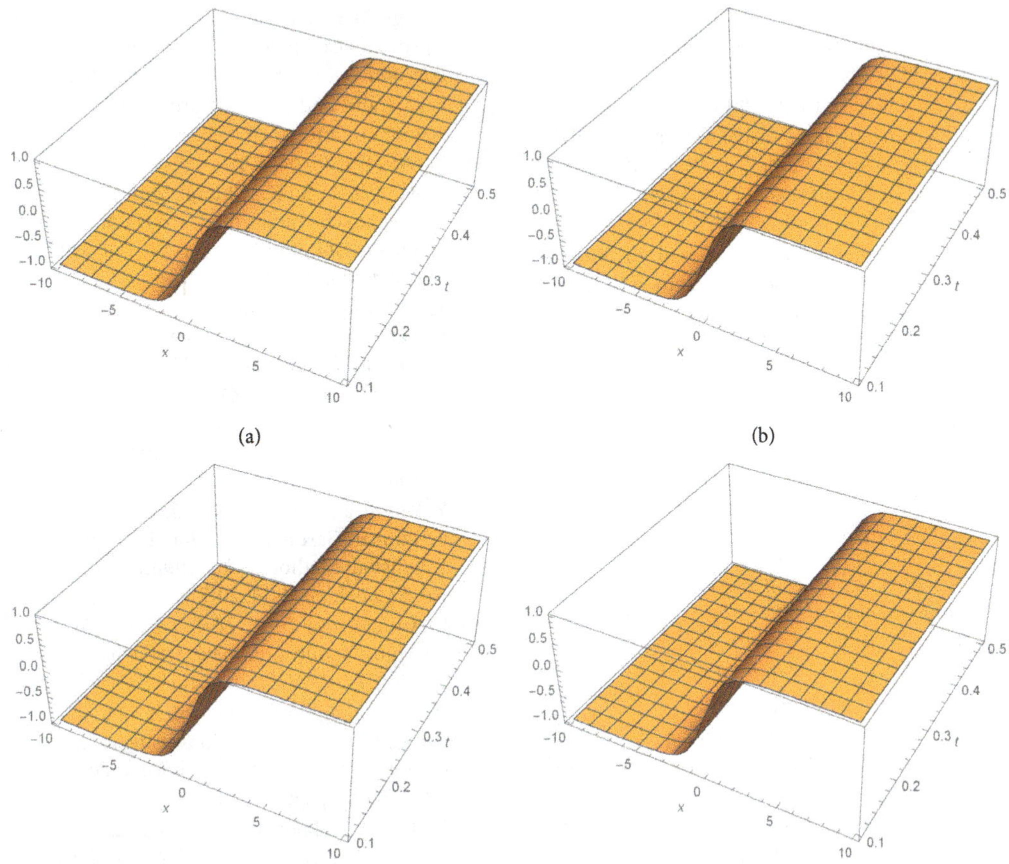

(a) (b) (c) (d)

FIGURE 4: The approximate solution for fractional Chan-Hilliard equation: (a) $\alpha = 1$, (b) Exact solution, (c) $\alpha = .99$, and (d) $\alpha = .95$.

TABLE 1: The absolute errors $|u_{exact} - u_3|$ for Gardner equation when $t = .2, \varepsilon = 1, \alpha \longrightarrow 1$.

x	$\lvert u_{exact} - u_{RPS}\rvert$	$\lvert u_{exact} - u_{qHAM}\rvert$
.1	166.002×10^{-6}	166.002×10^{-6}
.2	162.707×10^{-6}	162.707×10^{-6}
.3	156.257×10^{-6}	156.257×10^{-6}
.4	146.917×10^{-6}	146.917×10^{-6}
.5	135.064×10^{-6}	135.064×10^{-6}

TABLE 2: The absolute errors $|u_{exact} - u_4|$ for Cahn-Hilliard when $t = .2, \alpha \longrightarrow 1$.

x	$\lvert u_{exact} - u_{RPS}\rvert$	$\lvert u_{exact} - u_{HPM}\rvert$
.1	25.5541×10^{-6}	4.68338×10^{-3}
.2	41.5291×10^{-6}	7.28902×10^{-3}
.3	54.2246×10^{-6}	9.6162×10^{-3}
.4	62.8898×10^{-6}	11.5931×10^{-3}
.5	67.2637×10^{-6}	13.174×10^{-3}

Conflicts of Interest

The authors declare that they have no conflicts of interest.

References

[1] L. Debnath, "Recent applications of fractional calculus to science and engineering," *International Journal of Mathematics and Mathematical Sciences*, no. 54, pp. 3413–3442, 2003.

[2] M. Rahimy, "Applications of fractional differential equations," *Applied Mathematical Sciences*, vol. 4, no. 49-52, pp. 2453–2461, 2010.

[3] A. A. Arafa and S. Z. Rida, "Numerical solutions for some generalized coupled nonlinear evolution equations," *Mathematical and Computer Modelling*, vol. 56, no. 11-12, pp. 268–277, 2012.

[4] A. A. Arafa, "Series solutions of time-fractional host-parasitoid systems," *Journal of Statistical Physics*, vol. 145, no. 5, pp. 1357–1367, 2011.

[5] N. H. Sweilam, M. M. Khader, and R. F. Al-Bar, "Numerical studies for a multi-order fractional differential equation," *Physics Letters A*, vol. 371, no. 1-2, pp. 26–33, 2007.

[6] A. Golbabai and K. Sayevand, "Fractional calculus- a new approach to the analysis of generalized fourth-order diffusion-wave equations," *Computers & Mathematics with Applications. An International Journal*, vol. 61, no. 8, pp. 2227–2231, 2011.

[7] K. A. Gepreel, "The homotopy perturbation method applied to the nonlinear fractional Kolmogorov-Petrovskii-PISkunov equations," *Applied Mathematics Letters*, vol. 24, no. 8, pp. 1428–1434, 2011.

[8] A. A. Arafa, S. Z. Rida, and H. Mohamed, "Approximate analytical solutions of Schnakenberg systems by homotopy

analysis method," *Applied Mathematical Modelling: Simulation and Computation for Engineering and Environmental Systems*, vol. 36, no. 10, pp. 4789–4796, 2012.

[9] A. A. Arafa, S. Z. Rida, and M. Khalil, "The effect of anti-viral drug treatment of human immunodeficiency virus type 1 (HIV-1) described by a fractional order model," *Applied Mathematical Modelling: Simulation and Computation for Engineering and Environmental Systems*, vol. 37, no. 4, pp. 2189–2196, 2013.

[10] M. Javidi, "A numerical solution of the generalized Burgers-Huxley equation by spectral collocation method," *Applied Mathematics and Computation*, vol. 178, no. 2, pp. 338–344, 2006.

[11] A.-M. Wazwaz, "Analytic study on Burgers, Fisher, Huxley equations and combined forms of these equations," *Applied Mathematics and Computation*, vol. 195, no. 2, pp. 754–761, 2008.

[12] X.-W. Zhou, "Exp-function method for solving Huxley equation," *Mathematical Problems in Engineering*, Art. ID 538489, 7 pages, 2008.

[13] A. A. Arafa, S. Z. Rida, A. A. Mohammadein, and H. M. Ali, "Solving nonlinear fractional differential equation by generalized Mittag-Leffler function method," *Communications in Theoretical Physics*, vol. 59, no. 6, pp. 661–663, 2013.

[14] M. Sari and G. Gürarslan, "Numerical solutions of the generalized Burgers-Huxley equation by a differential quadrature method," *Mathematical Problems in Engineering*, Art. ID 370765, 11 pages, 2009.

[15] O. Abu Arqub, "An iterative method for solving fourth-order boundary value problems of mixed type integro-differential equations," *Journal of Computational Analysis and Applications*, vol. 18, no. 5, pp. 857–874, 2015.

[16] O. Abu Arqub, "Fitted reproducing kernel Hilbert space method for the solutions of some certain classes of time-fractional partial differential equations subject to initial and Neumann boundary conditions," *Computers & Mathematics with Applications*, vol. 73, no. 6, pp. 1243–1261, 2017.

[17] A.-M. Wazwaz, "Solitons and singular solitons for the Gardner-KP equation," *Applied Mathematics and Computation*, vol. 204, no. 1, pp. 162–169, 2008.

[18] C. S. Gardner, J. M. Greene, M. D. Kruskal, and R. M. Miura, "Method for solving the Korteweg-deVries equation," *Physical Review Letters*, vol. 19, no. 19, pp. 1095–1097, 1967.

[19] A.-M. Wazwaz, "New solitons and kink solutions for the Gardner equation," *Communications in Nonlinear Science and Numerical Simulation*, vol. 12, no. 8, pp. 1395–1404, 2007.

[20] Z. Fu, S. Liu, and S. Liu, "New kinds of solutions to Gardner equation," *Chaos, Solitons & Fractals*, vol. 20, no. 2, pp. 301–309, 2004.

[21] G.-q. Xu, Z.-b. Li, and Y.-p. Liu, "Exact solutions to a large class of nonlinear evolution equations," *Chinese Journal of Physics*, vol. 41, no. 3, pp. 232–241, 2003.

[22] D. Baldwin, Ü. Göktas, W. Hereman, L. Hong, R. S. Martino, and J. C. Miller, "Symbolic computation of exact solutions expressible in hyperbolic and elliptic functions for nonlinear PDEs," *Journal of Symbolic Computation*, vol. 37, no. 6, pp. 669–705, 2004.

[23] W. Hereman and A. Nuseir, "Symbolic methods to construct exact solutions of nonlinear partial differential equations," *Mathematics and Computers in Simulation*, vol. 43, no. 1, pp. 13–27, 1997.

[24] Y. Ugurlu and D. g. Kaya, "Solutions of the Cahn-Hilliard equation," *Computers & Mathematics with Applications. An International Journal*, vol. 56, no. 12, pp. 3038–3045, 2008.

[25] J. W. Cahn and J. E. Hilliard, "Free energy of a nonuniform system. I. Interfacial free energy," *The Journal of Chemical Physics*, vol. 28, no. 2, pp. 258–267, 1958.

[26] S. M. Choo, S. K. Chung, and Y. J. Lee, "A conservative difference scheme for the viscous Cahn-Hilliard equation with a nonconstant gradient energy coefficient," *Applied Numerical Mathematics*, vol. 51, no. 2-3, pp. 207–219, 2004.

[27] M. E. Gurtin, "Generalized Ginzburg-Landau and Cahn-Hilliard equations based on a microforce balance," *Physica D: Nonlinear Phenomena*, vol. 92, no. 3-4, pp. 178–192, 1996.

[28] O. S. Iyiola and O. G. Olayinka, "Analytical solutions of time-fractional models for homogeneous Gardner equation and non-homogeneous differential equations," *Ain Shams Engineering Journal*, vol. 5, no. 3, pp. 999–1004, 2014.

[29] Y. Pandir and H. H. Duzgun, "New Exact Solutions of Time Fractional Gardner Equation by Using New Version of F-Expansion Method," *Communications in Theoretical Physics*, vol. 67, no. 1, pp. 9–14, 2017.

[30] J. Ahmad and S. T. Mohyud-Din, "An efficient algorithm for some highly nonlinear fractional PDEs in mathematical physics," *PLoS ONE*, vol. 9, no. 12, Article ID e109127, 2014.

[31] W. Li, H. Yang, and B. He, "Exact solutions of fractional Burgers and Cahn-Hilliard equations using extended fractional Riccati expansion method," *Mathematical Problems in Engineering*, Art. ID 104069, 9 pages, 2014.

[32] H. Jafari, H. Tajadodi, N. Kadkhoda, and D. Baleanu, "Fractional subequation method for Cahn-Hilliard and Klein-Gordon equations," *Abstract and Applied Analysis*, vol. 2013, Article ID 587179, 5 pages, 2013.

[33] M. S. Mohamed and K. S. Mekheimer, "Analytical approximate solution for nonlinear space-time fractional Cahn-Hilliard equation," *International Electronic Journal of Pure and Applied Mathematics*, vol. 7, no. 4, pp. 145–159, 2014.

[34] Z. Dahmani and M. Benbachir, "Solutions of the Cahn-Hilliard equation with time- and space-fractional derivatives," *International Journal of Nonlinear Science*, vol. 8, no. 1, pp. 19–26, 2009.

[35] D. Baleanu, Y. Ugurlu, M. Inc, and B. Kilic, "Improved (G ' / G) -Expansion Method for the Time-Fractional Biological Population Model and Cahn-Hilliard Equation," *Journal of Computational and Nonlinear Dynamics*, vol. 10, no. 5, p. 051016, 2015.

[36] A. Bouhassoun and M. Hamdi Cherif, "Homotopy Perturbation Method For Solving The Fractional Cahn-Hilliard Equation," *Journal of Interdisciplinary Mathematics*, vol. 18, no. 5, pp. 513–524, 2015.

[37] J. Ahmad and S. T. Mohyud-Din, "An efficient algorithm for nonlinear fractional partial differential equations," *Proceedings of the Pakistan Academy of Sciences*, vol. 52, no. 4, pp. 381–388, 2015.

[38] J. Manafian and M. Lakestani, "A new analytical approach to solve some of the fractional-order partial differential equations," *Indian Journal of Physics*, vol. 91, no. 3, pp. 243–258, 2017.

[39] S. Tuluce Demiray, Y. Pandir, and H. Bulut, "Generalized Kudryashov method for time-fractional differential equations," *Abstract and Applied Analysis*, Art. ID 901540, 13 pages, 2014.

[40] O. Abu Arqub, "Series solution of fuzzy differential equations under strongly generalized differentiability," *Journal of Advanced Research in Applied Mathematics*, vol. 5, no. 1, pp. 31–52, 2013.

[41] O. Abu Arqub, A. El-Ajou, A. S. Bataineh, and I. Hashim, "A representation of the exact solution of generalized Lane-Emden equations using a new analytical method," *Abstract and Applied Analysis*, Art. ID 378593, 10 pages, 2013.

[42] O. Abu Arqub, Z. Abo-Hammour, R. Al-Badarneh, and S. Momani, "A reliable analytical method for solving higher-order initial value problems," *Discrete Dynamics in Nature and Society*, vol. 2013, Article ID 673829, 12 pages, 2013.

[43] A. El-Ajou, O. Abu Arqub, and S. Momani, "Approximate analytical solution of the nonlinear fractional KdV-Burgers equation: a new iterative algorithm," *Journal of Computational Physics*, vol. 293, pp. 81–95, 2015.

[44] O. A. Arqub, A. El-Ajou, and S. Momani, "Constructing and predicting solitary pattern solutions for nonlinear time-fractional dispersive partial differential equations," *Journal of Computational Physics*, vol. 293, pp. 385–399, 2015.

[45] O. A. Arqub, A. El-Ajou, Z. A. Zhour, and S. Momani, "Multiple solutions of nonlinear boundary value problems of fractional order: A new analytic iterative technique," *Entropy*, vol. 16, no. 1, pp. 471–493, 2014.

[46] A. El-Ajou, O. Abu Arqub, S. Momani, D. Baleanu, and A. Alsaedi, "A novel expansion iterative method for solving linear partial differential equations of fractional order," *Applied Mathematics and Computation*, vol. 257, pp. 119–133, 2015.

[47] A. El-Ajou, O. Abu Arqub, Z. Al Zhour, and S. Momani, "New results on fractional power series: theories and applications," *Entropy. An International and Interdisciplinary Journal of Entropy and Information Studies*, vol. 15, no. 12, pp. 5305–5323, 2013.

[48] A. El-Ajou, O. Abu Arqub, and M. Al-Smadi, "A general form of the generalized Taylor's formula with some applications," *Applied Mathematics and Computation*, vol. 256, pp. 851–859, 2015.

8

Linear Analysis of an Integro-Differential Delay Equation Model

Anael Verdugo[1,2]

[1]Department of Mathematics, California State University, Fullerton, CA, USA
[2]Center for Computational and Applied Mathematics, California State University, Fullerton, CA, USA

Correspondence should be addressed to Anael Verdugo; averdugo@fullerton.edu

Academic Editor: Elena Braverman

This paper presents a computational study of the stability of the steady state solutions of a biological model with negative feedback and time delay. The motivation behind the construction of our system comes from biological gene networks and the model takes the form of an integro-delay differential equation (IDDE) coupled to a partial differential equation. Linear analysis shows the existence of a critical delay where the stable steady state becomes unstable. Closed form expressions for the critical delay and associated frequency are found and confirmed by approximating the IDDE model with a system of N delay differential equations (DDEs) coupled to N ordinary differential equations. An example is then given that shows how the critical delay for the DDE system approaches the results for the IDDE model as N becomes large.

1. Introduction

New genetic experiments [1–3] and mathematical approaches [4–6] have been developed to help us better understand how genes interact within a cell. Theoretically, the structure of these interactions or networks are represented by the various chemical reactions happening at a certain time. If the reactions under consideration only involve a few genes, then their dynamic behavior could be understood intuitively and, most likely, confirmed with a biochemical experiment [2, 3]. However, if the system is formed of dozens of reactions, then developing an intuitive understanding of the system's dynamics would be difficult. Nevertheless, current research in the computational sciences [7, 8] shows that the study of these large gene networks is an important step which will help us unravel some of the mysteries in the field of cell biology [5, 6].

An important and popular modeling technique in the applied sciences is based on differential equations in all its various forms: linear [9], nonlinear [6, 10], partial [11], stochastic [12, 13], and delayed [3, 6, 14]. In this study we focus our attention to a differential equation model with constant delay, where the delay arises naturally as the time lag associated with various intracellular processes, like movement within the cell, synthesis of proteins, and transcription of DNA, among many others. The model that motivated this work was studied previously by [4–6] and is given by the following set of delay differential equations (DDEs):

$$\frac{dm}{dt} = -\mu_m m(t) + H(p(t-T)),$$
$$\frac{dp}{dt} = m(t) - \mu_p p(t), \tag{1}$$

where the time dependent variables are the mRNA concentration, $m(t)$, and its associated protein concentration, $p(t)$, and the constants μ_m and μ_p are the decay rates of the mRNA and protein molecules, respectively. The function $H(p(t-T))$ is generally a Hill equation representing the rate of *delayed* production of mRNA, where the delay, T, is assumed to be a positive constant. The associated biochemical representation of the system is given in Figure 1(a) and the biological context is the following: a gene is copied onto mRNA in the nucleus, which is then translated into a protein in the cytoplasm of the cell. The protein then returns to the nucleus and acts as a negative feedback regulator by repressing production of mRNA (see [4–6] for more biological background).

In this paper, we analyze the steady state stability of a model motivated by (1) and previously studied by the author in [15]. The model is given by an integro-delay differential equation (IDDE) coupled to a partial differential equation

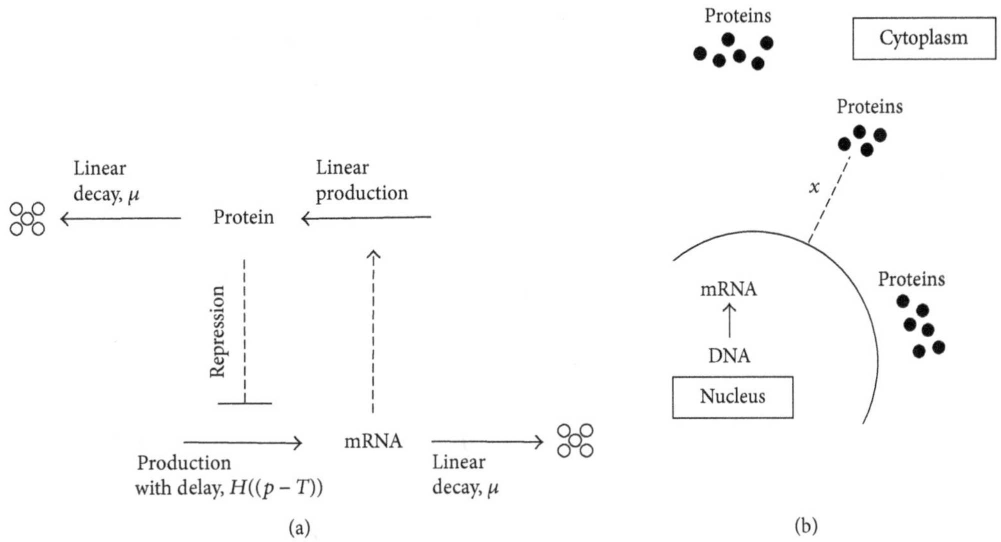

FIGURE 1: (a) Biological circuit diagram of (1). Protein production, protein decay, and mRNA decay are assumed to be linear processes. Production of mRNA is considered as a process affected by a delayed response of protein repression. Here the arrow (↑) represents activation and the perpendicular symbol (⊥) represents repression. Solid and dashed lines represent direct chemical reactions and indirect regulatory signals, respectively. Five small circles represent degradation byproducts. (b) Spatial distribution of protein production in the cytoplasm. Protein synthesis happens at various locations from the nucleus. The distance from the nucleus is represented here by the variable x, where $0 \le x \le 1$.

(PDE) and is characterized by an exponential "weighting" function that regulates the net *repression* effect on mRNA based on protein synthesis location. The model is given by

$$\frac{dm}{dt} = -\mu m + \int_0^1 e^{-|x-\bar{x}|} H(p(\bar{x}, t-T))\, d\bar{x}, \qquad (2)$$

$$\frac{dp}{dt} = m - \mu p, \qquad (3)$$

where $m = m(x,t)$, $p = p(x,t)$, and $e^{-|x-\bar{x}|}$ is a weighting function that accounts for a "stronger" mRNA repression for proteins being synthesized closer to the nucleus than more distant ones. The latter is due to the spatial distribution of protein production within the cytoplasm, which occurs after mRNA exits the nucleus and diffuses into the cytoplasm where it is caught, read, and translated into a protein. The exact location from the nucleus where this process occurs is arbitrary and here we quantify it with a *distance* variable $0 \le x \le 1$ as explained in Figure 1(b). The latter yields the integral term in (2) which represents the total sum of the repression effect that newly synthesized proteins have on mRNA.

The current work extends our previous study [15] in two different ways. First, the biological setup explained above sets our model (2)-(3) on firmer modeling grounds from our previous study [15]. Here we assume the variable x is "distance" from the nucleus, as opposed to x being a variable that represents gene sites in the DNA as argued in [15]. This is a crucial difference that yields a better understanding of our computational results. Second, the results from the analysis of the steady state and its associated stability are now confirmed via MATLAB's *dde23.m*, which provides more accurate approximations and numerical simulations for the associated $2N$-dimensional system. The latter was not accomplished in [15] and thus presented here for the first time.

The rest of the paper is organized as follows. In Section 2, we present the associated linear stability analysis of (2)-(3) and characterize the steady state solutions. Linear analysis reveals the existence of a critical delay where the stable steady state becomes unstable and thus closed form expressions for the critical delay, T_{cr}, and associated frequency ω are found. In Section 3, we construct a system of N DDEs coupled to N ordinary differential equations (ODEs) and use these to confirm the results obtained in Section 2. A numerical example is then given in Section 4, which shows how the critical delay for the DDE system approaches the results for the IDDE model as N becomes large. In Section 5, we discuss our findings and conclusions.

2. Linear Stability Analysis

In this section, we consider the steady state behavior of (2) and (3). Setting $dm/dt = dp/dt = 0$, we see from (3) that at steady state $m^* = \mu p^*$, where $(m^*(x), p^*(x))$ represents the steady state solution. Substituting the latter into (2) gives

$$\mu^2 p^*(x) = \int_0^1 e^{-|x-\bar{x}|} H(p^*(\bar{x}))\, d\bar{x}. \qquad (4)$$

Splitting the integration limits

$$\mu^2 p^*(x) = e^{-x} \int_0^x e^{\bar{x}} H(p^*(\bar{x}))\, d\bar{x} + e^{x} \int_x^1 e^{-\bar{x}} H(p^*(\bar{x}))\, d\bar{x}, \qquad (5)$$

and differentiating twice, we obtain an equivalent second-order two-point boundary value problem (BVP) for the equilibrium solution $p^* = p^*(x)$

$$\frac{d^2 p^*}{dx^2} = p^* - \frac{2}{\mu^2} H(p^*), \tag{6}$$

where the boundary conditions (BCs) are given by

$$\begin{aligned} p^*(0) &= \frac{1}{\mu^2} \int_0^1 e^{-\overline{x}} H(p^*(\overline{x}))\, d\overline{x}, \\ p^*(1) &= \frac{1}{e\mu^2} \int_0^1 e^{\overline{x}} H(p^*(\overline{x}))\, d\overline{x}. \end{aligned} \tag{7}$$

The BVP (6)–(7) has a unique solution as long as the right hand side (RHS) of (6) has bounded, positive, and continuous partial derivatives with respect to p^*. For the rest of this work, we let

$$H(p(x,t)) = 1 - p(x,t), \tag{8}$$

which allows mathematical tractability for the stability analysis presented below. Notice that since (8) satisfies all three aforementioned BVP conditions, then we are guaranteed the existence of a unique solution, which can be approximated using a numerical technique for BVPs, such as finite differences or a shooting method. See Section 4 for an example.

To study the stability of the steady state solution $(m^*(x), p^*(x))$, we set $p(x,t) = p^*(x) + \eta(x,t)$ and $m(x,t) = m^*(x) + \xi(x,t)$, substitute these into (2)-(3), and linearize the resulting equations in $\eta(x,t)$ and $\xi(x,t)$ to obtain

$$\frac{d\xi}{dt} = -\mu\xi - \int_0^1 e^{-|x-\overline{x}|} \eta_d(\overline{x})\, d\overline{x}, \tag{9}$$

$$\frac{d\eta}{dt} = \xi - \mu\eta. \tag{10}$$

Setting $\xi(x,t) = \phi(x)e^{\lambda t}$ and $\eta(x,t) = \psi(x)e^{\lambda t}$ gives

$$\begin{aligned} -e^{\lambda T}(\lambda+\mu)\phi(x) &= \int_0^1 e^{-|x-\overline{x}|}\psi(\overline{x})\, d\overline{x}, \\ (\lambda+\mu)\psi(x) &= \phi(x), \end{aligned} \tag{11}$$

which yields

$$r\psi(x) = \int_0^1 e^{-|x-\overline{x}|}\psi(\overline{x})\, d\overline{x}, \tag{12}$$

where the RHS has a *symmetric* integral kernel, $\psi(x)$ is an eigenfunction, and r is the associated eigenvalue given by

$$r = -e^{\lambda T}(\lambda+\mu)^2. \tag{13}$$

Since (12) is a self-adjoint operator of the form

$$L(\cdot) = \int_0^1 K(x,\overline{x})(\cdot)\, d\overline{x}, \tag{14}$$

then the eigenvalue problem (12) has real eigenvalues $r \in \mathbb{R}$. To compute r, we transform the integral equation (12) to the following equivalent second-order BVP:

$$\frac{d^2\psi}{dx^2} + \rho\psi = 0, \tag{15}$$

with solutions

$$\psi(x) = c_1 \sin(\rho x) + c_2 \cos(\rho x), \tag{16}$$

where c_1 and c_2 are constants and $\rho = \sqrt{2/r - 1}$. The endpoint BCs are obtained from (12) as follows:

$$\begin{aligned} \psi(0) &= \frac{\rho^2+1}{2} \int_0^1 e^{-\overline{x}}\psi(\overline{x})\, d\overline{x}, \\ \psi(1) &= \frac{\rho^2+1}{2e} \int_0^1 e^{\overline{x}}\psi(\overline{x})\, d\overline{x}. \end{aligned} \tag{17}$$

Substituting the solution (16) into the BCs (17) gives the system of equations

$$\begin{bmatrix} \rho\sin\rho - \cos\rho - e & -\sin\rho - \rho\cos\rho + e\rho \\ e\rho\sin\rho - e\cos\rho - 1 & -e\sin\rho - e\rho\cos\rho + \rho \end{bmatrix} \begin{bmatrix} c_1 \\ c_2 \end{bmatrix} = 0, \tag{18}$$

which yields the condition on ρ for nontrivial solutions

$$(\rho^2 - 1)\sin\rho - 2\rho\cos\rho = 0. \tag{19}$$

Using a numerical root finding technique on (19), we obtain $\rho = 1.30654, 3.67319, 6.58462, \ldots$ which gives the corresponding values for $r = 2/(1+\rho^2) = 0.73881, 0.13800, 0.04509, \ldots$. Thus to determine λ from r we have two cases:

(i) For $T = 0$, (13) gives $\lambda = -\mu \pm \sqrt{-r}$ and since $r = 2/(1+\rho^2) > 0$ then $\mathrm{Re}(\lambda) = -\mu < 0$ for $\mu > 0$. The latter shows that the equilibrium solution is stable when there is no delay.

(ii) For $T = T_{cr}$ and $\lambda = i\omega$, (13) becomes $r = -e^{i\omega T_{cr}}(i\omega + \mu)^2$ which gives the two real equations

$$\begin{aligned} r &= 2\mu\omega\sin(\omega T_{cr}) + (\omega^2 - \mu^2)\cos(\omega T_{cr}), \\ 0 &= (\omega^2 - \mu^2)\sin(\omega T_{cr}) - 2\mu\omega\cos(\omega T_{cr}). \end{aligned} \tag{20}$$

Solving (20) for $\sin(\omega T_{cr})$ and $\cos(\omega T_{cr})$, and using the identity $\sin^2(\omega T_{cr}) + \cos^2(\omega T_{cr}) = 1$, we obtain

$$\omega = \sqrt{r - \mu^2}. \tag{21}$$

Dividing the expressions for $\sin(\omega T_{cr})$ and $\cos(\omega T_{cr})$ and solving for T_{cr}, we obtain

$$T_{cr} = \frac{1}{\sqrt{r-\mu^2}} \arctan\left(\frac{2\mu\sqrt{r-\mu^2}}{r - 2\mu^2}\right), \tag{22}$$

which gives the value of the delay where the equilibrium solution loses stability. The smallest value for T_{cr} is obtained by setting $r = 0.73881$ to obtain an expression in terms of the decay rate μ. In Section 4, we present a numerical example to confirm these results.

3. Approximating the IDDE-PDE Equations with a DDE-ODE System

To check our previous results, we "discretize" the variables $\xi(x,t)$ and $\eta(x,t)$ in (9) and (10) with a set of $2N$ variables $\xi_i(t)$ and $\eta_i(t)$ for $i = 1,2,\dots,N$. This replaces the original model (9) and (10) with a $2N$-dimensional system of N DDEs coupled to N ODEs and replaces the integral in (9) with a sum of N terms as follows:

$$\dot{\xi}_i = -\mu\xi_i - \frac{1}{N}\sum_{j=1}^{N} e^{-|i-j|/N}\eta_j(t-T), \qquad \dot{\eta}_i = \xi_i - \mu\eta_i, \tag{23}$$

where $i = 1,2,\dots,N$. By assuming solutions of the form $\xi_i = \phi_i e^{\lambda t}$ and $\eta_i = \psi_i e^{\lambda t}$ and substituting them into (23), we obtain

$$-e^{\lambda T}(\lambda+\mu)\phi_i = \frac{1}{N}\sum_{j=1}^{N} e^{-|i-j|/N}\psi_j, \qquad (\lambda+\mu)\psi_i = \phi_i, \tag{24}$$

which yields the following eigenvalue problem:

$$c\psi_i = \sum_{j=1}^{N} e^{-|i-j|/N}\psi_j, \tag{25}$$

where $c = Nr$ and $r = -e^{\lambda T}(\lambda+\mu)^2$. For nontrivial solutions, system (25) of N equations must satisfy $\det(K - cI) = 0$, where K is the $N \times N$ matrix $K = [e^{-|i-j|/N}]$ and c is its associated eigenvalue. Since K is a symmetric matrix, then all of its eigenvalues are real. Furthermore, K is positive definite because $\det(M_{ii}) > 0$ for $i = 1,2,\dots,N$, where M_{ii} is the ith minor of K along the main diagonal. Hence K is a symmetric positive definite matrix, which shows that c is a positive real number. The steady state stability results are thus summarized as follows:

(i) For $T = 0$, we have that $\lambda = -\mu \pm \sqrt{-c/N}$, where $\mu, N, c > 0$. This shows that $\mathrm{Re}(\lambda) < 0$ and so the equilibrium solution with no delay is stable.

(ii) For $T = T_{cr}$, we take the smallest value of c for any given N and use (21) and (22) to obtain values for ω and T_{cr} where we set $r = c/N$. A numerical example of this case is presented in the following section.

4. Numerical Example

In this section, we present a numerical example to compare and confirm our previous results. From (6) and (8), we obtain

$$\frac{d^2p^*}{dx^2} - \gamma p^* = 1 - \gamma, \tag{26}$$

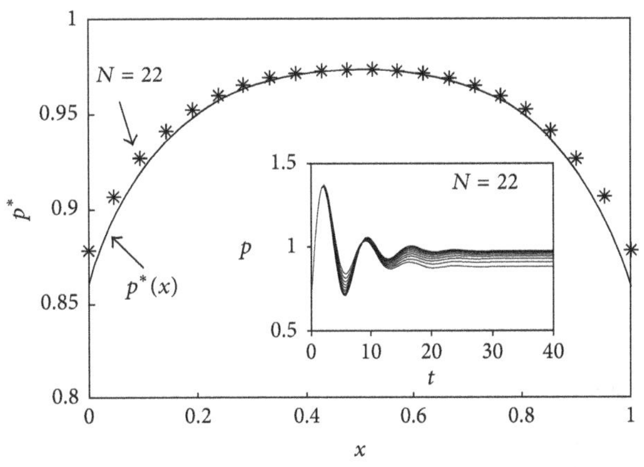

FIGURE 2: Numerical comparison between the steady state solutions for (29) and the 44-dimensional system given by (30). Here we also present a time course simulation for $N = 22$, which shows the short transient response to equilibrium. The asterisk marks (∗) are the numerical values extracted from $t = 40$ in the time course simulation for the 22 variables p_i for $i = 1,2,\dots,22$.

where $\gamma = 1 + 2/\mu^2 > 0$ gives the following solution:

$$p^*(x) = c_1\sinh(\sqrt{\gamma}x) + c_2\cosh(\sqrt{\gamma}x) + \frac{2}{\mu^2\gamma}. \tag{27}$$

Substituting (27) into the BCs (7), we obtain

$$c_1 = \left(1 - e^{\sqrt{\gamma}}\right)\cdot\left(\frac{1-\sqrt{\gamma}-(1+\sqrt{\gamma})e^{\sqrt{\gamma}}}{\gamma\left[(\mu^2\sqrt{\gamma}+\mu^2+1)e^{2\sqrt{\gamma}}+\mu^2\sqrt{\gamma}-\mu^2-1\right]}\right),$$
$$c_2 = \left(1 + e^{\sqrt{\gamma}}\right)\cdot\left(\frac{1-\sqrt{\gamma}-(1+\sqrt{\gamma})e^{\sqrt{\gamma}}}{\gamma\left[(\mu^2\sqrt{\gamma}+\mu^2+1)e^{2\sqrt{\gamma}}+\mu^2\sqrt{\gamma}-\mu^2-1\right]}\right). \tag{28}$$

Letting $\mu = 0.2$, we obtain

$$p^*(x) = 0.12\sinh(7.14x) - 0.12\cosh(7.14x) + 0.98, \tag{29}$$

which we have plotted in Figure 2 (solid). To confirm and compare this result, we numerically integrate the system

$$\dot{m}_i = -\mu m_i + \frac{1}{N}\sum_{j=1}^{N} e^{-|i-j|/N}\left(1 - p_j(t-T)\right), \qquad \dot{p}_i = m_i - \mu p_i, \tag{30}$$

for $N = 22$, $\mu = 0.2$, and $T = 0$ using MATLAB's built-in function *dde23.m*. Figure 2 shows a summary of the comparison between (29) and the 44-dimensional system (30), where we can see that good agreement was found between both systems as N becomes large. In addition, Figure 2 also presents a time course simulation for $N = 22$,

TABLE 1: Numerical results for $\mu = 0.2$ and various N.

N	c	ω	T_{cr}
2	1.6065	0.87365	0.51518
5	3.7453	0.84206	0.55386
10	7.4137	0.83748	0.55982
15	11.0992	0.83663	0.56094
22	**16.2655**	**0.83627**	**0.56142**
30	22.1729	0.83612	0.56161
50	36.9457	0.83601	0.56175
100	73.8836	0.83596	0.56181
200	147.7634	0.83595	0.56183
1000	738.8111	0.83595	0.56184

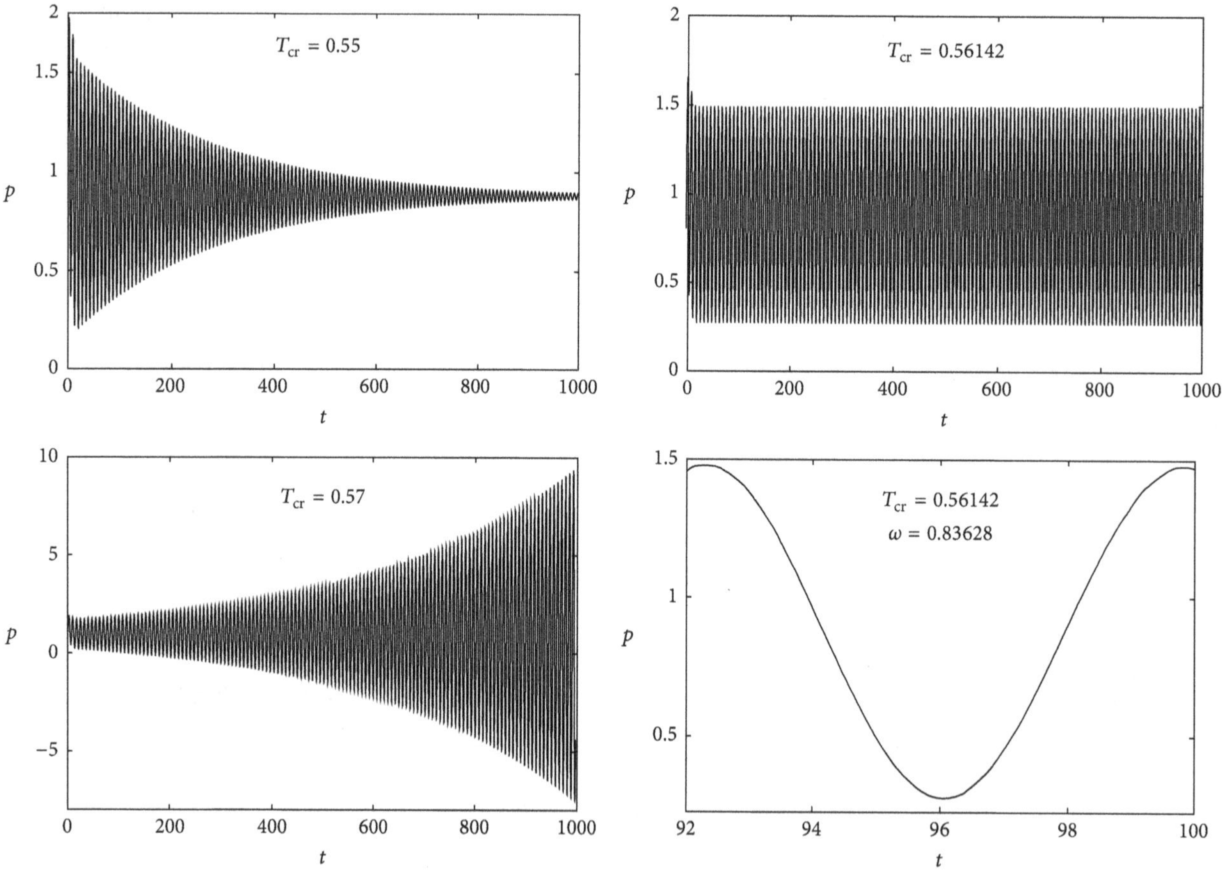

FIGURE 3: Numerical simulations for $\mu = 0.2$, $N = 22$, and different delay values. Here we plotted p_1, which serves as a representative for the other 21 p_i's, since they all exhibit the same time course simulations. We can see that the equilibrium solution is stable for $T = 0.55 < T_{cr}$ and unstable for $T = 0.57 > T_{cr}$. For $T_{cr} = 0.56142$ the system exhibits oscillations with a frequency $\omega = 2\pi/\text{period} = 2\pi/(99.794 - 92.281) = 0.83628$. These are the simulations associated with the $N = 22$ case in Table 1.

where we exhibit the short transients to equilibrium for the 22 variables p_i for $i = 1, 2, \ldots, 22$.

Now we use (21) and (22) to compute the critical delay and frequency where the steady state $p^*(x)$ loses its stability. For the IDDE system, setting $\mu = 0.2$ in (29) gives the values $\omega = 0.83595$ and $T_{cr} = 0.56184$, which we show as the limiting value for the DDE system when N becomes large. Table 1 shows the results for $\mu = 0.2$ for various values of N and Figure 3 presents the numerical simulations for $\mu = 0.2$, $N = 22$, and various delay values using MATLAB's *dde23.m*. For the case $N = 22$, Figure 3 shows that the equilibrium solution is stable for $T = 0.55 < T_{cr}$ and unstable for $T = 0.57 > T_{cr}$. For $T_{cr} = 0.56142$, the system exhibits oscillations with a frequency $\omega = 2\pi/\text{period} = 2\pi/(99.794 - 92.281) = 0.83628$ as predicted. Notice that in Figure 3 we only plotted p_1, which we use as one of the representatives for the other 21 p_i's, since they all exhibit the same time course simulation. Table 1 also shows the approach to the limiting value $T_{cr} = 0.56184$ which was approximately achieved in a system of 2000 equations.

5. Conclusions

In this work, we investigated the equilibrium solutions and their associated stability of a biological model with negative feedback and time delay. The model is formed of an IDDE coupled to a PDE having time, t, and distance, x, as independent variables. The study considers linear production and degradation rates of mRNA and protein and an exponential weighting function that models the net repression of all proteins due to spatial distribution in the cytoplasm. Our steady state analysis was accomplished by transforming the steady state integral equation into a second-order two-point boundary value problem, and it showed that the equilibrium solution, p^*, depends on the distance, x. Stability analysis then revealed that the nondelayed system is stable and that there exists a critical value for the delay where the equilibrium loses its stability.

We confirmed our results by "discretizing" our original model and approximating it with a system of N DDEs coupled to N ODEs. This resulted in a $2N$-dimensional system with delay where numerical evaluations for different N were performed and good agreement was found with the "continuous" IDDE counterpart as N became large. In particular, Table 1 shows that $T_{cr}^{\text{discrete}} \rightarrow T_{cr}^{\text{continuous}} = 0.56184$ as $N \rightarrow \infty$, which was confirmed using MATLAB's built-in function *dde23.m* on the full DDE model in a system of 2000 equations. Unfortunately, there are no numerical routines available in MATLAB for the simulation of IDDEs, but our results confirm that it is possible to dissect and understand the dynamics of such complicated equations via a discretization approach, as the one presented in Section 3. The current work corrects and extends our previous study [15] via MATLAB's *dde23.m* and thus providing more accurate and reliable approximations (and numerical simulations) for the associated $2N$-dimensional system used to confirm our results. Table 1 summarizes and corrects our previous results [15] by showcasing the $N = 22$ numerical simulations and their transition from stable to unstable behavior as seen in Figure 3. It is thus hoped that our approach will be useful to researchers in the field of computational mathematics and gene networks trying to model physical or biological systems characterized by IDDEs and PDEs. Future possible directions for this work include choosing $H(p)$ nonlinear, multiple delays, or a detailed bifurcation study proving that the system undergoes a Hopf bifurcation when $T = T_{cr}$.

Conflicts of Interest

The author declares that there are no conflicts of interest regarding the publication of this paper.

References

[1] I. Palmeirim, D. Henrique, D. Ish-Horowicz, and O. Pourquié, "Avian hairy gene expression identifies a molecular clock linked to vertebrate segmentation and somitogenesis," *Cell*, vol. 91, no. 5, pp. 639–648, 1997.

[2] M. B. Elowitz and S. Leibier, "A synthetic oscillatory network of transcriptional regulators," *Nature*, vol. 403, no. 6767, pp. 335–338, 2000.

[3] D. Bratsun, D. Volfson, L. S. Tsimring, and J. Hasty, "Delay-induced stochastic oscillations in gene regulation," *Proceedings of the National Acadamy of Sciences of the United States of America*, vol. 102, no. 41, pp. 14593–14598, 2005.

[4] J. Lewis, "Autoinhibition with transcriptional delay: a simple mechanism for the zebrafish somitogenesis oscillator," *Current Biology*, vol. 13, no. 16, pp. 1398–1408, 2003.

[5] N. A. M. Monk, "Oscillatory expression of Hesl, p53, and NF-κB driven by transcriptional time delays," *Current Biology*, vol. 13, no. 16, pp. 1409–1413, 2003.

[6] A. Verdugo and R. Rand, "Hopf bifurcation in a DDE model of gene expression," *Communications in Nonlinear Science and Numerical Simulation*, vol. 13, no. 2, pp. 235–242, 2008.

[7] H. de Jong, "Modeling and simulation of genetic regulatory systems: a literature review," *Journal of Computational Biology*, vol. 9, no. 1, pp. 67–103, 2002.

[8] T. Schlitt and A. Brazma, "Current approaches to gene regulatory network modelling," *BMC Bioinformatics*, vol. 8, no. 6, article no. S9, 2007.

[9] H. De Jong, J.-L. Gouzé, C. Hernandez, M. Page, T. Sari, and J. Geiselmann, "Qualitative simulation of genetic regulatory networks using piecewise-linear models," *The Bulletin of Mathematical Biology*, vol. 66, no. 2, pp. 301–340, 2004.

[10] A. Mochizuki, "Structure of regulatory networks and diversity of gene expression patterns," *Journal of Theoretical Biology*, vol. 250, no. 2, pp. 307–321, 2008.

[11] S. Turner, J. A. Sherratt, K. J. Painter, and N. J. Savill, "From a discrete to a continuous model of biological cell movement," *Physical Review E*, vol. 69, 2004.

[12] J. Goutsias and S. Kim, "Stochastic transcriptional regulatory systems with time delays: A mean-field approximation," *Journal of Computational Biology*, vol. 13, no. 5, pp. 1049–1076, 2006.

[13] A. Ribeiro, R. Zhu, and S. A. Kauffman, "A general modeling strategy for gene regulatory networks with stochastic dynamics," *Journal of Computational Biology*, vol. 13, no. 9, pp. 1630–1639, 2006.

[14] R. Edwards, P. Van Den Driessche, and L. Wang, "Periodicity in piecewise-linear switching networks with delay," *Journal of Mathematical Biology*, vol. 55, no. 2, pp. 271–298, 2007.

[15] A. Verdugo and R. H. Rand, "DDE Model of Gene Expression: A Continuum Approach," in *ASME Proceedings of IMECE*, pp. 1112–1120, 2008.

9

Existence and Uniqueness of Solution of Stochastic Dynamic Systems with Markov Switching and Concentration Points

Taras Lukashiv and Igor Malyk

Department of the System Analysis and Insurance and Financial Mathematics, Yuriy Fedkovych Chernivtsi National University, 28 Unversitetska St., Chernivtsi 58012, Ukraine

Correspondence should be addressed to Taras Lukashiv; t.lukashiv@gmail.com

Academic Editor: Elena Braverman

In this article the problem of existence and uniqueness of solutions of stochastic differential equations with jumps and concentration points are solved. The theoretical results are illustrated by one example.

1. Introduction

First, consider the works that are relevant to this subject. Note that number of these works are very small, since the existence of points of condensation is not very often encountered in real processes. However, the relevant equations can significantly enhance the understanding of the dynamics of real processes. In addition, this mathematical model can be a very good comparison for the classical model, the ordinary differential equations, stochastic differential equations, functional differential equations, and impulse equations. On the other hand, equations with concentration points cannot be considered as equations with Poison integral, because for these equations points of condensation do not exist with probability 1.

In paper [1] differential equations with delay in simplest form

$$\dot{x}(t) + ax(t-\tau) = \sum_{j=1}^{\infty} b_j x(t_j-)\,\delta(t-t_j) \qquad (1)$$

with infinity impulses are considered. However, in Theorems 2.1 and 3.1 [1] it is supposed that impulses satisfy the inequality $t_j - t_{j-1} > T = \text{const}$; that is, $t_j \to \infty$, if $j \to \infty$. According to Theorem 2.1 [1], one of the exponential stability conditions is

$$1 + |b_j| \le M \quad \text{for } j = 1, 2, 3, \ldots \qquad (2)$$

whence influence of impulses in determining case is obvious. In Section 3 of this paper authors consider determining differential-difference system with one delay and perturbation $x(t_i+) - x(t_i-) = b_i x(t_i-)$. Then one of the conditions of oscillation is

$$\lim_{i\to\infty} \sup\,(1+b_i)^{-1} \int_{t_i}^{t_i=\min\{\tau,T\}} p(s)\,ds > 1. \qquad (3)$$

On the other hand, Theorem 3.3 [1] consists of sufficient conditions about condition on the value of jumps $b_i > 0$, $i = 1, 2, 3, \ldots$, and $\sum_{i=1}^{\infty} d_i < \infty$.

In [2] delay depends on the time for differential equations with delay, and there is a condition on impulses $\lim_{k\to\infty} t_k = \infty$. Under the given conditions the boundedness of the solutions by exponential functions $k \cdot e^{\gamma t}$ is proved (Theorems 3.1, 4.1 [2]). Differential equations with impulses are used in a lot of application problems (however, besides delay effect, impulsive effect likewise exists in a wide variety of evolutionary processes in which states are changed abruptly at certain moments of time, involving such fields as medicine and biology, economics, mechanics, electronics, and telecommunications; artificial electronic systems, neural networks such as Hopfield neural networks, bidirectional neural networks, and recurrent neural networks often are subject to impulsive perturbations which can affect dynamical behaviors of the systems just as time delays). Authors note influence of the jumps on the capability analysis; namely, the condition on the jump's moments is $(\ln(\gamma_k)/(t_k - t_{k-1})) \le \gamma$, where γ and γ_k are described in Section 3. Thus, authors use weaker condition in comparison with $t_k - t_{k-1} > \text{const} > 0$, but existence of concentration points is not considered.

A significant contribution to the study of impulse systems of differential equations was made by Ukrainian academician Samoilenko. The monograph [3] studies impulse systems of ordinary differential equations; in the monograph [4] authors consider counted systems of ordinary differential equation in the next form

$$\frac{dx}{dt} = A(t)\,x \tag{4}$$

with impulse perturbations

$$\Delta x(t_j) = B_j x(t_j - 0). \tag{5}$$

For these systems problems of existence and uniqueness of the solution, limited periodicity is considered. In paper [5] the main problems of ordinary differential equations with stochastic parameters and perturbations are considered. As in papers [3, 4], authors suppose that $t_i - t_{i-1} > T$. In papers [3–5] the focus is on the existence of periodic solutions of the system and authors prove that existence of impulses changes qualitative characteristic of solutions (stability and periodicity). Beside this, in these papers there is a supposition about continuity on (t_i, t_{i+1}), as opposed to continuity on $[t_i, t_{i+1})$, as in this paper and in [6–8].

In [9] authors consider the second-order system, which describes behavior of $[y, \dot{y}]$ for the solution of 2nd order differential equation with impulse perturbations. By contrast, in this paper the delay process is found such that solution is non-Markov process in classic perception, but the condition on impulses is the same: $|t_i - t_{i-1}| > T$. The feature of this paper is transition of non-Markov process of perturbation to Markov process using additional variables. Based on this approach we can build finite-dimensional distributions.

Problem of existence and uniqueness of solution of impulse systems without the concentration points is considered in [10]. Also the problem of stability of solution using discontinuous impulses is considered.

All above-listed papers do not contain concentration points and cannot be used for describing systems with increase on the short time interval resonance.

The problem of existence and uniqueness of solution of dynamic systems with concentration points for determinate dynamic differential equations is solved in [11]. This paper is one of the first papers where the concentration points are considered. The examples of real processes, which are described by impulse differential equations, for which the condition $t_k - t_{k-1} > \text{const} > 0$ does not hold are considered in the paper [12]. Existence and uniqueness of random stochastic dynamic systems with permanent delay in the absence of the concentration points are considered in [8].

The sufficient conditions of existence and uniqueness of the solution of the systems of stochastic differential-difference equations with Markov switching with concentration points are shown in this paper. Thus the paper is actual and timely.

2. Problem Definition

Consider stochastic differential-difference equation

$$\begin{aligned} dx(t) &= a(t, \xi(t), x(t), x(t-r))\,dt \\ &\quad + b(t, \xi(t), x(t), x(t-r))\,dw(t), \end{aligned} \tag{6}$$

for $t \in \mathbb{R}_+ \setminus \mathbf{T}$ with Markov's switching

$$\begin{aligned} \Delta x(t_k) &= x(t_k) - x(t_k-) \\ &= g(t_k-, \xi(t_k-), \eta_k, x(t_k-)), \\ &\qquad t_k \in \mathbf{T} := \{t_k \uparrow,\ k = 1, 2, \ldots\}, \end{aligned} \tag{7}$$

and the initial conditions

$$\begin{aligned} x(t) &= \varphi(t), \\ -r &\le t \le 0, \\ r &> 0, \\ \varphi &\in \mathbf{D}, \\ \xi(0) &= y \in \mathbf{Y}, \\ \eta_0 &= h \in \mathbf{H}. \end{aligned} \tag{8}$$

Here $\xi(t)$ is the Markov process with values in the measured space $(\mathbf{Y}, \mathscr{Y})$ with generator Q; η_k, $k \ge 0$ is the Markov chain with values in the measured space $(\mathbf{H}, \mathscr{H})$, which is described by the matrix of transition probabilities $P(y, A) = P\{\eta_k \in A \mid \eta_{k-1} = y\}$, $y \in \mathbf{H}$, $A \in \mathscr{H}$; $w(t)$ is one-dimensional Wiener process. It should be noted that $w(t), \xi(t)$, $t \ge 0$, and η_k, $k \ge 0$, are independent [13]; $\mathbf{D} \equiv \mathbf{D}([-r, 0], \mathbb{R}^m)$ is the Skorokhod space of the right continuous functions with left-hand limits [14] with the norm

$$\|\varphi\| = \sup_{-r \le \theta \le 0} |\varphi(\theta)|. \tag{9}$$

Let measured mapping $a : \mathbb{R}_+ \times \mathbf{Y} \times \mathbb{R}^m \times \mathbb{R}^m \to \mathbb{R}^m$; $b : \mathbb{R}_+ \times \mathbf{Y} \times \mathbb{R}^m \times \mathbb{R}^m \to \mathbb{R}^m$; $g : \mathbb{R}_+ \times \mathbf{Y} \times \mathbf{H} \times \mathbb{R}^m \to \mathbb{R}^m$ satisfies the bounded condition and Lipschitz condition $\forall t \in [0, T]$, $y \in \mathbf{Y}$, $h \in \mathbf{H}$:

$$\begin{aligned} &|a(t, y, \phi_1, \phi_2)|^2 + |b(t, y, \phi_1, \phi_2)|^2 + |g(t, y, h, \phi_3)|^2 \\ &\quad \le C\left(1 + |\phi_1|^2 + |\phi_2|^2 + |\phi_3|^2\right), \\ &\qquad \forall t \ge 0,\ y \in \mathbf{Y},\ h \in \mathbf{H},\ \varphi_i \in \mathbb{R}^m,\ i = 1, 2, 3; \\ &|a(t, y, \phi_1, \phi_2) - a(t, y, \psi_1, \psi_2)|^2 \\ &\quad + |b(t, y, \phi_1, \phi_2) - b(t, y, \psi_1, \psi_2)|^2 \\ &\quad \le L\left(|\phi_1 - \psi_1|^2 + |\phi_2 - \psi_2|^2\right) \\ &\qquad \forall t \ge 0,\ y \in \mathbf{Y},\ \phi_i, \psi_i \in \mathbb{R}^m,\ i = 1, 2; \\ &|g(t_k, y, h, \phi_3) - g(t_k, y, h, \psi_3)|^2 \le l_k |\phi_3 - \psi_3|^2, \\ &\qquad \phi_3, \psi_3 \in \mathbb{R}^m,\ \sum_{k=1}^{\infty} l_k < \infty. \end{aligned} \tag{10}$$

Consider the case when the concentration point

$$\lim_{k\to\infty} t_k = t^* \in [0,T], \tag{11}$$

presents on $[0,T]$, where we learn existence and uniqueness of Cauchy problem (6)–(8) solution.

2.1. The Main Result. In this part of the work consider the problem of existence and uniqueness of solutions of stochastic differential-difference equations with impulse perturbations. Note that the conditions considered in this theorem are not elementary, since they have a quite complicated form.

Theorem 1. *Let*

(1) *the condition (10) hold;*

(2) $\sum_{k=1}^{\infty} \gamma_k < \infty$, $\gamma_k = \sup_{x\in\mathbb{R}^m, y\in\mathbf{Y}, h\in\mathbf{H}} |g(t_k, y, h, x)|$;

(3) *the condition*

$$\lim_{\varepsilon\downarrow 0}\left(\ln\varepsilon + N_\varepsilon \sum_{k=1}^{N_\varepsilon} l_k\right) = -\infty, \tag{12}$$

where

$$N_\varepsilon := \inf\left\{k \ge 1 : \sum_{m=k}^{\infty} \gamma_m < \varepsilon\right\}, \tag{13}$$

hold.

Then exists a unique solution of the Cauchy's problems (6)–(8).

Proof.

(I) Existing. Let us determine the following process:

$$\begin{aligned}\tilde{x}(t) &= \tilde{x}(t_k) + \int_{t_k}^{t} a(s,\xi(s),\tilde{x}(s),\tilde{x}(s-r))\,ds \\ &\quad + \int_{t_k}^{t} b(s,\xi(s),\tilde{x}(s),\tilde{x}(s-r))\,dw(s), \\ &\qquad t\in(t_k,t_{k+1}), \\ \tilde{x}(t_k) &= \tilde{x}(t_k-) + g(t_k-,\xi(t_k-),\eta_k,\tilde{x}(t_k-)), \\ &\qquad k\ge 0,\end{aligned} \tag{14}$$

which determines the initial condition

$$\begin{aligned} x(\tau) &= \varphi(\tau), \\ -r &\le \tau \le 0, \\ \xi(0) &= y \in \mathbf{Y}, \\ \eta_0 &= h \in \mathbf{H}. \end{aligned} \tag{15}$$

Consider $T' < t^*$. Then, for interval $[0,T']$, we can use classical theorem of existence and uniqueness [8]. Besides this, in the paper [15] it is proved that $x \in L_2[0,T']$. The next inequality

$$\begin{aligned}\mathbf{E}|x(t)| &\le \mathbf{E}|x(0)| \\ &\quad + \int_0^{T'} \mathbf{E}\left|a(s,\xi(s),x(s),x(s-r))\right| ds \\ &\quad + \sqrt{\int_0^{T'} \mathbf{E}\left|b(s,\xi(s),x(s),x(s-r))\right|^2 ds} \\ &\quad + \sum_{k=1}^{k'} \mathbf{E}\left|g(t_k-,\xi(t_k-),\eta_k,x(t_k-))\right|\end{aligned} \tag{16}$$

also holds. Using condition (7), we get

$$\lim_{T'\to t^*} \mathbf{E}|x(t)| < \infty. \tag{17}$$

This inequality proves the existence of the first moment for x; it means $x \in L_1[0,t^*]$ or $x \in L_1[0,T]$. Existence is proved.

(II) Uniqueness. Let two solutions $x^{(1)}(t)$, $x^{(2)}(t)$, $t \ge 0$ exist. Consider estimation $\mathbf{E}|x^{(1)}(t) - x^{(2)}(t)|^2$, $t \ge 0$, using its integral form and Lipschitz condition

$$\begin{aligned} I(t) &= \mathbf{E}\left|x^{(1)}(t) - x^{(2)}(t)\right|^2 \le K\int_0^t \mathbf{E}\left|x^{(1)}(s) - x^{(2)}(s)\right|^2 ds \\ &\quad + \mathbf{E}\left(\sum_{k=1}^{N(t)} \left|g\left(t_k-,\xi(t_k-),\eta_k,x^{(1)}(t_k-)\right) - g\left(t_k-,\xi(t_k-),\eta_k,x^{(2)}(t_k-)\right)\right|\right)^2 \\ &\le K\int_0^t \mathbf{E}\left|x^{(1)}(s) - x^{(2)}(s)\right|^2 ds \\ &\quad + 2\mathbf{E}\left(\sum_{k=1}^{N_\varepsilon(t)} \left|g\left(t_k-,\xi(t_k-),\eta_k,x^{(1)}(t_k-)\right) - g\left(t_k-,\xi(t_k-),\eta_k,x^{(2)}(t_k-)\right)\right|\right)^2 \\ &\quad + 2\mathbf{E}\left(\sum_{k=N_\varepsilon(t)+1}^{\infty} \left|g\left(t_k-,\xi(t_k-),\eta_k,x^{(1)}(t_k-)\right) - g\left(t_k-,\xi(t_k-),\eta_k,x^{(2)}(t_k-)\right)\right|\right)^2, \end{aligned} \tag{18}$$

where $N(t) = \sup\{k \in \mathbb{N} \mid t_k < t\} + 1$, $N_\varepsilon(t) = N(t) \wedge N_\varepsilon$, and $K = K(T,L)$. Let us use the Lipschitz condition for the second term of the last part of inequality and limited condition for the third one. Then we get

$$I(t) \leq K \int_0^t \mathbf{E}\left|x^{(1)}(s) - x^{(2)}(s)\right|^2 ds + 2N_\varepsilon \sum_{k=1}^{N_\varepsilon(t)} l_k \mathbf{E}\left|x^{(1)}(t_k) - x^{(2)}(t_k)\right|^2 + 2\varepsilon. \tag{19}$$

According to [11]

$$I(t) \leq 2\varepsilon \prod_{k=1}^{N_\varepsilon(t)} (1 + 2N_\varepsilon l_k) e^{KT} \leq e^{KT + 2N_\varepsilon \sum_{k=1}^{N_\varepsilon(t)} l_k + \ln \varepsilon + \ln 2}. \tag{20}$$

Then, according to the theorem's condition (3)

$$\lim_{\varepsilon \downarrow 0} \left(KT + 2N_\varepsilon \sum_{k=1}^{N_\varepsilon(t)} l_k + \ln \varepsilon + \ln 2 \right) = -\infty; \tag{21}$$

that is, $I(t) = 0$. This completes the proof of uniqueness. Theorem is proved. □

2.2. Model Example. Consider linear stochastic differential-difference equation

$$dx(t) = -a(\xi(t)) x(t) dt - b(\xi(t)) x(t - r) dt + \sigma(\xi(t)) dw(t), \quad t \geq 0, \ r > 0, \tag{22}$$

with impulse contagion

$$\Delta x\left(2 - \frac{1}{k}\right) = x\left(2 - \frac{1}{k}-\right) + e^{-\alpha k \eta_k} \left(\left| x\left(2 - \frac{1}{k}-\right) \right| \wedge 1 \right), \quad k \longrightarrow \infty, \tag{23}$$

and initial condition

$$x(\theta) = 1, \quad -r \leq \theta \leq 0, \quad \xi(0) = y_0 \in \mathbf{Y}, \quad \eta_0 = 1. \tag{24}$$

Here a, b, σ are constants that depend on Markov process ξ with generator $Q = \begin{pmatrix} -1 & 1 \\ 1 & -1 \end{pmatrix}$, where η_k, $k \geq 0$ is Markov chain with two nonabsorbing states $h_1 = 1$ i $h_2 = 2$ and transition matrix $P = \begin{pmatrix} 0.5 & 0.5 \\ 0.5 & 0.5 \end{pmatrix}$.

Let us define the values of the parameter α that the solution of the systems (22)–(24) exists.

Define the value of N_ε using equality

$$\sum_{m=k}^{\infty} \gamma_m = \sum_{m=k}^{\infty} e^{-\alpha m} = \frac{e^{-\alpha k}}{1 - e^{-\alpha}}. \tag{25}$$

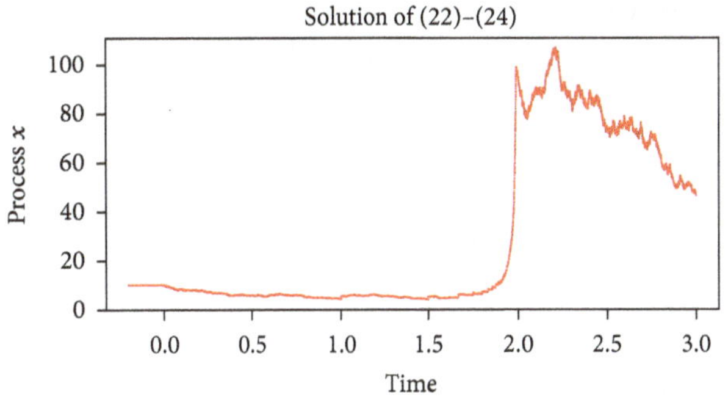

Figure 1

So,

$$N_\varepsilon = \left[-\frac{\ln \varepsilon (1 - e^{-\alpha})}{\alpha} + 1 \right], \tag{26}$$

and the theorem's condition (3) as $\alpha > 1$ is

$$\lim_{\varepsilon \downarrow 0} \left(\ln \varepsilon + \left[-\frac{\ln \varepsilon (1 - e^{-\alpha})}{\alpha} + 1 \right] \sum_{k=1}^{N_\varepsilon} l_k \right) = \lim_{\varepsilon \downarrow 0} \left(\ln \varepsilon + \left[-\frac{\ln \varepsilon (1 - e^{-\alpha})}{\alpha} + 1 \right] \frac{1}{1 - e^{-\alpha}} \right) = -\infty, \tag{27}$$

as $\alpha(1 - e^{-\alpha}) > 1$.

Let us give the R-realization of the problems (22)–(24) solution, using the following values:

If $\xi = 1$: $a = -1$, $b = -0.3$, $\sigma = 0.3$;

If $\xi = 2$: $a = 0.5$, $b = 0.04$, $\sigma = 2.1$;

$\eta_k \in \{1, 2\}$;

$\alpha = 1.673$, $h = 0.0001$, $r = 0.2$, $\varphi \equiv 10$.

As shown in Figure 1, concentration is in the point $t = 2$, and then process's behavior is continuous and stable for $t > 2$.

3. Conclusion

Usually, in mathematical describing of real processes with short-term perturbations evolution one supposes that perturbations are momentary and mathematical model is dynamic system with discontinuous trajectories. In this case the important class of the systems with impulse impact frequency increasing is lost. This paper is one of the important steps in learning of such systems and development of the quality persistence theory and learning the stabilization problem.

On the other hand, this paper significantly expands the class of equalities, for which we can consider the conditions of resistance, existence of periodic and quasi-periodic solutions, and tasks of the optimal control.

Additional Points

Annotation. The sufficient conditions of existence and uniqueness of the strong solution of stochastic dynamic systems with random structure with Markov switching and concentration points are proved in the paper.

Conflicts of Interest

The authors declare that there are no conflicts of interest regarding the publication of this article.

References

[1] K. Gopalsamy and B. G. Zhang, "On delay differential equations with impulses," *Journal of Mathematical Analysis and Applications*, vol. 139, no. 1, pp. 110–122, 1989.

[2] D. Xu and Z. Yang, "Impulsive delay differential inequality and stability of neural networks," *Journal of Mathematical Analysis and Applications*, vol. 305, no. 1, pp. 107–120, 2005.

[3] A. M. Samoilenko and N. A. Perestyuk, *Impulsive Differential Equations*, World Scientific, Singapore, 1995.

[4] A. M. Samoilenko and Y. V. Teplinskii, *CoUntable Systems of Differential Equations*, VSP, Boston, 2003.

[5] A. M. Samoilenko and O. Stanzhytskyi, *Qualitative and Asymptotic Analysis of Differential Equations with Random Perturbations*, World Scientific, Singapore, 2011.

[6] T. O. Lukashiv, I. V. Yurchenko, and V. K. Yasinskii, "Lyapunov function method for investigation of stability of stochastic Ito random-structure systems with impulse Markov switchings. I. General theorems on the stability of stochastic impulse systems," *Cybernetics and Systems Analysis*, vol. 45, no. 3, pp. 464–476, 2009.

[7] T. O. Lukashiv, V. K. Yasinskiy, and E. V. Yasinskiy, "Stabilization of stochastic diffusive dynamical systems with impulse markov switchings and parameters. part I. stability of impulse stochastic systems with markov parameters," *Journal of Automation and Information Sciences*, vol. 41, no. 2, pp. 1–24, 2009.

[8] T. O. Lukashiv and V. K. Yasynskyy, "Probabilistic stability in the whole of the stochastic dynamical systems of the random structure with constant delay," *Volyn' Mathematical Visnyk. Applied Mathematics*, vol. 10, no. 191, pp. 140–151, 2013.

[9] R. Iwankiewicz, "Equation for probability density of the response of a dynamic system to Erlang reneval random impulse processes," in *Advances in Reliability and Optimization of Structural Systems*, pp. 107–113, Taylor & Francis, Aalborg, Denmark, 2005.

[10] V. S. Denyssenko, "Stability of fuzzy impulsive Takagi-Sugeno' systems: method of linear matrix inequalities," *Reports of the National Academy of Sciences of Ukraine*, vol. 11, pp. 66–73, 2008.

[11] Å. P. Trofymchuk and Å. P. Trofymchuk, "Switching systems with fixed moments shocks the general location: existence, uniqueness of the solution and the correctness of the Cauchy problem," *Ukrainian Mathematical Journal*, vol. 42, no. 2, pp. 230–237, 1990.

[12] R. F. Nagayev, in *Mechanical processes with repeated damped collisions*, Nauka, Moskow, Russia, 1985.

[13] J. L. Doob, *Stochastic Processes*, John Wiley & Sons, New York, NY, USA, 1953.

[14] À. V. Skorokhod, *Asymptotic Methods in the Theory of Stochastic Differential Equations*, Naukova dumka, Êiev, 1987.

[15] V. K. Yasynskyy and I. V. Malyk, "Analysis of fluctuations of a parametric vacuum tube oscillator with delayed feedback," *Cybernetics and Systems Analysis*, vol. 51, no. 3, pp. 400–409, 2015.

10

The Morbidity of Multivariable Grey Model MGM$(1, m)$

Haixia Wang,[1] Lingdi Zhao,[1] and Mingzhao Hu[2]

[1]*School of Economics, Ocean University of China, Qingdao, Shandong 266100, China*
[2]*University of California Santa Barbara, Santa Barbara, CA 93106-3110, USA*

Correspondence should be addressed to Lingdi Zhao; lingdizhao512@163.com

Academic Editor: Tuncay Candan

This paper proposes the morbidity of the multivariable grey prediction MGM$(1, m)$ model. Based on the morbidity of the differential equations, properties of matrix, and Gerschgorin Panel Theorem, we analyze the factors that affect the morbidity of the multivariable grey model and give a criterion to justify the morbidity of MGM$(1, m)$. Finally, an example is presented to illustrate the practicality of our results.

1. Introduction

In recent decades, grey system theory, as well as fuzzy set theory [1] and rough set theory [2], is one of the most widely used theories to study uncertain problems. The grey system theory which was introduced by Deng [3], characterized by few data and poor information, has been successfully utilized in uncertain problems. On account of their enormous applications in agriculture, economics, management, and engineering, the grey system attracts many scientific research workers and scholars devoted to various aspect of those fields.

Grey forecasting models, an important part of grey systems, have been widely adopted to predict practical problems due to their simple calculating process and higher forecasting accuracy [4, 5]. However, some researchers put forward that the tiny changes of the initial data can result in the estimation errors, which is called the morbidity of the grey models. The research on morbidity and stability problems occupies an important part in grey forecasting system. Zheng et al. [6] pointed out that there existed morbidity in grey prediction models and analyzed the reasons in earlier times. Dang et al. [7] showed the possibility of the morbidity problem could only exist in GM$(1, 1)$ when the first item of original sequence was unequal to zero while other items were equal to zero approximatively. Wei [8] resolved the morbidity problem for the grey model with the accumulating method based on the condition number theory. Xiao and Li [9] studied the effects of the multiple transformation to the condition number of the non-equigap GM$(1, 1)$ model.

Except for the research on the morbidity of GM$(1, 1)$ model, there are also some studies concentrating on the morbidity of other grey models. Xiao and Guo [10] and Zeng and Xiao [11] researched on the morbidity problem of GM$(2, 1)$ which had two characteristic values. Wang et al. [12] summarized the main factor that affected the morbidity of GM$(1, 1, t^{\alpha})$ and suggested that there existed morbidity in some cases. Cui et al. [13, 14] found that there was no morbidity in NGM$(1, 1, k)$ and grey Verhulst model; the solution of those models will not make significant drift for the original data series of systems if there exist minor errors in collecting process.

Compared to the morbidity of grey models group, there is a little attention on the morbidity of multivariate grey prediction model MGM$(1, m)$. The MGM$(1, m)$ model was proposed by Zhai et al. [15] and has been developed rapidly and caught the attention of many researchers. Zou [16] applied a step by step optimum new information modeling method to build multivariable nonequidistance information grey model. Xiong et al. [17] optimized the background value and set the multiple linear regression model based on MGM in order to eliminate the fluctuations or random errors of the original data. Guo et al. [18] constructed SMGM$(1, m)$ through coupling self-memory principle of dynamic system to MGM; examples showed that it had superior predictive performance over other traditional grey prediction models.

Does the possibility of the morbidity in MGM exist? How to identify the morbidity of the multivariable grey model has become an important aspect in the process of constructing the MGM$(1, m)$ model. This paper discusses the possibility of the MGM$(1, m)$ model and the remainder of the paper is organized as follows: Section 2 introduces the morbidity of matrix equations and analyzes the factors that affect the condition number of special matrix. Section 3 provides a criterion to justify the morbidity of MGM$(1, m)$. Section 4 gives an example to illustrate the practicality of our results. Some conclusions are presented in Section 5.

2. The Morbidity of Equations

Considering the differential equation $Ax = b$, A is nonsingular matrix, b is the constant variable, and x is the solution of the equation.

Definition 1 (see [19]). If A or b has a small change and causes a larger change in the solution of the equation $Ax = b$, the equation is said to be morbidity equation.

Definition 2 (see [19]). Suppose that A is a square matrix with full rank. The condition number of A is

$$\operatorname{cond}(A)_v = \left\|A^{-1}\right\|_v \cdot \|A\|_v, \quad v = 1, 2, \ldots, \infty. \tag{1}$$

If A is a real symmetric matrix, then the condition number of A is

$$\operatorname{cond}(A) = \frac{|\lambda_{\max}(A)|}{|\lambda_{\min}(A)|}, \tag{2}$$

where $\lambda_{\max}$ is the maximal eigenvalue of the matrix and $\lambda_{\min}$ is the minimal eigenvalue of matrix. If cond$(A) \in (1, 10)$, A is well conditioned. If cond$(A) \in [10, 100)$, A is slightly ill-conditioned. If cond$(A) \in [100, 1000)$, A is moderately ill-conditioned. If cond$(A) \in [1000, \infty)$, A is strongly ill-conditioned.

In the process of parameters identification of multivariable grey model, we usually use the least square method to estimate the parameters, so there exist least square problems in the parameters.

Assuming that $C \in R^{m\times n}$, $y \in R^m$, and C is the parameters matrix of the grey model. If there exists a vector $x_0 \in R^n$, making $\|Cx - y\|_2$ achieve the minimum of the function, which is

$$\|Cx_0 - y\|_2 = \min_{x\in R^n} \|Cx - y\|_2, \tag{3}$$

then x_0 is the solution of the linear equation $Cx = y$, which is the estimated parameter of the grey model.

Suppose that $f(x) = \|Cx - y\|^2 = (Cx - y)^T(Cx - y) = x^TC^TCx - x^TD^Ty - y^TDx + y^Ty$. By the extremum condition of the equation, we have

$$\frac{df(x)}{dx} = 2C^TCx - 2C^Ty = 0. \tag{4}$$

Then we obtain the solution $C^TCx = C^Ty$, which is also the least square solution of the equation $Cx = y$.

In the multivariable grey prediction models, the data matrix C is usually the long matrix; it is not easy to solve its condition number. It should be noted that C^TC is a real symmetric matrix, the condition number is easy to obtain. Therefore, we often justify the morbidity of the multivariable grey model by the condition number of C^TC.

3. The Morbidity of MGM$(1, m)$

3.1. Grey MGM$(1, m)$ Model. The multiple variable grey prediction model abbreviated as MGM$(1, m)$ is one of the frequently used grey forecasting models. The MGM$(1, m)$ model constructing process is presented below.

Definition 3. Assume that the data sequence

$$X_j^{(0)} = \left(x_j^{(0)}(1), x_j^{(0)}(2), \ldots, x_j^{(0)}(m)\right)^T, \quad j = 1, 2, \ldots, m \tag{5}$$

is the original nonnegative data matrix. The data matrix

$$X_j^{(1)} = \left(x_j^{(1)}(1), x_j^{(1)}(2), \ldots, x_j^{(1)}(n)\right)^T, \quad j = 1, 2, \ldots, m \tag{6}$$

is the first-order accumulated generating matrix of $X^{(0)}$, where

$$x_j^{(1)}(k) = \sum_{i=1}^{k} x_j^{(0)}(i). \tag{7}$$

The adjacent neighbour average sequence of $X^{(1)}$ is

$$Z_j^{(1)} = \left(z_j^{(1)}(1), z_j^{(1)}(2), \ldots, z_j^{(1)}(n)\right), \tag{8}$$

where $z_j^{(1)}(k) = 0.5(x_j^{(1)}(k) + x_j^{(1)}(k-1))$, $k = 2, 3, \ldots, n$.

The first-order differential equations of the multivariable grey model MGM$(1, m)$ are as follows:

$$\frac{dx_1^{(1)}}{dt} = \alpha_{11}x_1^{(1)} + \alpha_{12}x_2^{(1)} + \cdots + \alpha_{1m}x_m^{(1)} + \beta_1$$

$$\frac{dx_2^{(1)}}{dt} = \alpha_{21}x_1^{(1)} + \alpha_{22}x_2^{(2)} + \cdots + \alpha_{2m}x_m^{(1)} + \beta_2$$

$$\vdots$$

$$\frac{dx_m^{(1)}}{dt} = \alpha_{m1}x_1^{(1)} + \alpha_{m2}x_2^{(1)} + \cdots + \alpha_{mm}x_m^{(m)} + \beta_m. \tag{9}$$

Note that

$$A = \left(\alpha_{ij}\right)_{m\times m}, \quad \beta = \left(\beta_1, \beta_2, \ldots, \beta_m\right)^T, \tag{10}$$

and (9) can be noted as

$$\frac{dX^{(1)}(t)}{dt} = AX^{(1)}(t) + \beta. \tag{11}$$

Applying the least square method to the first-order differential equation

$$\frac{dX^{(1)}(t)}{dt} = AZ^{(1)}(t) + \beta, \tag{12}$$

we obtain the estimated parameters

$$\widehat{Q} = \begin{pmatrix} \widehat{A'} \\ \widehat{\beta'} \end{pmatrix} = \left(P^TP\right)^{-1} P^T \left(Y_1, Y_2, \ldots, Y_m\right), \tag{13}$$

where

$$\widehat{Q} = \begin{pmatrix} \widehat{\alpha}_{11} & \widehat{\alpha}_{21} & \cdots & \widehat{\alpha}_{m1} \\ \widehat{\alpha}_{12} & \widehat{\alpha}_{22} & \cdots & \widehat{\alpha}_{m2} \\ \vdots & \vdots & \ddots & \vdots \\ \widehat{\alpha}_{1m} & \widehat{\alpha}_{2m} & \cdots & \widehat{\alpha}_{mm} \\ \widehat{\beta}_1 & \widehat{\beta}_2 & \cdots & \widehat{\beta}_m \end{pmatrix},$$

$$P = \begin{pmatrix} z_1^{(1)}(2) & z_2^{(1)}(2) & \cdots & z_m^{(1)}(2) & 1 \\ z_1^{(1)}(3) & z_2^{(1)}(3) & \cdots & z_m^{(1)}(3) & 1 \\ \vdots & \vdots & \ddots & \vdots & \vdots \\ z_1^{(1)}(n) & z_2^{(1)}(n) & \cdots & z_m^{(1)}(n) & 1 \end{pmatrix}, \tag{14}$$

and $Y_j = (x_j^{(0)}(2), x_j^{(0)}(3), \ldots, x_j^{(0)}(n))^T$, $j = 1, 2, \ldots, m$.

3.2. The Morbidity of MGM. In this part, we give a criterion to justify the morbidity of MGM$(1, m)$.

Lemma 4 (Gerschgorin Panel Theorem). *If $A \in C^{n\times n}$ and $A = (a_{ij})$, then every eigenvalue of A is contained in the plane, which is*

$$\lambda \in \bigcup_{i=1}^{n} D_i, \tag{15}$$

where D_i is the panel centred by a_{ii} in the complex plane and

$$D_i = \left\{ z \in C \mid \left|z - a_{ii}\right| \le \sum_{j=1, j\ne i}^{n} \left|a_{ij}\right| \right\}, \quad i = 1, 2, \ldots, n. \tag{16}$$

Theorem 5. *Suppose that $X_j^{(0)}(1), X_j^{(0)}(2), \ldots, X_j^{(0)}(n)$ are data vectors, and $X_j^{(1)}(n)$ is the first-order accumulated generating vector. If every consecutive neighbour $z_j^{(1)}(k) \ge 1$ ($j = 1, 2, \ldots, m$), then the multivariable grey model MGM$(1, m)$ is morbidity.*

Proof. In the process of estimating the parameters of A, β, by least square method, we calculate the matrix of P^TP, which is

$$P^TP = \begin{pmatrix} \sum_{k=2}^{n}\left(z_1^{(1)}(k)\right)^2 & \cdots & \sum_{k=2}^{n} z_1^{(1)}(k)\, z_m^{(1)}(k) & \sum_{k=2}^{n} z_1^{(1)}(k) \\ \sum_{k=2}^{n} z_1^{(1)}(k)\, z_2^{(1)}(k) & \cdots & \sum_{k=2}^{n} z_2^{(1)}(k)\, z_m^{(1)}(k) & \sum_{k=2}^{n} z_2^{(1)}(k) \\ \vdots & \ddots & \vdots & \vdots \\ \sum_{k=2}^{n} z_m^{(1)}(k)\, z_1^{(1)}(k) & \cdots & \sum_{k=2}^{n}\left(z_m^{(1)}(k)\right)^2 & \sum_{k=2}^{n} z_m^{(1)}(k) \\ \sum_{k=2}^{n} z_1^{(1)}(k) & \cdots & \sum_{k=2}^{n} z_m^{(1)}(k) & n-1 \end{pmatrix}. \tag{17}$$

From

$$\left(P^TP\right)^T = P^TP, \tag{18}$$

we know P^TP is a symmetric matrix; since P is invertible, we deduce that all the eigenvalues of the matrix P^TP are positive real numbers and P^TP is positive definite matrix. Therefore, the condition number of matrix P^TP can be represented by the maximal eigenvalue and minimal eigenvalue of the matrix.

Set $\lambda_1, \lambda_2, \ldots, \lambda_{n-1}$ as the eigenvalues of P^TP. By Gerschgorin Panel Theorem, we have

$$D_1 = \left\{ \lambda_1 \in R^+ \mid \left|\lambda_1 - \sum_{k=2}^{n}\left(z_1^{(1)}(k)\right)^2\right| \le \sum_{j=2}^{m}\left(\sum_{k=2}^{n} z_1^{(1)}(k)\, z_j^{(1)}(k)\right) + \sum_{k=2}^{n} z_1^{(1)}(k) \right\},$$

$$D_2 = \left\{ \lambda_2 \in R^+ \mid \left|\lambda_2 - \sum_{k=2}^{n}\left(z_2^{(1)}(k)\right)^2\right| \le \sum_{j=1, j\ne 2}^{m}\left(\sum_{k=2}^{n} z_2^{(1)}(k)\, z_j^{(1)}(k)\right) + \sum_{k=2}^{n} z_2^{(1)}(k) \right\},$$

$$\vdots$$

$$D_{n-1} = \left\{ \lambda_{n-1} \in R^{+} \mid \left|\lambda_{n-1} - (n-1)\right| \right.$$

$$\left. \leq \sum_{j=1}^{m} \sum_{k=2}^{n} z_j^{(1)}(k) \right\}. \tag{19}$$

It is easy to see that all the eigenvalues of $P^T P$ are contained in the $D_1 \cup D_2 \cup \cdots \cup D_{n-1}$; that is to say, every eigenvalue of $P^T P$ is contained in the panel.

If all the adjacent neighbour average sequences $z_i^{(1)}(k) \geq 1$ and the chosen sample is the minimal permitted data in grey system, then we conclude that $\sum_{k=2}^{n}(z_i^{(1)}(k))^2$ is larger than $n-1$, and the maximal eigenvalue and minimal eigenvalue are contained in different circles, and the centres of circles are far from each other. Therefore, the maximal eigenvalue and minimal eigenvalue are far away from each other on the number line. From the definition of the ill-conditioned matrix, we deduce that the multivariable grey model MGM$(1,m)$ is morbidity. This completes the proof. □

4. Example

In what follows, we give an example to illustrate the practicality of our results. The data are the price indexes of financial intermediation and real estate in 1981–1984, and data resource is the China statistical yearbook. Set $X_1^{(0)}$ and $X_2^{(0)}$ as the price index of financial intermediation and price index of the real estate, respectively; the data are shown in Table 1. As usual, we chose 4 group samples which are the minimum permitted data in grey models.

We construct MGM$(1,2)$ model to simulate and predict the data vectors. By the definition of P, we obtain

$$P = \begin{pmatrix} 1.339 & 1.133 & 1 \\ 1.7855 & 1.2125 & 1 \\ 2.3015 & 1.415 & 1 \end{pmatrix},$$

$$P^T P = \begin{pmatrix} 10.2778 & 6.9386 & 5.4260 \\ 6.9386 & 4.7561 & 3.7605 \\ 5.426 & 3.7605 & 3 \end{pmatrix}. \tag{20}$$

By Theorem 5, there exists morbidity in MGM$(1,2)$ model. In fact, all the eigenvalues of $P^T P$ are

$$\lambda_1 = 0.0011,$$
$$\lambda_2 = 0.1249, \tag{21}$$
$$\lambda_3 = 17.9079.$$

TABLE 1: The data vectors.

	$k=1$	$k=2$	$k=3$	$k=4$
$X_1^{(0)}$	1.102	1.576	1.995	2.608
$X_2^{(0)}$	1.084	1.182	1.243	1.587

It is clear that $\text{cond}(A) \geq 1000$, and there exists morbidity in the model. It proves that our criterion is a useful way to justify the morbidity of MGM$(1,m)$ model.

5. Conclusions

This paper discusses the morbidity of the multivariable grey model. From the morbidity of the differential equations, we analyze the factors that affect the morbidity of MGM$(1,m)$ model. By Gerschgorin Panel Theorem and the knowledge of matrix, we give a criterion to justify the morbidity of MGM$(1,m)$. An example is given to illustrate the maneuverability of our results.

Conflicts of Interest

The authors declare that they have no conflicts of interest.

Acknowledgments

This research is supported by National Natural Science Foundation of China (Grant no. 71473233).

References

[1] L. A. Zadeh, "Fuzzy algorithms," *Information and Control*, vol. 12, no. 2, pp. 94–102, 1968.
[2] Z. Pawlak, "Rough sets," *International Journal of Computer & Information Science*, vol. 11, no. 5, pp. 341–356, 1982.
[3] J. L. Deng, "Control problems of grey systems," *Systems Control Letters*, vol. 1, no. 5, pp. 288–294, 1982.
[4] Y. H. Wang, K. Qu, and Z. H. Wang, "A kind of nonlinear strengthening operators for predicting the output value of China's marine electric power industry," *The Journal of Grey System*, vol. 28, no. 2, pp. 35–52, 2016.
[5] S. F. Liu, J. Forrest, and Y. J. Yang, "Advances in grey systems research," *The Journal of Grey System*, vol. 25, no. 2, pp. 1–18, 2013.
[6] Z. N. Zheng, Y. Y. Wu, and H. L. Bao, "Morbidity problem in grey model," *Chinese Journal of Management Science*, vol. 9, no. 5, pp. 38–44, 2001.
[7] Y. G. Dang, Z. X. Wang, and S. F. Liu, "Study on morbidity problem in grey model," *Systems Engineering Theory Practice*, vol. 28, no. 1, pp. 156–160, 2008.
[8] Y. Wei, "Morbidity research on grey forecast model," *Communications in Computer and Information Science*, vol. 224, no. 1, pp. 294–298, 2011.
[9] X. P. Xiao and F. Q. Li, "Research on the stability of non-equigap grey control model under multiple transformations," *Kybernetes*, vol. 38, no. 10, pp. 1701–1708, 2009.
[10] X. P. Xiao and J. H. Guo, "The morbidity peoblem of GM(2, 1) model based on vector transformation," *The Journal of Grey System*, vol. 26, no. 3, pp. 1-11, 2014.

[11] X. Y. Zeng and X. P. Xiao, "Research on morbidity problem of accumulating method," *Journal of Systems Engineering and Electronics*, vol. 28, no. 4, pp. 542–572, 2006.

[12] Z. X. Wang, Y. G. Dang, and S. F. Liu, "The morbidity of GM(1, 1) power model," *System Engineering Theory Practice*, vol. 33, no. 7, 1859.

[13] J. Cui, S. F. Liu, N. M. Xie, and B. Zeng, "Study on morbidity of grey Verhulst forecasting model," *Systems Engineering—Theory Practice*, vol. 34, no. 2, pp. 416–420, 2014.

[14] J. Cui, Y. G. Dang, and S. F. Liu, "Study on morbidity of NGM(1, 1, *k*) model based on conditions of matrix," *Control and Decision*, vol. 25, no. 7, pp. 1050–1054, 2010.

[15] J. Zhai, J. M. Sheng, and Y. J. Feng, "The grey model MGM(1, *n*) and its application," *Systems Engineering—Theory Practice*, vol. 17, no. 5, pp. 109–113, 1997.

[16] R. B. Zou, "The Non-equidistant new information optimizing MGM(1, *n*) based on a step by step optimum constructing background value," *Applied Mathematics & Information Sciences*, vol. 6, no. 3, pp. 745–750, 2012.

[17] P. P. Xiong, Y. G. Dang, X. H. Wu, and X. M. Li, "Combined model based on optimized multi-variable grey model and multiple linear regression," *Journal of Systems Engineering and Electronics*, vol. 22, no. 4, pp. 615–620, 2011.

[18] X. J. Guo, S. F. Liu, L. F. Wu, Y. B. Gao, and Y. J. Yang, "A multi-variable grey model with a self-memory component and its application on engineering prediction," *Engineering Applications of Artificial Intelligence*, vol. 42, pp. 82–93, 2015.

[19] J. L. Chen and X. H. Chen, *Special Matrices*, Tsinghua university press, Beijing, China, 2001.

11

Existence of Weak Solutions for Fractional Integrodifferential Equations with Multipoint Boundary Conditions

Haide Gou and Baolin Li

College of Mathematics and Statistics, Northwest Normal University, Lanzhou 730070, China

Correspondence should be addressed to Baolin Li; ghdzxh@163.com

Academic Editor: Yuji Liu

By combining the techniques of fractional calculus with measure of weak noncompactness and fixed point theorem, we establish the existence of weak solutions of multipoint boundary value problem for fractional integrodifferential equations.

1. Introduction

In recent years, fractional differential equations in Banach spaces have been studied and a few papers consider fractional differential equations in reflexive Banach spaces equipped with the weak topology. As long as the Banach space is reflexive, the weak compactness offers no problem since every bounded subset is relatively weakly compact and therefore the weak continuity suffices to prove nice existence results for differential and integral equations [1, 2]. De Blasi [3] introduced the concept of measure of weak noncompactness and proved the analogue of Sadovskiis fixed point theorem for the weak topology (see also [4]). As stressed in [5], in many applications, it is always not possible to show the weak continuity of the involved mappings, while the sequential weak continuity offers no problem. This is mainly due to the fact that Lebesgues dominated convergence theorem is valid for sequences but not for nets. Recall that a mapping between two Banach spaces is sequentially weakly continuous if it maps weakly convergent sequences into weakly convergent sequences.

The theory of boundary value problems for nonlinear fractional differential equations is still in the initial stages and many aspects of this theory need to be explored. There are many papers dealing with multipoint boundary value problems both on resonance case and on nonresonance case; for more details see [6–11]. However, as far as we know, few results can be found in the literature concerning multipoint boundary value problems for fractional differential equations in Banach spaces and weak topologies. Zhou *et al.* [12] discuss the existence of solutions for nonlinear multipoint boundary value problem of integrodifferential equations of fractional order as follows:

$$\begin{gathered} {}^{c}D_{0+}^{\alpha}x(t) = f(t, x(t), (Hx)(t), (Kx)(t)), \\ t \in [0,1], \quad \alpha \in (1,2], \\ a_1 x(0) - b_1 x'(0) = d_1 x(\xi_1), \\ a_2 x(1) + b_2 x'(1) = d_2 x(\xi_2), \end{gathered} \tag{1}$$

with respect to strong topology, where ${}^{c}D_{0+}^{\alpha}$ denotes the fractional Caputo derivative and the operators given by

$$\begin{aligned} (Hx)(t) &= \int_0^t g(t,s)\,x(s)\,ds, \\ (Kx)(t) &= \int_0^t h(t,s)\,x(s)\,ds. \end{aligned} \tag{2}$$

Moreover, theory for boundary value problem of integrodifferential equations of fractional order in Banach spaces endowed with its weak topology has been few studied until now. In [13], we discussed the existence theorem of weak solutions nonlinear fractional integrodifferential equations in nonreflexive Banach spaces E:

$$\begin{gathered}{}^{c}D^{\alpha}_{0+}x(t) = f(t, x(t), (Tx)(t), (Sx)(t)), \\ t \in [0,1], \ \alpha \in (1,2], \\ a_1 x(0) - a_2 x'(0) = \gamma_1, \\ b_1 x(1) + b_2 x'(1) = \gamma_2,\end{gathered} \tag{3}$$

and obtain a new result by using the techniques of measure of weak noncompactness and Henstock-Kurzweil-Pettis integrals, where ${}^{c}D^{\alpha}_{0+}$ denotes the fractional Caputo derivative and the operators given by

$$\begin{aligned}(Tx)(s) &= \int_0^s k_1(s,\tau)\, g(\tau, x(\tau))\, d\tau, \\ (Sx)(s) &= \int_0^1 k_2(s,\tau)\, h(\tau, x(\tau))\, d\tau.\end{aligned} \tag{4}$$

Our analysis relies on the Krasnoselskii fixed point theorem combined with the technique of measure of weak noncompactness.

Motivated by the above works, in this paper, we use the techniques of measure of weak noncompactness combine with the fixed point theorem to discuss the existence theorem of weak solutions for a class of nonlinear fractional integrodifferential equations of the form

$$\begin{gathered}{}^{c}D^{\alpha}_{0+}x(t) = f(t, x(t), (Tx)(t), (Sx)(t)), \\ t \in [0,1], \ \alpha \in (1,2], \\ a_1 x(0) - b_1 x'(0) = d_1 x(\xi_1), \\ a_2 x(1) + b_2 x'(1) = d_2 x(\xi_2),\end{gathered} \tag{5}$$

where T and S are two operators defined by

$$\begin{aligned}(Tu)(t) &= \int_0^t k_1(t,s)\, g(s, u(s))\, ds, \\ (Su)(t) &= \int_0^a k_2(t,s)\, h(s, u(s))\, ds,\end{aligned} \tag{6}$$

E is a nonreflexive Banach space, ${}^{c}D^{\alpha}_{0+}$ denotes the fractional Caputo derivative, $k_1 \in C(D, R^+)$, $k_2 \in C(D_0, R^+)$, $D = \{(t,s) \in R^2 : 0 \le s \le t \le 1\}$, $D_0 = \{(t,s) \in R^2 : 0 \le t, s \le 1\}$, $a_1, b_1, d_1, a_2, b_2, d_2$ are real numbers, $0 < \xi_1, \xi_2 < 1$, $f : I \times E^3 \longrightarrow E, g, h : I \times E \longrightarrow E$ are given functions satisfying some assumptions that will be specified later, the integral is understood to be the Henstock-Kurzweil-Pettis, and solutions to (5) will be sought in $E = C(I, E_\omega)$.

The problems of our research are different between this paper and paper [13]. In paper [13], we studied two point boundary value problem by using the corresponding Green's function and fixed point theorems; moreover, we get some good results. In this paper, we use the techniques of measure of weak noncompactness and Henstock-Kurzweil-Pettis integrals to discuss the existence theorem of weak solutions for a class of the multipoint boundary value problem of fractional integrodifferential equations equipped with the weak topology. Our results generalized some classical results and improve the assumptions conditions, so our results improve the results in [13].

The paper is organized as follows: In Section 2 we recall some basic known results. In Section 3 we discuss the existence theorem of weak solutions for problem (5).

2. Preliminaries

Throughout this paper, we introduce notations, definitions, and preliminary results which will be used.

Let $I = [0,1]$ be the real interval, let E be a real Banach space with norm $\|\cdot\|$, its dual space E^* also $B(E^*)$ denotes the closed unit ball in E^*, and $E_\omega = (E, \omega) = (E, \sigma(E, E^*))$ denotes the space E with its weak topology. Denote by $C(I, E_\omega) = (C(I,E), \omega)$ the space of all continuous functions from I to E endowed with the weak topology and the usual supremum norm $\|x\| = \sup_{t\in I}|x(t)|$.

Let Ω_E be the collection of all nonempty bounded subsets of E, and let $\mathscr{W}_E$ be the subset of Ω_E consisting of all weakly compact subsets of E. Let B_r denote the closed ball in E centered at 0 with radius $r > 0$. The De Blasi [14] measure of weak noncompactness is the map $\beta : \Omega_E \longrightarrow [0, \infty)$ defined by

$$\begin{gathered}\beta(A) = \inf \{r > 0 : \text{there exists a set } W \\ \in \mathscr{W}_E \text{ such that } A \subseteq W + B_r\}\end{gathered} \tag{7}$$

for all $A \in \Omega_E$. The fundamental tool in this paper is the measure of weak noncompactness; for some properties of $\beta(A)$ and more details see [3].

Now, for the convenience of the reader, we recall some useful definitions of integrals.

Definition 1 (see [15]). A function $u : I \longrightarrow E$ is said to be Henstock-Kurzweil integrable on I if there exists an $J \in E$ such that, for every $\varepsilon > 0$, there exists $\delta(\xi) : I \longrightarrow \mathbb{R}^+$ such that, for every δ-fine partition $D = \{(I_i, \xi_i)\}_{i=1}^n$, we have

$$\left\| \sum_{i=1}^{n} u(\xi_i)\, \mu(I_i) - J \right\| < \varepsilon, \tag{8}$$

and we denote the Henstock-Kurzweil integral J by (HK) $\int_a^b u(s)ds$.

Definition 2 (see [15]). A function $f : I \longrightarrow E$ is said to be Henstock-Kurzweil-Pettis integrable or simply HKP-integrable on I, if there exists a function $g : I \longrightarrow E$ with the following properties:

(i) $\forall x^* \in E^*$, $x^* f$ is Henstock-Kurzweil integrable on I;

(ii) $\forall t \in I$, $\forall x^* \in E^*$, $x^* g(t) = \text{(HK)} \int_0^t x^* f(s)ds$.

This function g will be called a primitive of f and be denote by $g(t) = \int_0^t f(t)dt$ the Henstock-Kurzweil-Pettis integral of f on the interval I.

Definition 3 (see [16]). A family $\mathcal{M}$ of functions $f : S \longrightarrow E$ is called HK-equi-integrable if each $f \in \mathcal{M}$ is HK-integrable and for every $\varepsilon > 0$ there exists a gauge δ on S such that, for every δ-fine HK-partition π of S, we have

$$\left\| \sum_{(I,s)\in\pi} f(s)\lambda_m(I) - (\text{HK})\int_S \right\| \leq \varepsilon, \quad (9)$$

for all $f \in \mathcal{M}$.

Theorem 4 (see [16]). *Let (f_n) be a pointwise bounded sequence of HKP integrable functions $f_n : S \longrightarrow E$ and let $f : S \longrightarrow E$ be a function. Assume that,*

(i) *for every $x^* \in E^*$, $x^*(f_n(t)) \longrightarrow x^*(f(t))$ a.e. on S,*

(ii) *for every sequence $(x_k^*) \subset B(E^*)$, the sequence $(x_k^*(f_n))_{k,n}$ is HK-equi-integrable, then f is HKP-integrable and for every $I \in \mathcal{I}$, and we have*

$$\lim_{n\longrightarrow\infty} F_n(I) = F(I) \quad (10)$$

in the weak topology $\sigma(E, E^*)$, where F is the HKP-primitive of f and S is a fixed compact nondegenerate interval in $\mathbb{R}^n$. Denote by $\mathcal{I}$ the family of all closed nondegenerate subintervals of S.

Lemma 5 (see [17]). *If $B \subset C(I, E)$ is equicontinuous, $u_0 \in C(I, E)$, then $\overline{co}\{B, u_0\}$ is also equicontinuous in $C(I, E)$.*

Lemma 6 (see [17, 18]). *Let E be a Banach space, and let $B \subset C(I, E)$ be bounded and equicontinuous. Then $\beta(B(t))$ is continuous on I, and $\beta(B) = \max_{t\in I}\beta(B(t))$.*

Lemma 7 (see [14, 19]). *Let E be a Banach space and let $B \subset C(I, E)$ be bounded and equicontinuous. Then the map $t \longrightarrow \beta(B(t))$ is continuous on I and*

$$\beta(B) = \sup_{t\in I}\beta(B(t)) = \beta(B(I)), \quad (11)$$

where $B(t) = \{b(t) : b \in B\}$ and $B(I) = \bigcup_{t\in I}\{b(t) : h \in B\}$.

Lemma 8 (see [17]). *Let $B \subset C(I, E)$ be bounded and equicontinuous. Then $\beta(B(t))$ is continuous on I and*

$$\beta\left(\int_I B(s)\,ds\right) \leq \int_I \beta(B(s))\,ds. \quad (12)$$

We give the fixed point theorem, which play a key role in the proof of our main results.

Lemma 9 (see [20]). *Let E be a Banach space and β a regular and set additive measure of weak noncompactness on E. Let C be a nonempty closed convex subset of E, $x_0 \in C$, and n_0 a positive integer. Suppose $F : C \longrightarrow C$ is β-convex power condensing about x_0 and n_0. If F is weakly sequentially continuous and F(C) is bounded, then F has a fixed point in C.*

The following we recall the definition of the Caputo derivative of fractional order.

Definition 10. Let $x : I \longrightarrow E$ be a function. The fractional HKP-integral of the function x of order $\alpha \in \mathbb{R}_+$ is defined by

$$I_{0+}^{\alpha}x(t) := \int_0^t \frac{(t-s)^{\alpha-1}}{\Gamma(\alpha)}x(s)\,ds. \quad (13)$$

In the above definition the sign "$\int$" denotes the HKP-integral integral.

Definition 11. The Riemann-Liouville derivative of order α with the lower limit zero for a function $f : [0,\infty) \longrightarrow R$ can be written as

$$D_{0^+}^{\alpha}f(t) = \frac{1}{\Gamma(n-\alpha)}\frac{d^n}{dt^n}\int_0^t \frac{f(s)}{(t-s)^{\alpha+1-n}}ds, \quad t > 0,\ n-1 < \alpha < n. \quad (14)$$

Definition 12. The Caputo fractional derivative of order α for a function $f : [0,\infty) \longrightarrow E$ can be written as

$$^{c}D_{0+}^{\alpha}f(t) = D_{0^+}^{\alpha}\left[f(t) - \sum_{k=0}^{n-1}\frac{t^k}{k!}f^{(k)}(0)\right], \quad t > 0,\ n-1 < \alpha < n, \quad (15)$$

where $n = [\alpha] + 1$ and $[\alpha]$ denotes the integer part of α.

3. Main Results

In this section, we present the existence of solutions to problem (5) in the space $C(I, E_\omega)$.

Definition 13. A function $x \in C(I, E_w)$ is said to be a solution of problem (5) if x satisfies the equation $^{c}D_{0+}^{\alpha}x(t) = f(t, x(t), (Tx)(t), (Sx)(t))$ on I and satisfies the conditions $a_1x(0) - b_1x'(0) = d_1x(\xi_1)$, $a_2x(1) + b_2x'(1) = d_2x(\xi_2)$.

Lemma 14 (see [21]). *Let $\alpha > 0$. If one assumes $u \in C(0,1) \cap L(0,1)$, then the differential equation*

$$^{c}D_{0+}^{\alpha}u(t) = 0 \quad (16)$$

has solution $u(t) = c_0 + c_1t + c_2t^2 + \cdots + c_nt^{n-1}$, $c_i \in \mathbb{R}$, $i = 0, 1, \ldots, n$, $n = [\alpha] + 1$.

From the lemma above, we deduce the following statement.

Lemma 15 (see [21]). *Assume that $u \in C(0,1) \cap L(0,1)$ with a fractional derivative of order $\alpha > 0$ that belongs to $C(0,1) \cap L(0,1)$. Then*

$$I_{0+}^{\alpha}\left(^{c}D_{0+}^{\alpha}u(t)\right) = u(t) + c_0 + c_1t + c_2t^2 + \cdots + c_nt^{n-1} \quad (17)$$

for some $c_i \in \mathbb{R}$, $i = 0, 1, \ldots, n$, $n = [\alpha] + 1$.

The following we give the corresponding Greens function for problem (5).

Lemma 16. *Let* $\Delta \neq 0, \rho \in C(I, E_w)$ *and* $\alpha \in (1,2]$, *then the unique solution of*

$$\begin{aligned} &{}^{c}D_{0+}^{\alpha} x(t) = \rho(t), \quad t \in I, \\ &a_1 x(0) - b_1 x'(0) = d_1 x(\xi_1), \\ &a_2 x(1) + b_2 x'(1) = d_2 x(\xi_2) \end{aligned} \tag{18}$$

is given by

$$x(t) = \int_0^1 G(t,s)\,\rho(s)\,ds, \tag{19}$$

where the Green function G is given by

$$G(t,s) = \begin{cases} \dfrac{(t-s)^{\alpha-1}}{\Gamma(\alpha)} + \dfrac{d_1[a_2(1-t)+b_2+d_2(t-\xi_2)](\xi_1-s)^{\alpha-1}}{\Delta\Gamma(\alpha)} - \dfrac{a_2[(b_1+d_1\xi_1)+t(a_1-d_1)](1-s)^{\alpha-1}}{\Delta\Gamma(\alpha)} - \dfrac{b_2[(b_1+d_1\xi_1)+t(a_1-d_1)](1-s)^{\alpha-2}}{\Delta\Gamma(\alpha-1)} + \dfrac{d_2[(b_1+d_1\xi_1)+t(a_1-d_1)](\xi_2-s)^{\alpha-1}}{\Delta\Gamma(\alpha)}, & s \le \xi_1,\ s \le t; \\ \dfrac{d_1[a_2(1-t)+b_2+d_2(t-\xi_2)](\xi_1-s)^{\alpha-1}}{\Delta\Gamma(\alpha)} - \dfrac{a_2[(b_1+d_1\xi_1)+t(a_1-d_1)](1-s)^{\alpha-1}}{\Delta\Gamma(\alpha)} - \dfrac{b_2[(b_1+d_1\xi_1)+t(a_1-d_1)](1-s)^{\alpha-2}}{\Delta\Gamma(\alpha-1)} + \dfrac{d_2[(b_1+d_1\xi_1)+t(a_1-d_1)](\xi_2-s)^{\alpha-1}}{\Delta\Gamma(\alpha)}, & s \le \xi_1,\ t \le s; \\ \dfrac{(t-s)^{\alpha-1}}{\Gamma(\alpha)} - \dfrac{a_2[(b_1+d_1\xi_1)+t(a_1-d_1)](1-s)^{\alpha-1}}{\Delta\Gamma(\alpha)} - \dfrac{b_2[(b_1+d_1\xi_1)+t(a_1-d_1)](1-s)^{\alpha-2}}{\Delta\Gamma(\alpha-1)} + \dfrac{d_2[(b_1+d_1\xi_1)+t(a_1-d_1)](\xi_2-s)^{\alpha-1}}{\Delta\Gamma(\alpha)}, & \xi_1 \le s \le \xi_2,\ s \le t; \\ -\dfrac{a_2[(b_1+d_1\xi_1)+t(a_1-d_1)](1-s)^{\alpha-1}}{\Delta\Gamma(\alpha)} - \dfrac{b_2[(b_1+d_1\xi_1)+t(a_1-d_1)](1-s)^{\alpha-2}}{\Delta\Gamma(\alpha-1)} + \dfrac{d_2[(b_1+d_1\xi_1)+t(a_1-d_1)](\xi_2-s)^{\alpha-1}}{\Delta\Gamma(\alpha)}, & \xi_1 \le s \le \xi_2,\ t \le s; \\ \dfrac{(t-s)^{\alpha-1}}{\Gamma(\alpha)} - \dfrac{a_2[(b_1+d_1\xi_1)+t(a_1-d_1)](1-s)^{\alpha-1}}{\Delta\Gamma(\alpha)} - \dfrac{b_2[(b_1+d_1\xi_1)+t(a_1-d_1)](1-s)^{\alpha-2}}{\Delta\Gamma(\alpha-1)}, & \xi_2 \le s,\ s \le t; \\ -\dfrac{a_2[(b_1+d_1\xi_1)+t(a_1-d_1)](1-s)^{\alpha-1}}{\Delta\Gamma(\alpha)} - \dfrac{b_2[(b_1+d_1\xi_1)+t(a_1-d_1)](1-s)^{\alpha-2}}{\Delta\Gamma(\alpha-1)}, & \xi_2 \le s,\ t \le s. \end{cases} \tag{20}$$

Proof. Based on the idea of paper [7], assuming that $x(t)$ satisfies (18), by Lemma 15, we formally put

$$\begin{aligned} x(t) &= I_{0+}^{\alpha}\rho(t) - c_1 - c_2 t \\ &= \frac{1}{\Gamma(\alpha)} \int_0^t (t-s)^{\alpha-1} \rho(t)\,ds - c_1 - c_2 t \end{aligned} \tag{21}$$

for some constants $c_1, c_2 \in \mathbb{R}$.

On the other hand, by the relations $D_{0+}^{\alpha} I_{0+}^{\alpha} x(t) = x(t)$ and $I_{0+}^{\alpha} I_{0+}^{\beta} x(t) = I_{0+}^{\alpha+\beta} x(t)$, for $\alpha, \beta > 0,\ x \in C(I, E_w)$, we get

$$x'(t) = \frac{1}{\Gamma(\alpha-1)} \int_0^t (t-s)^{\alpha-2} \rho(s)\,ds - c_2. \tag{22}$$

By the boundary conditions of (18), we have

$$\begin{aligned} &(d_1 - a_1)c_1 + (b_1 + d_1\xi_1)c_2 \\ &\quad = d_1 I_{0+}^{\alpha}\rho(\xi_1) + b_1 I_{0+}^{\alpha-1}\rho(0) - a_1 I_{0+}^{\alpha}\rho(0), \\ &(d_2 - a_2)c_1 + (-a_2 - b_2 + d_2\xi_2)c_2 \\ &\quad = d_2 I_{0+}^{\alpha}\rho(\xi_2) - b_2 I_{0+}^{\alpha-1}\rho(1) - a_2 I_{0+}^{\alpha}\rho(1), \end{aligned} \tag{23}$$

By the proof of paper [12], we get

$$\begin{aligned} c_1 &= -\frac{d_1(a_2+b_2-d_2\xi_2)}{\Delta\Gamma(\alpha)} \int_0^{\xi_1} (\xi_1-s)^{\alpha-1}\rho(s)\,ds \\ &\quad + \frac{(b_1+d_1\xi_1)}{\Delta}\left[a_2 \int_0^1 \frac{(1-s)^{\alpha-1}}{\Gamma(\alpha)}\rho(s)\,ds \right. \\ &\quad + b_2 \int_0^1 \frac{(1-s)^{\alpha-2}}{\Gamma(\alpha-1)}\rho(s)\,ds \\ &\quad \left. - d_2 \int_0^{\xi_2} \frac{(\xi_2-s)^{\alpha-1}}{\Gamma(\alpha)}\rho(s)\,ds\right], \\ c_2 &= \frac{d_1(a_2-d_2)}{\Delta\Gamma(\alpha)} \int_0^{\xi_1} (\xi_1-s)^{\alpha-1}\rho(s)\,ds \\ &\quad + \frac{(a_1-d_1)}{\Delta}\left[a_2 \int_0^1 \frac{(1-s)^{\alpha-1}}{\Gamma(\alpha)}\rho(s)\,ds \right. \\ &\quad + b_2 \int_0^1 \frac{(1-s)^{\alpha-2}}{\Gamma(\alpha-1)}\rho(s)\,ds \\ &\quad \left. - d_2 \int_0^{\xi_2} \frac{(\xi_2-s)^{\alpha-1}}{\Gamma(\alpha)}\rho(s)\,ds\right], \end{aligned} \tag{24}$$

where $\Delta = [(b_1+d_1\xi_1)(a_2-d_2)+(a_2+b_2-d_2\xi_2)(a_1-d_1)] \neq 0$. Substituting the values of c_1 and c_2 in (21), we get

$$\begin{aligned} x(t) &= \frac{1}{\Gamma(\alpha)} \int_0^t (t-s)^{\alpha-1}\rho(s)\,ds \\ &\quad + \frac{d_1[a_2(1-t)+b_2+d_2(t-\xi_2)]}{\Delta\Gamma(\alpha)} \\ &\quad \cdot \int_0^{\xi_1} (\xi_1-s)^{\alpha-1}\rho(s)\,ds \\ &\quad - \frac{a_2[(b_1+d_1\xi_1)+t(a_1-d_1)]}{\Delta\Gamma(\alpha)} \\ &\quad \cdot \int_0^1 (1-s)^{\alpha-1}\rho(s)\,ds \\ &\quad - \frac{b_2[(b_1+d_1\xi_1)+t(a_1-d_1)]}{\Delta\Gamma(\alpha-1)} \\ &\quad \cdot \int_0^1 (1-s)^{\alpha-2}\rho(s)\,ds \\ &\quad + \frac{d_2[(b_1+d_1\xi_1)+t(a_1-d_1)]}{\Delta\Gamma(\alpha)} \\ &\quad \cdot \int_0^{\xi_2} (\xi_2-s)^{\alpha-1}\rho(s)\,ds = \int_0^1 G(t,s)\,\rho(s)\,ds. \end{aligned} \tag{25}$$

This completes the proof. □

Let $D_r = \{z \in C(I, E_w), \|z\| \le r\}$, $BV(I, \mathbb{R})$ denote the space of real bounded variation functions with its classical norm $\| \cdot \|_{BV}$.

Problem (5) will be studied under the following assumptions:

(1) For each weakly continuous function $x : I \longrightarrow E$, the functions $k_1(t,\cdot)g(\cdot, x(\cdot)), k_2(t,\cdot)h(\cdot, x(\cdot))$, $f(\cdot, x(\cdot), T(x)(\cdot), S(x)(\cdot))$ are HKP-integrable, $f : I \times E^3 \longrightarrow E, g, h : I \times E \longrightarrow E$ are weakly-weakly continuous function, and $\int_0^t g(s, x(s))ds$, $\int_0^1 h(s, x(s))ds$ are bounded.

(2)

(i) For any $r > 0$, there exist a HK-integrable function $m : I \longrightarrow \mathbb{R}^+$ and nondecreasing continuous functions $\psi_1 : [0, +\infty) \longrightarrow (0, \infty), \psi_2 : [0, +\infty) \longrightarrow [0, +\infty), \psi_3 : [0, +\infty) \longrightarrow [0, +\infty)$, ψ_2, ψ_3 satisfying $\psi_2(\lambda x) \le \lambda\psi_2(x)$, $\psi_3(\lambda x) \le \lambda\psi_3(x)$ for $\lambda > 0$ such that

$$\begin{aligned} &\|f(s,x,y,z)\| \\ &\quad \le m(s)\left[\psi_1(\|x\|) + \psi_2(|y|) + \psi_3(|z|)\right], \\ &\psi_2(|g(s,x)|) \le \psi_2(|x|), \\ &\psi_3(|h(s,x)|) \le \psi_3(|x|) \end{aligned} \tag{26}$$

for all $s \in I, (x, y, z) \in D_r \times D_r \times D_r$ with

$$\int_0^1 M(s)\,ds < \int_0^\infty \frac{dr}{\sum_{i=1}^3 \psi_i(s)}. \tag{27}$$

(ii) For each bounded set $X, Y, Z \subset D_r$, and each for each closed interval $J \subset I, t \in I$, there exists positive constant $l \ge 0$ such that

$$\begin{aligned} &\beta(k_1(J,J)\,g(J,Y) \le k_1^*\beta(Y(J)), \\ &\beta(k_2(J,J)\,h(J,Z) \le k_2^*\beta(Z(J)) \\ &\beta(f(t,X,Y,Z)) \le l\max\{\beta(X), \beta(Y), \beta(Z)\}, \end{aligned} \tag{28}$$

where $M(s) = G^* m(s) \max\{1, ak_1^*, ak_2^*\}, k_1^* = \sup_{t\in I}\|k_1(t,\cdot)\|_{BV}, k_2^* = \sup_{t\in I}\|k_2(t,\cdot)\|_{BV}$.

(3) For each $t \in I, G(t, .), k_i(t, \cdot) \in BV(I, \mathbb{R}), i = 1, 2$ are continuous; i.e., the maps $t \longmapsto G(t, .)$ and $t \longmapsto k_i(t, .)$ are $\|.\|_{BV}$-continuous.

(4) The family $\{x^* f(\cdot, x(\cdot), T(x)(\cdot), S(x)(\cdot)) : x^* \in E^*, \|x^*\| \le 1\}$ is uniformly HK-integrable over I for every $x \in D_r$.

Remark 17. From assumption (3) and the expression of function $G(t, s)$, it is obvious that it is bounded and let $G^* = \sup_{t\in I}\|G(t,\cdot)\|_{BV}$.

Now, we present the existence theorem for problem (5).

Theorem 18. *Assume that conditions (5)-(20). Then problem (5) has a solution $x \in C(I, E_w)$.*

Proof. Let $m = \max\{\sup_{t\in I}\|k_i(t,\cdot)\|_{BV}, i = 1, 2\}$ and $k_0 = \max\{\sup_{t\in I}|\int_0^t g(s, x(s))ds|, \sup_{t\in I}|\int_0^1 h(s, x(s))ds|\}$. Let $0 < k_0 < \min(r_0, r_0/m)$, for $x \in D_{r_0}$ and $x^* \in E^*$ such that $\|x\|^* \le 1$; we have

$$\begin{aligned} |x^*(Tx(s))| &= \left|(\mathrm{HK})\int_0^t x^*(k_1(t,s)\,g(s,x(s)))\,ds\right| \\ &\le \|x^*\| \sup_{t\in I}\|k_1(t,\cdot)\|_{BV}\int_0^1 \|g(s,x(s))\|\,ds \\ &\le m\cdot k_0 \le r_0, \end{aligned} \tag{29}$$

and also

$$\sup\{|x^*Tx| : x \in E^*, \|x^*\| \le 1\} \le r_0. \tag{30}$$

So $Tx \in D_{r_0}$. Similarly, we prove $Sx \in D_{r_0}$.

Defining the set

$$\begin{aligned} Q := \Big\{&x \in D_{r_0} : \|x(\cdot)\| \le r_0,\ \|x(t) - x(s)\| \\ &\le \frac{r_0}{G^*}\|G(t_2,\cdot) - G(t_1,\cdot)\|_{BV},\ t_1, t_2 \in I\Big\}, \end{aligned} \tag{31}$$

it is clear that the convex closed and equicontinuous subset $Q \subset D_{r_0} \subset C(I, E_w)$, where

$$\begin{aligned} b(t) &= I^{-1}\left(\int_0^t M(s)\,ds\right) \quad \text{and} \\ I(z) &= \int_0^z \frac{ds}{\sum_{i=1}^3 \psi_i(s)}. \end{aligned} \tag{32}$$

Clearly,

$$\begin{aligned} b'(t) &= M(t)\left(\sum_{i=1}^3 \psi_i(b(t))\right), \quad \text{and} \\ b(0) &= 0 \end{aligned} \tag{33}$$

for all $t \in I$. Also notice that Q is a closed, convex, bounded, and equicontinuous subset of $C(I, E_w)$. We define the operator $F : C(I, E_w) \longrightarrow C(I, E_w)$ by

$$Fx(t) = \int_0^1 G(t,s)\,f(s,x(s),(Tx)(s),(Sx)(s))\,ds, \quad t \in I, \tag{34}$$

where $G(\cdot, \cdot)$ is Green's function defined by (20). Clearly the fixed points of the operator F are solutions of problem (5). Since for $t \in I$ the function $s \longmapsto G(t, s)$ is of bounded variation, then by the proof of Theorem 3.1 in [13] and assumption (4), the function $G(t,\cdot)f(\cdot, x(\cdot), T(x)(\cdot), S(x)(\cdot))$ is HKP-integrable on I and thus the operator F makes sense.

We will show that F satisfies the assumptions of Lemma 8; the proof will be given in three steps.

Step 1. We shall show that the operator F maps into itself. To see this, let $x \in Q, t \in I$. Without loss of generality, assume that $Fx(t) \neq 0$. By Hahn-Banach theorem, there exists $x^* \in E^*$ with $\|x^*\| = 1$ and $\|Fx(t)\| = |x^*(Fx(t))|$. Thus

$$\begin{aligned}
\|Fx(t)\| &= \left|x^*(Fx(t))\right| \\
&\leq x^*\left(\int_0^1 G(t,s) f(s, x(s), (Tx)(s), (Sx)(s))\, ds\right) \\
&\leq \sup_{t\in I} \|G(t,\cdot)\|_{BV} \int_0^1 m(s) \\
&\quad \cdot \left[\psi_1(b(s)) + ak_1^*\psi_2(b(s)) + ak_2^*\psi_3(b(s))\right] ds \\
&\leq \int_0^1 b'(s)\, ds \leq I^{-1}\left(\int_0^1 M(s)\, ds\right) = r_0.
\end{aligned} \tag{35}$$

Then $\|Fx\| = \sup_{t\in I}|Fx(t)| \leq r_0$. Hence $F : Q \longrightarrow Q$.

Let $0 < t_1 < t_2 \leq 1$, without loss of generality; assume that $Fx(t_2) - Fx(t_1) \neq 0$. By Hahn-Banach theorem, there exists $x^* \in E^*$ with $\|x^*\| = 1$ and

$$\begin{aligned}
&\left\|Fx(t_2) - Fx(t_1)\right\| = x^*(Fx(t_2) - Fx(t_1)) \\
&\quad \leq \int_0^1 \left|G(t_2,s) - G(t_1,s)\right| \\
&\quad \cdot \left|x^*(f(s, x(s), (Tx)(s), (Sx)(s)))\right| ds \\
&\quad \leq \left\|G(t_2,\cdot) - G(t_1,\cdot)\right\|_{BV} \int_0^1 m(s) \\
&\quad \cdot \left[\psi_1(b(s)) + ak_1^*\psi_2(|b(s)) + ak_2^*\psi_3(b(s))\right] ds \\
&\quad \leq \frac{1}{G^*}\left\|G(t_2,\cdot) - G(t_1,\cdot)\right\|_{BV} \int_0^1 b'(s)\, ds \\
&\quad \leq \frac{1}{G^*}\left\|G(t_2,\cdot) - G(t_1,\cdot)\right\|_{BV} I^{-1}\left(\int_0^1 M(s)\, ds\right) \\
&\quad = \frac{r_0}{G^*}\left\|G(t_2,\cdot) - G(t_1,\cdot)\right\|_{BV},
\end{aligned} \tag{36}$$

and this estimation shows that F maps Q into itself.

Step 2. We will show that the operator F is weakly sequentially continuous. In order to be simple, we denote $Tx(t) = \phi(x)(t) = \int_0^1 k_1(t,s)g(s,x(s))ds$, $Sx(t) = \varphi(x)(t) = \int_0^1 k_2(t,s)h(s,x(s))ds$. To see this, by Lemma 9 of [22], a sequence $x_n(\cdot)$ weakly convergent to $x(\cdot) \in Q$ if and only if $x_n(\cdot)$ tends weakly to $x(t)$ for each $t \in I$. From Dinculeanu ([23, p. 380]) $(C(I,E))^* = M(I,E^*)$, $M(I,E^*)$ is the set of all bounded regular vector measures from I to E^* which are of bounded variation). Let $x^* \in E^*$, $t \in I$. Put $P_t = x^*\delta_t$, where δ_t is the Dirac measure concentrated at the point t. Then $P_t \in M(I,E^*)$. Since x_n converges weakly to $x \in Q$, then we have

$$\lim_{n\longrightarrow\infty} \langle p_t, x_n - x\rangle = 0 \tag{37}$$

which means that

$$\lim_{n\longrightarrow\infty} \langle x^*, x_n - x\rangle = 0. \tag{38}$$

Thus, for each $t \in I$, $x_n(t)$ converges weakly to $x(t) \in E$. Since $g(s,\cdot)$, $h(s,\cdot)$ are weakly-weakly sequentially continuous, then $g(s,x_n(s))$ and $h(s,x_n(s))$ converge weakly to $g(s,x(s))$ and $h(s,x(s))$, respectively. Hence, and by Theorem 4 and assumptions (1), we have

$$\lim_{n\longrightarrow\infty} \int_0^1 \left(\int_0^t (k_1(t,s)\, g(s,x_n(s)) - k_1(t,s)\, g(s,x(s)))\, ds\right) dm(s) = 0, \quad \forall m \in M(I,E^*). \tag{39}$$

This relation is equivalent to

$$\begin{aligned}
&\lim_{n\longrightarrow\infty} (m, \phi(x_n) - \phi(x)) \\
&\quad = \lim_{n\longrightarrow\infty} \int_0^1 (\phi(x_n)(t) - \phi(x)(t))\, dm(t) = 0, \quad \forall m \in M(I,E^*).
\end{aligned} \tag{40}$$

Similarly, we have

$$\lim_{n\longrightarrow\infty} \int_0^1 \left(\int_0^1 (k_2(t,s)\, h(s,x_n(s)) - k_2(t,s)\, h(s,x(s)))\, ds\right) dm(s) = 0, \quad \forall m \in M(I,E^*). \tag{41}$$

This relation is equivalent to

$$\begin{aligned}
&\lim_{n\longrightarrow\infty} (m, \varphi(x_n) - \varphi(x)) \\
&\quad = \lim_{n\longrightarrow\infty} \int_0^1 (\varphi(x_n)(t) - \varphi(x)(t))\, dm(t) = 0, \quad \forall m \in M(I,E^*).
\end{aligned} \tag{42}$$

Therefore, the operators T, S are weakly sequentially continuous in Q.

Moreover, because f is weakly-weakly sequentially continuous, we have that $f(s,x_n(s),(Tx_n)(s),(Sx_n)(s))$ converges weakly to $f(s,x(s),(Tx)(s),(Sx)(s))$ in E. By assumption (4), for every weakly convergent $(x_n)_n \subset D_{r_0}$, the set

$$\{x^* f(\cdot,x_n(\cdot),T(x_n)(\cdot),S(x_n)(\cdot)) : n \in N, x^* \in B(E^*)\} \tag{43}$$

is HK-equi-integrable. Since for $t \in I$ the function $s \longmapsto G(t,s)$ is of bounded variation, and by the proof of Theorem 3.1 in [13], the function $G(t,\cdot)f(\cdot,x_n(\cdot),(Tx_n)(\cdot),(Sx_n)(\cdot))$ is HKP-integrable on I for every $n \geq 1$, and by Theorem 4, we have that $\int_0^1 G(t,s)f(s,x_n(s),(Tx_n)(s),(Sx_n)(s))ds$ converges weakly to $\int_0^1 G(t,s)f(s,x(s),(Tx)(s),(Sx)(s))ds$ in E which means that

$$\lim_{n\longrightarrow\infty} \int_0^1 \left(\int_0^1 (G(t,s)\left[f(s,x_n(s),(Tx_n)(s),(Sx_n)(s)) - f(s,x(s),(Tx)(s),(Sx)(s))\right] ds \right) dm(s) = 0, \tag{44}$$

for all $m \in M(I,E^*)$. This relation is equivalent to

$$\lim_{n\longrightarrow\infty} (m, F(x_n) - F(x)) = \lim_{n\longrightarrow\infty} \int_0^1 (F(x_n)(t) - F(x)(t))\,dm(t) = 0, \quad \forall m \in M(I,E^*). \tag{45}$$

Therefore F is weakly-weakly sequentially continuous.

Step 3. We show that there is an integer n_0 such that the operator F is β-power-convex condensing about 0 and n_0. To see this, notice that, for each bounded set $H \subseteq Q$ and for each $t \in I$,

$$\begin{aligned}\beta\left(F^{(1,0)}(H)(t)\right) &= \beta(F(H)(t)) = \beta\left(\left\{\int_0^t G(t,s) \cdot f(s,x(s),(Tx)(s),(Sx)(s))\,ds : x \in H\right\}\right) \\ &\leq \beta(G^* t\overline{co}\{f(s,x(s),(Tx)(s),(Sx)(s)) : x \in H, s \in I\}) = G^* t\beta(\overline{co}\{f(s,x(s),(Tx)(s),(Sx)(s)) : x \in H, s \in I\}) \leq G^* t\beta(f(I \times H(I) \times T(H)(I) \times S(H)(I))) \leq G^* t \cdot \max\{1,k_1^*,k_2^*\} \cdot l\beta(H(I)).\end{aligned} \tag{46}$$

Let $\tau = G^* \cdot \max\{1,k_1^*,k_2^*\} \cdot l > 0$. Lemma 7 implies (since H is equicontinuous) that

$$\beta\left(F^{(1,0)}(H)(t)\right) \leq t\tau\beta(H). \tag{47}$$

Since $F^{(1,0)}(H)$ is equicontinuous, it follows from Lemma 5 that $F^{(2,0)}(H)$ is equicontinuous. Using (47), we get

$$\begin{aligned}\beta\left(F^{(2,0)}(H)(t)\right) &= \beta\left(\left\{\int_0^t G(t,s) \cdot f(s,x(s),(Tx)(s),(Sx)(s))\,ds : x \in \overline{co}\left(F^{(1,0)}(H) \cup \{0\}\right)\right\}\right) \\ &\leq \beta\left(\left\{\int_0^t G(t,s) \cdot f(s,x(s),(Tx)(s),(Sx)(s))\,ds : x \in V\right\}\right),\end{aligned} \tag{48}$$

where $V = \overline{co}(F^{(1,0)}(H) \cup \{0\})$; it is clear that V is equicontinuous set. By Lemma 8, we get

$$\beta(V(s)) = \beta\left(F^{(1,0)}(H)(s)\right) \leq s\tau\beta(H), \tag{49}$$

and therefore,

$$\int_0^t \beta(V(s))\,ds \leq s\tau\frac{t^2}{2}\beta(H). \tag{50}$$

Thus,

$$\beta\left(F^{(2,0)}(H)(t)\right) \leq \frac{(\tau t)^2}{2}\beta(H). \tag{51}$$

By induction, we get

$$\beta\left(F^{(n,0)}(H)(t)\right) \leq \frac{(\tau t)^n}{n!}\beta(H). \tag{52}$$

And by Lemma 7, we have

$$\beta\left(F^{(n,0)}(H)\right) \leq \frac{(\tau T)^n}{n!}\beta(H). \tag{53}$$

Since $\lim_{n\longrightarrow\infty}((\tau t)^n/n!) = 0$, then there exist an n_0 with $(\tau t)^{n_0}/n_0! < 1$, and we have

$$\beta\left(F^{(n_0,0)}(H)\right) \leq \beta(H). \tag{54}$$

Consequently, F is β-power-convex condensing about 0 and n_0, by Lemma 8, then problem (5) has a solution $x \in C(I, E_w)$. □

4. Conclusions

In this paper, we use the techniques of measure of weak noncompactness and Henstock-Kurzweil-Pettis integrals to discuss the existence theorem of weak solutions for a class of the multipoint boundary value problem of fractional integrodifferential equations equipped with the weak topology. Our results generalized some classical results.

Conflicts of Interest

The authors declare that they have no conflicts of interest.

Authors' Contributions

All authors contributed equally to the writing of this paper. All authors read and approved the final manuscript.

Acknowledgments

This work is supported by National Natural Science Foundation of China (Grant no. 11061031).

References

[1] D. O'Regan, "Existence results for nonlinear integral equations," *Journal of Mathematical Analysis and Applications*, vol. 192, no. 3, pp. 705–726, 1995.

[2] D. O'Regan, "Singular integral equations arising in draining and coating flows," *Applied Mathematics and Computation*, vol. 205, no. 1, pp. 438–441, 2008.

[3] F. S. De Blasi, "On a property of the unit sphere in a Banach space," *Bulletin mathematiques de la Societe des sciences mathematiques de Roumanie*, vol. 21(69), no. 3-4, pp. 259–262, 1977.

[4] G. Emmanuele, "Measure of weak noncompactness and fixed point theorems," *Bulletin mathematiques de la Societe des sciences mathematiques de Roumanie*, vol. 25, no. 4, pp. 353–358, 1981.

[5] O. Arino, S. Gautier, and J.-P. Penot, "A fixed point theorem for sequentially continuous mappings with application to ordinary differential equations," *Funkcialaj Ekvacioj. Serio Internacia*, vol. 27, no. 3, pp. 273–279, 1984.

[6] Z. Bai, "On positive solutions of a nonlocal fractional boundary value problem," *Nonlinear Analysis: Theory, Methods & Applications*, vol. 72, no. 2, pp. 916–924, 2010.

[7] H. A. Salem, "On the fractional order m-point boundary value problem in reflexive Banach spaces and weak topologies," *Journal of Computational and Applied Mathematics*, vol. 224, no. 2, pp. 565–572, 2009.

[8] W. Y. Zhong and W. Lin, "Nonlocal and multiple-point boundary value problem for fractional differential equations," *Computers & Mathematics with Applications*, vol. 59, no. 3, pp. 1345–1351, 2010.

[9] Z. Bai, "On solutions of some fractional m-point boundary value problems at resonance," *Electronic Journal of Qualitative Theory of Differential Equations*, pp. 1–15, 2010.

[10] Q. Song, X. Dong, Z. Bai, and B. Chen, "Existence for fractional Dirichlet boundary value problem under barrier strip conditions," *Journal of Nonlinear Sciences and Applications. JNSA*, vol. 10, no. 7, pp. 3592–3598, 2017.

[11] Z. Bai, S. Zhang, S. Sun, and C. Yin, "Monotone iterative method for fractional differential equations," *Electronic Journal of Differential Equations*, vol. 2016, article 6, 2016.

[12] W. Zhou and Y. Chu, "Existence of solutions for fractional differential equations with multi-point boundary conditions," *Communications in Nonlinear Science and Numerical Simulation*, vol. 17, no. 3, pp. 1142–1148, 2012.

[13] B. Li and H. Gou, "Weak solutions nonlinear fractional integrodifferential equations in nonreflexive Banach spaces," *Boundary Value Problems*, Paper No. 209, 13 pages, 2016.

[14] N. Hussain and M. A. Taoudi, "Krasnosel'skii-type fixed point theorems with applications to Volterra integral equations," *Fixed Point Theory and Applications*, vol. 2013, no. 1, article 196, 2013.

[15] M. Cicho'n, I. Kubiaczyk, and A. Sikorska, "The Henstock-Kurzweil-Pettis integrals and existence theorems for the Cauchy problem," *Czechoslovak Mathematical Journal*, vol. 54(129), no. 2, pp. 279–289, 2004.

[16] S. B. Kaliaj, A. D. Tato, and F. D. Gumeni, "Controlled convergence theorem for the Henstock-Kurwzeil-Pettis integrals on m-dimiensional compact intervals," *Czechoslovak Mathematical Journal*, vol. 62, no. 1, pp. 243–255, 2012.

[17] D. Guo, V. Lakshmikantham, and X. Liu, *Nonlinear Integral Equations in Abstract Spaces*, vol. 373 of *Mathematics and its Applications*, Kluwer Academic Publishers Group, Dordrecht, The Netherlands, 1996.

[18] J. Banaś and K. Goebel, *Measures of Noncompactness*, vol. 60 of *Lecture Notes in Pure and Applied Mathematics*, Marcel Dekker, New York, NY, USA, 1980.

[19] A. R. Mitchell and C. Smith, "An existence theorem for weak solutions of differential equations in Banach spaces," in *Nonlinear equations in abstract spaces (Proc. Internat. Sympos., Univ. TEXas, Arlington, TEX., 1977)*, pp. 387–403, Academic Press, New York, 1978.

[20] R. P. Agarwal, D. O'Regan, and M.-A. Taoudi, "Fixed point theorems for convex-power condensing operators relative to the weak topology and applications to Volterra integral equations," *Journal of Integral Equations and Applications*, vol. 24, no. 2, pp. 167–181, 2012.

[21] Z. Bai and H. Lü, "Positive solutions for boundary value problem of nonlinear fractional differential equation," *Journal of Mathematical Analysis and Applications*, vol. 311, no. 2, pp. 495–505, 2005.

[22] A. R. Mitchell and C. Smith, "An existence theorem for weak solutions of differential equations in Banach spaces," in *Nonlinear Equations in Abstract Spaces*, pp. 387–403, Orlando, 1978.

[23] J. Garcia-Falset, "Existence of fixed points and measures of weak noncompactness," *Nonlinear Analysis. Theory, Methods & Applications. An International Multidisciplinary Journal*, vol. 71, no. 7-8, pp. 2625–2633, 2009.

12

Stability Analysis of Additive Runge-Kutta Methods for Delay-Integro-Differential Equations

Hongyu Qin,[1] Zhiyong Wang,[2] Fumin Zhu,[3] and Jinming Wen[4]

[1]*Wenhua College, Wuhan 430074, China*
[2]*School of Mathematical Sciences, University of Electronic Science and Technology of China, Sichuan 611731, China*
[3]*College of Economics, Shenzhen University, Shenzhen 518060, China*
[4]*Department of Electrical and Computer Engineering, University of Toronto, Toronto, Canada M5S3G4*

Correspondence should be addressed to Fumin Zhu; zhufumin@szu.edu.cn

Academic Editor: Gaston Mandata N'guérékata

This paper is concerned with stability analysis of additive Runge-Kutta methods for delay-integro-differential equations. We show that if the additive Runge-Kutta methods are algebraically stable, the perturbations of the numerical solutions are controlled by the initial perturbations from the system and the methods.

1. Introduction

Spatial discretization of many nonlinear parabolic problems usually gives a class of ordinary differential equations, which have the stiff part and the nonstiff part; see, e.g., [1–5]. In such cases, the most widely used time-discretizations are the special organized numerical methods, such as the implicit-explicit numerical methods [6, 7], the additive Runge-Kutta methods [8–12], and the linearized methods [13, 14]. When applying the split numerical methods to numerically solve the equations, it is important to investigate the stability of the numerical methods.

In this paper, it is assumed that the spatial discretization of time-dependent partial differential equations yields the following nonlinear delay-integro-differential equations:

$$\begin{aligned} y'(t) &= f^{[1]}(t, y(t)) \\ &\quad + f^{[2]}\left(t, y(t), y(t-\tau), \int_{t-\tau}^{t} g(t,s,y(s))\,ds\right), \quad (1) \\ &\qquad t > 0, \\ y(t) &= \psi(t), \quad -\tau \le t \le 0. \end{aligned}$$

Here τ is a positive delay term, $\psi(t)$ is continuous, $f^{[1]}$: $[t_0, +\infty] \times X \rightarrow X$, and $f^{[2]}$: $[t_0, +\infty] \times X \times X \times X \rightarrow X$, such that problem (1) owns a unique solution, where X is a real or complex Hilbert space. Particularly, when $g \equiv 0$, problem (1) is reduced to the nonlinear delay differential equations. When the delay term $\tau = 0$, problem (1) is reduced to the ordinary differential equations.

The investigation on stability analysis of different numerical methods for problem (1) has fascinated generations of researchers. For example, Torelli [15] considered stability of Euler methods for the nonautonomous nonlinear delay differential equations. Hout [16] studied the stability of Runge-Kutta methods for systems of delay differential equations. Baker and Ford [17] discussed stability of continuous Runge-Kutta methods for integrodifferential systems with unbounded delays. Zhang and Vandewalle [18] discussed the stability of the general linear methods for integrodifferential equations with memory. Li and Zhang obtained the stability and convergence of the discontinuous Galerkin methods for nonlinear delay differential equations [19, 20]. More references for this topic can be found in [21–30]. However, few works have been found on the stability of splitting methods for the proposed methods.

In the present work, we present the additive Runge-Kutta methods with some appropriate quadrature rules to numerically solve the nonlinear delay-integrodifferential equations (1). It is shown that if the additive Runge-Kutta methods are algebraically stable, the obtained numerical solutions are globally and asymptotically stable under the given assumptions, respectively. The rest of the paper is organized as follows. In Section 2, we present the numerical methods for problems (1). In Section 3, we consider stability analysis of the numerical schemes. Finally, we present some extensions in Section 4.

2. The Numerical Methods

In this section, we present the additive Runge-Kutta methods with the appropriate quadrature rules to numerically solve problem (1).

The coefficients of the additive Runge-Kutta methods can be organized in Buther tableau as follows (cf. [31]):

$$\begin{array}{c|c|c} c & A^{[1]} & A^{[2]} \\ \hline & \left(b^{[1]}\right)^T & \left(b^{[2]}\right)^T \end{array}, \tag{2}$$

where $c = [c_1, \cdots, c_s]^T$, $b^{[k]} = [b_1^{[k]}, \cdots, b_s^{[k]}]^T$, and $A^{[k]} = (a_{ij}^{[k]})_{i,j=1}^s$ for $k = 1, 2$.

Then, the presented ARKMs for problem (1) can be written by

$$\begin{aligned} y_{n+1} &= y_n + h\sum_{j=1}^s b_j^{[1]} f^{[1]}\left(t_n + c_j h, y_j^{(n)}\right) \\ &\quad + h\sum_{j=1}^s b_j^{[2]} f^{[2]}\left(t_n + c_j h, y_j^{(n)}, \tilde{y}_j^{(n)}\right), \\ y_i^{(n)} &= y_n + h\sum_{j=1}^s a_{ij}^{[1]} f^{[1]}\left(t_n + c_j h, y_j^{(n)}\right) \\ &\quad + h\sum_{j=1}^s a_{ij}^{[2]} f^{[2]}\left(t_n + c_j h, y_j^{(n)}, y_j^{(n-m)}, \tilde{y}_j^{(n)}\right), \\ &\qquad i = 1, 2, \cdots, s, \end{aligned} \tag{3}$$

where y_n and $y_i^{(n)}$ are approximations to the analytic solution $y(t_n)$ and $y(t_n + c_i h)$, respectively, $y_n = \psi(t_n)$ for $n \le 0$, $y_i^{(n)} = \psi(t_n + c_i h)$ for $t_n + c_i h \le 0$, and $\tilde{y}_i^{(n)}$ denotes the approximation to $\int_{t_n+c_ih-\tau}^{t_n+c_ih} g(t_n + c_i h, \xi, y(\xi))d\xi$, which can be computed by some appropriate quadrature rules

$$\tilde{y}_i^{(n)} = h\sum_{k=0}^m p_k g\left(t_n + c_i h, t_{n-k} + c_i h, y_i^{(n-k)}\right), \qquad i = 1, 2, \cdots, s. \tag{4}$$

For example, we usually adopt the repeated Simpson's rule or Newton-Cotes rule, etc., according to the requirement of the convergence of the method (cf. [18]).

3. Stability Analysis

In this section, we consider the numerical stability of the proposed methods. First, we introduce a perturbed problem, whose solution satisfies

$$\begin{aligned} z'(t) &= f^{[1]}(t, z(t)) \\ &\quad + f^{[2]}\left(t, z(t), z(t-\tau), \int_{t-\tau}^t g(t, s, z(s))\, ds\right), \quad t > 0, \\ y(t) &= \phi(t), \quad -\tau \le t \le 0. \end{aligned} \tag{5}$$

It is assumed that there exist some inner product $<\cdot,\cdot>$ and the induced norm $\|\cdot\|$ such that

$$\begin{aligned} &\operatorname{Re}\left\langle y - z, f^{[1]}(t, y) - f^{[1]}(t, z)\right\rangle \le \alpha \|y - z\|^2, \\ &\operatorname{Re}\left\langle y - z, f^{[2]}(t, y, u_1, v_1) - f^{[2]}(t, z, u_2, v_2)\right\rangle \\ &\quad \le \beta_1 \|y - z\|^2 + \beta_2 \|u_1 - u_2\|^2 + \gamma \|v_1 - v_2\|^2, \\ &\|g(t, v, s_1) - g(t, v, s_2)\| \le \theta \|s_1 - s_2\|, \end{aligned} \tag{6}$$

where $\alpha < 0$, $\beta_1 < 0$, $\beta_2 > 0$, $\gamma > 0$, and $\theta > 0$ are constants. It is remarkable that the conditions can be equivalent to the assumptions in [32, 33] (see. [34] *Remark* 2.1).

Definition 1 (cf. [9]). An additive Runge-Kutta method is called algebraically stable if the matrices

$$\begin{aligned} B_\nu &:= \operatorname{diag}\left(b_1^{[\nu]}, \cdots, b_s^{[\nu]}\right), \quad \nu = 1, 2, \\ M_{\nu\mu} &:= B_\nu A^{[\mu]} + A^{[\nu]T} B_\mu - b^{[\nu]} b^{[\mu]T} \end{aligned} \tag{7}$$

are nonnegative.

Theorem 2. *Assume an additive Runge-Kutta method is algebraically stable and $\beta_1 + \beta_2 + 4\gamma\tau^2\eta^2\theta^2 < 0$, where $\eta = \max\{p_1, p_2, \cdots, p_k\}$. Then, it holds that*

$$\begin{aligned} \|y_n - z_n\| &\le \sqrt{\left(1 + 2\sum_{i=1}^s \tau b_i^{[2]} \beta_2 + 4\gamma\tau^2\eta^2\theta^2\right)} \\ &\quad \cdot \max_{-\tau \le t \le 0} \|\psi(t) - \phi(t)\|, \end{aligned} \tag{8}$$

where y_n and z_n are numerical approximations to problems (1) and (5), respectively.

Proof. Let $\{y_n, y_i^{(n)}, \widetilde{y}_i^{(n)}\}$ and $\{z_n, z_i^{(n)}, \widetilde{z}_i^{(n)}\}$ be two sequences of approximations to problems (1) and (5), respectively, by ARKMs with the same stepsize h and write

$$\begin{aligned} U_i^{(n)} &= y_i^{(n)} - z_i^{(n)}, \\ \widetilde{U}_i^{(n)} &= \widetilde{y}_i^{(n)} - \widetilde{z}_i^{(n)}, \\ U_0^{(n)} &= y_n - z_n, \\ W_i^{[1]} &= h\left[f^{[1]}\left(t_n + c_i^{[1]}h, y_i^{(n)}\right) - f^{[1]}\left(t_n + c_i^{[1]}h, z_i^{(n)}\right)\right], \\ W_i^{[2]} &= h\left[f^{[2]}\left(t_n + c_i^{[2]}h, y_i^{(n)}, y_i^{(n-m)}, \widetilde{y}_i^{(n)}\right) - f^{[2]}\left(t_n + c_i^{[2]}h, z_i^{(n)}, z_i^{(n-m)}, \widetilde{z}_i^{(n)}\right)\right]. \end{aligned} \tag{9}$$

With the notation, the ARKMs for (1) and (5) yield

$$\begin{aligned} U_0^{(n+1)} &= U_0^n + \sum_{\mu=1}^{2}\sum_{j=1}^{s} b_j^{[\mu]} W_j^{[\mu]}, \\ U_i^{(n)} &= U_0^{(n)} + \sum_{\mu=1}^{2}\sum_{j=1}^{s} a_{ij}^{[\mu]} W_j^{[\mu]}, \quad i = 1, 2, \cdots, s. \end{aligned} \tag{10}$$

Thus, we have

$$\begin{aligned} \left\|U_0^{(n+1)}\right\|^2 &= \left\langle U_0^{(n)} + \sum_{\mu=1}^{2}\sum_{j=1}^{s} b_j^{[\mu]} W_j^{[\mu]}, U_0^{(n)} + \sum_{\nu=1}^{2}\sum_{i=1}^{s} b_i^{[\nu]} W_i^{[\nu]} \right\rangle = \left\|U_0^{(n)}\right\|^2 + 2\sum_{\mu=1}^{2}\sum_{i=1}^{s} b_i^{[\mu]} \\ &\cdot \operatorname{Re}\left\langle U_0^{(n)}, W_i^{[\mu]}\right\rangle + \sum_{\mu,\nu=1}^{2}\sum_{i,j=1}^{s} b_i^{[\mu]} b_j^{[\nu]} \left\langle W_i^{[\mu]}, W_j^{[\nu]}\right\rangle \\ &= \left\|U_0^{(n)}\right\|^2 + 2\sum_{\mu=1}^{2}\sum_{i=1}^{s} b_i^{[\mu]} \cdot \operatorname{Re}\left\langle U_i^{(n)} - \sum_{\nu=1}^{2}\sum_{j=1}^{s} a_{ij}^{[\nu]} W_j^{[\nu]}, W_i^{[\mu]} \right\rangle \\ &+ \sum_{\mu,\nu=1}^{2}\sum_{i,j=1}^{s} b_i^{[\mu]} b_j^{[\nu]} \left\langle W_i^{[\mu]}, W_j^{[\nu]}\right\rangle = \left\|U_0^{(n)}\right\|^2 + 2\sum_{\mu=1}^{2}\sum_{i=1}^{s} b_i^{[\mu]} \operatorname{Re}\left\langle U_i^{(n)}, W_i^{[\mu]}\right\rangle \\ &- \sum_{\mu,\nu=1}^{2}\sum_{i,j=1}^{s} \left(b_i^{[\mu]} a_{ij}^{[\nu]} + a_{ji}^{[\mu]} b_j^{[\nu]} - b_i^{[\mu]} b_j^{[\nu]}\right) \cdot \left\langle W_i^{[\mu]}, W_j^{[\nu]}\right\rangle. \end{aligned} \tag{11}$$

Since that the matrix $\mathscr{M}$ is a nonnegative matrix, we obtain

$$-\sum_{\mu,\nu=1}^{2}\sum_{i,j=1}^{s} \left(b_i^{[\mu]} a_{ij}^{[\nu]} + a_{ji}^{[\mu]} b_j^{[\nu]} - b_i^{[\mu]} b_j^{[\nu]}\right) \left\langle W_i^{[\mu]}, W_j^{[\nu]}\right\rangle \le 0. \tag{12}$$

Furthermore, by conditions (6), we find

$$\operatorname{Re}\left\langle U_i^{(n)}, W_i^{[1]}\right\rangle \le \alpha h \left\|U_i^{(n)}\right\|^2, \tag{13}$$

and

$$\operatorname{Re}\left\langle U_i^{(n)}, W_i^{[2]}\right\rangle \le \beta_1 h \left\|U_i^{(n)}\right\|^2 + \beta_2 h \left\|U_i^{(n-m)}\right\|^2 + \gamma h \left\|\widetilde{U}_i^{(n)}\right\|^2. \tag{14}$$

Together with (11), (12), (13), and (14), we get

$$\begin{aligned} \left\|U_0^{(n+1)}\right\|^2 &\le \left\|U_0^{(n)}\right\|^2 + 2\sum_{i=1}^{s} h b_i^{[1]} \alpha \left\|U_i^{(n)}\right\|^2 + 2\sum_{i=1}^{s} h b_i^{[2]} \left(\beta_1 \left\|U_i^{(n)}\right\|^2 + \beta_2 \left\|U_i^{(n-m)}\right\|^2 + \gamma \left\|\widetilde{U}_i^{(n)}\right\|^2\right) \\ &\le \left\|U_0^{(n)}\right\|^2 + 2\sum_{i=1}^{s} h b_i^{[2]} \left(\beta_1 \left\|U_i^{(n)}\right\|^2 + \beta_2 \left\|U_i^{(n-m)}\right\|^2 + \gamma \left\|\widetilde{U}_i^{(n)}\right\|^2\right). \end{aligned} \tag{15}$$

Note that

$$\left\|\widetilde{U}_i^{(n)}\right\|^2 = \left\| h\sum_{k=0}^{m} p_k \left[g\left(t_n + c_i h, t_{n-k} + c_i h, y_i^{n-k}\right) - g\left(t_n + c_i h, t_{n-k} + c_i h, z_i^{n-k}\right)\right]\right\|^2 \le (m+1) \cdot \eta^2 \theta^2 h^2 \sum_{k=0}^{m} \left\|U_i^{(n-k)}\right\|^2. \tag{16}$$

Then, we obtain

$$\begin{aligned} \left\|U_0^{(n+1)}\right\|^2 &\le \left\|U_0^{(n)}\right\|^2 + 2\sum_{i=1}^{s} h b_i^{[2]} \left(\beta_1 \left\|U_i^{(n)}\right\|^2 + \beta_2 \left\|U_i^{(n-m)}\right\|^2 + \gamma (m+1) \eta^2 \theta^2 h^2 \sum_{k=0}^{m} \left\|U_i^{(n-k)}\right\|^2\right) \\ &\le \left\|U_0^{(0)}\right\|^2 + 2\sum_{j=0}^{n}\sum_{i=1}^{s} h b_i^{[2]} \left(\beta_1 \left\|U_i^{(j)}\right\|^2 + \beta_2 \left\|U_i^{(j-m)}\right\|^2 + \gamma (m+1) \eta^2 \theta^2 h^2 \sum_{k=0}^{m} \left\|U_i^{(j-k)}\right\|^2\right) \end{aligned}$$

$$\begin{aligned}
&\leq \left\|U_0^{(0)}\right\|^2 + 2\sum_{j=0}^{n}\sum_{i=1}^{s} hb_i^{[2]}\left(\beta_1\left\|U_i^{(j)}\right\|^2 + \beta_2\left\|U_i^{(j)}\right\|^2\right.\\
&\left.+ \gamma(m+1)^2h^2\eta^2\theta^2\left\|U_i^{(j)}\right\|^2\right)\\
&+ 2\sum_{j=-m}^{-1}\sum_{i=1}^{s} hb_i^{[2]}\left(\beta_2\left\|U_i^{(j)}\right\|^2\right.\\
&\left.+ \gamma(m+1)^2h^2\eta^2\theta^2\left\|U_i^{(j)}\right\|^2\right) \leq \left\|U_0^{(0)}\right\|^2\\
&+ 2\sum_{j=0}^{n}\sum_{i=1}^{s} hb_i^{[2]}\left(\beta_1+\beta_2+4\gamma\tau^2\eta^2\theta^2\right)\left\|U_i^{(j)}\right\|^2\\
&+ 2\sum_{j=-m}^{-1}\sum_{i=1}^{s} hb_i^{[2]}\left(\beta_2+4\gamma\tau^2\eta^2\theta^2\right)\left\|U_i^{(j)}\right\|^2 \leq \left\|U_0^{(0)}\right\|^2\\
&+ 2\sum_{j=-m}^{-1}\sum_{i=1}^{s} hb_i^{[2]}\left(\beta_2+4\gamma\tau^2\eta^2\theta^2\right)\left\|U_i^{(j)}\right\|^2 \leq \left\|U_0^{(0)}\right\|^2\\
&+ 2\sum_{i=1}^{s} mhb_i^{[2]}\left(\beta_2+4\gamma\tau^2\eta^2\theta^2\right)\max_{-m\leq j\leq -1}\left\|U_i^{(j)}\right\|^2 .
\end{aligned} \tag{17}$$

Hence,

$$\left\|U_0^{(n+1)}\right\|^2 \leq C\max_{-\tau\leq t\leq 0}\left\|\psi(t)-\phi(t)\right\|^2, \tag{18}$$

where $C = [(1 + 2\sum_{i=1}^{s}\tau b_i^{[2]}\beta_2 + 4\gamma\tau^2\eta^2\theta^2)]$. This completes the proof. □

Theorem 3. *Assume an additive Runge-Kutta method is algebraically stable and $\beta_1 + \beta_2 + 4\gamma\tau^2\eta^2\theta^2 < 0$. Then, it holds that*

$$\lim_{n\to\infty}\left\|U_0^{(n)}\right\| = 0. \tag{19}$$

Proof. Similar to the proof of Theorem 2, it holds that

$$\begin{aligned}
&\left\|U_0^{(n+1)}\right\|^2\\
&\quad\leq \left\|U_0^{(0)}\right\|^2\\
&\quad\quad+ 2\sum_{j=0}^{n}\sum_{i=1}^{s} hb_i^{[2]}\left(\beta_1+\beta_2+4\gamma\tau^2\eta^2\theta^2\right)\left\|U_i^{(j)}\right\|^2\\
&\quad\quad+ 2\sum_{j=-m}^{-1}\sum_{i=1}^{s} hb_i^{[2]}\left(\beta_2+4\gamma\tau^2\eta^2\theta^2\right)\left\|U_i^{(j)}\right\|^2 .
\end{aligned} \tag{20}$$

Note that $\beta_1 + \beta_2 + 4\gamma\tau^2\eta^2\theta^2 < 0$ and $b_i^{[2]} > 0$; we have

$$\lim_{n\to\infty}\sum_{i=1}^{s} b_i^{[2]}\left\|U_i^{(n)}\right\| = 0. \tag{21}$$

On the other hand,

$$\begin{aligned}
\left\|W_i^{[1]}\right\| &= \left\|h\left[f^{[1]}\left(t_n + c_i^{[1]}h, y_i^{(n)}\right)\right.\right.\\
&\quad\left.\left.- f^{[1]}\left(t_n + c_i^{[1]}h, z_i^{(n)}\right)\right]\right\| \leq L_1\left\|U_i^{(n)}\right\|
\end{aligned} \tag{22}$$

$$\begin{aligned}
\left\|W_i^{[2]}\right\| &= \left\|h\left[f^{[2]}\left(t_n + c_i^{[2]}h, y_i^{(n)}, y_i^{(n-m)}, \tilde{y}_i^{(n)}\right)\right.\right.\\
&\quad\left.\left.- f^{[2]}\left(t_n + c_i^{[2]}h, z_i^{(n)}, z_i^{(n-m)}, \tilde{z}_i^{(n)}\right)\right]\right\| \leq L_2\left(\left\|U_i^{(n)}\right\|\right.\\
&\quad\left.+ \left\|U_i^{(n-m)}\right\| + \left\|\tilde{y}_i^{(n)} - \tilde{z}_i^{(n)}\right\|\right).
\end{aligned} \tag{23}$$

Now, in view of (10), (21), (22), and (23), we obtain

$$\lim_{n\to\infty}\left\|U_0^{(n)}\right\| = 0. \tag{24}$$

This completes the proof. □

Remark 4. In [35], Yuan et al. also discussed nonlinear stability of additive Runge-Kutta methods for multidelay-integro-differential equations. However, the main results are different. The main reason is that the results in [35] imply that the perturbations of the numerical solutions tend to infinity when the time increase, while the stability results in present paper indicate that the perturbations of the numerical solutions are independent of the time. Besides, the asymptotical stability of the methods is also discussed in the present paper.

4. Conclusion

The additive Runge-Kutta methods with some appropriate quadrature rules are applied to solve the delay-integro-differential equations. It is shown that if the additive Runge-Kutta methods are algebraically stable, the obtained numerical solutions can be globally and asymptotically stable, respectively. In the future works, we will apply the methods to solve more real-world problems.

Conflicts of Interest

The authors declare that they have no conflicts of interest.

Acknowledgments

This work is supported in part by the National Natural Science Foundation of China (71601125).

References

[1] V. Thomée, *Galerkin Finite Element Methods for Parabolic Problems*, Springer, Berlin, Germany, 1997.

[2] J. Wu, *Theory and Applications of Partial Functional-Differential Equations*, Springer, New York, NY, USA, 1996.

[3] J. R. Cannon and Y. Lin, "Non-classical H1 projection and Galerkin methods for non-linear parabolic integro-differential equations," *Calcolo*, vol. 25, pp. 187–201, 1988.

[4] D. Li and J. Wang, "Unconditionally optimal error analysis of crank-nicolson galerkin fems for a strongly nonlinear parabolic system," *Journal of Scientific Computing*, vol. 72, no. 2, pp. 892–915, 2017.

[5] B. Li and W. Sun, "Error analysis of linearized semi-implicit galerkin finite element methods for nonlinear parabolic equations," *International Journal of Numerical Analysis & Modeling*, vol. 10, no. 3, pp. 622–633, 2013.

[6] U. M. Ascher, S. J. Ruuth, and B. T. Wetton, "Implicit-explicit methods for time-dependent partial differential equations," *SIAM Journal on Numerical Analysis*, vol. 32, no. 3, pp. 797–823, 1995.

[7] G. Akrivis and B. Li, "Maximum norm analysis of implicit-explicit backward difference formulas for nonlinear parabolic equations," *SIAM Journal on Numerical Analysis*, 2017.

[8] I. Higueras, "Strong stability for additive Runge-Kutta methods," *SIAM Journal on Numerical Analysis*, vol. 44, no. 4, pp. 1735–1758, 2006.

[9] A. Araujo, "A note on B-stability of splitting methods," *Computing and Visualization in Science*, vol. 26, no. 2-3, pp. 53–57, 2004.

[10] C. A. Kennedy and M. H. Carpenter, "Additive Runge-Kutta schemes for convection-diffusion-reaction equations," *Applied Numerical Mathematics*, vol. 44, no. 1-2, pp. 139–181, 2003.

[11] T. Koto, "Stability of IMEX Runge-Kutta methods for delay differential equations," *Journal of Computational and Applied Mathematics*, vol. 211, pp. 201–212, 2008.

[12] H. Liu and J. Zou, "Some new additive Runge-Kutta methods and their applications," *Journal of Computational and Applied Mathematics*, vol. 190, no. 1-2, pp. 74–98, 2006.

[13] D. Li, C. Zhang, and M. Ran, "A linear finite difference scheme for generalized time fractional Burgers equation," *Applied Mathematical Modelling: Simulation and Computation for Engineering and Environmental Systems*, vol. 40, no. 11-12, pp. 6069–6081, 2016.

[14] D. Li, J. Wang, and J. Zhang, "Unconditionally convergent L1-Galerkin FEMs for nonlinear time-fractional Schrödinger equations," *SIAM Journal on Scientific Computing*, vol. 39, no. 6, pp. A3067–A3088, 2017.

[15] L. Torelli, "Stability of numerical methods for delay differential equations," *Journal of Computational and Applied Mathematics*, vol. 25, no. 1, pp. 15–26, 1989.

[16] K. J. in't Hout, "Stability analysis of Runge-Kutta methods for systems of delay differential equations," *IMA Journal of Numerical Analysis*, vol. 17, no. 1, pp. 17–27, 1997.

[17] C. T. Baker and A. Tang, "Stability analysis of continuous implicit Runge-Kutta methods for Volterra integro-differential systems with unbounded delays," *Applied Numerical Mathematics*, vol. 24, no. 2-3, pp. 153–173, 1997.

[18] C. Zhang and S. Vandewalle, "General linear methods for Volterra integro-differential equations with memory," *SIAM Journal on Scientific Computing*, vol. 27, no. 6, pp. 2010–2031, 2006.

[19] D. Li and C. Zhang, "Nonlinear stability of discontinuous Galerkin methods for delay differential equations," *Applied Mathematics Letters*, vol. 23, no. 4, pp. 457–461, 2010.

[20] D. Li and C. Zhang, "L∞ error estimates of discontinuous Galerkin methods for delay differential equations," *Applied Numerical Mathematics*, vol. 82, pp. 1–10, 2014.

[21] V. K. Barwell, "Special stability problems for functional differential equations," *BIT*, vol. 15, pp. 130–135, 1975.

[22] A. Bellen and M. Zennaro, "Strong contractivity properties of numerical methods for ordinary and delay differential equations," *Applied Numerical Mathematics*, vol. 9, no. 3-5, pp. 321–346, 1992.

[23] K. Burrage, "High order algebraically stable Runge-Kutta methods," *BIT*, vol. 18, no. 4, pp. 373–383, 1978.

[24] K. Burrage and J. C. Butcher, "Nonlinear stability of a general class of differential equation methods," *BIT*, vol. 20, no. 2, pp. 185–203, 1980.

[25] G. J. Cooper and A. Sayfy, "Additive Runge-Kutta methods for stiff ordinary differential equations," *Mathematics of Computation*, vol. 40, no. 161, pp. 207–218, 1983.

[26] K. Dekker and J. G. Verwer, *Stability of Runge-Kutta Methods for Stiff Nonlinear Differential Equations*, North-Holland Publishing, Amsterdam, The Netherlands, 1984.

[27] L. Ferracina and M. N. Spijker, "Strong stability of singly-diagonally-implicit Runge-Kutta methods," *Applied Numerical Mathematics*, vol. 58, no. 11, pp. 1675–1686, 2008.

[28] K. J. in't Hout and M. N. Spijker, "The θ-methods in the numerical solution of delay differential equations," in *The Numerical Treatment of Differential Equations*, K. Strehmel, Ed., vol. 121, pp. 61–67, 1991.

[29] M. Zennaro, "Asymptotic stability analysis of Runge-Kutta methods for nonlinear systems of delay differential equations," *Numerische Mathematik*, vol. 77, no. 4, pp. 549–563, 1997.

[30] D. Li, C. Zhang, and W. Wang, "Long time behavior of non-Fickian delay reaction-diffusion equations," *Nonlinear Analysis: Real World Applications*, vol. 13, no. 3, pp. 1401–1415, 2012.

[31] B. Garcia-Celayeta, I. Higueras, and T. Roldan, "Contractivity/monotonicity for additive Range-kutta methods: Inner product norms," *Applied Numerical Mathematics*, vol. 56, no. 6, pp. 862–878, 2006.

[32] C. Huang, "Dissipativity of one-leg methods for dynamical systems with delays," *Applied Numerical Mathematics*, vol. 35, no. 1, pp. 11–22, 2000.

[33] C. Zhang and S. Zhou, "Nonlinear stability and D-convergence of Runge-Kutta methods for delay differential equations," *Journal of Computational and Applied Mathematics*, vol. 85, no. 2, pp. 225–237, 1997.

[34] C. Huang, S. Li, H. Fu, and G. Chen, "Nonlinear stability of general linear methods for delay differential equations," *BIT Numerical Mathematics*, vol. 42, no. 2, pp. 380–392, 2002.

[35] H. Yuan, J. Zhao, and Y. Xu, "Nonlinear stability and D-convergence of additive Runge-Kutta methods for multidelay-integro-differential equations," *Abstract and Applied Analysis*, vol. 2012, Article ID 854517, 22 pages, 2012.

13

Analysis of a Predator-Prey Model with Switching and Stage-Structure for Predator

T. Suebcharoen

Center of Excellence in Mathematics and Applied Mathematics, Department of Mathematics, Faculty of Science, Chiang Mai University, Chiang Mai 50200, Thailand

Correspondence should be addressed to T. Suebcharoen; teeranush.s@gmail.com

Academic Editor: Yuji Liu

This paper studies the behavior of a predator-prey model with switching and stage-structure for predator. Bounded positive solution, equilibria, and stabilities are determined for the system of delay differential equation. By choosing the delay as a bifurcation parameter, it is shown that the positive equilibrium can be destabilized through a Hopf bifurcation. Some numerical simulations are also given to illustrate our results.

1. Introduction

The predator-prey system is important in dynamical population models and has been discussed by many authors [1–15].

In the related studies, a switching predator-prey model which has the switching property of predator was introduced by [7]. It was assumed that the predators catch prey in an abundant habitat. After a decrease in prey species population, the predator moves to another abundant habitat. In [8], the authors investigated a switching model of a two-prey one-predator system and they have shown that the system undergoes a Hopf bifurcation. They used the carrying capacity of prey as the bifurcations parameter. More examples on switching models can be found in [9–11]. Saito and Takeuchi [12] proposed a stage-structure model of a species' growth consisting of immature and mature individuals. It is assumed that the predators are divided into two-stage groups: juveniles and adults. Only the adult predators are able to catch prey species. As for the juvenile predators, they live with the adult predators. It is assumed that juveniles survive on prey already caught by adults. They live on a different resource which is available in the abundant habitat from the adult predators. Consequently, stage-structure model is more realistic than the model without stage-structure. In [14], it was further assumed that the time from juveniles to adults is itself state dependent. Qu and Wei [15] studied the asymptotic behavior of a predator-prey model with stage-structure. They found that an orbitally asymptotically stable periodic orbit exists in that model.

The purpose of the present paper is to study nonlinear delayed differential equations each of which describes a switching and stage structured predator-prey model. The present paper is organized as follows. In the next section, the main mathematical model is formulated and the positivity and boundedness of solutions are presented. In Section 3, we discuss the local stability of equilibria by analyzing the corresponding characteristic equations and we prove the existence of Hopf bifurcations for the model. Finally, numerical results and a brief discussion are provided.

2. Model

In this paper, we extend the switching predator-prey model in [8] by introducing stage structured with time delay into the model. We consider the switching with stage-structure predator-prey model of the following form:

$$\frac{dx_1}{dt} = rx_1\left(1 - \frac{x_1}{k}\right) + pqx_2 - \frac{\beta x_1 x_2 y}{x_1 + x_2}$$

$$\frac{dx_2}{dt} = rx_2\left(1 - \frac{x_2}{k}\right) + pqx_1 - \frac{\beta x_1 x_2 y}{x_1 + x_2}$$

$$\frac{dy}{dt} = 2\delta\beta e^{-\gamma\tau}\frac{x_1(t-\tau)x_2(t-\tau)y(t-\tau)}{x_1(t-\tau)+x_2(t-\tau)} - \mu y$$
$$\frac{dy_j}{dt} = 2\delta\beta\frac{x_1x_2y}{x_1+x_2} - 2\delta\beta e^{-\gamma\tau}\frac{x_1(t-\tau)x_2(t-\tau)y(t-\tau)}{x_1(t-\tau)+x_2(t-\tau)} - \gamma y_j \quad (1)$$

with initial conditions

$$x_1(\theta), x_2(\theta), y(\theta), y_j(\theta) \geq 0 \quad \text{continuous on } [-\tau, 0), \quad x_1(0), x_2(0), y(0) > 0, \quad y_j(0) > 0. \quad (2)$$

The model is formulated under the following assumptions:

(1) It is assumed that two-prey species, denoted by x_1 and x_2, respectively, can be modelled by a logistic equation when the predator is absent. The parameter r is the prey intrinsic growth rate and k is its carrying capacity.

(2) The prey lives in two different habitats and each prey is able to migrate among two different habitats. The parameter p is the probability of successful transition from each habitat and q is inverse barrier strength in going out of the first habitat and the second habitat.

(3) The functions $\beta x_1/(x_1 + x_2)$ and $\beta x_2/(x_1 + x_2)$ have a characteristic property of a switching mechanism, where β is capturing rate.

(4) The parameter δ is the rate of conversion of prey to predator and μ is the death rate of predator.

(5) The predators are derived into two-stage groups: juveniles and adults, which are divided by age τ, and they are denoted by $y_j(t)$ and $y(t)$, respectively. It is assumed that juveniles take τ units of time to mature and $e^{-\gamma\tau}$ is the surviving rate of juveniles to adults. Notice, we assume that the juveniles suffer a mortality rate of γ.

For ecological reasons, we always assume that the initial data $x_1(\theta)$, $x_2(\theta)$, $y(\theta)$, $y_j(\theta) \geq 0$ continuous on $[-\tau, 0)$, and $x_1(0)$, $x_2(0)$, $y(0)$, $y_j(0) > 0$. If $(x_1(t), x_2(t), y(t), y_j(t))$ is a solution of system (1) through that initial data, it is easy to verify that $(x_1(t), x_2(t), y(t), y_j(t))$ is positive on the maximum existence interval of solution. Such solutions will be called positive solution. Moreover, if such a solution is bounded above and below, it is called a positive solution. Furthermore, we discuss the bounded positive solutions of system (1) which implies a natural restriction; that is, our system (1) must have a bounded positive solution. The following theorem guarantees that our stage-structure predator-prey model (1) with initial condition (2) always has a bounded solution. Therefore, every solution to system (1) is positive and bounded.

Theorem 1. *Every solution of system (1) with initial condition (2) is bounded for all $t \geq 0$ and all of these solutions are ultimately bounded.*

Proof. Let $V(t) = \gamma(\delta x_1 + \delta x_2 + y + y_j)$. By calculating the derivative of $V(t)$ with respect to t along the positive solution of the system of system (1), we have

$$\begin{aligned}\dot{V}(t) &= \gamma\delta\dot{x}_1 + \gamma\delta\dot{x}_2 + \gamma\dot{y} + \gamma\dot{y}_j \\ &= \gamma\delta\left(rx_1 - \frac{r}{k}x_1^2 + pqx_2\right) \\ &\quad + \gamma\delta\left(rx_2 - \frac{r}{k}x_2^2 + pqx_1\right) - \gamma\mu y - \gamma^2 y_j.\end{aligned} \quad (3)$$

Let $\gamma > \mu$. We have

$$\begin{aligned}\dot{V}(t) + \mu V(t) &= (\gamma\delta r + pq + \gamma\mu\delta)(x_1 + x_2) \\ &\quad - \frac{\gamma\delta r}{k}(x_1^2 + x_2^2) - \gamma(\gamma - \mu)y_j \\ &< (\gamma\delta r + pq + \gamma\mu\delta)(x_1 + x_2) \\ &\quad - \frac{\gamma\delta r}{k}(x_1^2 + x_2^2).\end{aligned} \quad (4)$$

Hence, there exists a positive constant C, such that

$$\dot{V}(t) + \mu V(t) \leq C. \quad (5)$$

Thus, we get

$$V \leq \left(V(0) - \frac{C}{\mu}\right)e^{-\mu t} + \frac{C}{\mu}. \quad (6)$$

Therefore, $V(t)$ is ultimately bounded; that is, each solution of system (1) is ultimately bounded. □

3. Local Stability and Existence of Hopf Bifurcation

The main goal in this section is to investigate the stability of a positive equilibrium and the existence of a Hopf bifurcation.

Because of the last equation of system (1), $y_j(t)$ is completely determined by $x_1(t)$, $x_2(t)$, $y(t)$. Therefore, in the rest of this paper, we will study the following system:

$$\frac{dx_1}{dt} = rx_1\left(1 - \frac{x_1}{k}\right) + pqx_2 - \frac{\beta x_1x_2y}{x_1+x_2}$$
$$\frac{dx_2}{dt} = rx_2\left(1 - \frac{x_2}{k}\right) + pqx_1 - \frac{\beta x_1x_2y}{x_1+x_2}$$

$$\frac{dy}{dt} = 2\delta\beta e^{-\gamma\tau} \frac{x_1(t-\tau)x_2(t-\tau)y(t-\tau)}{x_1(t-\tau)+x_2(t-\tau)} - \mu y \tag{7}$$

with the initial conditions $x_1(\theta)$, $x_2(\theta)$, $y(\theta) \geq 0$ continuous on $[-\tau, 0)$ and $x_1(0)$, $x_2(0)$, $y(0) > 0$.

Before we proceed further, let us scale (7) by putting

$$\begin{aligned}
\overline{x}_1 &= \frac{x_1}{k}, \\
\overline{x}_2 &= \frac{x_2}{k}, \\
\overline{y} &= e^{\gamma\tau} y \\
\alpha &= \frac{2\delta\beta}{pq} k e^{-\gamma\tau} \\
g &= \frac{r}{pq}, \\
b &= \frac{e^{-\gamma\tau}\beta}{pq}, \\
d &= \frac{\mu}{pq}, \\
\overline{t} &= pqt, \\
\overline{\tau} &= pq\tau,
\end{aligned} \tag{8}$$

and dropping the bars for the sake of simplicity. We obtain the following system containing dimensionless quantities:

$$\begin{aligned}
\frac{dx_1}{dt} &= gx_1(1-x_1) + x_2 - \frac{bx_1x_2y}{x_1+x_2} \\
\frac{dx_2}{dt} &= gx_2(1-x_2) + x_1 - \frac{bx_1x_2y}{x_1+x_2} \\
\frac{dy}{dt} &= \alpha \frac{x_1(t-\tau)x_2(t-\tau)y(t-\tau)}{x_1(t-\tau)+x_2(t-\tau)} - dy.
\end{aligned} \tag{9}$$

Next, we find equilibria of system (9) by equating the derivatives on the left-hand sides to zero. The equilibria are solutions of the system

$$\begin{aligned}
gx_1(1-x_1) + x_2 - \frac{bx_1x_2y}{x_1+x_2} &= 0 \\
gx_2(1-x_2) + x_1 - \frac{bx_1x_2y}{x_1+x_2} &= 0 \\
\alpha \frac{x_1(t-\tau)x_2(t-\tau)y(t-\tau)}{x_1(t-\tau)+x_2(t-\tau)} - dy &= 0.
\end{aligned} \tag{10}$$

This gives two possible equilibria which are

(i) boundary equilibrium $E_1 = (x_1^*, gx_1^*(x_1^* - 1), 0)$, which is corresponding to extinction of the predator, where $x_1^* > 1$ is a real positive root of the cubic equation

$$g^3 x_1^{*3} - 2g^3 x_1^{*2} + (g^3 - g^2) x_1^* + (g^2 - 1) = 0. \tag{11}$$

(ii) positive equilibrium $E_2 = (\overline{x}_1, \overline{x}_2, \overline{y})$, which is corresponding to coexistence of prey and predator and

$$\begin{aligned}
\overline{x}_1 &= \frac{d}{\alpha}(\overline{x}+1) \\
\overline{x}_2 &= \frac{d}{\alpha\overline{x}}(\overline{x}+1) \\
\overline{y} &= \frac{\overline{x}+1}{b\overline{x}}\left(g(1-\overline{x}_2)+\overline{x}\right),
\end{aligned} \tag{12}$$

Here $\overline{x} = \overline{x}_1/\overline{x}_2$ is a real positive root of the cubic equation

$$gd\overline{x}^3 + (gd - g\alpha + \alpha)\overline{x}^2 + (g\alpha - \alpha - gd)\overline{x} - gd = 0 \tag{13}$$

or

$$(\overline{x} - 1)\left(gd\overline{x}^2 + (2gd - g\alpha + \alpha)\overline{x} + gd\right) = 0. \tag{14}$$

Obviously, $\overline{x} = 1$ is the one real positive root of (13). The other two values of $\overline{x}$ will be real and positive if

$$g > \frac{\alpha}{\alpha - 4d}. \tag{15}$$

We now analyze the stability of each equilibrium.

Let $E = (\hat{x}_1, \hat{x}_2, \hat{y})$ be any arbitrary equilibrium. The characteristic equation about E is given by

$$\begin{vmatrix}
g - 2g\hat{x}_1 - \dfrac{b\hat{y}\hat{x}_2^2}{(\hat{x}_1+\hat{x}_2)^2} - \lambda & 1 - \dfrac{b\hat{y}\hat{x}_1^2}{(\hat{x}_1+\hat{x}_2)^2} & -\dfrac{b\hat{x}_1\hat{x}_2}{\hat{x}_1+\hat{x}_2} \\
1 - \dfrac{b\hat{y}\hat{x}_2^2}{(\hat{x}_1+\hat{x}_2)^2} & g - 2g\hat{x}_2 - \dfrac{b\hat{y}\hat{x}_1^2}{(\hat{x}_1+\hat{x}_2)^2} - \lambda & -\dfrac{b\hat{x}_1\hat{x}_2}{\hat{x}_1+\hat{x}_2} \\
\dfrac{\alpha\hat{y}\hat{x}_2^2}{(\hat{x}_1+\hat{x}_2)^2} e^{-\lambda\tau} & \dfrac{\alpha\hat{y}\hat{x}_1^2}{(\hat{x}_1+\hat{x}_2)^2} e^{-\lambda\tau} & -d + \dfrac{\alpha\hat{x}_1\hat{x}_2}{\hat{x}_1+\hat{x}_2} e^{-\lambda\tau} - \lambda
\end{vmatrix} = 0 \tag{16}$$

The next lemma gives conditions for the stability of equilibrium $E_1 = (x_1^*, x_2^*, 0)$.

Theorem 2. *The equilibrium* $E_1 = (x_1^*, x_2^*, 0)$ *is*

(i) *unstable if* $d < g\alpha x_1^*(x_1^* - 1)/(1 + g(x_1^* - 1))$;

(ii) *locally asymptotically stable if* $d > g\alpha x_1^*(x_1^* - 1)/(1 + g(x_1^* - 1))$.

Proof. We consider the characteristic equation of (16) at the equilibrium E_1. It follows that

$$\left(\lambda + d - g\alpha \frac{x_1^*(x_1^* - 1)}{1 + g(x_1^* - 1)} e^{-\lambda\tau}\right) \cdot \left((g - 2gx_1^* - \lambda)\left(g - 2g^2x_1^*(x_1^* - 1) - \lambda\right) - 1\right) = 0. \tag{17}$$

Hence, one characteristic root is the solution of the equation

$$f_1(\lambda) \equiv \lambda + d - g\alpha \frac{x_1^*(x_1^* - 1)}{1 + g(x_1^* - 1)} e^{-\lambda\tau} = 0. \tag{18}$$

If $d < g\alpha x_1^*(x_1^* - 1)/(1 + g(x_1^* - 1))$, then $f_1(0) = d - g\alpha(x_1^*(x_1^* - 1)/(1 + g(x_1^* - 1))) < 0$, and $f_1(+\infty) = \infty$. Therefore, $f_1(\lambda)$ has at least one positive root and the equilibrium E_1 is unstable.

On the other hand, let $d > g\alpha x_1^*(x_1^* - 1)/(1 + g(x_1^* - 1))$; that is,

$$d - \frac{g\alpha x_1^*(x_1^* - 1)}{1 + g(x_1^* - 1)} > 0. \tag{19}$$

Then $f_1(-\infty) = -\infty$ and $f_1(0) > 0$. Thus, a root of $f_1(\lambda)$ has negative real part. Hence, the other characteristic roots are the solution of the equation

$$(g - 2gx_1^* - \lambda)\left(g - 2g^2x_1^*(x_1^* - 1) - \lambda\right) - 1 = 0; \tag{20}$$

that is,

$$f_2(\lambda) \equiv \lambda^2 + (x_1^* - 1)\left(2g + 2g^2x_1^*\right)\lambda + 4g^3(x_1^*)^3 - 6g^3(x_1^*)^2 + 2\left(g^3 - g^2\right)x_1^* + \left(g^2 - 1\right) = 0. \tag{21}$$

Since $x_1^* > 1$ is a real positive root of the cubic equation $g^3x_1^{*3} - 2g^3x_1^{*2} + (g^3 - g^2)x_1^* + (g^2 - 1) = 0$, we have $(x_1^* - 1)(2g + 2g^2x_1^*) > 0$. We, then, consider the last few terms from (21)

$$\begin{aligned} 4g^3(x_1^*)^3 &- 6g^3(x_1^*)^2 + 2\left(g^3 - g^2\right)x_1^* + g^2 - 1 \\ &= \left(g^3(x_1^*)^3 - 2g^3(x_1^*)^2 + \left(g^3 - g^2\right)x_1^* + g^2 - 1\right) \\ &\quad + 3g^3(x_1^*)^3 + \left(g^3 - g^2\right)x_1^* - 4g^3(x_1^*)^2 \\ &= \left(g^3(x_1^*)^3 - 2g^3(x_1^*)^2 + \left(g^3 - g^2\right)x_1^*\right) \\ &\quad + 2g^3(x_1^*)^2(x_1^* - 1) \\ &= -\left(g^2 - 1\right) + 2g^3(x_1^*)^2(x_1^* - 1) \\ &= g^2(2x_1^*x_2^* - 1) + 1 > 0. \end{aligned} \tag{22}$$

Thus, all the roots of characteristic equation have negative real part. The equilibrium E_1 is locally asymptotically stable. □

Now, we analyze the stability of positive equilibrium $E_2(\overline{x}_1, \overline{x}_2, \overline{y})$. The associated characteristic equation is

$$G(\lambda) = \lambda^3 + a_1\lambda^2 + a_2\lambda + \left(a_3\lambda^2 + a_4\lambda + a_5\right)e^{-\lambda\tau} + a_6 = 0, \tag{23}$$

where

$$\begin{aligned} a_1 &= -b_3 - b_4 + d, \\ a_2 &= b_3b_4 - \left(1 - b_1\overline{x}^2\right)(1 - b_1) - d(b_3 + b_4), \\ a_3 &= -d, \\ a_4 &= d\left(b_1 + b_1\overline{x}^2 + b_3 + b_4\right), \\ a_5 &= d\left(1 - b_1^2\overline{x}^2 - b_3b_4 - b_1b_4 - b_1b_3\overline{x}^2\right), \\ a_6 &= d\left(b_3b_4 - \left(1 - b_1\overline{x}^2\right)(1 - b_1)\right), \end{aligned} \tag{24}$$

$$b_1 = \frac{b\overline{y}}{(\overline{x} + 1)^2} > 0,\ b_2 = \frac{2b\overline{x}\overline{y}}{\overline{x} + 1} > 0,\ b_3 = g + b_2\overline{x} - 2\overline{x}^2 - 2g\overline{x} - b_1,\ b_4 = -g + b_2 - 2\overline{x} - b_1\overline{x}^2.$$

In the following, we study the Hopf bifurcation for system (9), using the time delay τ as the bifurcation parameter. We assume that $\lambda = i\omega$ $(\omega > 0)$ is a root of the characteristic equation (23). Then we get

$$-\omega^3 i - \omega^2 a_1 + a_2 \omega i + \left(-\omega^2 a_3 + a_4 \omega i + a_5\right)(\cos\omega\tau - i\sin\omega\tau) + a_6 = 0. \quad (25)$$

By separating real part and imaginary part, we obtain

$$\begin{aligned}\left(a_5 - a_3\omega^2\right)\cos\omega\tau + a_4\omega\sin\omega\tau &= a_1\omega^2 - a_6\\ (a_4\omega)\cos\omega\tau + \left(a_3\omega^2 - a_5\right)\sin\omega\tau &= \omega^3 - a_2\omega.\end{aligned} \quad (26)$$

By squaring both sides of the equations and using the property that $\sin^2\omega\tau + \cos^2\omega\tau = 1$, we can simplify the above equation. As a result,

$$\omega^6 + \left(a_1^2 - 2a_2 - a_3^2\right)\omega^4 + \left(a_2^2 - 2a_1a_6 + 2a_3a_5 - a_4^2\right)\omega^2 + \left(a_6^2 - a_5^2\right) = 0. \quad (27)$$

Denote $v = \omega^2$, $e_1 = a_1^2 - 2a_2 - a_3^2$, $e_2 = a_2^2 - 2a_1a_6 + 2a_3a_5 - a_4^2$, and $e_3 = a_6^6 - a_5^2$. Then (27) becomes

$$h(v) = v^3 + e_1 v^2 + e_2 v + e_3. \quad (28)$$

By the Routh-Hurwitz criterion, we conclude that if

$$\begin{aligned}a_1 + a_3 &> 0,\\ a_5 + a_6 &> 0\\ (a_1 + a_3)(a_2 + a_4) &> a_5 + a_6,\end{aligned} \quad (29)$$

(23) has no positive real roots. Therefore, we get the following results.

Theorem 3. *Suppose conditions in (29) hold and $e_1, e_2 > 0$, $e_3 \geq 0$. Then the equilibrium E_2 is locally asymptotically stable.*

Proof. For $h(v)$ defined in (28), we have

$$\frac{dh(v)}{dv} = 3v^2 + 2e_1 v + e_2, \quad (30)$$

and the zeros of (30) are

$$v_{1,2} = \frac{-e_1 \pm \sqrt{e_1^2 - 3e_2}}{3}. \quad (31)$$

If $e_1, e_2 > 0$, then $\sqrt{e_1^2 - 3e_2} < e_1$. Hence, v_1 and v_2 are negative. Thus, $dh(v)/dv = 0$ has no positive root. Since $h(0) = e_3 \geq 0$, it follows that $h(v) = 0$ has no positive roots. Therefore, the equilibrium E_2 is locally asymptotically stable. □

Theorem 4. *Suppose that conditions in (29) hold and that*

(i) *either $e_3 < 0$,*

(ii) *or $e_3 \geq 0, e_2 < 0$, and $2\omega_0^6 + (a_1^2 - 2a_2 - 2a_3^2)\omega_0^4 + 2a_5^2 - a_6^2 \neq 0$,*

where ω_0 satisfies $G(i\omega_0) = 0$ with G given in (23). Then the equilibrium E_2 is locally asymptotically stable if $\tau < \tau_0$ and is unstable if $\tau > \tau_0$, where

$$\tau_0 = \frac{1}{\omega_0} \cdot \cos^{-1}\left(\frac{(a_4 - a_1a_3)\omega_0^4 + (a_1a_5 + a_3a_6 - a_2a_4)\omega_0^2 - a_5a_6}{a_4^2\omega_0^2 + (a_5 - a_3\omega_0^2)^2}\right). \quad (32)$$

Furthermore, when $\tau = \tau_0$, a Hopf bifurcation occurs; that is, a family of periodic solutions are bifurcated from E_2 as τ passes through the critical value τ_0.

Proof. If $e_3 < 0$, then it follows from (28) that $h(0) < 0$ and $\lim_{v\to\infty} h(v) = \infty$. Thus, (27) has at least one positive root. If $e_2 < 0$, then $v_1 = (-e_1 + \sqrt{e_1^2 - 3e_2})/3$ is one positive root of $dh(v)/dv = 0$. Since $h(0) = e_3 \geq 0$, it follows that $h(v) = 0$ has at least one positive root. As a consequence, (27) has a positive root ω_0. This implies that the characteristic equation (23) has a pair of purely imaginary roots.

Let $u(\tau) = \eta(\tau) + i\omega(\tau)$ be the eigenvalue of (23) such that $\eta(\tau_0) = 0$ and $\omega(\tau_0) = \omega_0$. If there exists $\omega_0 > 0$, such that $G(i\omega) = 0$. Then by the first equation of (26), we have

$$\cos\left(\omega_0\tau_j\right) = \frac{(a_4 - a_1a_3)\omega_0^4 + (a_1a_5 + a_3a_6 - a_2a_4)\omega_0^2 - a_5a_6}{a_4^2\omega_0^2 + (a_5 - a_3\omega_0^2)^2}, \quad (33)$$

and then

$$\tau_j = \cos^{-1}\frac{(a_4 - a_1a_3)\omega_0^4 + (a_1a_5 + a_3a_6 - a_2a_4)\omega_0^2 - a_5a_6}{a_4^2\omega_0^2 + (a_5 - a_3\omega_0^2)^2} + \frac{2\pi j}{\omega_0}, \quad j = 0, 1, 2, \ldots. \quad (34)$$

By taking the derivative of the characteristic equation (23) with respect to τ, we have

$$\frac{d\lambda(\tau)}{d\tau} = \frac{\left(a_3\lambda^3 + a_4\lambda^2 + a_5\lambda\right)e^{-\lambda\tau}}{(3\lambda^2 + 2a_1\lambda + a_2) - (a_3\lambda^2 + a_4\lambda + a_5)\tau e^{-\lambda\tau} + (2a_3\lambda + a_4)e^{-\lambda\tau}}. \quad (35)$$

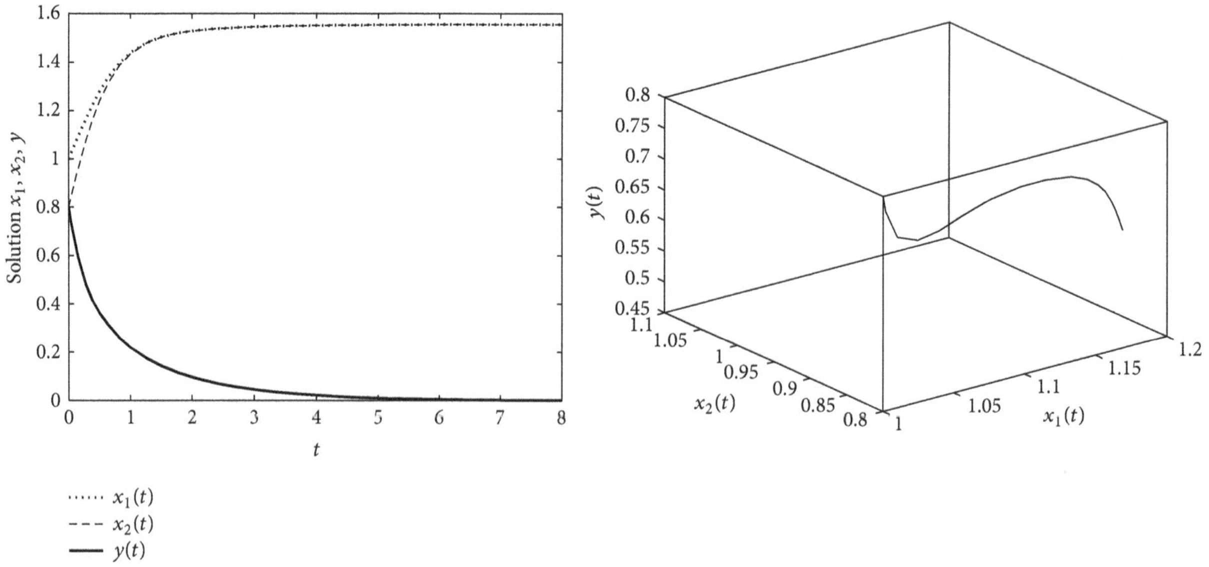

FIGURE 1: The behavior of x_1, x_2, and y with respect to t for Example 5.

Thus,

$$\left(\frac{d\lambda(\tau)}{d\tau}\right)^{-1} = \frac{\left(3\lambda^2 + 2a_1\lambda + a_2\right) + \left(2a_3\lambda + a_4\right)e^{-\lambda\tau}}{\lambda\left(a_3\lambda^2 + a_4\lambda + a_5\right)\tau e^{-\lambda\tau}} - \frac{\tau}{\lambda}. \tag{36}$$

We can also verify the following transversality condition [16]:

$$\begin{aligned}\left(\left.\frac{d\mathbf{Re}\lambda(\tau)}{d\tau}\right|_{\tau=\tau_0}\right)^{-1} &= \mathbf{Re}\left(\frac{\left(3\lambda^2 + 2a_1\lambda + a_2\right) + \left(2a_3\lambda + a_4\right)e^{-\lambda\tau}}{\lambda\left(a_3\lambda^2 + a_4\lambda + a_5\right)\tau e^{-\lambda\tau}} - \frac{\tau}{\lambda}\right)\Bigg|_{\tau=\tau_0} \\ &= \frac{2\omega_0^6 + \left(a_1^2 - 2a_2 - 2a_3^2\right)\omega_0^4 + 2a_5^2 - a_6^2}{\omega_0^2\left(\left(a_5 - a_3\omega_0^2\right)^2 + \left(a_4\omega_0\right)^2\right)} \neq 0.\end{aligned} \tag{37}$$

Therefore, if $\tau = \tau_0$, then a Hopf bifurcation occurs; that is, a family of periodic solutions appear as τ passes through the critical value τ_0. □

4. Numerical Simulations and Discussion

In this section, we present some numerical simulation of system (9) at different parameters to illustrate our analytic results.

Example 5. Let $g = 1.8$ $b = 0.6$ $\alpha = 2$ $d = 3$ and we consider the following system:

$$\begin{aligned}\frac{dx_1}{dt} &= 1.8x_1\left(1 - x_1\right) + x_2 - \frac{0.6x_1x_2y}{x_1 + x_2} \\ \frac{dx_2}{dt} &= 1.8x_2\left(1 - x_2\right) + x_1 - \frac{0.6x_1x_2y}{x_1 + x_2} \\ \frac{dy}{dt} &= 2\frac{x_1(t-\tau)x_2(t-\tau)y(t-\tau)}{x_1(t-\tau) + x_2(t-\tau)} - 3y.\end{aligned} \tag{38}$$

In this case, we obtain only one boundary equilibrium $E_1 = (1.556, 1.557, 0)$, and the conditions of (ii) in Theorem 2 are satisfied. Therefore, the equilibrium E_1 is locally asymptotically stable. The behaviors of x_1, x_2, and y with respect to t are shown in Figure 1. According to the graph in Figure 1, the predator population decreases and eventually the predator species becomes extinct. As for prey species, the population of both species reaches the equilibrium as the predator population approaches zero.

Example 6. As an example, consider the following system:

$$\begin{aligned}\frac{dx_1}{dt} &= 1.8x_1\left(1 - x_1\right) + x_2 - \frac{0.6x_1x_2y}{x_1 + x_2} \\ \frac{dx_2}{dt} &= 1.8x_2\left(1 - x_2\right) + x_1 - \frac{0.6x_1x_2y}{x_1 + x_2} \\ \frac{dy}{dt} &= 2\frac{x_1(t-\tau)x_2(t-\tau)y(t-\tau)}{x_1(t-\tau) + x_2(t-\tau)} - 0.3y.\end{aligned} \tag{39}$$

There is a positive equilibrium $E_2 = (0.3, 0.3, 7.53)$. By direct calculation, we have $e_3 = -0.01903$, $\omega_0 = 0.639$, and $2\omega_0^6 + (a_1^2 - 2a_2 - 2a_3^2)\omega_0^4 + 2a_5^2 - a_6^2 = 0.2234 \neq 0$. From Theorem 4, there is a critical value $\tau_0 = 1.1071$, and the

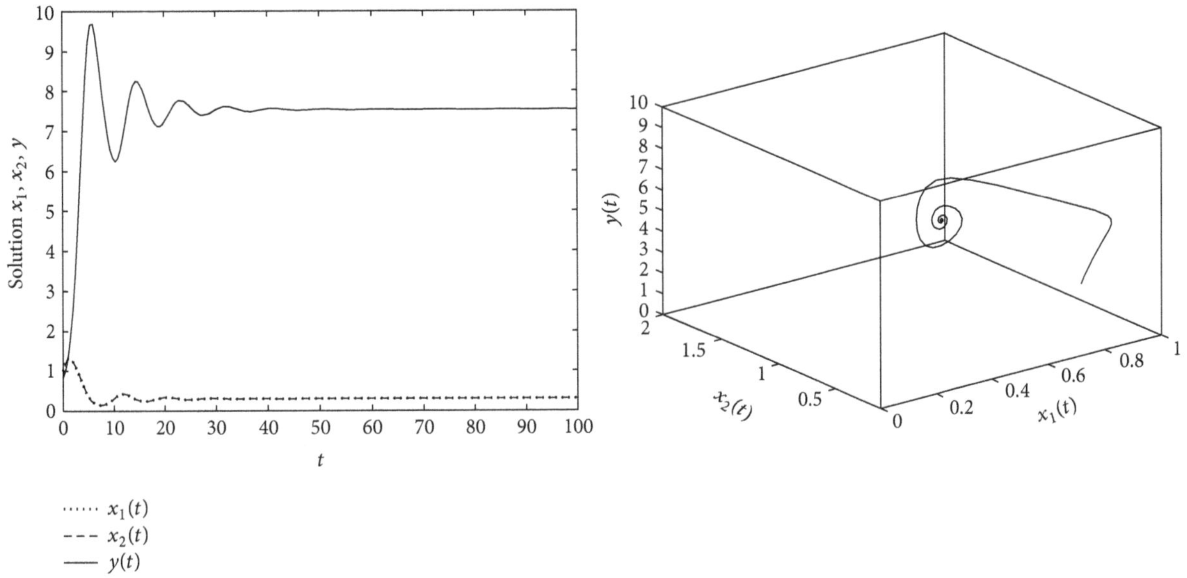

FIGURE 2: The behavior of x_1, x_2, and y with respect to t for Example 6 with $\tau = 0.5$.

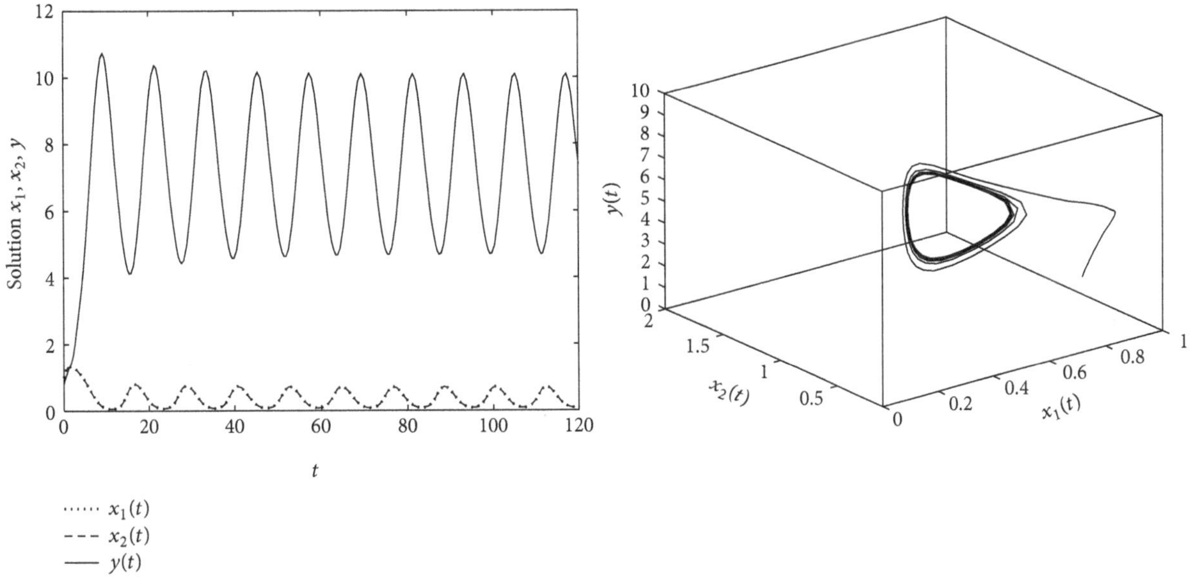

FIGURE 3: The behavior of x_1, x_2, and y with respect to t for Example 6 with $\tau = 1.8$.

equilibrium E_2 is locally asymptotically stable as $\tau < \tau_0 = 1.1071$. A Hopf bifurcation occurs as $\tau = \tau_0 = 1.1071$ and the equilibrium becomes unstable and stable periodic solutions exist for $\tau > \tau_0 = 1.1071$. Figures 2 and 3 show the solutions of that system corresponding to $\tau = 0.5$ and $\tau = 1.8$. Furthermore, a bifurcation diagram for Example 6 is shown in Figure 4. This is an example when the predator and prey coexist permanently. If the time that juvenile takes to be mature is less than τ_0, then both predators and prey population reach the nonzero equilibrium. They can coexist permanently. On the other hand, if the time that juvenile predators takes to become mature and ready to hunt is longer than τ_0, then the population of both predator and prey species becomes unstable and periodic.

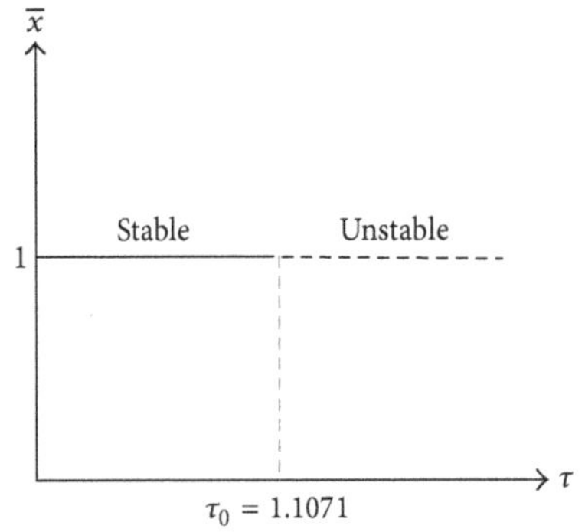

FIGURE 4: Bifurcation diagram for Example 6.

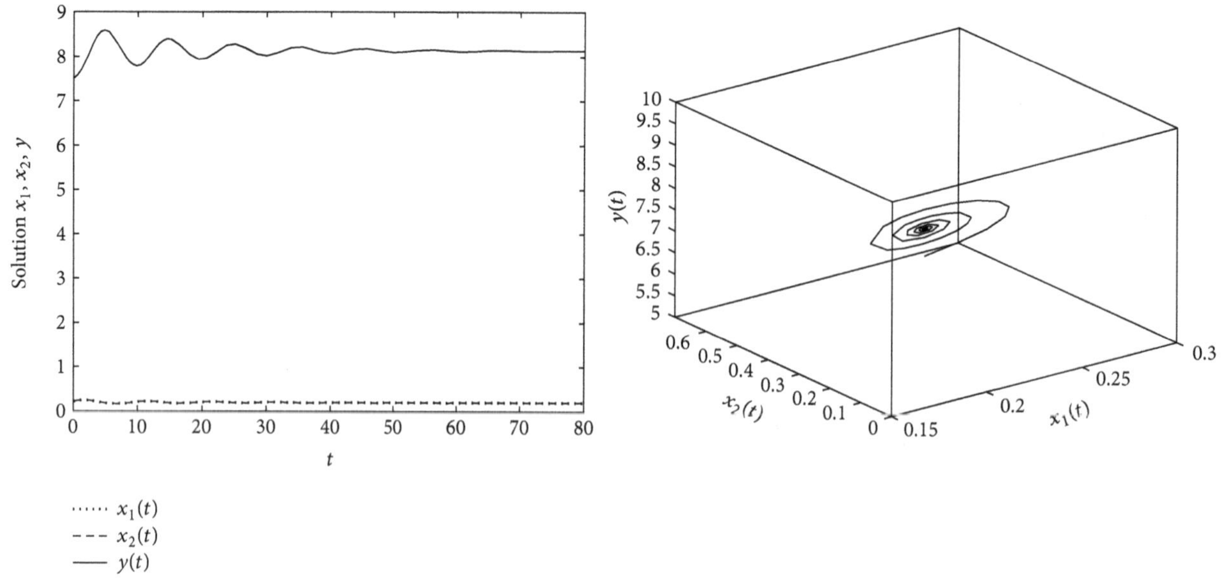

FIGURE 5: The behavior of x_1, x_2, and y with respect to t for equilibrium E_2^1 with $\tau = 0.6$.

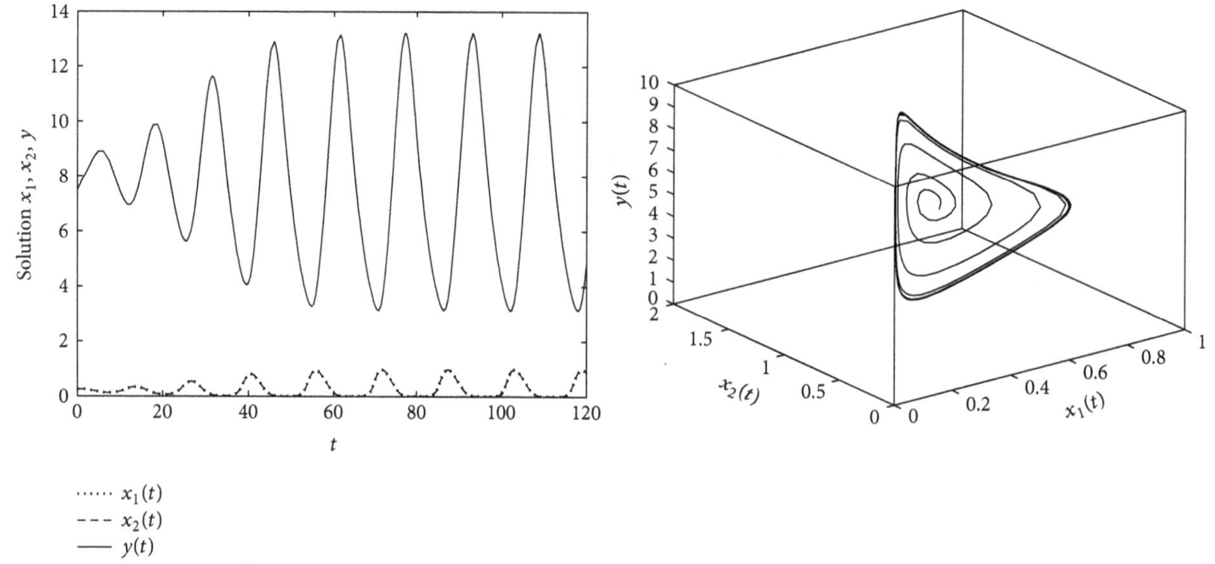

FIGURE 6: The behavior of x_1, x_2, and y with respect to t for equilibrium E_2^1 with $\tau = 2$.

Example 7. As an example, consider the following system:

$$\begin{aligned}
\frac{dx_1}{dt} &= 1.8x_1\left(1 - x_1\right) + x_2 - \frac{0.6x_1x_2y}{x_1 + x_2} \\
\frac{dx_2}{dt} &= 1.8x_2\left(1 - x_2\right) + x_1 - \frac{0.6x_1x_2y}{x_1 + x_2} \\
\frac{dy}{dt} &= 2\frac{x_1(t-\tau)x_2(t-\tau)y(t-\tau)}{x_1(t-\tau) + x_2(t-\tau)} - 0.2y.
\end{aligned} \tag{40}$$

In this case, we obtain three positive equilibria $E_2^1 = (0.2, 0.2, 8.133)$, $E_2^2 = (0.1519, 0.2927, 8.7420)$, and $E_2^3 = (0.2925, 0.1519, 8.7410)$. By Theorem 4, we know that the positive equilibrium E_2^1 is locally asymptotically stable when $\tau < \tau_0 = 1.8227$ and unstable when $\tau > \tau_0 = 1.8227$, and the system can also undergo a Hopf bifurcation at the equilibrium E_2^1 when τ crosses through the critical value $\tau > \tau_0 = 1.8227$; see Figures 5 and 6. Similarly, at the positive equilibrium E_2^2, a Hopf bifurcation occurs as $\tau = \tau_0 = 2.4353$. Hence, the positive equilibrium E_2^2 is locally asymptotically stable when $\tau < \tau_0 = 2.4353$ and unstable when $\tau > \tau_0 = 2.4353$; see Figures 7 and 8. For equilibrium E_2^3, (27) has no positive real root. Hence, equilibrium E_2^3 is locally asymptotically stable and no stability switches can occur; see Figure 9. On this last

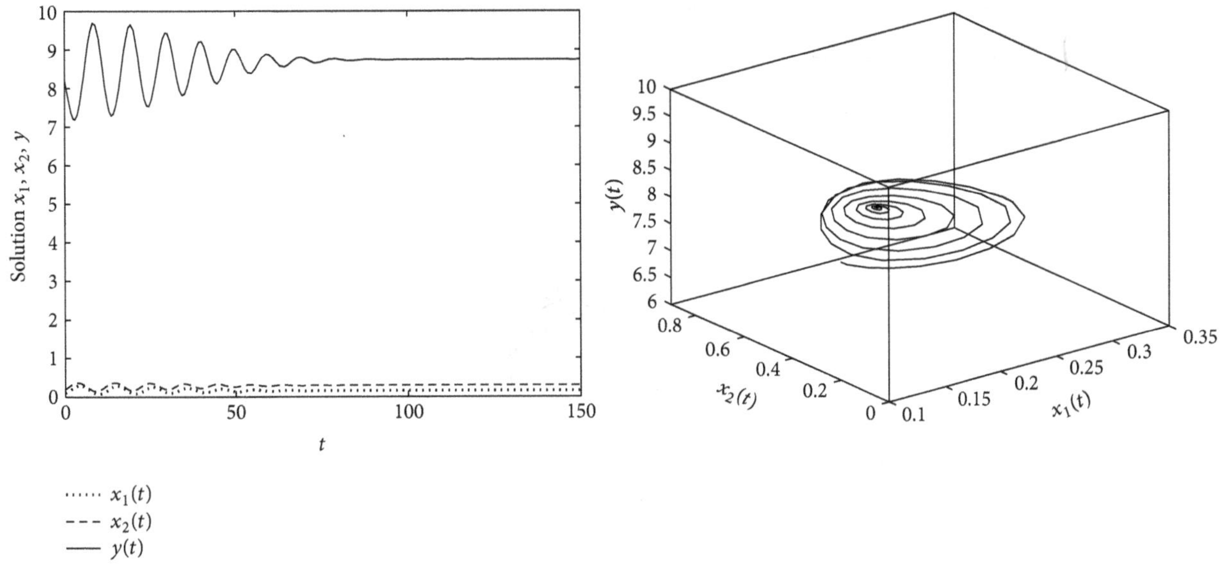

FIGURE 7: The behavior of x_1, x_2, and y with respect to t for equilibrium E_2^2 with $\tau = 1.8$.

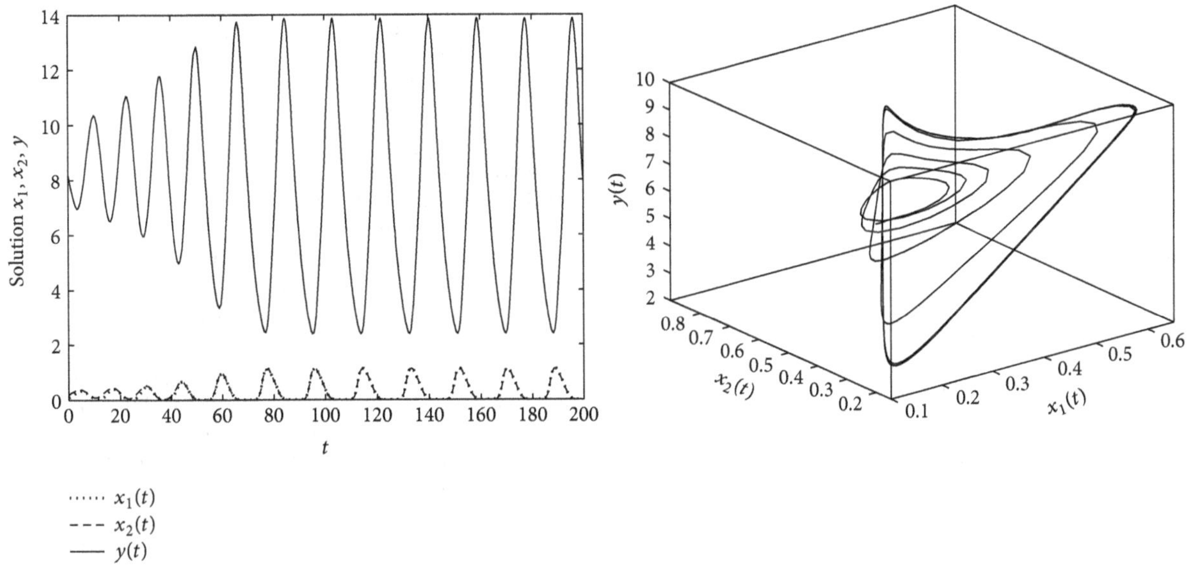

FIGURE 8: The behavior of x_1, x_2, and y with respect to t for equilibrium E_2^2 with $\tau = 2.6$.

example, what occurs at the first two equilibria E_2^1 and E_2^2 is similar to the previous example where Hopf bifurcations occur at τ_0's and the stable limit cycle exists. The predator and prey species coexist. As for the other equilibrium E_2^3, the system is locally asymptotically stable where predator and prey species also coexist. Finally, the bifurcation diagram for Example 7 is shown in Figure 10.

5. Concluding Remarks

In this paper, we find that system (7) has complex dynamics behavior. By Theorem 2, our results show that the predator and prey coexist permanently if $d < g\alpha x_1^*(x_1^* - 1)/(1 + g(x_1^* - 1))$; that is, the adult predators' reproductive rate at the peak of prey abundance is larger than its death rate. On the other hand, the predator faces extinction, if $d > g\alpha x_1^*(x_1^* - 1)/(1 + g(x_1^* - 1))$, which implies that the predator's possible highest reproductive rate is less than its death rate. We also find the stability switches of the positive equilibrium E_2 due to the increase of τ. Our results show that when there is no time delay or the time delay is very small, the positive equilibrium E_2 is locally asymptotically stable. As the time delay increases to the critical value, it can cause a stable equilibrium to become unstable and Hope bifurcation can occur.

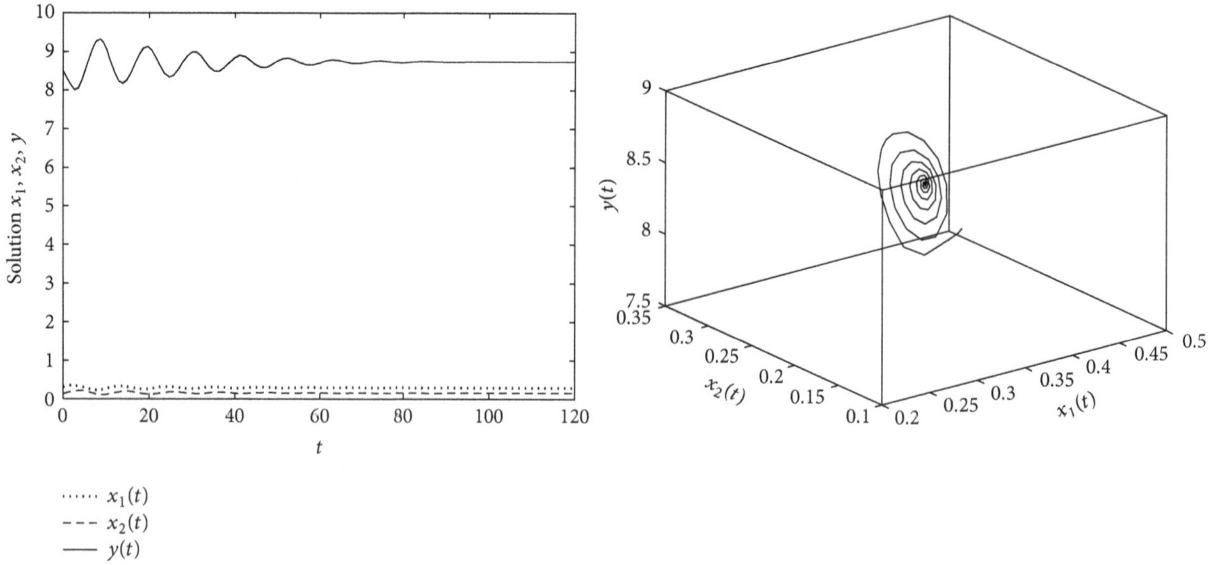

FIGURE 9: The behavior of x_1, x_2, and y with respect to t for equilibrium E_2^3 with $\tau = 2.6$.

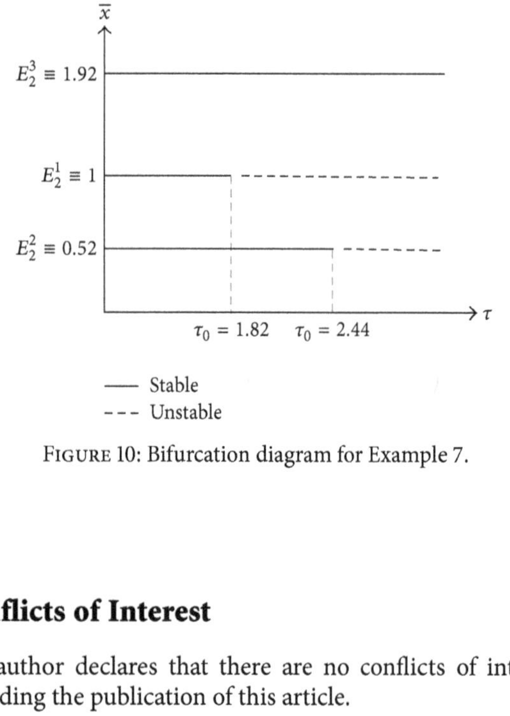

FIGURE 10: Bifurcation diagram for Example 7.

Conflicts of Interest

The author declares that there are no conflicts of interest regarding the publication of this article.

Acknowledgments

This research was supported by Chiang Mai University, Thailand.

References

[1] C. Celik, "The stability and Hopf bifurcation for a predator-prey system with time delay," *Chaos, Solitons and Fractals*, vol. 37, no. 1, pp. 87–99, 2008.

[2] H. I. Freedman and S. G. Ruan, "Hopf bifurcation in three-species food chain models with group defense," *Mathematical Biosciences*, vol. 111, no. 1, pp. 73–87, 1992.

[3] S. B. Hsu and T. W. Huang, "Global stability for a class of predator-prey systems," *SIAM Journal on Applied Mathematics*, vol. 55, no. 3, pp. 763–783, 1995.

[4] S.-B. Hsu and T.-W. Hwang, "Hopf bifurcation analysis for a predator-prey system of Holling and Leslie type," *Taiwanese Journal of Mathematics*, vol. 3, no. 1, pp. 35–53, 1999.

[5] Z. Lu and X. Liu, "Analysis of a predator-prey model with modified Holling-Tanner functional response and time delay," *Nonlinear Analysis. Real World Applications. An International Multidisciplinary Journal*, vol. 9, no. 2, pp. 641–650, 2008.

[6] Y. Song and S. Yuan, "Bifurcation analysis in a predator-prey system with time delay," *Nonlinear Analysis. Real World Applications. An International Multidisciplinary Journal*, vol. 7, no. 2, pp. 265–284, 2006.

[7] M. Tansky, "Switching effect in prey-predator system," *Journal of Theoretical Biology*, vol. 70, no. 3, pp. 263–271, 1978.

[8] Q. J. Khan, E. Balakrishnan, and G. C. Wake, "Analysis of a predator-prey system with predator switching," *Bulletin of Mathematical Biology*, vol. 66, no. 1, pp. 109–123, 2004.

[9] W. W. Murdoch, "Switching in general predators: experiments on predator specificity and stability of prey populations," *Ecological Monographs*, vol. 39, no. 4, pp. 335–364, 1969.

[10] Prajneshu and P. Holgate, "A prey-predator model with switching effect," *Journal of Theoretical Biology*, vol. 125, no. 1, pp. 61–66, 1987.

[11] E. Teramoto, K. Kawasaki, and N. Shigesada, "Switching effect of predation on competitive prey species," *Journal of Theoretical Biology*, vol. 79, no. 3, pp. 303–315, 1979.

[12] Y. Saito and Y. Takeuchi, "A time-delay model for prey-predator growth with stage structure," *Canadian Applied Mathematics Quarterly*, vol. 11, no. 3, pp. 293–302, 2003.

[13] W. G. Aiello and H. I. Freedman, "A time-delay model of single-species growth with stage structure," *Mathematical Biosciences*, vol. 101, no. 2, pp. 139–153, 1990.

[14] W. G. Aiello, H. I. Freedman, and J. Wu, "Analysis of a model representing stage-structured population growth with state-dependent time delay," *SIAM Journal on Applied Mathematics*, vol. 52, no. 3, pp. 855–869, 1992.

[15] Y. Qu and J. Wei, "Bifurcation analysis in a time-delay model for prey-predator growth with stage-structure," *Nonlinear Dynamics. An International Journal of Nonlinear Dynamics and Chaos in Engineering Systems*, vol. 49, no. 1-2, pp. 285–294, 2007.

[16] M. Y. Li and H. Shu, "Global dynamics of a mathematical model for HTLV-I infection of $CD4^{+}T$ cells with delayed CTL response," *Nonlinear Analysis: Real World Applications*, vol. 13, no. 3, pp. 1080–1092, 2012.

14

The Existence of Strong Solutions for a Class of Stochastic Differential Equations

Junfei Zhang

School of Statistics and Mathematics, Central University of Finance and Economics, Beijing 100081, China

Correspondence should be addressed to Junfei Zhang; zhangfei851115@163.com

Academic Editor: José A Langa

In this paper, we will consider the existence of a strong solution for stochastic differential equations with discontinuous drift coefficients. More precisely, we study a class of stochastic differential equations when the drift coefficients are an increasing function instead of Lipschitz continuous or continuous. The main tools of this paper are the lower solutions and upper solutions of stochastic differential equations.

1. Introduction

There are many works [1–3] about the existence and uniqueness of strong or weak solutions for the following stochastic differential equation (denoted briefly by SDE):

$$dX_t = b(t, X_t)\,dt + \sigma(t, X_t)\,dW_t \quad t \geq 0, \tag{1}$$

where $b(t,x) : \mathbb{R}_+ \times \mathbb{R} \longrightarrow \mathbb{R}$ and $\sigma(t,x) : \mathbb{R}_+ \times \mathbb{R} \longrightarrow \mathbb{R}$ are called drift and diffusion coefficients, respectively. W_t is standard Brownian motion. Usually, the drift and diffusion coefficients are Lipschitz or local Lipschitz continuous or at least are continuous with respect to x when the existence and uniqueness of solutions are investigated. In fact, the solutions of stochastic differential equations may exist when their drift and diffusion coefficients are discontinuous with respect to x. Therefore, many authors discussed the existence of solutions for SDE with discontinuous coefficients. For example, L. Karatzas and S. E. Shreve [1] (Proposition 3.6 of §5.3) considered the existence of a weak solution when the drift coefficient of SDE need not be continuous with respect to x. A. K. Zvonkin [4] considered the following stochastic differential equation with a discontinuous diffusion coefficient:

$$X_t = \int_0^t \operatorname{sgn}(X_s)\,dW_s; \quad 0 \leq t < \infty, \tag{2}$$

where

$$\operatorname{sgn}(x) = \begin{cases} 1, & x > 0; \\ -1, & x \leq 0. \end{cases} \tag{3}$$

The weak solution of this stochastic differential equation exists, but there is not the strong solution. N. V. Ktylov [5] and N. V. Ktylov and R. Liptser [6] also discussed existence issues of SDE when their diffusion coefficients are discontinuous with respect to x. And many authors also considered the approximation solutions of SDE with discontinuous coefficients, such as [7–11].

In this paper, we will consider the existence of a strong solution of SDE (1) when the drift coefficient $b(t,x)$ is an increasing function but need not be continuous with respect to x and the diffusion coefficient $\sigma(t, X_t)$ satisfies (C_σ) condition. Section 1 is an introduction. In Section 2, we will show a comparison theorem by using the upper and lower solutions of SDE. We will prove our main result by using the above comparison theorem in Section 3.

2. The Setup and a Comparison Theorem

In our paper, we just consider a 1-dimensional case. We always assume that $(\Omega, \mathcal{F}, \mathbf{P})$ is a completed probability space, $W =: \{W_t : t \geq 0\}$ is a real-valued Brownian motion defined on

$(\Omega, \mathscr{F}, \mathbf{P})$, and $\{\mathscr{F}_t : t \geq 0\}$ is natural filtration generated by the Brownian motion W; i.e., for any $t \geq 0$

$$\mathscr{F}_t = \sigma\{W_s : s \leq t\}. \quad (4)$$

We consider SDE (1) with coefficients $b(t,x) : \mathbb{R}^+ \times \mathbb{R} \longrightarrow \mathbb{R}$ and $\sigma(t,x) : \mathbb{R}^+ \times \mathbb{R} \longrightarrow \mathbb{R}$, where $\mathbb{R}^+$ and $\mathbb{R}$ are a positive real number and real number, respectively. And we use $\|\cdot\|$ to denote norm of $\mathbb{R}$. The following is the definition of a strong solution for SDE.

Definition 1. An adapted continuous process X_t defined on $(\Omega, \mathscr{F}, \mathbf{P})$ is said to be a strong solution for SDE (1) if it satisfies that

$$X_t = X_0 + \int_0^t b(s, X_s)\, ds + \int_0^t \sigma(s, X_s)\, dW_s, \quad t \geq 0, \quad (5)$$

holds with probability 1.

Moreover, X_t and $\widetilde{X}_t$ are two strong solutions of SDE (1); then $P[X_t = \widetilde{X}_t;\ 0 \leq t < \infty] = 1$. Under this condition, the solution of SDE (1) is said to be unique.

The following is the conception of upper and lower solutions for stochastic differential equations, which are given by N. Halidias and P. E. Kloeden [12]. Many authors discussed the upper and lower solutions of the stochastic differential equation by using the other name which is the solutions of the stochastic differential inequality, for example, S. Assing and R. Manthey [13] and X. Ding and R. Wu [14].

Definition 2. An adapted continuous stochastic process U_t (resp., L_t) is an upper (resp., lower) solution of SDE (1) if the inequalities

(1) $U_t \geq U_s + \int_s^t b(u, U_u)du + \int_s^t \sigma(u, U_u)dW_u,\ t \geq s \geq 0$;

(2) $L_t \leq L_s + \int_s^t b(u, L_u)du + \int_s^t \sigma(u, L_u)dW_u,\ t \geq s \geq 0$,

hold with probability 1.

Remark 3. It is not an easy thing to calculate the exact upper and lower solution of the general stochastic differential equations. However, one can discuss the existence of upper and lower solutions. S. Assing and R. Manthey [13] discussed the "maximal/minimal solution" of the stochastic differential inequality. They proved the existence of a "maximal/minimal solution" under some conditions. However, it is easy to show there exist the upper solutions of stochastic differential equations if the minimal solution of the stochastic differential inequality exists. In fact, the minimal solution is special upper solutions of stochastic differential equations. Similarly, we can show the existence of the lower solution by using the maximal solution of the stochastic differential inequality.

Usually, the existence and uniqueness of solutions of SDE (1) are investigated under the conditions in which the diffusion coefficient satisfies Lipschitz condition and liner growth condition. In fact, the Lipschitz condition can be generalized. In this paper, the diffusion coefficient satisfies the (C_σ) condition.

(C_σ): For $N > 0$, there exist an increasing function $\rho_N : \mathbb{R}_+ \longrightarrow \mathbb{R}_+$ and a predictable process $G_N(t, \omega)$ such that

$$|\sigma(t,\omega,x) - \sigma(t,\omega,y)| \leq G_N(t,\omega)\, \rho_N(|x-y|),$$
$$\int_0^t G_N(t,\omega)\, dt < \infty \ a.s., \quad (6)$$
$$\int_{0^+} \rho_N^{-2}(u)\, du = \infty,$$

for all $t \geq 0$, and $x, y \in \mathbb{R}$ with $\|x\|, \|y\| \leq N$.

Note that the Lipschitz condition satisfies the (C_σ) condition. The following lemma is an important tool of this paper and had to be proved in proposition 2.3 of X. Ding and R. Wu [14].

Lemma 4. *In SDE (1), we assume σ satisfies (C_σ) and b satisfies that, for each $N > 0$, there exists a measurable process $L_N(t,\omega)$ such that*

$$\|b(t,\omega,x) - b(t,\omega,y)\| \leq L_N(t,\omega)\, \|x-y\|,$$
$$\int_0^t L_N(t,\omega)\, dt < \infty, \ a.s., \quad (7)$$

for all $t \geq 0$ and $x, y \in \mathbb{R}$ with $\|x\|, \|y\| \leq N$. Then SDE (1) has a unique local (explosion in the finite time) strong solution.

Remark 5. Moreover, if b and σ satisfy the liner growth condition (cf. J. Jacod and J. Memin [15])

$$\|b(t,\omega,x)\| + \|\sigma(t,\omega)\| \leq H(t,\omega)(1 + \|x\|), \quad (8)$$

where $H(t,\omega)$, $t \geq 0$, is a predictable process such that $\int_0^t H^2(s,\omega)ds < \infty$, *a.s.* Then SDE (1) has a unique global strong solution.

The following theorem can be considered as a comparison theorem, and we will use it to arrive at our main result.

Theorem 6. *Let $b : \mathbb{R}^+ \times \Omega \longrightarrow \mathbb{R}$ be predictable such that $\int_0^t b^2(s,\omega)ds < \infty$, a.s. for any $t \geq 0$, and let $\sigma : \mathbb{R}^+ \times \Omega \times \mathbb{R} \longrightarrow \mathbb{R}$ be predictable. Suppose that σ satisfies (C_σ) and there exists a predictable process $H(t,\omega)$, $t \geq 0$ such that*

$$\|\sigma(t,\omega)\| \leq H(t,\omega)(1 + \|x\|), \quad (9)$$

where $\int_0^t H^2(s,\omega)ds < \infty$, a.s. And suppose that U_t and L_t are upper and lower solutions of the following SDE:

$$X_t = X_0 + \int_0^t b(s,\omega)\, ds + \int_0^t \sigma(s, X_s)\, dW_s, \quad t \geq 0, \quad (10)$$

such that $L_0 \leq X_0 \leq U_0$, a.s.

Then there is a unique strong solution X_t which satisfies that $L_t \leq X_t \leq U_t$ for any $t \geq 0$ holds with probability 1.

Proof. Obviously, we have that SDE (10) has a unique strong solution X_t by using Lemma 4 and Remark 5. In the following we will show

$$\mathbf{P}\{L_t \leq X_t \leq U_t,\ \forall t \geq 0\} = 1. \quad (11)$$

We only prove $\mathbf{P}\{X_t \le U_t,\ \forall t \ge 0\} = 1$, because we can prove $\mathbf{P}\{L_t \le X_t,\ \forall t \ge 0\} = 1$ by using the similar way.

Define the stopping time

$$T_N =: \inf\{t \in [0,\infty) : |X_t| \vee |L_t| \vee t > N\} \wedge N. \quad (12)$$

Obviously, $T_N \longrightarrow \infty$ when $N \longrightarrow \infty$. And define the stopping time $\tau =: \inf\{t \in [0,\infty) : X_t < L_t\}$. If $\mathbf{P}\{\tau < T_N\} = 0$ for $N \ge 1$, then $\mathbf{P}\{\tau < \infty\} = 0$; that is, $\mathbf{P}\{L_t \le X_t,\ \forall t \ge 0\} = 1$. Indeed, $\forall q \in Q^+$ and $N \ge 1$, we define $\alpha =: (\tau + q) \wedge T_N$ and $\Omega_\alpha =: \{X_\alpha < L_\alpha\}$. Note that

$$\mathbf{P}\{\Omega_\alpha\} = 0, \quad \forall q \in Q^+,\ N \ge 1 \Longrightarrow \mathbf{P}\{\tau < T_N\} = 0. \quad (13)$$

In fact, by $\mathbf{P}\{\Omega_\alpha\} = 0$ and X, L being continuous and the denseness of the rational number in $\mathbb{R}$, we have

$$X_{(\tau+t)\wedge T_N} \ge L_{(\tau+t)\wedge T_N} \ a.s. \text{ on } \{\tau < T_N\} \quad (14)$$

for all $t \ge 0$. That is for $a.s.\ \omega \in \{\tau < T_N\}$ and $t \in [\tau(\omega), T_N(\omega)]$ one has $X_t \ge L_t$. However, by the definition of τ and $L_\tau \le X_\tau, a.s.$ we have $\mathbf{P}\{\tau < T_N\} = 0$.

In the following we shall prove $\mathbf{P}\{\Omega_\alpha\} = 0,\ \forall q \in Q^+,\ N \ge 1$. Set $\beta =: \sup\{t \in [0,\alpha) : L_t \le X_t\}$. By continuity of X and L we have $X_\beta \ge L_\beta,\ a.s.$ Obviously, $\{X_\alpha \ge L_\alpha\} = \{\beta = \alpha\}$. So, we have $\Omega_\alpha =: \{X_\alpha < L_\alpha\} = \{\beta < \alpha\}$. Hence, for $\omega \in \Omega_\alpha$ and $t \in (\beta(\omega), \alpha(\omega)]$ we have $X_t < L_t$. Using L as a lower solution of SDE (10), we have

$$L_t - X_t \le \int_\beta^t [\sigma(s, L_s) - \sigma(s, X_s)]\, dW_s =: M_t. \quad (15)$$

Hence,

$$[L_t - X_t]\, I_{\Omega_\alpha} I_{(\beta,\alpha]}(t) \le M_t I_{\Omega_\alpha} I_{(\beta,\alpha]}(t). \quad (16)$$

Let us take $M^+ =: \max\{M, 0\}$. By the Tanaka formula (refer to [3]) we have

$$\begin{aligned} M_t^+ I_{\Omega_\alpha} &= M_\beta^+ I_{\Omega_\alpha} + I_{\Omega_\alpha} \int_\beta^t I_{\{M_s>0\}} dM_s \\ &\quad + \frac{1}{2} I_{\Omega_\alpha} \left[L_t^0(M) - L_\beta^0(M)\right], \end{aligned} \quad (17)$$

where $L_t^x(M)$ denotes local time at the point x for M. By the definition of local time, one can prove easily that $L_t^0(M) - L_\beta^0(M) = 0$, for $t \in (\beta, \alpha]$ on Ω_α. So, by $M_\beta^+ I_{\Omega_\alpha} = 0$ (using the definition M) we have

$$\begin{aligned} M^+ I_{\Omega_\alpha} &= \int_\beta^t I_{\{M_s>0\}} I_{\Omega_\alpha} [\sigma(s, L_s) - \sigma(s, X_s)]\, dW_s \\ &=: N_t. \end{aligned} \quad (18)$$

Since for $\omega \in \Omega_\alpha$ and $t \in (\beta(\omega), \alpha(\omega)]$ we have $X_t < L_t$, by (18) we have

$$M^+ I_{\Omega_\alpha} \le N_t + \int_\beta^t I_{\{M_s>0\}} I_{\Omega_\alpha} [L_s - U_s]\, ds. \quad (19)$$

Using (16), we have

$$M^+ I_{\Omega_\alpha} \le N_t + \int_\beta^t I_{\Omega_\alpha} M^+ ds. \quad (20)$$

By the stochastic Gronwall inequality (e.g., Lemma 2.1 [14]), we have

$$I_{\Omega_\alpha} M_\alpha^+ e^{-t} \le N_\beta e^{-t} + \int_\beta^\alpha e^{-t} dN_s. \quad (21)$$

By $N_\beta = 0$ we have

$$E\left(I_{\Omega_\alpha} M_\alpha^+ e^{-t}\right) \le E \int_\beta^\alpha e^{-t} dN_s = 0. \quad (22)$$

So, using (16) once again we have

$$I_{\Omega_\alpha} [L_\alpha - X_\alpha] \le I_{\Omega_\alpha} M_\alpha^+ = 0 \ a.e. \quad (23)$$

That is $L_\alpha \le X_\alpha$ on Ω_α a.s. Hence, $\mathbf{P}\{\Omega_\alpha\} = 0$. The proof is completed. □

3. Existence of Strong Solutions

In this section, we will show the existence of the solution for SDEs with discontinuous drift coefficients. The method of the proof of our main result is based on Amann's fixed point theorem (e.g., Theorem 11.D [16]), so we introduce it in the following.

Lemma 7. *Suppose that*

(1) the mapping $f : X \longrightarrow X$ is monotone increasing on an ordered set X

(2) every chain in X has a supremum

(3) there is an element $x_0 \in X$ for which $x_0 \le f(x_0)$

Then f has a smallest fixed point in the set $\{x \in X : x_0 \le x\}$.

The following theorem is our main result.

Theorem 8. *Let $b, \sigma : \mathbb{R}^+ \times \Omega \times \mathbb{R} \longrightarrow \mathbb{R}$ be predictable. Suppose that b is an increasing function in x and σ satisfies (C_σ) and there exists a predictable process $H(t,\omega)$, $t \ge 0$, such that*

$$\|b(t,\omega,x)\| + \|\sigma(t,\omega,x)\| \le H(t,\omega)(1 + \|x\|), \quad (24)$$

where $\int_0^t H^2(s,\omega)ds < \infty$, a.s. Moreover, suppose that U_t and L_t are upper and lower solutions of the SDE

$$X_t = X_0 + \int_0^t b(s, X_s)\, ds + \int_0^t \sigma(s, X_s)\, dW_s, \quad t \ge 0, \quad (25)$$

such that $L_0 \le X_0 \le U_0$, a.s.

Then there is at least a strong solution X_t which satisfies that $L_t \le X_t \le U_t$ for $t \ge 0$ holds with probability 1.

Proof. Let $\mathcal{X}$ be a space of adapted and continuous processes and define the order relation $\preceq$:

$$X \preceq Y \Longleftrightarrow \mathbf{P}\{X_t \le Y_t,\ \forall t \ge 0\} = 1, \quad (26)$$

for $X, Y \in \mathscr{X}$. We consider a subset of the space $(\mathscr{X}, \preceq)$

$$\begin{aligned} \mathscr{D} &=: [L, U] \\ &=: \{X \in \mathscr{X} : \mathbf{P}\{L_t \le X_t \le U_t,\ \forall t \ge 0\} = 1\}. \end{aligned} \tag{27}$$

For arbitrary fixed $Z \in \mathscr{D}$, we consider the following equation:

$$X_t = X_0 + \int_0^t b(s, Z_s)\, ds + \int_0^t \sigma(s, X_s)\, dW_s; \tag{28}$$

by Theorem 6 there exists a unique strong solution X_t^*. Define a mapping $S : \mathscr{D} \longrightarrow \mathscr{X}$ and $S(Z) = X^*$. To complete the proof it is enough to show S has a fixed point.

Since b is an increasing function and U is an upper solution of SDE (25), we have that

$$U_t \ge U_s + \int_s^t b(u, Z_u)\, du + \int_s^t \sigma(u, U_u)\, dW_u \tag{29}$$

holds with probability 1 for $t \ge s \ge 0$. Then U is also an upper solution of SDE (28). Similarly, we have that

$$L_t \le L_s + \int_s^t b(u, Z_u)\, du + \int_s^t \sigma(u, L_u)\, dW_u \tag{30}$$

holds with probability 1 for $t \ge s \ge 0$ such that L is also a lower solution of SDE (28). Hence, using Theorem 6 we have

$$\mathbf{P}\{L_t \le S(Z_t) \le U_t,\ \forall t \ge 0\} = 1. \tag{31}$$

Since Z is arbitrary, we have $S : \mathscr{D} \longrightarrow \mathscr{D}$ and $L \preceq S(L)$ and $S(U) \preceq U$. If S is an increasing mapping, by Lemma 7 S has a fixed point on $\mathscr{D}$. In fact, take $Z^1, Z^2 \in \mathscr{D}$ and $Z^1 \preceq Z^2$ and set $X^i =: S(Z^i)$; that is,

$$X_t^i = X_0 + \int_0^t b(s, Z_s^i)\, ds + \int_0^t \sigma(s, X_s^i)\, dW_s, \quad i = 1, 2. \tag{32}$$

Since b is an increasing function, we have that

$$X_t^2 \ge X_s + \int_s^t b(u, Z_u^1)\, du + \int_s^t \sigma(u, X_u^2)\, dW_u \tag{33}$$

holds with probability 1 for $t \ge s \ge 0$. Hence X^2 is an upper solution of the following equation:

$$X_t = X_0 + \int_0^t b(s, Z_s^1)\, ds + \int_0^t \sigma(s, X_s)\, dW_s. \tag{34}$$

And by (29) U is an upper solution of (34). Using Theorem 6 again, we have

$$\mathbf{P}\{S(Z_t^1) \le S(Z_t^2) \le U_t,\ t \ge 0\} = 1; \tag{35}$$

that is, $S(Z_t^1) \preceq S(Z_t^2)$. Hence S is an increasing function. The proof is completed.

Example 9. We consider the following SDE:

$$dX_t = \operatorname{sgn}(X_t)\, dt + dW_t, \quad \forall t \ge 0, \tag{36}$$

with initial value X_0. Obviously, $X_0 - t + W_t \le X_0 + \int_0^t \operatorname{sgn}(X_s)ds + W_t \le X_0 + t + W_t$. By Theorem 8, there exists at least one solution X_t such that $X_0 - t + W_t \le X_t \le X_0 + t + W_t, t \ge 0$ holds with probability 1.

Example 10. We have the SDE

$$dX_t = f(X_t, t)\, dt + \sigma dW_t, \quad \forall t \ge 0, \tag{37}$$

with initial value X_0, where $f(x,t)$ is a bounded function and is defined as

$$f(x,t) = \begin{cases} M+1, & x \ge M; \\ x+1, & 0 \le x < M; \\ x-1, & -M \le x < 0; \\ -M-1, & x \le -M. \end{cases} \tag{38}$$

It is easy to show $X_t = X_0 - (M+1)t + \sigma W_t$ and $X_t = X_0 + (M+1)t + \sigma W_t$ are the lower solution and upper solution of (37), respectively. And $f(x,t)$ is an increasing function in x but is not continuous in x, so we have that SDE (37) has a strong solution by using Theorem 8.

Conflicts of Interest

The author declares that they have no conflicts of interest.

Acknowledgments

This paper was supported by the Fundamental Research Funds for the Central Universities and the School of Statistics and Mathematics of CUFE.

References

[1] I. Karatzas and S. E. Shreve, *Brownian Motion and Stochastic Calculus*, Springer, New York, NY, USA, 2nd edition, 1991.

[2] X. Mao, *Stochastic Differential Equations and Applications*, Horwood, Chichester, UK, 2nd edition, 1997.

[3] P. Protter, *Stochastic Integration and Differential Equations*, Springer, Berlin, Germany, 1990.

[4] A. K. Zvonkin, "A transformation of the phase space of a diffusion process that will remove the drift," *Mathematics of the USSR-Sbornik*, vol. 22, pp. 129–149, 1974.

[5] N. V. Krylov, "On weak uniqueness for some diffusions with discontinuous coefficients," *Stochastic Processes and Their Applications*, vol. 113, no. 1, pp. 37–64, 2004.

[6] N. V. Krylov and R. Liptser, "On diffusion approximation with discontinuous coefficients," *Stochastic Processes and Their Applications*, vol. 102, no. 2, pp. 235–264, 2002.

[7] N. Halidias and P. E. Kloeden, "A note on the Euler-Maruyama scheme for stochastic differential equations with a discontinuous monotone drift coefficient," *BIT Numerical Mathematics*, vol. 48, no. 1, pp. 51–59, 2008.

[8] A. Kohatsu-Higa, A. Lejay, and K. Yasuda, "On weak approximation of stochastic differential equations with discontinuous drift coefficient," *Journal of Mathematical Economics*, vol. 1788, pp. 94–106, 2012.

[9] H.-L. Ngo and D. Taguchi, "Strong convergence for the Euler-CMaruyama approximation of stochastic differential equations with discontinuous coefficients," *Statistics and Probability Letters*, vol. 125, pp. 55–63, 2017.

[10] G. Leobacher and M. Szölgyenyi, "Convergence of the Euler-Maruyama method for multidimensional SDEs with discontinuous drift and degenerate diffusion coefficient," *Numerische Mathematik*, vol. 138, no. 1, pp. 219–239, 2018.

[11] P. Przybylowicz, "Optimality of Euler-type algorithms for approximation of stochastic differential equations with discontinuous coefficients," *International Journal of Computer Mathematics*, vol. 91, no. 7, pp. 1461–1479, 2014.

[12] N. Halidias and P. E. Kloeden, "A note on strong solutions of stochastic differential equations with a discontinuous drift coefficient," *Journal of Applied Mathematics and Stochastic Analysis*, pp. 1–6, 2006.

[13] S. Assing and R. Manthey, "The behavior of solutions of stochastic differential inequalities," *Probability Theory and Related Fields*, vol. 103, no. 4, pp. 493–514, 1995.

[14] X. Ding and R. Wu, "A new proof for comparison theorems for stochastic differential inequalities with respect to semimartingales," *Stochastic Processes and Their Applications*, vol. 78, no. 2, pp. 155–171, 1998.

[15] J. Jacod and J. Memin, "Weak and strong solutions of stochastic differential equations: existence and stability," *Stochastic Integrals*, vol. 851, pp. 169–212, 1980.

[16] E. Zeidler, *Nonliner Functional Analysis and Its Applications I: Fixed-Point Theorems*, Springer, Berlin, Germany, 1986.

15

Convergent Power Series of sech(x) and Solutions to Nonlinear Differential Equations

U. Al Khawaja[1] **and Qasem M. Al-Mdallal**[2]

[1]*Physics Department, United Arab Emirates University, P.O. Box 15551, Al Ain, UAE*
[2]*Department of Mathematical Sciences, United Arab Emirates University, P.O. Box 15551, Al Ain, UAE*

Correspondence should be addressed to U. Al Khawaja; u.alkhawaja@uaeu.ac.ae

Academic Editor: Jaume Giné

It is known that power series expansion of certain functions such as sech(x) diverges beyond a finite radius of convergence. We present here an iterative power series expansion (IPS) to obtain a power series representation of sech(x) that is convergent for all x. The convergent series is a sum of the Taylor series of sech(x) and a complementary series that cancels the divergence of the Taylor series for $x \ge \pi/2$. The method is general and can be applied to other functions known to have finite radius of convergence, such as $1/(1+x^2)$. A straightforward application of this method is to solve analytically nonlinear differential equations, which we also illustrate here. The method provides also a robust and very efficient numerical algorithm for solving nonlinear differential equations numerically. A detailed comparison with the fourth-order Runge-Kutta method and extensive analysis of the behavior of the error and CPU time are performed.

1. Introduction

It is well-known that the Taylor series of some functions diverge beyond a finite radius of convergence [1]. For instance, by way of example not exhaustive enumeration, the Taylor series of sech(x) and $1/(1+x^2)$ diverge for $x \ge \pi/2$ and $x \ge 1$, respectively. Increasing the number of terms in the power series does not increase the radius of convergence; it only makes the divergence sharper. The radius of convergence can be increased only slightly via some functional transforms [2]. Among the many different methods of solving nonlinear differential equations [3–9], the power series is the most straightforward and efficient [10]. It has been used as a powerful numerical scheme for many problems [11–19] including chaotic systems [20–23]. Many numerical algorithms and codes have been developed based on this method [10–12, 20–24]. However, the above-mentioned finiteness of radius of convergence is a serious problem that hinders the use of this method to wide class of differential equations, in particular the nonlinear ones. For instance, the nonlinear Schrödinger equation (NLSE) with cubic nonlinearity has the sech(x) as a solution. Using the power series method to solve this equation produces the power series of a sech(x), which is valid only for $x < \pi/2$.

A review of the literature reveals that the power series expansion was exploited by several researchers [10–12, 20–24] to develop powerful numerical methods for solving nonlinear differential equations. Therefore, this paper is motivated by a desire to extend these attempts to a develop a numerical scheme with systematic control on the accuracy and error. Specifically, two main advances are presented in this paper: (1) a method of constructing a convergent power series representation of a given function with an arbitrarily large radius of convergence and (2) a method of obtaining analytic power series solution of a given nonlinear differential equation that is free from the finite radius of convergence. Through this paper, we show robustness and efficiency of the method via a number of examples including the chaotic Lorenz system [25] and the NLSE. Therefore, solving the problem of finite radius of convergence will open the door wide for applying the power series method to much larger class of differential equations, particularly the nonlinear ones.

It is worth mentioning that the literature includes several semianalytical methods for solving nonlinear differential equations; such as homotopy analysis method (HAM), homotopy perturbation method (HPM), and Adomian decomposition method (ADM); for more details see [26–29] and the references therein. Essentially, these methods generate iteratively a series solution for the nonlinear systems where we have to solve a linear differential equation at each iteration. Although these methods prove to be effective in solving *most* of nonlinear differential equations and in obtaining a convergent series solution, they have few disadvantages such as the large number of terms in the solution as the number of iterations increases. One of the most important advantages of the present technique is the simplicity in transforming the nonlinear differential equation into a set of simple algebraic difference equations which can be easily solved.

The paper is thus divided into two, seemingly separated, but actually connected main parts. In the first (Section 2), we show, for a given function, how a convergent power series is constructed out of the nonconverging one. In the second part (Section 3.1), we essentially use this idea to solve nonlinear differential equations. In Section 3.2, we investigate the robustness and efficiency of the method by studying the behavior of its error and CPU time versus the parameters of the method. We summarise our results in Section 4.

2. Iterative Power Series Method

This section describes how to obtain a convergent power series for a given function that is otherwise not converging for all x. In brief, the method is described as follows. We expand the function $f(x)$ in a power series as usual, say around $x = 0$. Then we reexpress the coefficients, $f^{(n)}(x)$, in terms of $f(x)$. This establishes a recursion relation between the higher-order coefficients, $f^{(n)}(0)$, and the lowest order ones, $f^{(0)}(0)$ and $f^{(1)}(0)$, and thus the power series is written in terms of only these two coefficients. Then the series and its derivative are calculated at $x = \Delta$, where Δ is much less than the radius of convergence of the power series. A new power series expansion of $f(x)$ is then performed at $x = \Delta$. Similarly, the higher-order coefficients are reexpressed in terms of the lowest order coefficients $f^{(0)}(\Delta)$ and $f^{(1)}(\Delta)$. The value of the previous series and its derivative calculated at $x = \Delta$ are then given to $f^{(0)}(\Delta)$ and $f^{(1)}(\Delta)$, respectively. Then a new expansion around 2Δ is performed with the lowest order coefficients being taken from the previous series, and so on. This iterative process is repeated N times. The final series will correspond to a convergent series at $x = N\Delta$.

Here is a detailed description of the method. The function $f(x)$ is expanded in a Taylor series, $T^0(x)$, around $x = 0$. The infinite Taylor series is an exact representation of $f(x)$ for $x < R$ where R is the radius of convergence. For $x \geq R$ the series diverges. We assume that x is divided into N small intervals $\Delta = x/N$ such that $\Delta < R$. Expanding $f(x)$ around the beginning of each interval we obtain N convergent Taylor series representing $f(x)$ in each interval

$$T^j(y) = \sum_{n=0}^{\infty} \frac{1}{n!} f^{(n)}(j\Delta)(y - j\Delta)^n, \qquad j\Delta \leq y < (j+1)\Delta,\ j = 0, 1, 2, \ldots, N, \tag{1}$$

where $T^j(y)$ denotes the Taylor series expansion of $f(y)$ around $y = j\Delta$ and $f^{(n)}(j\Delta)$ is the nth derivative of $f(y)$ calculated at $y = j\Delta$. It is noted that we use $y \in [(j-1)\Delta, j\Delta]$ as the independent variable for the nth Taylor series expansion to distinguish it from $x = N\Delta$. However, these series can not be combined in a single series since their ranges of applicability are different and do not overlap. To obtain a single convergent power series out of the set of series T^j, we put forward two new ideas, which constitute the basis of our method; namely:

(1) Reexpress $f^{(n)}(y)$ in terms of $f(y)$ as $f^{(n)}(y) = F_n[f(y)]$, where the functional $F_n[f(y)]$ is determined by direct differentiation of $f(y)$ for n times and then reexpressing the result in terms of $f(y)$ only. We conjecture that this is possible for a wide class of functions if not all. At least for the two specific functions considered here, this turned out to be possible. Equation (1) then takes the form

$$T^j(y) = \sum_{n=0}^{\infty} a_n\left(a_0^j\right)(y - j\Delta)^n, \tag{2}$$

where we have renamed $f(j\Delta)$ by a_0^j and $F_n[f(j\Delta)]/n!$ by $a_n(a_0^j)$ for a reason to be obvious in the next section. Thus, the coefficients a_n for all n are determined only by a_0^j.

(2) Calculate a_0^j from T^{j-1} at $j\Delta$

$$a_0^j = T^{j-1}(j\Delta) = \sum_{n=0}^{\infty} a_n\left(a_0^{j-1}\right)\Delta^n, \quad j = 1, 2, \ldots, N, \tag{3}$$

which amounts to assigning the value of the Taylor series at the end of an interval to a_0^j of the consecutive one. Equation (3) captures the essence of the recursive feature of our method; a_0^N is calculated recursively from a_0^0 by repeated action of the right-hand-side on a_0^0. While T^j represents the function f within an interval of width Δ, the sequence a_0^j corresponds to the values of the function at the end of the intervals. In the limit $N \to \infty$, or equivalently $\Delta \to 0$, the discrete set of a_0^j values and $j\Delta$ render to the continuous function $f(x)$ and its independent variable x, respectively. Formally, the convergent power series expansion of $f(x)$ around $x = 0$ will thus be given by

$$f(x) = \lim_{N\to\infty} S^N, \tag{4}$$

where S^N denotes the Nth iteration of

$$S[f(0)] = \sum_{n=0}^{\infty} a_n(f(0))\left(\frac{x}{N}\right)^n. \tag{5}$$

As an illustrative example, we apply the method to $f(x) = \operatorname{sech}(x)$. The infinite Taylor series expansion of this function diverges sharply to infinity at $x = \pi/2$. The first step is to determine the coefficients a_n^j, which are the coefficients of the T^j series

$$\begin{aligned}T^j(y) &= \operatorname{sech}(y)\big|_{y=j\Delta} + [-\operatorname{sech}(y)\tanh(y)]_{y=j\Delta}(y - j\Delta) + \frac{1}{2!}\left[-\operatorname{sech}^3(y) + \operatorname{sech}(y)\tanh^2(y)\right]_{y=j\Delta} \\ &\cdot (y-j\Delta)^2 + \frac{1}{3!}\left[\left(5\operatorname{sech}^2(y) - \tanh^2(y)\right) \cdot \operatorname{sech}(y)\tanh(y)\right]_{y=j\Delta}(y-j\Delta)^3 + \frac{1}{4!}\left[5\operatorname{sech}^5(y) - 18\operatorname{sech}^3(y)\tanh^2(y) + \operatorname{sech}(y)\tanh^4(y)\right]_{y=j\Delta} \\ &\cdot (y-j\Delta)^4 + \cdots. \end{aligned} \tag{6}$$

The next step is to reexpress the higher-order coefficients, a_n^j, in terms of the zeroth-order coefficient $a_0^j = \operatorname{sech}(y)|_{y=j\Delta}$. The property $\operatorname{sech}^2(y) + \tanh^2(y) = 1$ is used to that end. It is noticed, however, that while it is possible to express the even-n coefficients in terms of a_0^j only, the odd-n coefficients can only be expressed terms of both a_0^j and $a_1^j = -\operatorname{sech}(y)\tanh(y)|_{y=j\Delta}$. In the context of solving differential equations using the power series method, this reflects the fact that the solution is expressed in terms of two independent parameters (initial conditions). The sech function is indeed a solution of a second-order differential equation, which is solved using this method in the next section. Equation (6) then takes the form

$$\begin{aligned}T^j(y) &= a_0^j + a_1^j(y-j\Delta) + \frac{1}{2!}\left[1 - 2\left(a_0^j\right)^2\right] \cdot a_0^j(y-j\Delta)^2 + \frac{1}{3!}\left[1 - 6\left(a_0^j\right)^2\right] \cdot a_1^j(y-j\Delta)^3 \\ &+ \frac{1}{4!}\left[1 - 8\left(a_0^j\right)^2 + 8\left(a_0^j\right)^4 - 12\left(a_1^j\right)^2\right] \cdot a_0^j(y-j\Delta)^4 + \cdots. \end{aligned} \tag{7}$$

Calculating $T^j(y)$ series at the end of its interval of applicability, $y = (j+1)\Delta$, we get

$$T^j((j+1)\Delta) = a_0^j + a_1^j\Delta + a_2^j\left(a_0^j\right)\Delta^2 + a_3^j\left(a_0^j, a_1^j\right)\Delta^3 + a_4^j\left(a_0^j\right)\Delta^4 + \cdots, \tag{8}$$

where the "recursion" coefficients are given by

$$\begin{aligned} a_2^j\left(a_0^j\right) &= \frac{1}{2!}\left[1 - 2\left(a_0^j\right)^2\right]a_0^j, \\ a_3^j\left(a_0^j, a_1^j\right) &= \frac{1}{3!}\left[1 - 6\left(a_0^j\right)^2\right]a_1^j, \\ a_4^j\left(a_0^j\right) &= \frac{1}{4!}\left[1 - 8\left(a_0^j\right)^2 + 8\left(a_0^j\right)^4 - 12\left(a_1^j\right)^2\right]a_0^j. \end{aligned} \tag{9}$$

Finally, we assign $T^j((j+1)\Delta)$ to a_0^{j+1}

$$a_0^{j+1} = a_0^j + a_1^j\Delta + a_2^j\left(a_0^j\right)\Delta^2 + a_3^j\left(a_0^j, a_1^j\right)\Delta^3 + a_4^j\left(a_0^j\right)\Delta^4 + \cdots. \tag{10}$$

The second independent coefficient a_1^{j+1} is determined by the derivative of $T^j(y)$ calculated at $y = (j+1)\Delta$

$$a_1^{j+1} = a_1^j + 2a_2^j\left(a_0^j\right)\Delta + 3a_3^j\left(a_0^j, a_1^j\right)\Delta^2 + 4a_4^j\left(a_0^j\right)\Delta^3 + \cdots. \tag{11}$$

While, in the limit $N \to \infty$, a_0^j corresponds to the function $f(x)$, the sequence a_1^j corresponds to $f'(x)$. Therefore, the power series expansion of sech(x) and its first derivative are given by

$$\begin{pmatrix} f(x) \\ f'(x) \end{pmatrix} = \lim_{N\to\infty} \begin{pmatrix} a_0 + a_1\left(\frac{x}{N}\right) + a_2(a_0)\left(\frac{x}{N}\right)^2 + a_3(a_0, a_1)\left(\frac{x}{N}\right)^3 + a_4(a_0)\left(\frac{x}{N}\right)^4 + \cdots \\ a_1 + 2a_2(a_0)\left(\frac{x}{N}\right) + 3a_3(a_0, a_1)\left(\frac{x}{N}\right)^2 + 4a_4(a_0)\left(\frac{x}{N}\right)^3 + \cdots \end{pmatrix}^N, \tag{12}$$

where the superscript of the matrix on the right-hand-side, N, denotes the Nth iteration of the matrix. The superscript j has been removed since the functional form of the recursion coefficients does not depend on j. The procedure of calculating the power series recursively is described as follows. First, $a_0 = \operatorname{sech}(0) = 1$ and $a_1 = \operatorname{sech}'(0) = 0$ are substituted in the right-hand-side of the last equation. Then the result of the upper element is taken as the updated value of a_0, and, similarly, the lower element updates a_1. The two updated values are then resubstituted back in the right-hand-side. The process is repeated N times. To obtain an explicit form of the series we truncate the Taylor series at $n_{max} = 4$ and use $N = 4$ iterations. The resulting expansion takes the form

$$\operatorname{sech}(x) = 1 - \frac{1}{2}x^2 + \frac{5}{24}x^4 - 0.0806681x^6 + 0.0302048x^8 + \cdots + 1.4434798 \times 10^{-461}x^{624}. \quad (13)$$

It is noted that the higher-order coefficients, which correspond to ratios of large integers, are represented in real numbers for convenience. Already with such a small number of iterations, $N = 4$, the number of terms equals 313. By inspection, we find that the number of terms equals $((n_{\max} + 1)^N + 1)/2$. Here, $n_{\max}$ is even due to the fact that sech(x) is an even function.

It is also noted that the series (13) is composed of the Taylor expansion of sech(x) around zero, represented by the first three terms, and a series of higher-order terms generated from the nonlinearity in the recursion relations of a_n. In fact, we prove in the next section that this property holds for any $n_{\max}$, N, and function $f(x)$, provided that the Taylor series of the later exists. Therefore, the power series expansion of sech(x), given by (12), can be put in the suggestive form

$$\operatorname{sech}(x) = T + \lim_{N\to\infty} C(N), \quad (14)$$

where T is the infinite Taylor series of $f(x)$ about $x = 0$ and $C(N)$ is a complementary series. It turns out that the complementary series increases the radius of convergence of T for $x \geq \pi/2$. For finite N, the effect of $C(N)$ is to shift the radius of convergence, R, to a larger value such that $R \to \infty$ for $N \to \infty$. In Figure 1 we plot the convergent power series obtained by the present method as given by (12) using $n_{\max} = 4$ and $N = 100$. The curve is indistinguishable from the plot of sech(x). Both the Taylor series expansion, T, and the complementary series, C, diverge sharply at $x = \pi/2$. Since C is essentially zero for $x < \pi/2$, it will not affect the sum $T + C$. However, its major role is to cancel the divergency for $x \geq \pi/2$. In the limit $N \to \infty$, T will be an exact representative of sech(x) for $x < \pi/2$ and C will equal zero in the same interval. For $x \geq \pi/2$, the divergences in T and C cancel each other with a remainder that is an exact representative of sech(x). In this manner, $T + C$ will represent sech(x) for all x.

For finite values of $n_{\max}$ and N, the series $T + C$ is an approximate representative of sech(x). Truncating the Taylor series at $n_{\max}$ introduces an error of order $\Delta^{n_{\max}+1}$. This error will be magnified N times due the recursive substitutions. The total error is then estimated by

$$\text{error} = \left(\frac{x}{N}\right)^{n_{\max}+1} N. \quad (15)$$

For the parameters used in Figure 1, this error is of order 10^{-6} at $x = 5$. This can be reduced to extremely small values such as 10^{-131} with $n_{\max} = 100$. However, the number of terms in the series $T + C$ will be of order 10^{200} which is extremely large and hinders any analytical manipulations.

As another example, we consider $f(x) = 1/(1 + x^2)$ with Taylor series diverging at $x = 1$. Much of the formulation we followed for the previous case holds here and the specifics of the function alter only the recursion relations, (9):

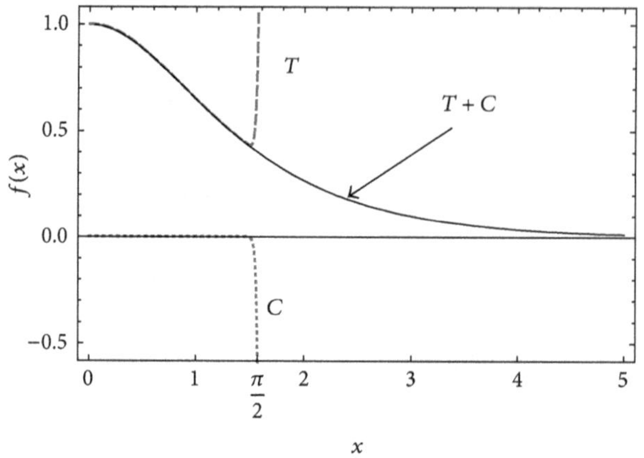

Figure 1: The solid curve corresponds to the convergent power series obtained by the present method as given by (12) using $n_{\max} = 4$ and $N = 100$. It is indistinguishable from the solid curve of $f(x) =$ sech(x). Dashed curve corresponds to the Taylor series expansion, T, and the dotted curve corresponds to the complementary series, C.

$$a_2^j\left(a_0^j\right) = \left[3 - 4a_0^j\right]\left(a_0^j\right)^2, \quad (16)$$

$$a_3^j\left(a_0^j, a_1^j\right) = 2\left[1 - 2a_0^j\right]a_0^j a_1^j, \quad (17)$$

$$a_4^j\left(a_0^j\right) = \left[16 - 64a_0^j + 85\left(a_0^j\right)^2 - 52\left(a_0^j\right)^3 + 16\left(a_0^j\right)^4\right]a_0^j. \quad (18)$$

The convergent power series is obtained by using these recursion relations in (12). Plots similar to those of Figure 1 are obtained.

We present now a proof that the convergent power series produced by the recursive procedure always regenerates the Taylor series in addition to a complementary one.

Proposition 1. *If we expand $f(x)$ in a Taylor series, T, around $x = 0$ truncated at $n_{\max}$ and use the recursive procedure, as described above, the resulting convergent power series always takes the form $T + C$ where C is a power series of orders larger than $n_{\max}$. This is true for any number of iterations, N, maximum power of the Taylor series, $n_{\max}$, and for all functions that satisfy the general differential equation*

$$f''(x) = F\left[f(x)\right], \quad (19)$$

where $F[\cdot]$ is an analytic real functional that does not contain derivatives.

Proof. It is trivial to prove this for a specific case, such as sech(x). For the general case, we prove this only for $n_{\max} = 4$ and $N = 2$. The Taylor series expansion of sech(x) around $x = 0$ is

$$\operatorname{sech}(x) = 1 - \frac{1}{2}x^2 + \frac{5}{24}x^4 + \cdots. \quad (20)$$

In our recursive procedure, this is put in the equivalent form

$$\begin{pmatrix} \operatorname{sech}(\Delta) \\ \operatorname{sech}'(\Delta) \end{pmatrix} \simeq \begin{pmatrix} a_0 + a_1\Delta + \frac{1-2a_0^2}{2}\Delta^2 + \frac{1}{3!}a_1\left(1-6a_0^2\right)\Delta^3 + \frac{1}{4!}a_0\left(1-8a_0^2+12a_0^4-12a_1^2\right)\Delta^4 \\ \left(1-2a_0^2\right)\Delta + \frac{1}{2}a_1\left(1-6a_0^2\right)\Delta^2 + \frac{1}{6}a_0\left(1-8a_0^2+12a_0^4-12a_1^2\right)\Delta^3 \end{pmatrix}^N, \tag{21}$$

where the approximation stems from using finite N and $n_{\max}$, and $\Delta = x/N$. For $N = 1$, $a_0 = 1$, and $a_1 = 0$, (20) is regenerated. However, in our recursive procedure a_0 and a_1 are kept as variables since they will be substituted for at each recursive step. Only at the last step are their numerical values inserted. For $N = 2$, we resubstitute in the last equation for a_0 and a_1 by their updated expressions, as follows:

$$\begin{pmatrix} a_0 \\ a_1 \end{pmatrix} \longrightarrow \begin{pmatrix} a_0 + a_1\Delta + \frac{1-2a_0^2}{2}\Delta^2 + \frac{1}{3!}a_1\left(1-6a_0^2\right)\Delta^3 + \frac{1}{4!}a_0\left(1-8a_0^2+12a_0^4-12a_1^2\right)\Delta^4 \\ \left(1-2a_0^2\right)\Delta + \frac{1}{2}a_1\left(1-6a_0^2\right)\Delta^2 + \frac{1}{6}a_0\left(1-8a_0^2+12a_0^4-12a_1^2\right)\Delta^3 \end{pmatrix}. \tag{22}$$

Substituting the updated expressions for a_0 and a_1 in (21), we get

$$\begin{aligned} \operatorname{sech}(x) \simeq{}& a_0 + a_1\left(\frac{x}{N}\right) + 2a_0\left(1-2a_0^2\right)\left(\frac{x}{N}\right)^2 \\ &+ \frac{4}{3}a_1\left(1-6a_0^2\right)\left(\frac{x}{N}\right)^3 \\ &+ \frac{2}{3}a_0\left(1-8a_0^2+12a_0^4-12a_1^2\right)\left(\frac{x}{N}\right)^4. \end{aligned} \tag{23}$$

Clearly for $N = 2$, the last equation gives the Taylor expansion, that is, (21) with $N = 1$. The complimentary series, C, is absent here since we have terminated the expansions at $n = n_{\max} = 4$. For $N = 3$, another step of resubstituting updated expressions is needed, and so on.

Now, we present the proof for the more general case, namely, when $f(x)$ is unspecified but is a solution to (19). We start with the following Taylor series expansion of $f(x)$ and its derivative

$$\begin{pmatrix} f(\Delta) \\ f'(\Delta) \end{pmatrix} = \begin{pmatrix} a_0 + a_1\Delta + a_2(a_0,a_1)\Delta^2 + a_3(a_0,a_1)\Delta^3 + a_4(a_0,a_1)\Delta^4 \\ a_1 + 2a_2(a_0,a_1)\Delta + 3a_3(a_0,a_1)\Delta^2 + 4a_4(a_0,a_1)\Delta^3 \end{pmatrix}. \tag{24}$$

Substituting on the right-hand-side for a_0 and a_1 by $f(\Delta)$ and $f'(\Delta)$, respectively, we get

$$\begin{aligned} f(\Delta) ={}& a_0 + a_1\Delta + 4a_2\Delta^2 + \left(5a_3 + 2a_2\frac{\partial a_2}{\partial a_1} + a_1\frac{\partial a_2}{\partial a_0}\right)\Delta^3 + \left(6a_4 + 3a_3\frac{\partial a_2}{\partial a_1} + 2a_2\frac{\partial a_3}{\partial a_1} \right. \\ &\left. + 2a_2^2\frac{\partial^2 a_2}{\partial a_1^2} + a_2\frac{\partial a_2}{\partial a_0} + a_1\frac{\partial a_3}{\partial a_0} + 2a_1a_2\frac{\partial^2 a_2}{\partial a_1 \partial a_0} + \frac{1}{2}a_1^2\frac{\partial^2 a_2}{\partial a_0^2}\right)\Delta^4. \end{aligned} \tag{25}$$

The partial derivatives can not be calculated unless the functional forms of the recursion coefficients are known. One possibility is to specify the function being expanded, $f(x)$, as we did at the start of this proof. The other possibility is to exploit the differential equation that $f(x)$ is a solution for, namely, (19). Substituting the Taylor expansion of $f(\Delta)$, from (24) in (19) and expanding up to the fourth power in Δ, we obtain the following relations:

$$\begin{aligned} a_2 &= -\frac{F}{2}, \\ a_3 &= -\frac{a_1F'}{6}, \\ a_4 &= -\frac{a_2F'}{12} - \frac{a_1^2F''}{24}, \end{aligned} \tag{26}$$

which lead to

$$\begin{aligned} \frac{\partial a_2}{\partial a_0} &= -\frac{F'}{2}, \\ \frac{\partial a_2}{\partial a_1} &= -\frac{\Delta F'}{2}, \\ \frac{\partial^2 a_2}{\partial a_0^2} &= -\frac{F''}{2}, \\ \frac{\partial^2 a_2}{\partial a_1^2} &= -\frac{\Delta^2 F''}{2}, \end{aligned}$$

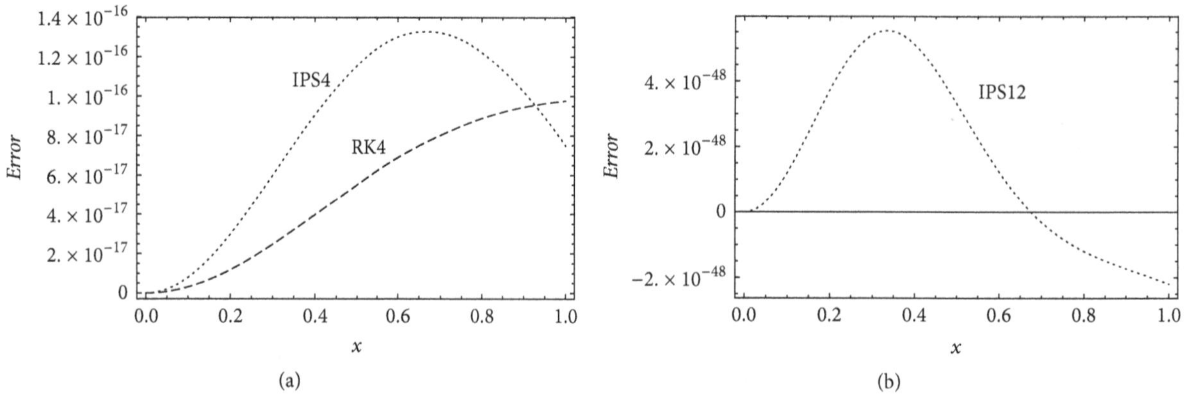

FIGURE 2: Error defined by the difference between the numerical and exact soliton solution of the NLSE, (29). (a) Dashed line corresponds to RK4 and dotted line corresponds to IPS4. (b) IPS12. Parameters used: $N = 1000$, number of digits $N_d = 50$, $f(0) = 1$, and $f'(0) = 0$.

$$\frac{\partial^2 a_2}{\partial a_0 \partial a_1} = -\frac{\Delta F''}{2},$$
$$\frac{\partial a_3}{\partial a_1} = -\frac{F'}{6}. \tag{27}$$

Using these equations in (28), we obtain

$$f(\Delta) = a_0 + a_1\Delta + 4a_2\Delta^2 + 8a_3(a_0, a_1)\Delta^3 + 16a_4(a_0, a_1)\Delta^4. \tag{28}$$

For $N = 2$, the last equation regenerates the Taylor series of $f(x)$, namely, (24) with $N = 1$, and this completes the proof. □

3. Application to Nonlinear Differential Equations

The method described in the previous section can be used as a powerful solver and integrator of nonlinear differential equations both analytically and numerically. In Section 3.1, we apply the method on a number of well-known problems. In Section 3.2, we show the power of the method in terms of detailed analysis of the error and CPU time.

3.1. Examples. For the sake of demonstration, we consider the following nonlinear differential equation:

$$\frac{1}{2}f(x) - \frac{1}{2}f''(x) - f^3(x) = 0. \tag{29}$$

The reason for selecting this equation is that $f(x) = \text{sech}(x)$ is one of its exact solutions. Substituting the power series expansion $f(x) = \sum_{n=0}^{\infty} a_n x^n$, we obtain, as usual, the recursion relations

$$a_2(a_0) = \frac{1}{2!}\left[1 - 2(a_0)^2\right]a_0,$$
$$a_3(a_0, a_1) = \frac{1}{3!}\left[1 - 6(a_0)^2\right]a_1,$$
$$a_4(a_0) = \frac{1}{4!}\left[1 - 8a_0^2 + 12a_0^4 - 12a_1^2\right]a_0, \tag{30}$$

where a_0 and a_1 turn out to be independent parameters which in the present case correspond to the initial conditions on the solution and its first derivative. It is not surprising that these recursion relations are identical with those we found for the sech(x) in the previous section, (9). Therefore, substituting the above recursion relations in $f(x) = \sum_{n=0}^{\infty} a_n x^n$ we obtain the Taylor series expansion, T, of $f(x) = \text{sech}(x)$. Removing the divergency in T follows exactly the same steps as in the previous section, and thus an exact solution in terms of a convergent power series is obtained, as also plotted in Figure 1.

For $f(x) = 1/(1 + x^2)$, the relevant differential equation is

$$f''(x) + f^3(x) - 6f^2(x) = 0. \tag{31}$$

Substituting the power series expansion in this equation, the recursion relations will be given by (16)–(18). Similarly, the convergent series solution will be obtained, as in the previous section.

3.2. Numerical Method. As a numerical method, the power series is very powerful and efficient [10]. The power series method with $N_{max} = 4$, denoted by IPS4, is used to solve the NLSE, (29) and the error is calculated as the difference between the numerical solution and the exact solution, namely, sech(x). The equation is then resolved using the fourth-order Runge-Kutta (RK4) method. In Figure 2, we plot the error of both methods which turn out to be of the same order. Using the iterative power series method with $N_{max} = 12$, (IPS12), the error drops to infinitesimally low values. Neither the CPU time nor the memory requirements for IPS12 are much larger than those for IPS4; it is straight forward upgrade to higher orders which leads to ultrahigh efficiency. This is verified by the set of tables, Tables 1–4, where we compute sech(1) using both the RK4 and the iterative power series method and show the corresponding CPU time. For the same N, Tables 1 and 3 show that both RK4 and

TABLE 1: RK4 sech(x) solution of the NLSE, (29), computed at $x = 1$.

N	RK4 sech(1)	CPU time
1000	0.6480542736639463753683598987682775440961395993015	0.087257
2000	0.6480542736638892102461002212645001242748400651983	0.166534
3000	0.6480542736638861522791729310382098086291071707752	0.229016
4000	0.6480542736638856377319630763886301512099763161750	0.314480
5000	0.6480542736638854971232600221160524143929321411426	0.427610
Exact	0.6480542736638853995749773532261503231084893120719	

TABLE 2: IPS4 sech(x) solution of the NLSE, (29), computed at $x = 1$.

N	IPS4 sech(1)	CPU time
1000	0.6480542736639323955233350367786700400715255548373	0.056793
2000	0.6480542736638883277492809223389308596155602276884	0.137941
3000	0.6480542736638859773823119964459740185046024606302	0.215395
4000	0.6480542736638855823023023050565759829475356280281	0.304501
5000	0.6480542736638854743968579026604641270130595226186	0.372284
Exact	0.6480542736638853995749773532261503231084893120719	

TABLE 3: IPS12 sech(x) solution of the NLSE, (29), computed at $x = 1$.

N	IPS12 sech(1)	CPU time
1000	0.6480542736638853995749773532261503231079594354079	0.280617
2000	0.6480542736638853995749773532261503231084891816361	0.595639
3000	0.6480542736638853995749773532261503231084893110634	0.891370
4000	0.6480542736638853995749773532261503231084893120400	1.080357
5000	0.6480542736638853995749773532261503231084893120697	1.366386
Exact	0.6480542736638853995749773532261503231084893120719	

TABLE 4: IPS12 sech(x) solution of the NLSE, (29), computed at $x = 1$, but with much less number of iterations than in Table 3.

N	IPS12 sech(1)	CPU time
3	0.6480542794079665629469114154348980055814088430953	0.000782
6	0.6480542736643770346283969779587807646256058100135	0.001429
9	0.6480542736638872007452856567074697922787489590238	0.002123
12	0.6480542736638854259793573015747954030451932565027	0.002768
15	0.6480542736638854001495023257702479909741369016589	0.003594
Exact	0.6480542736638853995749773532261503231084893120719	

IPS4 produce the first 16 digits of the exact value (underlined numbers in the last raw) and consume almost the same CPU time. Table 3 shows that, for the same N, IPS12, reproduces the first 49 digits of the exact value. The CPU time needed for such ultrahigh accuracy is just about 3 times that of the RK4 and IPS4. Of course the accuracy can be arbitrarily increased by increasing N or more efficiently N_{max}. For IPS12 to produce only the first 16 digits, as in RK4 and IPS4, only very small number of iterations is needed, as shown in Table 4. The CPU time in this case is about 100 times less than that of RK4 and IPS4, highlighting the high efficiency of the power series method.

A more challenging test on the power series method is the chaotic Lorenz system [25] given by

$$\begin{aligned} \dot{z}_1 &= -\sigma z_1 + \sigma z_2, \\ \dot{z}_2 &= -z_1 z_3 + r z_1 - z_2, \\ \dot{z}_3 &= z_1 z_2 - b z_3, \end{aligned} \quad (32)$$

where we take the usual values $\sigma = 10$, $b = -8/3$, and $R = 28$ with initial conditions $z_1(0) = z_2(0) = 1$ and $z_3(0) = 20$. It is straight forward to generalise the method to three differential equations; therefore we do not show the

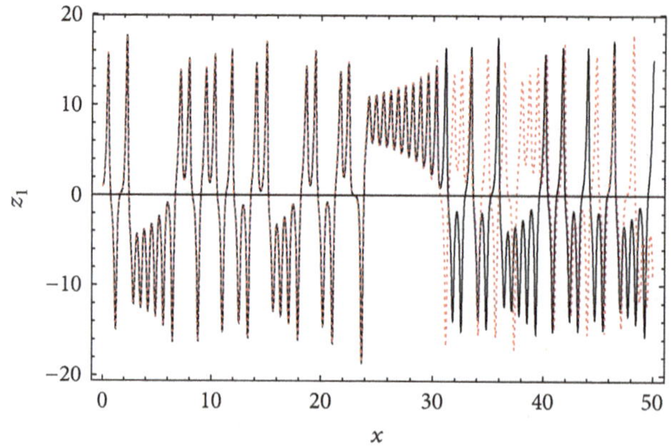

FIGURE 3: Numerical solution of the chaotic Lorenz system, (32), using the RK4 and IPS methods. Parameters: for IPS (solid black line), $N_{\max} = 16$, $N = 5000$. For RK4 (dashed red line), $N = 2 \times 10^5$. For both curves we used $N_d = 1000$, $z_1 = 1 = z_2 = 1$, $z_3 = 20$, $\sigma = 10$, $b = -8/3$, and $R = 28$.

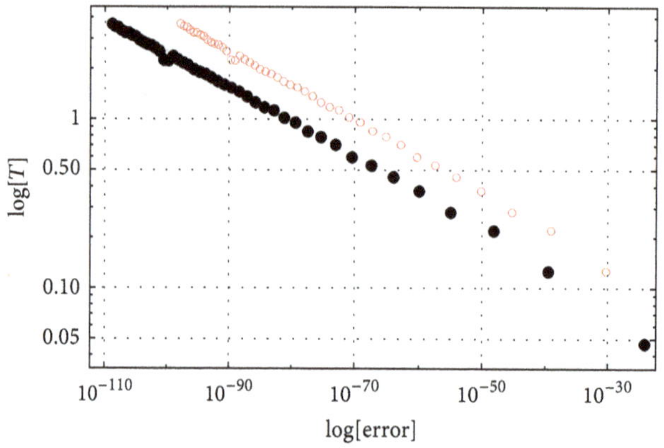

FIGURE 4: CPU time versus error on a log-log scale for the iterative power series solution of the NLSE, (29). Black filled circles correspond to error calculated from the difference between the power series solution and the exact solution. Red empty circles correspond to the error estimated by the formula (15). Parameters used: $N_{\max} = 50$, $n_d = 500$, and N ranges between 2 and 90.

details of the calculation. In Figure 3, the results of solving the Lorenz system using RK4 and IPS12 are shown. For the same parameters, namely discretization, RK4 reaches stability at about $x = 30$; that is, the curve for $x < 30$ is unchanged by increasing N, but for $x > 30$, the curve keeps changing by increasing N. In comparison, IPS12 reaches stability at about $x = 50$. In chaotic systems, it is quite challenging to go that deep in the chaotic region. Hence, the need for such high accuracy methods.

Achieving higher accuracy requires larger CPU time usage. Therefore, it is important to investigate how the CPU time, denoted here by T, depends on the main parameters of the method, namely N and $N_{\max}$. A typical plot is shown in Figure 4, where we plot on a log-log scale the CPU time versus the error. The linear relationship indicates $T \propto \text{error}^p$, where p is the slope of the line joining the points in Figure 4. The error can be calculated in two ways: (i) the difference between the numerical solution and (ii) theoretical estimate, (15). Both ways are shown in the figure and they have the same slope. However, as expected, error defined by (15), which is actually an upper limit on the error, is always larger than the first one. To find how the CPU time depends explicitly on N and $N_{\max}$, we argue that the dependence should be of the form $T \propto N N_{\max}^3$. This is justified by the fact that CPU time should be linearly proportional to the number of terms computed. The number of terms computed increases linearly with the number of iterations N. The number of terms in the power series is linearly proportional to $N_{\max}$. When substituted in the NLSE with cubic nonlinearity, the resulting number of terms, and thus T, will be proportional to $N_{\max}^3$. In Figure 5, it is shown indeed that the ratio $T/N N_{\max}^3$ saturates asymptotically to a constant for large N and $N_{\max}$ since the scaling behaviors mentioned here apply for large N and $N_{\max}$. The proportionality constant, c, is very small and corresponds to the CPU time of calculating one term. It is dependent on the machine, the programming optimization [10], and the number of digits used, N_d. In terms of the number of digits, the CPU time increases, as shown in Figure 6, where it is noticed that CPU time is almost constant for number of digits $N_d < 500$.

4. Conclusions

We have presented an iterative power series method that solves the problem of finite radius of convergence. We have

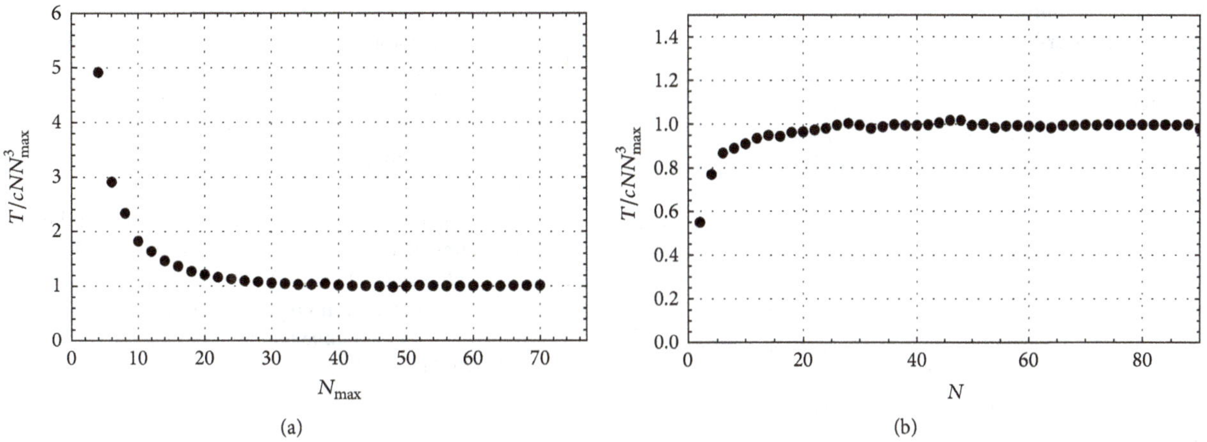

FIGURE 5: CPU time versus N_{max} (a) and N (b) for the NLSE, (29). Parameters used: for (a), $N = 400$. For (b), $N_{max} = 70$. For both plots, $N_d = 1000$, $c = 6.2 \times 10^{-7}$, $f(0) = 1$, $f'(0) = 0$, and $\Delta = 1/N$.

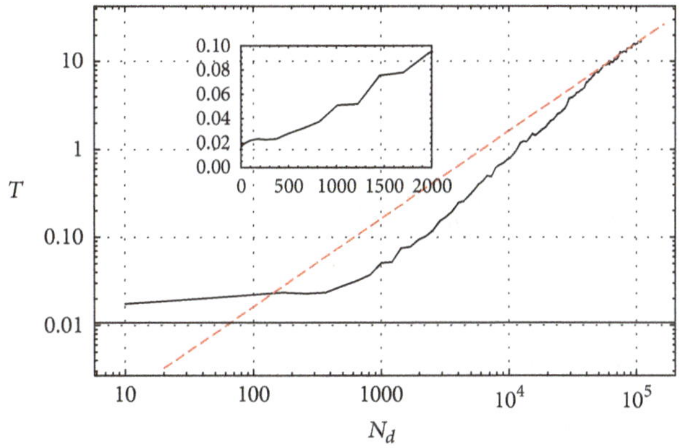

FIGURE 6: CPU time versus the number of digits for the power series solution of NLSE, (29). Parameters used: $N_{max} = 4$, $N = 400$, $f(0) = 1$, $f'(0) = 0$, and $\Delta = 1/N$. Inset shows a zoom on linear scale. Red dashed line is an asymptote of slope equal 1.

proved that the iterative power series is always composed of a sum of the typical power series of the function and a complementary series that cancels the divergency. The method is divided into two schemes where in the first we find a convergent power series for a given function and in the second we solve a given nonlinear differential equation. The result of the iterative power series expansion of sech(x) is remarkably convergent for arbitrary radius of convergence and accuracy, as shown by Figures 1 and 2 and Tables 1–4. Extremely high accuracy can be obtained by using higher-order iterative power series via increasing N_{max} with relatively low CPU time usage. Robustness and efficiency of the method have been shown by solving the chaotic Lorenz system and the NLSE. Extensive analysis of the error and CPU time characterising the method is performed. Although we have focused on the localised sech(x) solution of the NLSE, all other solitary wave solutions (conoidal waves) can be obtained using the present method, just by choosing the appropriate initial conditions.

The method can be generalised to partial and fractional differential equations making its domain of applicability even wider.

Conflicts of Interest

The authors declare that there are no conflicts of interest regarding the publication of this paper.

References

[1] M. Abramowitz and I. A. Stegun, *Handbook of Mathematical Functions with Formulas, Graphs, and Mathematical Tables*, vol. 7, Dover Publications, New York, NY, USA, 1965.

[2] R. E. Scraton, "A note on the summation of divergent power series," *Mathematical Proc. of the Camb. Phil. Soc*, vol. 66, pp. 109–114, 1969.

[3] A. D. Polyanin and V. F. Zaitsev, *Handbook of Nonlinear Partial Differential Equations*, Chapman and Hall, New York, NY, USA, 2003.

[4] C. S. Gardner, J. M. Greene, M. D. Kruskal, and R. M. Miura, "Method for solving the Korteweg-deVries equation," *Physical Review Letters*, vol. 19, no. 19, pp. 1095–1097, 1967.

[5] P. D. Lax, "Integrals of nonlinear equations of evolution and solitary waves," *Communications on Pure and Applied Mathematics*, vol. 21, pp. 467–490, 1968.

[6] V. B. Matveev and M. A. Salle, *Darboux Transformations and Solitons*, Springer-Verlag, Berlin, Gemany, 1991.

[7] R. Hirota, *Topics in Current Physics 17*, R. K. Eullough and P. I. Caudrey, Eds., Springer-Verlag, Berlin, Germany, 1980.

[8] G. Adomian, *Solving Frontier Problems of Physics: The Decomposition Method*, Kluwer Academic, Dordrecht, The Netherlands, 1994.

[9] S. Liao and Y. Tan, "A general approach to obtain series solutions of nonlinear differential equations," *Studies in Applied Mathematics*, vol. 119, no. 4, pp. 297–354, 2007.

[10] R. Barrio, M. Rodríguez, A. Abad, and F. Blesa, "Breaking the limits: the Taylor series method," *Applied Mathematics and Computation*, vol. 217, no. 20, pp. 7940–7954, 2011.

[11] G. Corliss and Y. F. Chang, "Solving ordinary differential equations using Taylor series," *ACM Transactions on Mathematical Software*, vol. 8, no. 2, pp. 114–144, 1982.

[12] Y. F. Chang and G. Corliss, "ATOMFT: solving ODEs and DAEs using Taylor series," *Computers & Mathematics with Applications*, vol. 28, no. 10-12, pp. 209–233, 1994.

[13] J. D. Pryce, "Solving high-index DAEs by Taylor series," *Numerical Algorithms*, vol. 19, no. 1-4, pp. 195–211, 1998.

[14] R. Barrio, "Performance of the Taylor series method for ODEs/DAEs," *Applied Mathematics and Computation*, vol. 163, no. 2, pp. 525–545, 2005.

[15] N. S. Nedialkov and J. D. Pryce, "Solving differential-algebraic equations by taylor series (I): Computing taylor coefficients," *BIT Numerical Mathematics*, vol. 45, no. 3, pp. 561–591, 2005.

[16] N. S. Nedialkov and J. D. Pryce, "Solving differential-algebraic equations by Taylor series. (II): Computing the system Jacobian," *BIT Numerical Mathematics*, vol. 47, no. 1, pp. 121–135, 2007.

[17] N. S. Nedialkov and J. D. Pryce, "Solving differential algebraic equations by Taylor series. (III): THE DAETS code," *JNAIAM. Journal of Numerical Analysis, Industrial and Applied Mathematics*, vol. 3, no. 1-2, pp. 61–80, 2008.

[18] N. S. Nedialkov, K. R. Jackson, and G. F. Corliss, "Validated solutions of initial value problems for ordinary differential equations," *Applied Mathematics and Computation*, vol. 105, no. 1, pp. 21–68, 1999.

[19] W. Tucker, "A Rigorous ODE Solver and Smale's 14th Problem," *Foundations of Computational Mathematics*, vol. 2, no. 1, pp. 53–117, 2002.

[20] R. Barrio and F. Blesa, "Systematic search of symmetric periodic orbits in 2DOF Hamiltonian systems," *Chaos, Solitons & Fractals*, vol. 41, no. 2, pp. 560–582, 2009.

[21] R. Barrio, F. Blesa, and S. Serrano, "Periodic, escape and chaotic orbits in the Copenhagen and the (n+1)-body ring problems," *Communications in Nonlinear Science and Numerical Simulation*, vol. 14, no. 5, pp. 2229–2238, 2009.

[22] R. Barrio, "Painting chaos: a gallery of sensitivity plots of classical problems," *International Journal of Bifurcation and Chaos*, vol. 16, no. 10, pp. 2777–2798, 2006.

[23] R. Barrio and S. Serrano, "A three-parametric study of the Lorenz model," *Physica D: Nonlinear Phenomena*, vol. 229, no. 1, pp. 43–51, 2007.

[24] Á. Jorba and M. Zou, "A software package for the numerical integration of ODEs by means of high-order Taylor methods," *Experimental Mathematics*, vol. 14, no. 1, pp. 99–117, 2005.

[25] E. Lorenz, "Deterministic nonperiodic flow," *Journal of Atomic and Molecular Sciences*, vol. 20, pp. 130–141, 1963.

[26] F. M. Allan, "Derivation of the Adomian decomposition method using the homotopy analysis method," *Applied Mathematics and Computation*, vol. 190, no. 1, pp. 6–14, 2007.

[27] Q. M. Al-Mdallal, M. I. Syam, and P. . Ariel, "Extended homotopy perturbation method and the axisymmetric flow past a porous stretching sheet," *International Journal for Numerical Methods in Fluids*, vol. 69, no. 5, pp. 909–925, 2012.

[28] M. A. Hajji and K. Al-Khaled, "Analytic studies and numerical simulations of the generalized Boussinesq equation," *Applied Mathematics and Computation*, vol. 191, no. 2, pp. 320–333, 2007.

[29] S. Liao, "On the homotopy analysis method for nonlinear problems," *Applied Mathematics and Computation*, vol. 147, no. 2, pp. 499–513, 2004.

16

Collocation Method based on Genocchi Operational Matrix for Solving Generalized Fractional Pantograph Equations

Abdulnasir Isah,[1,2] Chang Phang,[2] and Piau Phang[3]

[1]*Department of Mathematics, Ahmadu Bello University, Zaria, Nigeria*
[2]*Department of Mathematics and Statistics, Faculty of Science, Technology and Human Development, Universiti Tun Hussein Onn Malaysia, Batu Pahat, Malaysia*
[3]*Faculty of Computer Science and Information Technology, Universiti Malaysia Sarawak, Sarawak, Malaysia*

Correspondence should be addressed to Piau Phang; pphang@unimas.my

Academic Editor: Patricia J. Y. Wong

An effective collocation method based on Genocchi operational matrix for solving generalized fractional pantograph equations with initial and boundary conditions is presented. Using the properties of Genocchi polynomials, we derive a new Genocchi delay operational matrix which we used together with the Genocchi operational matrix of fractional derivative to approach the problems. The error upper bound for the Genocchi operational matrix of fractional derivative is also shown. Collocation method based on these operational matrices is applied to reduce the generalized fractional pantograph equations to a system of algebraic equations. The comparison of the numerical results with some existing methods shows that the present method is an excellent mathematical tool for finding the numerical solutions of generalized fractional pantograph equations.

1. Introduction

Fractional calculus, the calculus of derivative and integral of any order, is used as a powerful tool in science and engineering to study the behaviors of real world phenomena especially the ones that cannot be fully described by the classical methods and techniques [1]. Differential equations with proportional delays are usually referred to as pantograph equations or generalized pantograph equations. The name pantograph was originated from the study work of Ockendon and Tayler [2]. Many researchers have studied different applications of these equations in applied sciences such as biology, physics, economics, and electrodynamics [3–5]. Solutions of pantograph equations were also studied by many authors numerically and analytically. Bhrawy et al. proposed a new generalized Laguerre-Gauss collocation method for numerical solution of generalized fractional pantograph equations [1]. Tohidi et al. in [6] proposed a new collocation scheme based on Bernoulli operational matrix for numerical solution of generalized pantograph equation. Yusufoglu [7] proposed an efficient algorithm for solving generalized pantograph equations with linear functional argument. In [8], Yang and Huang presented a spectral-collocation method for fractional pantograph delay integrodifferential equations and in [9] Yüzbasi and Sezer presented an exponential approximation for solutions of generalized pantograph delay differential equations. Chebyshev and Bessel polynomials are, respectively, used in [10, 11] to obtain the solutions of generalized pantograph equations. Operational matrices of fractional derivatives and integration have become very important tool in the field of numerical solution of fractional differential equations. In this paper, a member of Appell polynomials called Genocchi polynomials is used; although this polynomial is not based on orthogonal functions, it possesses operational matrices of derivatives with high accuracy. It is very important to note that this polynomial shares some great advantages with Bernoulli and Euler polynomials for approximating an arbitrary function over some classical orthogonal polynomials; we refer the reader to [6] for these advantages. On top of that, we had successfully applied the operational matrix via Genocchi polynomials for solving integer-order delay differential equations [12] and fractional optimal control problems [13], and the numerical solutions obtained are comparable or even more accurate compared to some existing well-known methods. Motivated by these advantages, in this paper, we intend to extend the result for integer-order delay differential equations in [12] to fractional delay differential equations or so-called generalized fractional pantograph equations. To the best of our knowledge, this is the first time that the operational matrix based on Genocchi polynomials is applied to solve the fractional pantograph equations. On the other hand, some other types of polynomials were employed to solve some special type of fractional calculus problems; for example, Bessel polynomials were used for the solution of fractional-order logistic population model [14]; Bernstein polynomials were also used for the solution of Riccati type differential equations [15].

In this paper, we use the new operational matrix of fractional-order derivative via Genocchi polynomials to provide approximate solutions of the generalized fractional pantograph equations of the following form [1]:

$$D^{\alpha} y(t) = \sum_{j=0}^{J} \sum_{n=0}^{m-1} p_{j,n}(t) D^{\beta_n} y(\lambda_{j,n} t + \mu_{j,n}) + g(t), \quad 0 \le t \le 1 \tag{1}$$

subject to the following conditions:

$$\sum_{n=0}^{m-1} a_{n,i} y^{(n)}(0) = d_i, \quad i = 0, 1, \ldots, m-1, \tag{2}$$

where $a_{n,i}$, $\lambda_{j,n}$, and $\mu_{j,n}$ are real or complex coefficients; $m-1 < \alpha < m$, $0 < \beta_0 < \beta_1 < \cdots < \beta_{m-1} < \alpha$, while $p_{j,n}(t)$ and $g(t)$ are given continuous functions in the interval $[0, 1]$.

The rest of the paper is organized as follows: Section 2 introduces some mathematical preliminaries of fractional calculus. In Section 3, we discuss some important properties of Genocchi polynomials. In Section 4, we derive the Genocchi delay operational matrix and we apply the collocation method for solving fractional pantograph equation (1) using the Genocchi operational matrix of fractional derivative and the delay operational matrix in Section 5. In Section 6, the proposed method is applied to several examples and conclusion is given in Section 7.

2. Preliminaries

2.1. Fractional Derivative and Integration. We recall some basic definitions and properties of fractional calculus that we will use. There are various competing definitions for fractional derivatives [16, 17]. The Riemann-Liouville definition played a vital role in the development of the theory of fractional calculus. However, there are certain disadvantages of using this definition when modeling real world phenomena. To cope with these disadvantages, Caputo definition was introduced which is found to be more reliable in application. So we use this definition of fractional derivatives. We begin with the definition of Riemann-Liouville integral, in which the fractional integral operator I of a function $f(t)$ is defined as follows.

Definition 1. The Riemann-Liouville integral I of fractional-order α of $f(t)$ is given by

$$I^{\alpha} f(t) = \frac{1}{\Gamma(\alpha)} \int_0^t (t-\tau)^{\alpha-1} f(\tau)\, d\tau, \quad t > 0, \ \alpha \in \mathbb{R}^+, \tag{3}$$

where $\Gamma(\cdot)$ is the Gamma function. The fractional derivative of order $\alpha > 0$ due to Riemann-Liouville is defined by

$$(D_l^{\alpha} f)(t) = \left(\frac{d}{dt}\right)^m (I^{m-\alpha} f)(t), \quad (\alpha > 0, \ m-1 < \alpha < m). \tag{4}$$

The following are important properties of Riemann-Liouville fractional integral I^{α}:

$$I^{\alpha} I^{\beta} f(t) = I^{\alpha+\beta} f(t), \quad \alpha > 0, \ \beta > 0,$$
$$I^{\alpha} t^{\beta} = \frac{\Gamma(\beta+1)}{\Gamma(\beta+\alpha+1)} t^{\beta+\alpha}. \tag{5}$$

Definition 2. The Caputo fractional derivative D^{α} of a function $f(t)$ is defined as

$$D^{\alpha} f(t) = \frac{1}{\Gamma(n-\alpha)} \int_0^t \frac{f^{(n)}(\tau)}{(t-\tau)^{\alpha-n+1}} d\tau, \quad n-1 < \alpha \le n, \ n \in \mathbb{N}. \tag{6}$$

Some properties of Caputo fractional derivatives are as follows:

$$D^{\alpha} C = 0, \quad (C \text{ is constant}),$$
$$D^{\alpha} t^{\beta} = \begin{cases} 0, & \beta \in \mathbb{N} \cup \{0\},\ \beta < \lceil \alpha \rceil \\ \dfrac{\Gamma(\beta+1)}{\Gamma(\beta+1-\alpha)} t^{\beta-\alpha}, & \beta \in \mathbb{N} \cup \{0\},\ \beta \ge \lceil \alpha \rceil \text{ or } \beta \notin \mathbb{N},\ \beta > \lfloor \alpha \rfloor, \end{cases} \tag{7}$$

where $\lceil \alpha \rceil$ denotes the smallest integer greater than or equal to α and $\lfloor \alpha \rfloor$ denotes the largest integer less than or equal to α.

Similar to the integer-order differentiation, the Caputo fractional differential operator is a linear operator; that is,

$$D^{\alpha}\left(\lambda f(t)+\mu g(t)\right)=\lambda D^{\alpha}f(t)+\mu D^{\alpha}g(t) \tag{8}$$

for λ and μ constants.

3. Genocchi Polynomials and Some Properties

Genocchi polynomials and numbers have been extensively studied in many different contexts in branches of mathematics such as elementary number theory, complex analytic number theory, homotopy theory (stable homotopy groups of spheres), differential topology (differential structures on spheres), theory of modular forms (Eisenstein series), and quantum physics (quantum groups). The classical Genocchi polynomial $G_n(x)$ is usually defined by means of the exponential generating functions [18–20].

$$\frac{2te^{xt}}{e^t+1}=\sum_{n=0}^{\infty}G_n(x)\frac{t^n}{n!},\quad (|t|<\pi), \tag{9}$$

where $G_n(x)$ is the Genocchi polynomial of degree n and is given by

$$G_n(x)=\sum_{k=0}^{n}\binom{n}{k}G_{n-k}x^k. \tag{10}$$

G_{n-k} here is the Genocchi number.

Some of the important properties of these polynomials include

$$\int_0^1 G_n(x)G_m(x)\,dx=\frac{2(-1)^n n!m!}{(m+n)!}G_{m+n}\quad n,m\ge 1, \tag{11}$$

$$\frac{dG_n(x)}{dx}=nG_{n-1}(x),\quad n\ge 1, \tag{12}$$

$$G_n(1)+G_n(0)=0,\quad n>1. \tag{13}$$

Before we move to the next level, we need the following linear independence on which the rest of theoretical results are based.

Lemma 3. *The set $A=\{G_1(t),G_2(t),\ldots,G_N(t)\}\subset L^2[0,1]$ is a linearly independent set in $L^2[0,1]$.*

Proof. To show that A is the set of linearly independent elements of $L^2[0,1]$, it is enough to show that the Gram determinant is not zero. That is,

$$\text{Gram}(G_1,G_2,\ldots,G_N)\neq 0, \tag{14}$$

where

$$\text{Gram}(G_1,G_2,\ldots,G_N)=\begin{vmatrix}\langle G_1,G_1\rangle & \langle G_1,G_2\rangle & \cdots & \langle G_1,G_N\rangle\\ \langle G_2,G_1\rangle & \langle G_2,G_2\rangle & \cdots & \langle G_2,G_N\rangle\\ \vdots & \vdots & \cdots & \vdots\\ \langle G_n,G_1\rangle & \langle G_n,G_2\rangle & \cdots & \langle G_n,G_N\rangle\end{vmatrix}. \tag{15}$$

Now, to prove that this determinant is not equal to zero, we first reduce the Gram matrix to an upper triangular matrix by Gaussian elimination and it is not difficult to see that the elements of the diagonal of the reduced matrix are given by

$$a(n)=\frac{(n!(n+1)!)^2}{(2n!)(2n+1)!},\quad n\in\mathbb{N}. \tag{16}$$

Clearly, one can see that, for any $\in\mathbb{N}$, $a(n)\neq 0$. Consequently, the determinant given by

$$\prod_{n=1}^{N}a(n) \tag{17}$$

is not equal to zero. Therefore, the set A is the set of linearly independent sets. □

3.1. Function Approximation. Assume that $\{G_1(t),G_2(t),\ldots,G_N(t)\}\subset L^2[0,1]$ is the set of Genocchi polynomials and $Y=\text{Span}\{G_1(t),G_2(t),\ldots,G_N(t)\}$. Let $f(t)$ be arbitrary element of $L^2[0,1]$; since Y is a finite dimensional subspace of $L^2[0,1]$ space, $f(t)$ has a unique best approximation in Y, say $f^*(t)$, such that

$$\|f(t)-f^*(t)\|_2\le\|f(t)-y(t)\|_2,\quad\forall y(t)\in Y. \tag{18}$$

This implies that, $\forall y(t)\in Y$,

$$\langle f(t)-f^*(t),y(t)\rangle=0, \tag{19}$$

where $\langle\cdot\rangle$ denotes inner product. Since $f^*(t)\in Y$, there exist the unique coefficients $c_1,c_2,\ldots,c_N$ such that

$$f(t)\approx f^*(t)=\sum_{n=1}^{N}c_nG_n(t)=\mathbf{C}^T\mathbf{G}(t), \tag{20}$$

where $\mathbf{C}=[c_1,c_2,\ldots,c_N]^T$, $\mathbf{G}(t)=[G_1(t),G_2(t),\ldots,G_N(t)]^T$.

Using (19), we have

$$\left\langle f(t)-\mathbf{C}^T\mathbf{G}(t),G_i(t)\right\rangle=0\quad i=1,2,\ldots,N; \tag{21}$$

for simplicity, we write

$$\mathbf{C}^T\langle\mathbf{G}(t),\mathbf{G}(t)\rangle=\langle f(t),\mathbf{G}(t)\rangle, \tag{22}$$

where $\langle\mathbf{G}(t),\mathbf{G}(t)\rangle$ is an $N\times N$ matrix.

Let $W=\langle\mathbf{G}(t),\mathbf{G}(t)\rangle=\int_0^1\mathbf{G}(t)\mathbf{G}^T(t)dt$.

The entries of the matrix W can be calculated from (11). Therefore, any function $f(t)\in L^2[0,1]$ can be expanded by Genocchi polynomials as $f(t)=\mathbf{C}^T\mathbf{G}(t)$, where

$$\mathbf{C}=W^{-1}\langle f(t),\mathbf{G}(t)\rangle. \tag{23}$$

4. Genocchi Operational Matrix

In this section, we derive the operational matrices for the delay and that of fractional derivative based on Genocchi polynomials for the solution of fractional pantograph equations.

4.1. Genocchi Delay Operational Matrix. The Genocchi delay vector $\mathbf{G}(t-\mu)$ can be expressed as

$$\mathbf{G}(t-\mu) = \mathbf{R}\mathbf{G}(t), \tag{24}$$

where $\mathbf{R}$ is the $N \times N$ operational delay matrix given by

$$\mathbf{R} = W_1 W^{-1} = \begin{bmatrix} 1 & 0 & 0 & 0 & 0 & \cdots & 0 \\ -2\mu & 1 & 0 & 0 & 0 & \cdots & 0 \\ 3\mu^2 & -3\mu & 1 & 0 & 0 & \cdots & 0 \\ -4\mu^3 & 6\mu^2 & -4\mu & 1 & 0 & \cdots & 0 \\ 5\mu^4 & -10\mu^3 & 10\mu^2 & -5\mu & 1 & \cdots & 0 \\ \vdots & \vdots & \vdots & \cdots & \vdots & \vdots & \vdots \\ b_n(1) & b_n(2) & b_n(3) & b_n(4) & b_n(5) & \cdots & 1 \end{bmatrix}, \tag{25}$$

where $W_1 = \int_0^1 \mathbf{G}(t-\mu)\mathbf{G}^T(t)dt$ and $b_n(i) = (-1)^{n-i}(\binom{n}{i})\mu^{n-i}$, $i = 1, 2, \ldots, n$.

Also, for any delay function $f(t-\mu)$, we can express it in terms of Genocchi polynomials as shown in (26):

$$f(t-\mu) = \sum_{i=1}^{N} c_i G_i(t-\mu) = \mathbf{C}^T \mathbf{R}\mathbf{G}(t), \tag{26}$$

where $\mathbf{C}$ is given in (23).

The following lemma is also of great importance.

Lemma 4. *Let $G_i(t)$ be the Genocchi polynomials; then $D^\alpha G_i(t) = 0$, for $i = 1, \ldots, \lceil\alpha\rceil - 1$, $\alpha > 0$.*

The proof of this lemma is obvious; one can use (7) and (8) on (10).

4.2. Genocchi Operational Matrix of Fractional Derivative. If we consider the Genocchi vector $\mathbf{G}(t)$ given by $\mathbf{G}(t) = [G_1(t), G_2(t), \ldots, G_N(t)]$, then the derivative of $\mathbf{G}(t)$ with the aid of (12) can be expressed in the matrix form by

$$\frac{d\mathbf{G}(t)^T}{dt} = M\mathbf{G}(t)^T, \tag{27}$$

where

$$M = \begin{bmatrix} 0 & 0 & 0 & \cdots & 0 & 0 & 0 \\ 2 & 0 & 0 & \cdots & 0 & 0 & 0 \\ 0 & 3 & 0 & \cdots & 0 & 0 & 0 \\ 0 & 0 & 4 & \cdots & 0 & 0 & 0 \\ \vdots & \vdots & \vdots & \cdots & \vdots & \vdots & \vdots \\ 0 & 0 & 0 & \cdots & N-1 & 0 & 0 \\ 0 & 0 & 0 & \cdots & 0 & N & 0 \end{bmatrix}. \tag{28}$$

Thus, M is $N \times N$ operational matrix of derivative.

It is not difficult to show inductively that the kth derivative of $\mathbf{G}(t)$ can be given by

$$\frac{d^k\mathbf{G}(t)^T}{dt^k} = \mathbf{G}(t)\left(M^T\right)^k. \tag{29}$$

In the following theorem, the operational matrix of fractional-order derivative for the Genocchi polynomials is given.

Theorem 5 (see [21]). *Suppose that $\mathbf{G}(t)$ is the Genocchi vector given in (20) and let $\alpha > 0$. Then,*

$$D^\alpha \mathbf{G}(t)^T = P^\alpha \mathbf{G}(t)^T, \tag{30}$$

where P^α is $N \times N$ operational matrix of fractional derivative of order α in Caputo sense and is defined as follows:

$$P^{(\alpha)} = \begin{bmatrix} 0 & 0 & \cdots & 0 \\ \vdots & \vdots & \cdots & \vdots \\ 0 & 0 & \cdots & 0 \\ \sum_{k=\lceil\alpha\rceil}^{\lceil\alpha\rceil} \rho_{\lceil\alpha\rceil,k,1} & \sum_{k=\lceil\alpha\rceil}^{\lceil\alpha\rceil} \rho_{\lceil\alpha\rceil,k,2} & \cdots & \sum_{k=\lceil\alpha\rceil}^{\lceil\alpha\rceil} \rho_{\lceil\alpha\rceil,k,N} \\ \vdots & \vdots & \cdots & \vdots \\ \sum_{k=\lceil\alpha\rceil}^{i} \rho_{i,k,1} & \sum_{k=\lceil\alpha\rceil}^{i} \rho_{i,k,2} & \cdots & \sum_{k=\lceil\alpha\rceil}^{i} \rho_{i,k,N} \\ \vdots & \vdots & \cdots & \vdots \\ \sum_{k=\lceil\alpha\rceil}^{N} \rho_{N,k,1} & \sum_{k=\lceil\alpha\rceil}^{N} \rho_{N,k,2} & \cdots & \sum_{k=\lceil\alpha\rceil}^{N} \rho_{N,k,N} \end{bmatrix}, \tag{31}$$

where $\rho_{i,k,j}$ is given by

$$\rho_{i,k,j} = \frac{i!G_{i-k}}{(i-k)!\Gamma(k+1-\alpha)} c_j. \tag{32}$$

G_{i-k} is the Genocchi number and c_j can be obtained from (23).

Proof. For the proof, see [21].

4.3. Upper Bound of the Error for the Operational Matrix of Fractional Derivative P^{α}. We begin here by proving the upper bound of the error of arbitrary function approximation by Genocchi polynomials in the following Lemma.

Lemma 6. *Suppose that $f(t) \in C^{n+1}[0,1]$ and $Y = \text{Span}\{G_1(t), G_2(t), \ldots, G_N(t)\}$; if $\mathbf{C}^T\mathbf{G}(t)$ is the best approximation of $f(t)$ out of Y, then*

$$\left\| f(t) - \mathbf{C}^T\mathbf{G}(t) \right\| \le \frac{R}{(n+1)!\sqrt{2n+3}}, \tag{33}$$

where $R = \max_{t\in[0,1]} |f^{(n+1)}(t)|$.

To see this, we set $\{1, t, \ldots, t^n\}$ as a basis for the polynomial space of degree n.

Define $y_1(t) = f(0) + tf'(0) + (t^2/2!)f''(0) + \cdots + (t^n/n!)f^{(n)}(0)$.

From Taylor's expansion, one has $|f(t) - y_1(t)| = |(t^{n+1}/(n+1)!)f^{(n+1)}(\xi_t)|$, where $\xi_t \in (0,1)$.

Since $\mathbf{C}^T\mathbf{G}(t)$ is the best approximation of $f(t)$ out of Y and $y_1(t) \in Y$, from (18), one has

$$\begin{aligned} \left\| f(t) - \mathbf{C}^T\mathbf{G}(t) \right\|_2^2 &\le \left\| f(t) - y_1(t) \right\|_2^2 \\ &\le \int_0^1 \left| f(t) - y_1(t) \right|^2 dt \\ &= \int_0^1 \left(\frac{t^{n+1}}{(n+1)!} \right)^2 \left\| f^{(n+1)}(\xi_t) \right\|^2 dt \\ &\le \frac{R^2}{((n+1)!)^2} \int_0^1 t^{2n+2} dt = \frac{R^2}{((n+1)!)^2 (2n+3)}. \end{aligned} \tag{34}$$

Taking the square root of both sides, one has

$$\left\| f(t) - \mathbf{C}^T\mathbf{G}(t) \right\| \le \frac{R}{(n+1)!\sqrt{2n+3}} \tag{35}$$

which is the desired error bound.

We use the following theorem from [22].

Theorem 7 (see [22]). *Suppose that H is a Hilbert space and Y is a closed subspace of H such that $\dim Y < \infty$ and $y_1, y_2, \ldots, y_n$ is a basis for Y. Let f be an arbitrary element in H and let y_0 be the unique best approximation of f out of Y. Then,*

$$\left\| f - y_0 \right\|^2 = \frac{\text{Gram}(f, y_1, y_2, \ldots, y_n)}{\text{Gram}(y_1, y_2, \ldots, y_n)}, \tag{36}$$

where

$$\text{Gram}(y_1, y_2, \ldots, y_n) = \begin{vmatrix} \langle y_1, y_1 \rangle & \langle y_1, y_2 \rangle & \cdots & \langle y_1, y_n \rangle \\ \langle y_2, y_1 \rangle & \langle y_2, y_2 \rangle & \cdots & \langle y_2, y_n \rangle \\ \vdots & \vdots & \cdots & \vdots \\ \langle y_n, y_1 \rangle & \langle y_n, y_2 \rangle & \cdots & \langle y_n, y_n \rangle \end{vmatrix}. \tag{37}$$

Theorem 8. *Suppose that $f(t) \in L^2[0,1]$ is approximated by $f_n(t)$ as*

$$f_n(t) = \sum_{i=1}^{n} c_i G_i(t) = \mathbf{C}^T\mathbf{G}(t); \tag{38}$$

then,

$$\lim_{n\to\infty} \left\| f(t) - f_n(t) \right\| = 0. \tag{39}$$

The proof of this theorem obviously follows from Lemma 6.

The operational matrix error vector E^{α} is given by

$$E^{\alpha} = P^{\alpha}\mathbf{G}(t) - D^{\alpha}\mathbf{G}(t), \tag{40}$$

where

$$E^{\alpha} = \begin{bmatrix} e_1^{\alpha} \\ e_2^{\alpha} \\ \vdots \\ e_n^{\alpha} \end{bmatrix}; \tag{41}$$

from Theorem 7, we get

$$\begin{aligned} &\left\| f_0(t) - \sum_{j=1}^{N} c_j G_j(t) \right\| \\ &\quad = \left(\frac{\text{Gram}(f(t), G_1(t), \ldots, G_N(t))}{\text{Gram}(G_1(t), \ldots, G_N(t))} \right)^{1/2}. \end{aligned} \tag{42}$$

Thus, according to equations (29) and (30) in [21], one has

$$\begin{aligned} \left\| e_i^{\alpha} \right\| &= \Bigg| D^{\alpha}G_i(t) \\ &\quad - \sum_{j=1}^{N} \left(\sum_{k=\lceil\alpha\rceil}^{i} \frac{i! G_{i-k}}{(i-k)!\Gamma(k+1-\alpha)} c_j \right) G_j(t) \Bigg| \\ &\le \sum_{k=\lceil\alpha\rceil}^{i} \frac{i! G_{i-k}}{(i-k)!\Gamma(k+1-\alpha)} t^{k-\alpha} \Bigg| f_0(t) \\ &\quad - \sum_{j=1}^{N} c_j G_j(t) \Bigg| \le \sum_{k=\lceil\alpha\rceil}^{i} \frac{i! G_{i-k}}{(i-k)!\Gamma(k+1-\alpha)} t^{k-\alpha} \\ &\quad \cdot \left(\frac{\text{Gram}(f(t), G_1(t), \ldots, G_N(t))}{\text{Gram}(G_1(t), \ldots, G_N(t))} \right)^{1/2}. \end{aligned} \tag{43}$$

By considering Theorem 8 and (43), we can conclude that by increasing the number of the Genocchi bases the vector e_i^{α} tends to zero.

For comparison purpose in Table 1, we show below the errors of operational matrix of fractional derivative based on Genocchi polynomials and shifted Legendre polynomials derived in [23, 24] when $N = 10$ and $\alpha = 0.75$ at different

TABLE 1: Comparison of the operational matrix errors for the GPOMFD and SLPOMFD.

E^{α}	$x = 1$		$x = 0$		$x = 0.5$	
	GPOMFD	SLPOMFD	GPOMFD	SLPOMFD	GPOMFD	SLPOMFD
e_1	0.00000	0.00000	0.00000	0.00000	0.00000	0.00000
e_2	0.01288	0.01343	0.51657	0.51648	0.00353	0.00345
e_3	0.02204	0.03462	0.77767	1.56518	0.00531	0.01047
e_4	0.00004	0.02916	0.01241	3.17608	0.00008	0.02487
e_5	0.03944	0.17666	1.28048	5.36557	0.00911	0.02570
e_6	0.00292	0.33223	0.02450	8.00571	0.00001	0.05647
e_7	0.15747	2.61748	5.38767	15.06874	0.03839	0.04226
e_8	0.00748	11.8527	0.15766	11.80719	0.00060	0.92194
e_9	1.21090	38.04476	39.19581	54.74287	0.27897	15.64398
e_{10}	0.04902	806.14232	1.32181	629.08480	0.00519	20.30772

points on $[0, 1]$. From this table, it is clear that the accuracy of Genocchi polynomials operational matrix of fractional derivative (GPOMFD) is better than the shifted Legendre polynomials operational matrix of fractional derivatives (SLPOMFD). We believe that this is the case for any value of N because the Genocchi polynomials have smaller coefficients of individual terms compared to shifted Legendre polynomials.

5. Collocation Method Based on Genocchi Operational Matrices

In this section, we use the collocation method based on Genocchi operational matrix of fractional derivatives and Genocchi delay operational matrix to solve numerically the generalized fractional pantograph equation. We now derive an algorithm for solving (1). To do this, let the solution of (1) be approximated by the first N terms Genocchi polynomials. Thus, we write

$$y_N(t) \approx \sum_{n=1}^{N} c_n G_n(t) = G(t) C, \tag{44}$$

where the Genocchi coefficient vector C and the Genocchi vector $G(t)$ are given by

$$\begin{aligned} C^T &= [c_1, c_2, \ldots, c_N], \\ G(t) &= [G_1(t), G_2(t), \ldots, G_N(t)]; \end{aligned} \tag{45}$$

thus, $D^{\alpha} y_N(t)$ and $D^{\beta_n} y_N(t)$, $n = 0, 1, \ldots, m-1$, can be expressed, respectively, as follows:

$$\begin{aligned} D^{\alpha} y_N(t) &= G(t) \left(P^T\right)^{\alpha} C, \\ D^{\beta_n} y_N(t) &= G(t) \left(P^T\right)^{\beta_n} C, \quad n = 0, 1, \ldots, m-1. \end{aligned} \tag{46}$$

Substituting (44) and (46) in (1), we have

$$\begin{aligned} & G(t) \left(P^T\right)^{\alpha} C \\ & = \sum_{j=0}^{J} \sum_{n=0}^{m-1} p_{j,n}(t) G\left(\lambda_{j,n} t + \mu_{j,n}\right) \left(P^T\right)^{\beta_n} C + g(t), \end{aligned} \tag{47}$$

where $G(\lambda_{j,n} t + \mu_{j,n}) = [G_1(\lambda_{j,n} t + \mu_{j,n}), G_2(\lambda_{j,n} t + \mu_{j,n}), \ldots, G_N(\lambda_{j,n} t + \mu_{j,n})]$.

Also the initial condition will produce m other equations:

$$\sum_{n=0}^{m-1} a_{n,i} G(0) \left(P^T\right)^{i} C = d_i, \quad i = 0, 1, \ldots, m-1. \tag{48}$$

To find the solution $y_N(t)$ we collocate (47) at the collocation points $t_j = j/(N-m)$, $j = 1, 2, \ldots, N-m$, to obtain

$$\begin{aligned} & G\left(t_j\right) \left(P^T\right)^{\alpha} C \\ & = \sum_{j=0}^{J} \sum_{n=0}^{m-1} p_{j,n}\left(t_j\right) G\left(\lambda_{j,n} t_j + \mu_{j,n}\right) \left(P^T\right)^{\beta_n} C \\ & + g\left(t_j\right) \end{aligned} \tag{49}$$

for $j = 1, 2, \ldots, N-m$. Additionally, one can also use both the operational matrix of fractional derivative and delay operational matrix to solve problem (1). According to (44), we can approximate the delay function $y(\lambda_{j,n} t_j + \mu_{j,n})$ and its fractional derivative using the operational matrices $\mathbf{P}$ and $\mathbf{R}$ as follows:

$$\begin{aligned} y_N\left(\lambda_{j,n} t_j + \mu_{j,n}\right) &= \mathbf{R} C^T G(t), \\ D^{\beta_n} y_N\left(\lambda_{j,n} t_j + \mu_{j,n}\right) &= \mathbf{R} C^T P^{\beta_n} G(t). \end{aligned} \tag{50}$$

Putting this approximation together with (44) in (1), we have

$$G(t) \left(P^T\right)^{\alpha} C = \sum_{j=0}^{J} \sum_{n=0}^{m-1} p_{j,n}(t) \mathbf{R} C^T P^{\beta_n} G(t) + g(t). \tag{51}$$

TABLE 2: Comparison errors obtained by the present method and those obtained in [25] when $\alpha = 0.6$ and $\tau = 0.3$ for Example 1.

t	Error (new method) [25]	Error (FAM) [25]	Error (present method)
0.2	0.0781197	0.078155	$3.37154E-03$
0.4	0.129928	0.129978	$6.52102E-03$
0.6	0.190687	0.19076	$9.77309E-03$
0.8	0.248601	0.248694	$1.30349E-02$
1.0	0.307649	0.307763	$1.64103E-02$

Thus, collocating (51) at the same collocation point as that in (47), we get

$$G(t_j)(P^T)^\alpha C = \sum_{j=0}^{J}\sum_{n=0}^{m-1} p_{j,n}(t_j)\mathbf{R}C^T P^{\beta_n} G(t_j) + g(t_j). \tag{52}$$

Hence, (49) or (52) is $N - m$ nonlinear algebraic equation. Any of these equations together with (48) makes N algebraic equations which can be solved using Newton's iterative method. Consequently, $y_N(x)$ given in (44) can be calculated.

6. Numerical Examples

In this section, some numerical examples are given to illustrate the applicability and accuracy of the proposed method. All the numerical computations have been done using Maple 18.

Example 1. Consider the following example solved in [25]:

$$D^\alpha y(t) = \frac{2}{\Gamma(3-\alpha)} y^{1-\alpha/2}(t) + y(t-\tau) - y(t) + 2\tau\sqrt{y(t)} - \tau^2 \tag{53}$$

subject to

$$y(t) = 0, \quad t \le 0. \tag{54}$$

The exact solution for this example is given by $y(t) = t^2$. We solve the example when $\alpha = 0.6$ and $\tau = 0.3$. In Table 2, we compare the errors obtained by our method with those obtained using FAM and new approach in [25]. As reported in [25], the time required for the new method is 104.343750 seconds and for the FAM the time taken is 215.031250 seconds for completing the same task, whereas in our method we only need 38.080 seconds to complete the computations.

Example 2 (see [1]). Consider the following generalized fractional pantograph equation:

$$D^{5/2} y(t) = -y(t) - y(t-0.5) + g(t), \quad t \in [0,1] \tag{55}$$

subject to

$$y(0) = 0, \quad y'(0) = 0, \quad y''(0) = 0, \tag{56}$$

where

$$g(t) = \frac{\Gamma(4)}{\Gamma(3/2)} t^{1/2} + t^3 + (t-0.5)^3. \tag{57}$$

The exact solution of this problem is known to be $y(t) = t^3$. This problem is solved in [1] using generalized Laguerre-Gauss collocation scheme. We apply our technique with $N = 4$. Approximating (55) with Genocchi polynomials, we have

$$G(t)(P^T)^{5/2} C = -G(t)C + G(t-0.5)C + g(t). \tag{58}$$

Also from the initial conditions we have

$$G(0)C = 0, \quad G(0)(P^T)C = 0, \quad G(0)(P^T)^2 C = 0. \tag{59}$$

Thus, collocating (58) at $t = 0.267339$, we get

$$-3.507078326 + 15.27479180c_4 + 2c_1 + 0.2727646328c_3 - 1.930640400c_2 = 0, \tag{60}$$

and (59) gives

$$c_1 - c_2 + c_4 = 0, \quad 2c_1 - 3c_2 = 0, \quad 6c_3 - 12c_4 = 0. \tag{61}$$

Solving these equations, we have

$$c_1 = 0.4999969148, \quad c_2 = 0.7499953722, \quad c_3 = 0.4999969148, \quad c_4 = 0.2499984574. \tag{62}$$

Thus, $y(t) = G(x)C$ is calculated and we have $0.9999938296t^3$ which is almost the exact solution. In Table 3, we compare the absolute errors obtained by our method (with only few terms $N = 4$) and the absolute errors obtained in [1] when $N = 22$ with different Laguerre parameters β.

TABLE 3: Comparison of the absolute errors obtained by the present method and those obtained in [1] for Example 2.

t		$N = 22$ [1]		$N = 4$
	$\beta = 2$	$\beta = 3$	$\beta = 5$	Present method
0.1	$1.030E-04$	$1.019E-05$	$6.273E-06$	$6.17040E-09$
0.2	$6.510E-04$	$6.051E-05$	$3.892E-05$	$4.93630E-08$
0.3	$1.740E-03$	$1.495E-04$	$1.023E-04$	$1.66600E-07$
0.4	$3.283E-03$	$2.559E-04$	$1.901E-04$	$3.94910E-07$
0.5	$5.138E-03$	$3.546E-04$	$2.944E-04$	$7.71300E-07$
0.6	$7.175E-03$	$4.261E-04$	$4.088E-04$	$1.33280E-06$
0.7	$9.303E-03$	$4.592E-04$	$5.306E-04$	$2.11640E-06$
0.8	$1.147E-02$	$4.510E-04$	$6.597E-04$	$3.15920E-06$
0.9	$1.367E-02$	$4.055E-04$	$7.977E-04$	$4.49820E-06$
1.0	$1.589E-02$	$3.311E-04$	$9.468E-04$	$6.17040E-06$

TABLE 4: Comparison of the absolute errors obtained by the present method and those in [26] for Example 4.

t	Absolute error [26] $N = 4$	Absolute error (present method) $N = 3$
0.1	$0.100E-07$	$1.15000E-09$
0.2	$0.115E-07$	$2.00000E-09$
0.3	$0.115E-07$	$2.55000E-09$
0.4	$0.107E-07$	$2.80000E-09$
0.5	$0.967E-08$	$2.70000E-09$
0.6	$0.811E-08$	$2.40000E-09$
0.7	$0.641E-08$	$1.70000E-09$
0.8	$0.440E-08$	$8.00000E-10$
0.9	$0.223E-08$	$5.00000E-10$
1.0	$0.372E-09$	$2.00000E-09$

Example 3. Consider the following fractional pantograph equation:

$$D^{1/2}y(t) = 2y\left(\frac{3t}{2}\right) + \frac{8t^{3/2}}{3\sqrt{\pi}} - \frac{9t^2}{2}, \quad t \in [0,1] \tag{63}$$

subject to

$$y(0) = 0, \quad y(1) = 1. \tag{64}$$

The exact solution of this problem is known to be $y(t) = t^2$. We solve (63) using our technique with $N = 3$ only. As in Example 2, we obtained the values of the coefficients to be

$$c_1 = \frac{1}{2}, \quad c_2 = \frac{1}{2}, \quad c_3 = \frac{1}{3}. \tag{65}$$

Thus, $y_N(t) = G(t)C$ is calculated to be t^2 which is the exact solution and so there is nothing to compare for the error is zero.

Example 4. Consider the following fractional pantograph equation solved in [26]:

$$D^2y(t) + D^{3/2}y(t) + y(t) = y\left(\frac{t}{2}\right) + \frac{3t^2}{4} + 4\sqrt{\frac{t}{\pi}} + 2, \quad t \in [0,1] \tag{66}$$

subject to

$$y(0) = 0, \quad y(1) = 1. \tag{67}$$

The exact solution of this problem is known to be $y(t) = t^2$. As in Example 3, we solve (66) using our technique with $N = 3$ and the values of the coefficients obtained are

$$c_1 = 0.5000000011, \quad c_2 = 0.5000000011, \quad c_3 = 0.3333333384. \tag{68}$$

Thus, $y_N(t) = G(t)C$ is calculated and compared with the exact solution. This problem is solved using Taylor collocation method in [26] when $N = 4, 5$, and 6. In Table 4, we compare the absolute errors obtained by present method when $N = 3$ with the errors obtained when $N = 4$ in [26].

7. Conclusion

In this paper, a collocation method based on the Genocchi delay operational matrix and the operational matrix of fractional derivative for solving generalized fractional pantograph equations is presented. The comparison of the results shows that the present method is an excellent mathematical tool for finding the numerical solutions delay equation. The advantage of the method over others is that only few terms are needed and every operational matrix involves more numbers of zeroes; as such the method has less computational complexity and provides the solution at high accuracy.

Conflicts of Interest

The authors declare that there are no conflicts of interest regarding the publication of this manuscript.

Acknowledgments

This work was supported in part by MOE-UTHM FRGS Grant Vot 1433. The authors also acknowledge financial support from UTHM through GIPS U060. The third author would like to acknowledge Faculty of Computer Science and Information Technology, UNIMAS, for providing continuous support.

References

[1] A. H. Bhrawy, A. A. Al-Zahrani, Y. A. Alhamed, and D. Baleanu, "A new generalized laguerre-gauss collocation scheme for numerical solution of generalized fractional pantograph equations," *Romanian Journal of Physics*, vol. 59, no. 7-8, pp. 646–657, 2014.

[2] J. R. Ockendon and A. B. Tayler, "The dynamics of a current collection system for an electric locomotive," *Proceedings of the Royal Society of London A: Mathematical, Physical and Engineering Sciences*, vol. 322, pp. 447–468, 1971.

[3] W. G. Aiello, H. I. Freedman, and J. Wu, "Analysis of a model representing stage-structured population growth with state-dependent time delay," *SIAM Journal on Applied Mathematics*, vol. 52, no. 3, pp. 855–869, 1992.

[4] M. Dehghan and F. Shakeri, "The use of the decomposition procedure of Adomian for solving a delay differential equation arising in electrodynamics," *Physica Scripta*, vol. 78, no. 6, article 065004, 2008.

[5] Y. Kuang, *Delay Differential Equations with Applications in Population Dynamics*, Academic Press, New York, NY, USA, 1993.

[6] E. Tohidi, A. H. Bhrawy, and K. Erfani, "A collocation method based on Bernoulli operational matrix for numerical solution of generalized pantograph equation," *Applied Mathematical Modelling. Simulation and Computation for Engineering and Environmental Systems*, vol. 37, no. 6, pp. 4283–4294, 2013.

[7] E. Yusufoglu, "An efficient algorithm for solving generalized pantograph equations with linear functional argument," *Applied Mathematics and Computation*, vol. 217, no. 7, pp. 3591–3595, 2010.

[8] Y. Yang and Y. Huang, "Spectral-collocation methods for fractional pantograph delay—integrodifferential equations," *Advances in Mathematical Physics*, vol. 2013, Article ID 821327, 14 pages, 2013.

[9] S. Yüzbasi and M. Sezer, "An exponential approximation for solutions of generalized pantograph-delay differential equations," *Applied Mathematical Modelling. Simulation and Computation for Engineering and Environmental Systems*, vol. 37, no. 22, pp. 9160–9173, 2013.

[10] S. Sedaghat, Y. Ordokhani, and M. Dehghan, "Numerical solution of the delay differential equations of pantograph type via Chebyshev polynomials," *Communications in Nonlinear Science and Numerical Simulation*, vol. 17, no. 12, pp. 4815–4830, 2012.

[11] S. Yüzbasi, N. Sahin, and M. Sezer, "A Bessel collocation method for numerical solution of generalized pantograph equations," *Numerical Methods for Partial Differential Equations*, vol. 28, no. 4, pp. 1105–1123, 2012.

[12] A. Isah and C. Phang, "Operational matrix based on Genocchi polynomials for solution of delay differential equations," *Ain Shams Engineering Journal*, 2017.

[13] C. Phang, N. F. Ismail, A. Isah, and J. R. Loh, "A new efficient numerical scheme for solving fractional optimal control problems via a Genocchi operational matrix of integration," *Journal of Vibration and Control*, 2017.

[14] S. Yüzbasi, "A collocation method for numerical solutions of fractional-order logistic population model," *International Journal of Biomathematics*, vol. 9, no. 2, article 1650031, 14 pages, 2016.

[15] S. Yüzbasi, "Numerical solutions of fractional Riccati type differential equations by means of the Bernstein polynomials," *Applied Mathematics and Computation*, vol. 219, no. 11, pp. 6328–6343, 2013.

[16] A. A. Kilbas, H. M. Srivastava, and J. J. Trujillo, "Preface," *North-Holland Mathematics Studies*, vol. 204, pp. 7–10, 2006.

[17] I. Podlubny, *Fractional Differential Equations: An Introduction to Fractional Derivatives, Fractional Differential Equations, to Methods of Their Solution and Some of Their Applications*, vol. 198, Academic Press, San Diego, Calif, USA, 1999.

[18] S. Araci, "Novel identities for q-Genocchi numbers and polynomials," *Journal of Function Spaces and Applications*, vol. 2012, Article ID 214961, 13 pages, 2012.

[19] S. Araci, "Novel identities involving Genocchi numbers and polynomials arising from applications of umbral calculus," *Applied Mathematics and Computation*, vol. 233, pp. 599–607, 2014.

[20] T. Kim, "Some identities for the Bernoulli, the Euler and the Genocchi numbers and polynomials," *Advanced Studies in Contemporary Mathematics*, vol. 20, no. 1, pp. 23–28, 2010.

[21] A. Isah and C. Phang, "Genocchi Wavelet-like operational matrix and its application for solving non-linear fractional differential equations," *Open Physics*, vol. 14, no. 1, pp. 463–472, 2016.

[22] E. Kreyszig, *Introductory Functional Analysis with Applications*, vol. 81, John Wiley & Sons, New York, NY, USA, 1989.

[23] A. Saadatmandi and M. Dehghan, "A new operational matrix for solving fractional-order differential equations," *Computers & Mathematics with Applications*, vol. 59, no. 3, pp. 1326–1336, 2010.

[24] P. Chang and A. Isah, "Legendre Wavelet Operational Matrix of fractional Derivative through wavelet-polynomial transformation and its Applications in Solving Fractional Order Brusselator system," *Journal of Physics: Conference Series*, vol. 693, no. 1, article 012001, 2016.

[25] V. Daftardar-Gejji, Y. Sukale, and S. Bhalekar, "Solving fractional delay differential equations: a new approach," *Fractional Calculus and Applied Analysis*, vol. 18, no. 2, pp. 400–418, 2015.

[26] A. Anapali, Y. Öztürk, and M. Gülsu, "Numerical Approach for Solving Fractional Pantograph Equation," *International Journal of Computer Applications*, vol. 113, no. 9, pp. 45–52, 2015.

17

Existence of Asymptotically Almost Automorphic Mild Solutions of Semilinear Fractional Differential Equations

Junfei Cao,[1] Zaitang Huang,[2] and Gaston M. N'Guérékata[3]

[1]Department of Mathematics, Guangdong University of Education, Guangzhou 510303, China
[2]School of Mathematical Sciences, Guangxi Teachers Education University, Nanning 530023, China
[3]Department of Mathematics, Morgan State University, Baltimore, MD 21251, USA

Correspondence should be addressed to Gaston M. N'Guérékata; nguerekata@aol.com

Academic Editor: Patricia J. Y. Wong

This paper is concerned with the existence of asymptotically almost automorphic mild solutions to a class of abstract semilinear fractional differential equations $D_t^{\alpha}x(t) = Ax(t) + D_t^{\alpha-1}F(t, x(t), Bx(t))$, $t \in \mathbb{R}$, where $1 < \alpha < 2$, A is a linear densely defined operator of sectorial type on a complex Banach space X and B is a bounded linear operator defined on X, F is an appropriate function defined on phase space, and the fractional derivative is understood in the Riemann-Liouville sense. Combining the fixed point theorem due to Krasnoselskii and a decomposition technique, we prove the existence of asymptotically almost automorphic mild solutions to such problems. Our results generalize and improve some previous results since the (locally) Lipschitz continuity on the nonlinearity F is not required. The results obtained are utilized to study the existence of asymptotically almost automorphic mild solutions to a fractional relaxation-oscillation equation.

1. Introduction

The almost periodic function introduced seminally by Bohr in 1925 plays an important role in describing the phenomena that are similar to the periodic oscillations which can be observed frequently in many fields, such as celestial mechanics, nonlinear vibration, electromagnetic theory, plasma physics, engineering, and ecosphere. The concept of almost automorphy, which is an important generalization of the classical almost periodicity, was first introduced in the literature [1–4] by Bochner in relation to some aspects of differential geometry. Since then, this pioneer work has attracted more and more attention and has been substantially extended in several different directions. Many authors have made important contributions to this theory (see, for instance, [5–17] and the references therein). Especially, in [5, 6], the authors gave an important overview about the theory of almost automorphic functions and their applications to differential equations.

As a natural extension of almost automorphy, the concept of asymptotic almost automorphy, which is the central issue to be discussed in this paper, was introduced in the literature [18] by N'Guérékata in the early eighties. Since then, this notion has found several developments and has been generalized into different directions. Until now, the asymptotically almost automorphic functions as well as the asymptotically almost automorphic solutions for differential systems have been investigated by many mathematicians; see [19] by Bugajewski and N'Guérékata, [20] by Diagana, Hernández, and dos Santos, and [21] by Ding, Xiao, and Liang for the asymptotically almost automorphic solutions to integrodifferential equations, see [22] by Zhao, Chang, and N'Guérékata for the asymptotically almost automorphic solutions to the nonlinear delay integral equations, and see [23] by Chang and Tang and [24] by Zhao, Chang, and Nieto for the asymptotically almost automorphic solutions to stochastic differential equations, and the existence of asymptotically almost automorphic solutions has become one of the most attractive topics in the qualitative theory of differential equations due to its significance and applications in physics, mathematical biology, control theory, and so on. We refer the reader to the monographs of N'Guérékata [25] for the recently theory and applications of asymptotically almost automorphic functions.

With motivation coming from a wide range of engineering and physical applications, fractional differential equations have recently attracted great attention of mathematicians and scientists. This kind of equations is a generalization of ordinary differential equations to arbitrary noninteger orders. Fractional differential equations find numerous applications in the field of viscoelasticity, feedback amplifiers, electrical circuits, electro analytical chemistry, fractional multipoles, neuron modelling encompassing different branches of physics, chemistry, and biological sciences [26–32]. Many physical processes appear to exhibit fractional order behavior that may vary with time or space. In recent years, there has been a significant development in ordinary and partial differential equations involving fractional derivatives; we only enumerate here the monographs of Kilbas et al. [26, 27], Diethelm [28], Hilfer [29], Podlubny [30], Miller [31], and Zhou [32] and the papers of Agarwal et al. [33, 34], Benchohra et al. [35, 36], El-Borai [37], Lakshmikantham et al. [38–41], Mophou et al. [42–45], N'Guérékata [46], and Zhou et al. [47–50] and the reference therein.

The study of almost periodic and almost automorphic type solutions to fractional differential equations was initiated by Araya and Lizama [11]. In their work, the authors investigated the existence and uniqueness of an almost automorphic mild solution of the semilinear fractional differential equation

$$D_t^{\alpha} x(t) = Ax(t) + F(t, x(t)), \quad t \in \mathbb{R}, \ 1 < \alpha < 2, \tag{1}$$

when A is a generator of an α-resolvent family and D_t^{α} is the Riemann-Liouville fractional derivative. In [51], Cuevas and Lizama considered the fractional differential equation:

$$D_t^{\alpha} x(t) = Ax(t) + D_t^{\alpha-1} F(t, x(t)), \quad t \in \mathbb{R}, \ 1 < \alpha < 2, \tag{2}$$

where A is a linear operator of sectorial negative type on a complex Banach space X and the fractional derivative is understood in the Riemann-Liouville sense. Under suitable conditions on $F(t, x)$, the authors proved the existence and uniqueness of an almost automorphic mild solution to (2). Cuevas et al. [52, 53] studied, respectively, the pseudo almost periodic and pseudo almost periodic class infinity mild solutions to (2) assuming that $F : \mathbb{R} \times X \longrightarrow X$ and $(t, x) \longrightarrow F(t, x)$ is a pseudo almost periodic and pseudo almost periodic of class infinity function satisfying suitable conditions in $x \in X$. Agarwal et al. [54] studied the existence and uniqueness of a weighted pseudo almost periodic mild solution to equation (2). Ding et al. [55] investigated the existence and uniqueness of almost automorphic solution to (2) assuming that $F : \mathbb{R} \times X \longrightarrow X$ and $(t, x) \longrightarrow F(t, x)$ is Stepanov-like almost automorphic in $t \in \mathbb{R}$ satisfying some kind of Lipschitz conditions. Cuevas et al. [56] studied the existence of almost periodic (resp., pseudo almost periodic) mild solutions to equation (2) assuming that $F : \mathbb{R} \times X \longrightarrow X$ and $(t, x) \longrightarrow F(t, x)$ is Stepanov almost (resp., Stepanov-like pseudo almost) periodic in $t \in \mathbb{R}$ uniformly for $x \in X$. Chang et al. [57] studied the existence and uniqueness of weighted pseudo almost automorphic solution to equation (2) with Stepanov-like weighted pseudo almost automorphic coefficient. He et al. [58] studied also the existence and uniqueness of weighted Stepanov-like pseudo almost automorphic mild solution to (2). Cao et al. [59] studied the existence and uniqueness of antiperiodic mild solution to (2). In [60], Cuevas et al. showed sufficient conditions to ensure the existence and uniqueness of mild solution for (2) in the following classes of vector-valued function spaces: periodic functions, asymptotically periodic functions, pseudo periodic functions, almost periodic functions, asymptotically almost periodic functions, pseudo almost periodic functions, almost automorphic functions, asymptotically almost automorphic functions, pseudo almost automorphic functions, compact almost automorphic functions, asymptotically compact almost automorphic functions, pseudo compact almost automorphic functions, S-asymptotically ω-periodic functions, decay functions, and mean decay functions.

Recently, Xia et al. [61] established some sufficient criteria for the existence and uniqueness of (μ, ν)-pseudo almost automorphic solution to the semilinear fractional differential equation

$$D_t^{\alpha} x(t) = Ax(t) + D_t^{\alpha-1} F(t, Bx(t)), \quad t \in \mathbb{R}, \tag{3}$$

where $1 < \alpha < 2$, A is a sectorial operator of type $\omega < 0$ on a complex Banach space X and B is a bounded linear operator. The fractional derivative is understood in the Riemann-Liouville sense. Their discussion is divided into two cases, i.e., $F : \mathbb{R} \times X \longrightarrow X$, $(t, x) \longrightarrow F(t, x)$ is (μ, ν)-pseudo almost automorphic and $F : \mathbb{R} \times X \longrightarrow X$, and $(t, x) \longrightarrow F(t, x)$ is Stepanov-like (μ, ν)-pseudo almost automorphic. Kavitha et al. [62] studied weighted pseudo almost automorphic solutions of the fractional integrodifferential equation

$$D_t^{\alpha} x(t) = Ax(t) + D_t^{\alpha-1} F(t, x(t), Kx(t)), \quad t \in \mathbb{R}, \tag{4}$$

where $1 < \alpha < 2$ and

$$Kx(t) = \int_{-\infty}^{t} k(t-s) h(s, x(s)) \, ds, \tag{5}$$

A is a linear densely defined sectorial operator on a complex Banach space X, $F : \mathbb{R} \times X \times X \longrightarrow X$, and $(t, x, y) \longrightarrow F(t, x, y)$ is a weighted pseudo almost automorphic function in $t \in \mathbb{R}$ for each $x, y \in X$ satisfying suitable conditions. The fractional derivative is understood in the Riemann-Liouville sense. Mophou [63] investigated the existence and uniqueness of weighted pseudo almost automorphic mild solution to the fractional differential equation:

$$D_t^{\alpha} x(t) = Ax(t) + D_t^{\alpha-1} F(t, x(t), Bx(t)), \quad t \in \mathbb{R}, \ 1 < \alpha < 2, \tag{6}$$

where $A : D(A) \subset X \longrightarrow X$ is a linear densely operator of sectorial type on a complex Banach space X, $B : X \longrightarrow X$ is a bounded linear operator and $F : \mathbb{R} \times X \times X \longrightarrow X$, and $(t, x, y) \longrightarrow F(t, x, y)$ is a weighted pseudo almost automorphic function in $t \in \mathbb{R}$ for each $x, y \in X$ satisfying suitable conditions. The fractional derivative D_t^{α} is to be understood in Riemann-Liouville sense. Chang et al.

[64] investigated some existence results of μ-pseudo almost automorphic mild solutions to (6) assuming that $F : \mathbb{R} \times X \times X \longrightarrow X$ and $(t, x, y) \longrightarrow F(t, x, y)$ is a μ-pseudo almost automorphic function in $t \in \mathbb{R}$ for each $x, y \in X$ satisfying suitable conditions. For more on the almost periodicity and almost automorphy for fractional differential equations and related issues, we refer the reader to [65–67] and others.

Equation (6) is motivated by physical problems. Indeed, due to their applications in fields of science where characteristics of anomalous diffusion are presented, type (6) equations are attracting increasing interest (cf. [68–70] and references therein). For example, anomalous diffusion in fractals [69] or in macroeconomics [71] has been recently well studied in the setting of fractional Cauchy problems like (6). For this reason, (6) has gotten a considerable attention in recent years (cf. [51–64, 68–71] and the references therein).

To the best of our knowledge, much less is known about the existence of asymptotically almost automorphic mild solutions to (6) when the nonlinearity $F(t, x, y)$ as a whole loses the Lipschitz continuity with respect to x and y. Motivated by the abovementioned works, the purpose of this paper is to establish some new existence results of asymptotically almost automorphic mild solutions to (6). In our results, the nonlinearity $F : \mathbb{R} \times X \times X \longrightarrow X$, $(t, x, y) \longrightarrow F(t, x, y)$ does not have to satisfy a (locally) Lipschitz condition (see Remark 22). However, in many papers (for instance, [11, 51–64]) on almost periodic type and almost automorphic type solutions to fractional differential equations, to be able to apply the well-known Banach contraction principle, a (locally) Lipschitz condition for the nonlinearity of corresponding fractional differential equations is needed. As can be seen, our results generalize those as well as related research and have more broad applications. In particular, as application and to illustrate our main results, we will examine some sufficient conditions for the existence of asymptotically almost automorphic mild solutions to the fractional relaxation-oscillation equation given by

$$\partial_t^\alpha u(t,x) = \partial_x^2 u(t,x) - pu(t,x) + \partial_t^{\alpha-1}\Big[\mu a(t) \cdot \sin\left(\frac{1}{2+\cos t+\cos\sqrt{2}t}\right)[\sin u(t,x) + u(t,x)] + \nu e^{-|t|}[u(t,x) + \sin u(t,x)]\Big], \quad t \in \mathbb{R},\ x \in [0,\pi] \tag{7}$$

with boundary conditions $u(t, 0) = u(t, \pi) = 0$, $t \in \mathbb{R}$, where $a(t) \in BC(\mathbb{R}, \mathbb{R}^+)$ is a function and p, μ, and ν are positive constants.

The rest of this paper is organized as follows. In Section 2, some concepts, the related notations, and some useful lemmas are introduced and established. In Section 3, we prove the existence of asymptotically almost automorphic mild solutions to such problems. The results obtained are utilized to study the existence of asymptotically almost automorphic mild solutions to a fractional relaxation-oscillation equation given in Section 4.

2. Preliminaries

This section is concerned with some notations, definitions, lemmas, and preliminary facts which are used in what follows.

From now on, let $(X, \|\cdot\|)$ and $(Y, \|\cdot\|_Y)$ be two Banach spaces and $BC(\mathbb{R}, X)$ (resp., $BC(\mathbb{R} \times Y \times Y, X)$) is the space of all X-valued bounded continuous functions (resp., jointly bounded continuous functions $F : \mathbb{R} \times Y \times Y \longrightarrow X$). Furthermore, $C_0(\mathbb{R}, X)$ (resp., $C_0(\mathbb{R} \times Y \times Y, X)$) is the closed subspace of $BC(\mathbb{R}, X)$ (resp., $BC(\mathbb{R} \times Y \times Y, X)$) consisting of functions vanishing at infinity (vanishing at infinity uniformly in any compact subset of $Y \times Y$, in other words,

$$\lim_{|t|\longrightarrow+\infty} \|g(t,x,y)\| = 0 \quad \text{uniformly for } (x,y) \in \mathbb{K}, \tag{8}$$

where $\mathbb{K}$ is an any compact subset of $Y \times Y$). Let also $\mathbb{L}(X)$ be the Banach space of all bounded linear operators from X into itself endowed with the norm:

$$\|T\|_{\mathbb{L}(X)} = \sup\{\|Tx\| : x \in X,\ \|x\| = 1\}. \tag{9}$$

For a bounded linear operator $A \in \mathbb{L}(X)$, let $\rho(A)$ and $D(A)$ stand for the resolvent and domain of A, respectively.

First, let us recall some basic definitions and results on almost automorphic and asymptotically almost automorphic functions.

Definition 1 ((Bochner) [1] (N'Guérékata) [6]). A continuous function $F : \mathbb{R} \longrightarrow X$ is said to be almost automorphic if for every sequence of real numbers $\{s_n'\}$, there exists a subsequence $\{s_n\}$ such that

$$\Theta(t) = \lim_{n\longrightarrow\infty} F(t+s_n) \tag{10}$$

is well defined for each $t \in \mathbb{R}$ and

$$\lim_{n\longrightarrow\infty} \Theta(t-s_n) = F(t) \quad \text{for each } t \in \mathbb{R}. \tag{11}$$

Denote by $AA(\mathbb{R}, X)$ the set of all such functions.

Remark 2 (see [6]). By the point-wise convergence, the function $\Theta(t)$ in Definition 1 is measurable but not necessarily continuous. Moreover, if $\Theta(t)$ is continuous, then $F(t)$ is uniformly continuous (cf., e.g., [17], Theorem 2.6), and if the convergence in Definition 1 is uniform on $\mathbb{R}$, one gets almost periodicity (in the sense of Bochner and von Neumann). Almost automorphy is thus a more general concept than almost periodicity. There exists an almost automorphic function which is not almost periodic. The function $F : \mathbb{R} \longrightarrow \mathbb{R}$ given by

$$F(t) = \sin\left(\frac{1}{2+\cos t+\cos\sqrt{2}t}\right) \tag{12}$$

is an example of such functions [72].

Lemma 3 (see [5]). *$AA(\mathbb{R}, X)$ is a Banach space with the norm* $\|F\|_\infty = \sup_{t\in\mathbb{R}}\|F(t)\|$.

Definition 4 (see [6]). A continuous function $F : \mathbb{R}\times Y\times Y \longrightarrow X$ is said to be almost automorphic in $t \in \mathbb{R}$ uniformly for all $(x, y) \in K$, where K is any bounded subset of $Y\times Y$, if for every sequence of real numbers $\{s_n'\}$, there exists a subsequence $\{s_n\}$ such that

$$\lim_{n\longrightarrow\infty} F(t+s_n,x,y) = \Theta(t,x,y) \ \text{ exists} \qquad \text{for each } t\in\mathbb{R} \text{ and each } (x,y)\in K \tag{13}$$

and

$$\lim_{n\longrightarrow\infty} \Theta(t-s_n,x,y) = F(t,x,y) \ \text{ exists} \qquad \text{for each } t\in\mathbb{R} \text{ and each } (x,y)\in K. \tag{14}$$

The collection of those functions is denoted by $AA(\mathbb{R}\times Y\times Y, X)$.

Remark 5. The function $F : \mathbb{R}\times X\times X \longrightarrow X$ given by

$$F(t,x,y) = \sin\left(\frac{1}{2+\cos t+\cos\sqrt{2}t}\right)[\sin(x)+y] \tag{15}$$

is almost automorphic in $t \in \mathbb{R}$ uniformly for all $(x, y) \in K$, where K is any bounded subset of $X\times X$, $X = L^2[0, \pi]$.

Similar to Lemma 2.2 of [73] and Proposition 3.2 of [63], we have the following result on almost automorphic functions.

Lemma 6. *Let $F : \mathbb{R}\times X\times X \longrightarrow X$ be almost automorphic in $t \in \mathbb{R}$ uniformly for all $(x, y) \in K$, where K is any bounded subset of $X \times X$, and assume that $F(t, x, y)$ is uniformly continuous on K uniformly for $t \in \mathbb{R}$, that is, for any $\varepsilon > 0$, there exists $\delta > 0$ such that $x_1, x_2, y_1, y_2 \in K$ and $\|x_1 - y_1\| + \|x_2 - y_2\| < \delta$ imply that*

$$\left\|F(t,x_1,x_2) - F(t,y_1,y_2)\right\| < \varepsilon \quad \forall t\in\mathbb{R}. \tag{16}$$

Let $x, y : \mathbb{R} \longrightarrow X$ be almost automorphic. Then the function $\Upsilon : \mathbb{R} \longrightarrow X$ defined by $\Upsilon(t) = F(t, x(t), y(t))$ is almost automorphic.

Proof. Suppose that $\{s_n\}$ is a sequence of real numbers. Then by the definition of almost automorphic functions, we can extract a subsequence $\{\tau_n\}$ of $\{s_n\}$ such that

$$\begin{aligned}
&(P_1)\ \lim_{n\longrightarrow\infty} x(t+\tau_n) = \tilde{x}(t) \quad \text{for each } t\in\mathbb{R},\\
&(P_2)\ \lim_{n\longrightarrow\infty} \tilde{x}(t-\tau_n) = x(t) \quad \text{for each } t\in\mathbb{R},\\
&(P_3)\ \lim_{n\longrightarrow\infty} y(t+\tau_n) = \tilde{y}(t) \quad \text{for each } t\in\mathbb{R},\\
&(P_4)\ \lim_{n\longrightarrow\infty} \tilde{y}(t-\tau_n) = y(t) \quad \text{for each } t\in\mathbb{R},\\
&(P_5)\ \lim_{n\longrightarrow\infty} F(t+\tau_n,x,y) = \tilde{F}(t,x,y) \quad \text{for each } t\in\mathbb{R},\ x,y\in X,\\
&(P_6)\ \lim_{n\longrightarrow\infty} \tilde{F}(t-\tau_n,x,y) = F(t,x,y) \quad \text{for each } t\in\mathbb{R},\ x,y\in X.
\end{aligned} \tag{17}$$

Write

$$\tilde{\Upsilon}(t) := \tilde{F}(t,\tilde{x}(t),\tilde{y}(t)), \quad t\in\mathbb{R}. \tag{18}$$

Then

$$\begin{aligned}
&\left\|\Upsilon(t+\tau_n) - \tilde{\Upsilon}(t)\right\|\\
&= \left\|F(t+\tau_n,x(t+\tau_n),y(t+\tau_n)) - \tilde{F}(t,\tilde{x}(t),\tilde{y}(t))\right\|\\
&\le \left\|F(t+\tau_n,x(t+\tau_n),y(t+\tau_n)) - F(t+\tau_n,\tilde{x}(t),\tilde{y}(t))\right\| + \left\|F(t+\tau_n,\tilde{x}(t),\tilde{y}(t)) - \tilde{F}(t,\tilde{x}(t),\tilde{y}(t))\right\|.
\end{aligned} \tag{19}$$

Since $x(t)$ and $y(t)$ are almost automorphic, then $x(t)$, $y(t)$ and $\tilde{x}(t)$, and $\tilde{y}(t)$ are bounded. Therefore we can choose a bounded subset $K \subset X\times X$, such that

$$(x(t),y(t))\in K, \quad (\tilde{x}(t),\tilde{y}(t))\in K \quad \forall t\in\mathbb{R}. \tag{20}$$

By (P_1), (P_3), and the uniform continuity of $F(t, x, y)$ in $(x(t), y(t)) \in K$, we have

$$\lim_{n\longrightarrow\infty} \left\|F(t+\tau_n,x(t+\tau_n),y(t+\tau_n)) - F(t+\tau_n,\tilde{x}(t),\tilde{y}(t))\right\| = 0. \tag{21}$$

Moreover, by (P_5),

$$\lim_{n\longrightarrow\infty} \left\|F(t+\tau_n,\tilde{x}(t),\tilde{y}(t)) - \tilde{F}(t,\tilde{x}(t),\tilde{y}(t))\right\| = 0, \tag{22}$$

so remembering the above triangle inequality, we deduce that

$$\lim_{n\longrightarrow\infty} \left\|\Upsilon(t+\tau_n) - \tilde{\Upsilon}(t)\right\| = 0 \quad \text{for each } t\in\mathbb{R}. \tag{23}$$

Using the same argument we can prove that

$$\lim_{n\longrightarrow\infty} \left\|\tilde{\Upsilon}(t-\tau_n) - \Upsilon(t)\right\| = 0 \quad \text{for each } t\in\mathbb{R}. \tag{24}$$

This proves that $\Upsilon(t)$ is almost automorphic by the definition. □

Remark 7. If $F(t, x, y)$ satisfies a Lipschitz condition with respect to x and y uniformly in $t \in \mathbb{R}$, i.e., for each pair $x_1, x_2, y_1, y_2 \in X$,

$$\begin{aligned}
&\left\|F(t,x_1,x_2) - F(t,y_1,y_2)\right\|\\
&\le L\left(\left\|x_1-y_1\right\| + \left\|x_2-y_2\right\|\right)
\end{aligned} \tag{25}$$

uniformly in $t \in \mathbb{R}$, where $L > 0$ is called the Lipschitz constant for the function $F(t, x, y)$, then $F(t, x, y)$ is uniformly continuous on K uniformly for $t \in \mathbb{R}$, where K is any bounded subset of $X \times X$.

Remark 8. If $F(t, x, y)$ satisfies a local Lipschitz condition with respect to x and y uniformly in $t \in \mathbb{R}$, i.e., for each pair $x_1, x_2, y_1, y_2 \in X$, $t \in \mathbb{R}$,

$$\begin{aligned} &\|F(t, x_1, x_2) - F(t, y_1, y_2)\| \\ &\quad \leq L(t)(\|x_1 - y_1\| + \|x_2 - y_2\|), \end{aligned} \tag{26}$$

where $L(t) \in BC(\mathbb{R}, \mathbb{R}^+)$, then $F(t, x, y)$ is uniformly continuous on K uniformly for $t \in \mathbb{R}$, where K is any bounded subset of $X \times X$.

Definition 9 (see [6]). A continuous function $F : \mathbb{R} \longrightarrow X$ is said to be asymptotically almost automorphic if it can be decomposed as $F(t) = G(t) + \Phi(t)$, where

$$\begin{aligned} &G(t) \in AA(\mathbb{R}, X), \\ &\Phi(t) \in C_0(\mathbb{R}, X). \end{aligned} \tag{27}$$

Denote by $AAA(\mathbb{R}, X)$ the set of all such functions.

Remark 10. The function $F : \mathbb{R} \longrightarrow \mathbb{R}$ defined by

$$\begin{aligned} F(t) &= G(t) + \Phi(t) \\ &= \sin\left(\frac{1}{2 + \cos t + \cos\sqrt{2}t}\right) + e^{-|t|} \end{aligned} \tag{28}$$

is an asymptotically almost automorphic function with

$$\begin{aligned} &G(t) = \sin\left(\frac{1}{2 + \cos t + \cos\sqrt{2}t}\right) \in AA(\mathbb{R}, \mathbb{R}), \\ &\Phi(t) = e^{-|t|} \in C_0(\mathbb{R}, \mathbb{R}). \end{aligned} \tag{29}$$

Lemma 11 (see [6]). *$AAA(\mathbb{R}, X)$ is also a Banach space with the supremum norm $\|\cdot\|_\infty$.*

Definition 12 (see [6]). A continuous function $F : \mathbb{R} \times Y \times Y \longrightarrow X$ is said to be asymptotically almost automorphic if it can be decomposed as $F(t, x, y) = G(t, x, y) + \Phi(t, x, y)$, where

$$\begin{aligned} &G(t, x, y) \in AA(\mathbb{R} \times Y \times Y, X), \\ &\Phi(t, x, y) \in C_0(\mathbb{R} \times Y \times Y, X). \end{aligned} \tag{30}$$

Denote by $AAA(\mathbb{R} \times Y \times Y, X)$ the set of all such functions.

Remark 13. The function $F : \mathbb{R} \times X \times X \longrightarrow X$ given by

$$\begin{aligned} F(t, x, y) &= G(t, x, y) + \Phi(t, x, y) \\ &= \sin\left(\frac{1}{2 + \cos t + \cos\sqrt{2}t}\right)[\sin(x) + y] \\ &\quad + e^{-|t|}[x + \sin(y)] \end{aligned} \tag{31}$$

is asymptotically almost automorphic in $t \in \mathbb{R}$ uniformly for all $(x, y) \in K$, where K is any bounded subset of $X \times X$, $X = L^2[0, \pi]$ and

$$\begin{aligned} G(t, x, y) &= \sin\left(\frac{1}{2 + \cos t + \cos\sqrt{2}t}\right)[\sin(x) + y] \\ &\in AA(\mathbb{R} \times X \times X, X), \\ \Phi(t, x, y) &= e^{-|t|}[x + \sin(y)] \in C_0(\mathbb{R} \times X \times X, X). \end{aligned} \tag{32}$$

Next we give some basic definitions and properties of the fractional calculus theory which are used further in this paper.

Definition 14 (see [26]). The fractional integral of order $\alpha > 0$ with the lower limit t_0 for a function f is defined as

$$I^\alpha f(t) = \frac{1}{\Gamma(\alpha)} \int_{t_0}^{t} (t - s)^{\alpha - 1} f(s)\, ds, \quad t > t_0,\ \alpha > 0 \tag{33}$$

provided that the right-hand side is point-wise defined on $[t_0, \infty)$, where Γ is the Gamma function.

Definition 15 (see [26]). Riemann-Liouville derivative of order $\alpha > 0$ with the lower limit t_0 for a function $f : [t_0, \infty) \longrightarrow \mathbb{R}$ can be written as

$$\begin{aligned} D_t^\alpha f(t) &= \frac{1}{\Gamma(n - \alpha)} \frac{d^n}{dt^n} \int_{t_0}^{t} (t - s)^{-\alpha} f(s)\, ds, \\ &\qquad t > t_0,\ n - 1 < \alpha < n. \end{aligned} \tag{34}$$

The first and maybe the most important property of Riemann-Liouville fractional derivative is that, for $t > t_0$ and $\alpha > 0$, one has $D_t^\alpha(I^\alpha f(t)) = f(t)$, which means that Riemann-Liouville fractional differentiation operator is a left inverse to the Riemann-Liouville fractional integration operator of the same order α.

It is important to define sectorial operator for the definition of mild solution of any fractional abstract equations. So, let us now give the definitions of sectorial linear operators and their associated solution operators.

Definition 16 ([74] sectorial operator). A closed and linear operator A is said to be sectorial of type ω and angle θ if there exist $0 < \theta < \pi/2$, $M > 0$, and $\omega \in \mathbb{R}$ such that its resolvent $\rho(A)$ exists outside the sector $\omega + S_\theta := \{\omega + \lambda : \lambda \in \mathbb{C}, |\arg(-\lambda)| < \theta\}$ and

$$\left\|(\lambda - A)^{-1}\right\| \leq \frac{M}{|\lambda - \omega|}, \quad \lambda \notin \omega + S_\theta. \tag{35}$$

Sectorial operators are well studied in the literature, usually for the case $\omega = 0$. For a recent reference including several examples and properties we refer the reader to [74]. Note that an operator A is sectorial of type ω if and only if $\omega I - A$ is sectorial of type 0.

Definition 17 (see [75]). Let A be a closed and linear operator with domain $D(A)$ defined on a Banach space X. We call A

the generator of a solution operator if there are $\omega \in \mathbb{R}$ and a strongly continuous function $S_\alpha : \mathbb{R}^+ \longrightarrow \mathbb{L}(X)$ such that $\{\lambda^\alpha : \operatorname{Re}\lambda > \omega\} \subseteq \rho(A)$ and

$$\lambda^{\alpha-1}(\lambda^\alpha - A)^{-1} x = \int_0^\infty e^{-\lambda t} S_\alpha(t) x \, dt, \quad \operatorname{Re}\lambda > \omega, \ x \in X. \tag{36}$$

In this case, $S_\alpha(t)$ is called the solution operator generated by A.

Note that if A is sectorial of type ω with $0 \le \theta \le \pi(1-\alpha/2)$, then A is the generator of a solution operator given by

$$S_\alpha(t) := \frac{1}{2\pi i}\int_\gamma e^{-\lambda t}\lambda^{\alpha-1}(\lambda^\alpha - A)^{-1} d\lambda, \tag{37}$$

where γ is a suitable path lying outside the sector $\omega + \Sigma_\theta$ (cf. [74]).

Very recently, Cuesta in [74](Theorem 1) has proved that if A is a sectorial operator of type $\omega < 0$ for some $M > 0$ and $0 \le \theta < \pi(1-\alpha/2)$, then there exists $C > 0$ such that

$$\|S_\alpha(t)\|_{\mathbb{L}(X)} \le \frac{CM}{1+|\omega| t^\alpha} \quad \text{for } t \ge 0. \tag{38}$$

In the border case $\alpha = 1$, this is analogous to saying that A is the generator of a exponentially stable C_0-semigroup. The main difference is that in the case $\alpha > 1$ the solution family $S_\alpha(t)$ decays like $t^{-\alpha}$. Cuesta's result proves that $S_\alpha(t)$ is, in fact, integrable.

In the following, we present the following compactness criterion, which is a special case of the general compactness result of Theorem 2.1 in [76].

Lemma 18 (see [76]). *A set $D \subset C_0(\mathbb{R}, X)$ is relatively compact if*

(1) *D is equicontinuous;*

(2) *$\lim_{|t|\longrightarrow\infty} x(t) = 0$ uniformly for $x \in D$;*

(3) *the set $D(t) := \{x(t) : x \in D\}$ is relatively compact in X for every $t \in \mathbb{R}$.*

The following Krasnoselskii's fixed point theorem plays a key role in the proofs of our main results, which can be found in many books.

Lemma 19 (see [77]). *Let U be a bounded closed and convex subset of X and J_1, J_2 be maps of U into X such that $J_1x + J_2y \in U$ for every pair $x, y \in U$. If J_1 is a contraction and J_2 is completely continuous, then $J_1x + J_2x = x$ has a solution on U.*

3. Asymptotically Almost Automorphic Mild Solutions

In this section, we study the existence of asymptotically almost automorphic mild solutions for the semilinear fractional differential equations of the form

$$D_t^\alpha x(t) = Ax(t) + D_t^{\alpha-1}F(t, x(t), Bx(t)), \quad t \in \mathbb{R}, \ 1 < \alpha < 2, \tag{39}$$

where $A : D(A) \subset X \longrightarrow X$ is a linear densely defined operator of sectorial type of $\omega < 0$ on a complex Banach space X, $B : X \longrightarrow X$ is a bounded linear operator and $F : \mathbb{R} \times X \times X \longrightarrow X$, and $(t, x, y) \longrightarrow F(t, x, y)$ is a given function to be specified later. The fractional derivative D_t^α is to be understood in Riemann-Liouville sense.

We recall the following definition that will be essential for us.

Definition 20 (see [63]). Assume that A generates an integrable solution operator $S_\alpha(t)$. A continuous function $x : \mathbb{R} \longrightarrow X$ satisfying the integral equation

$$x(t) = \int_{-\infty}^t S_\alpha(t-\sigma) F(\sigma, x(\sigma), Bx(\sigma)) d\sigma, \quad t \in \mathbb{R} \tag{40}$$

is called a mild solution on $\mathbb{R}$ to (39).

In the proofs of our results, we need the following auxiliary result.

Lemma 21. *Given $Y(t) \in AA(\mathbb{R}, X)$ and $Z(t) \in C_0(\mathbb{R}, X)$, let*

$$\Phi_1(t) := \int_{-\infty}^t S_\alpha(t-s) Y(s) ds, \quad \Phi_2(t) := \int_{-\infty}^t S_\alpha(t-s) Z(s) ds, \quad t \in \mathbb{R}. \tag{41}$$

Then $\Phi_1(t) \in AA(\mathbb{R}, X)$, $\Phi_2(t) \in C_0(\mathbb{R}, X)$.

Proof. Firstly, note that

$$\int_0^\infty \frac{1}{1+|\omega| s^\alpha} ds = \frac{|\omega|^{-1/\alpha}\pi}{\alpha \sin(\pi/\alpha)} \quad \text{for } 1 < \alpha < 2. \tag{42}$$

Then

$$\begin{aligned} \|\Phi_1(t)\| &= \left\|\int_{-\infty}^t S_\alpha(t-s) Y(s) ds\right\| \\ &= \left\|\int_0^{+\infty} S_\alpha(\tau) Y(t-\tau) d\tau\right\| \\ &\le CM \|Y\|_\infty \int_0^\infty \frac{1}{1+|\omega|\tau^\alpha} d\tau \\ &= \frac{CM|\omega|^{-1/\alpha}\pi}{\alpha\sin(\pi/\alpha)} \|Y\|_\infty, \end{aligned} \tag{43}$$

which implies that $\Phi_1(t)$ is well defined and continuous on $\mathbb{R}$. Since $Y(t) \in AA(\mathbb{R}, X)$, then for any $\varepsilon > 0$ and every sequence of real numbers $\{s_n'\}$, there exist a subsequence $\{s_n\}$, a function $\widetilde{Y}(t)$, and $N \in \mathbb{N}$ such that

$$\left\|Y(s+s_n) - \widetilde{Y}(s)\right\| < \varepsilon \quad \text{for each } n > N \text{ and every } s \in \mathbb{R}. \tag{44}$$

Define

$$\widetilde{\Phi_1}(t) := \int_{-\infty}^{t} T(t-s)\widetilde{Y}(s)\,ds. \tag{45}$$

Then

$$\begin{aligned}
\left\|\Phi_1(t+s_n)-\widetilde{\Phi_1}(t)\right\| &= \left\|\int_{-\infty}^{t+s_n} S_\alpha(t+s_n-s)Y(s)\,ds - \int_{-\infty}^{t} S_\alpha(t-s)Y(s)\,ds\right\| \\
&= \left\|\int_{0}^{+\infty} S_\alpha(s)Y(t+s_n-s)\,ds - \int_{0}^{+\infty} S_\alpha(s)Y(t-s)\,ds\right\| \\
&\le CM\int_0^{\infty}\frac{1}{1+|\omega|s^\alpha}\left\|Y(s+s_n)-\widetilde{Y}(s)\right\|ds \\
&\le \frac{CM|\omega|^{-1/\alpha}\pi\varepsilon}{\alpha\sin(\pi/\alpha)}
\end{aligned} \tag{46}$$

for each $n > N$ and every $t \in \mathbb{R}$. This implies that

$$\widetilde{\Phi_1}(t) = \lim_{n\longrightarrow\infty}\Phi_1(t+s_n) \tag{47}$$

is well defined for each $t \in \mathbb{R}$.

By a similar argument one can obtain

$$\lim_{n\longrightarrow\infty}\widetilde{\Phi_1}(t-s_n) = \Phi_1(t) \quad \text{for each } t \in \mathbb{R}. \tag{48}$$

Thus $\Phi_1(t) \in AA(\mathbb{R}, X)$.

Since $Z(t) \in C_0(\mathbb{R}, X)$, one can choose an $N_1 > 0$ such that $\|Z(t)\| < \varepsilon$ for all $t > N_1$. This enables us to conclude that, for all $t > N_1$,

$$\begin{aligned}
\left\|\Phi_2(t)\right\| &\le \left\|\int_{-\infty}^{N_1} S_\alpha(t-s)Z(s)\,ds\right\| + \left\|\int_{N_1}^{t} S_\alpha(t-s)Z(s)\,ds\right\| \\
&\le CM\|Z\|_\infty\int_{-\infty}^{N_1}\frac{1}{1+|\omega|(t-s)^\alpha}ds + \varepsilon CM\int_{N_1}^{t}\frac{1}{1+|\omega|(t-s)^\alpha}ds \\
&\le \frac{CM\|Z\|_\infty}{|\omega|}\int_{-\infty}^{N_1}\frac{1}{(t-s)^\alpha}ds + \frac{CM|\omega|^{-1/\alpha}\pi\varepsilon}{\alpha\sin(\pi/\alpha)} \\
&\le \frac{CM\|Z\|_\infty}{|\omega|}\frac{1}{(\alpha-1)(t-N_1)^{\alpha-1}} + \frac{CM|\omega|^{-1/\alpha}\pi\varepsilon}{\alpha\sin(\pi/\alpha)},
\end{aligned} \tag{49}$$

which implies

$$\lim_{t\longrightarrow+\infty}\left\|\Phi_2(t)\right\| = 0. \tag{50}$$

On the other hand, from $Z(t) \in C_0(\mathbb{R}, X)$ it follows that there exists an $N_2 > 0$ such that $\|Z(t)\| < \varepsilon$ for all $t < -N_2$. This enables us to conclude that, for all $t < -N_2$,

$$\begin{aligned}
\left\|\Phi_2(t)\right\| &= \left\|\int_{-\infty}^{t} S_\alpha(t-s)Z(s)\,ds\right\| \\
&\le \int_{-\infty}^{t}\left\|S_\alpha(t-s)\right\|\left\|Z(s)\right\|ds \\
&\le CM\varepsilon\int_{-\infty}^{t}\frac{1}{1+|\omega|(t-s)^\alpha}ds \\
&= \frac{CM|\omega|^{-1/\alpha}\pi\varepsilon}{\alpha\sin(\pi/\alpha)},
\end{aligned} \tag{51}$$

which implies

$$\lim_{t\longrightarrow-\infty}\left\|\Phi_2(t)\right\| = 0. \tag{52}$$

□

Now we are in position to state and prove our first main result. To prove our main result, let us introduce the following assumptions:

(H_1) $F(t,x,y) = F_1(t,x,y) + F_2(t,x,y) \in AAA(\mathbb{R}\times X\times X, X)$ with

$$\begin{aligned}
&F_1(t,x,y) \in AA(\mathbb{R}\times X\times X, X), \\
&F_2(t,x,y) \in C_0(\mathbb{R}\times X\times X, X)
\end{aligned} \tag{53}$$

and there exists a constant $L > 0$ such that, for all $t \in \mathbb{R}$ and $x_1, x_2, y_1, y_2 \in X$,

$$\begin{aligned}
&\left\|F_1(t,x_1,x_2) - F_1(t,y_1,y_2)\right\| \\
&\quad \le L\left(\left\|x_1-y_1\right\| + \left\|x_2-y_2\right\|\right).
\end{aligned} \tag{54}$$

(H_2) There exist a function $\beta(t) \in C_0(\mathbb{R}, \mathbb{R}^+)$ and a nondecreasing function $\Phi : \mathbb{R}^+ \longrightarrow \mathbb{R}^+$ such that, for all $t \in \mathbb{R}$ and $x, y \in X$ with $\|x\| + \|y\| \le r$,

$$\begin{aligned}
&\left\|F_2(t,x,y)\right\| \le \beta(t)\Phi(r) \\
&\text{and } \liminf_{r\longrightarrow+\infty}\frac{\Phi(r)}{r} = \rho_1.
\end{aligned} \tag{55}$$

Remark 22. Assuming that $F(t,x,y)$ satisfies the assumption (H_1), it is noted that $F(t,x,y)$ does not have to meet the

Lipschitz continuity with respect to x and y. Such class of asymptotically almost automorphic functions $F(t,x,y)$ are more complicated than those with Lipschitz continuity with respect to x and y and little is known about them.

Let $\beta(t)$ be the function involved in assumption (H_2). Define

$$\sigma(t) := \int_{-\infty}^{t} \frac{\beta(s)}{1+|\omega|(t-s)^{\alpha}} ds, \quad t \in \mathbb{R}. \tag{56}$$

Lemma 23. $\sigma(t) \in C_0(\mathbb{R},\mathbb{R}^+)$.

Proof. Since $\beta(t) \in C_0(\mathbb{R},\mathbb{R}^+)$, one can choose a $T_1 > 0$ such that $\|\beta(t)\| < \varepsilon$ for all $t > T_1$. This enables us to conclude that, for all $t > T_1$,

$$\begin{aligned} \|\sigma(t)\| &\le \left\| \int_{-\infty}^{T_1} \frac{\beta(s)}{1+|\omega|(t-s)^{\alpha}} ds \right\| + \left\| \int_{T_1}^{t} \frac{\beta(s)}{1+|\omega|(t-s)^{\alpha}} ds \right\| \\ &\le \|\beta\|_{\infty} \int_{-\infty}^{T_1} \frac{1}{1+|\omega|(t-s)^{\alpha}} ds + \varepsilon \int_{T_1}^{t} \frac{1}{1+|\omega|(t-s)^{\alpha}} ds \\ &\le \frac{\|\beta\|_{\infty}}{|\omega|} \int_{-\infty}^{T_1} \frac{1}{(t-s)^{\alpha}} ds + \frac{|\omega|^{-1/\alpha} \pi\varepsilon}{\alpha \sin(\pi/\alpha)} \\ &\le \frac{\|\beta\|_{\infty}}{|\omega|} \frac{1}{(\alpha-1)(t-T_1)^{\alpha-1}} + \frac{|\omega|^{-1/\alpha} \pi\varepsilon}{\alpha \sin(\pi/\alpha)}, \end{aligned} \tag{57}$$

which implies

$$\lim_{t\longrightarrow+\infty} \|\sigma(t)\| = 0. \tag{58}$$

On the other hand, from $\beta(t) \in C_0(\mathbb{R},\mathbb{R}^+)$ it follows that there exists a $T_2 > 0$ such that $\|\beta(t)\| < \varepsilon$ for all $t < -T_2$. This enables us to conclude that, for all $t < -T_2$,

$$\begin{aligned} \|\sigma(t)\| &= \left\| \int_{-\infty}^{t} \frac{\beta(s)}{1+|\omega|(t-s)^{\alpha}} ds \right\| \\ &\le \varepsilon \int_{-\infty}^{t} \frac{1}{1+|\omega|(t-s)^{\alpha}} ds = \frac{|\omega|^{-1/\alpha} \pi\varepsilon}{\alpha \sin(\pi/\alpha)}, \end{aligned} \tag{59}$$

which implies

$$\lim_{t\longrightarrow-\infty} \|\sigma(t)\| = 0. \tag{60}$$

□

Theorem 24. *Assume that A is sectorial of type $\omega < 0$. Let $F : \mathbb{R} \times X \times X \longrightarrow X$ satisfy the hypotheses (H_1) and (H_2). Put $\rho_2 := \sup_{t\in\mathbb{R}} \sigma(t)$. Then (39) has at least one asymptotically almost automorphic mild solution provided that*

$$\frac{CML(1+\|B\|_{\mathbb{L}(X)})|\omega|^{-1/\alpha}\pi}{\alpha \sin(\pi/\alpha)} + CM(1+\|B\|_{\mathbb{L}(X)})\rho_1\rho_2 < 1. \tag{61}$$

Proof. The proof is divided into the following five steps.

Step 1. Define a mapping Λ on $AA(\mathbb{R},X)$ by

$$(\Lambda v)(t) = \int_{-\infty}^{t} S_{\alpha}(t-s) F_1(s, v(s), Bv(s)) ds, \quad t \in \mathbb{R} \tag{62}$$

and prove Λ has a unique fixed point $v(t) \in AA(\mathbb{R},X)$.

Firstly, since the function $s \longrightarrow F_1(s,v(s),Bv(s))$ is bounded in $\mathbb{R}$ and

$$\begin{aligned} \|[\Lambda v](t)\| &\le \int_{-\infty}^{t} \|S_{\alpha}(t-s)\| \|F_1(s,v(s),Bv(s))\| ds \\ &\le CM \int_{-\infty}^{t} \frac{1}{1+|\omega|(t-s)^{\alpha}} \|F_1(s,v(s),Bv(s))\| ds \\ &\le CM \|F_1\|_{\infty} \int_{-\infty}^{t} \frac{1}{1+|\omega|(t-s)^{\alpha}} ds \\ &= \frac{CML|\omega|^{-1/\alpha}\pi \|F_1\|_{\infty}}{\alpha \sin(\pi/\alpha)}, \end{aligned} \tag{63}$$

this implies that $(\Lambda v)(t)$ exists. Moreover from $F_1(t,x,y) \in AA(\mathbb{R}\times X\times X, X)$ satisfying (54), together with Lemma 6 and Remark 7, it follows that

$$F_1(\cdot, v(\cdot), Bv(\cdot)) \in AA(\mathbb{R},X) \quad \text{for every } v(\cdot) \in AA(\mathbb{R},X). \tag{64}$$

This, together with Lemma 21, implies that Λ is well defined and maps $AA(\mathbb{R},X)$ into itself.

In the sequel, we verify that Λ is continuous.

Let $v_n(t), v(t)$ be in $AA(\mathbb{R},X)$ with $v_n(t) \longrightarrow v(t)$ as $n \longrightarrow \infty$; then one has

$$\begin{aligned} &\|[\Lambda v_n](t) - [\Lambda v](t)\| = \left\| \int_{-\infty}^{t} S_{\alpha}(t-s) \cdot [F_1(s,v_n(s),Bv_n(s)) - F_1(s,v(s),Bv(s))] ds \right\| \le L \int_{-\infty}^{t} \|S_{\alpha}(t-s)\| \\ &\cdot [\|v_n(s) - v(s)\| + \|Bv_n(s) - Bv(s)\|] ds \\ &\le CML \int_{-\infty}^{t} \frac{1}{1+|\omega|(t-s)^{\alpha}} (1+\|B\|_{\mathbb{L}(X)}) \|v_n(s) - v(s)\| ds \le CML(1+\|B\|_{\mathbb{L}(X)}) \|v_n - v\|_{\infty} \end{aligned}$$

$$\cdot \int_{-\infty}^{t} \frac{1}{1 + |\omega| (t-s)^{\alpha}} ds = \frac{CML (1 + \|B\|_{\mathbb{L}(X)}) |\omega|^{-1/\alpha} \pi}{\alpha \sin (\pi/\alpha)} \|v_n - v\|_{\infty}. \quad (65)$$

Therefore, as $n \longrightarrow \infty$ and $\Lambda v_n \longrightarrow \Lambda v$, hence Λ is continuous.

Next, we prove that Λ is a contraction on $AA(\mathbb{R}, X)$ and has a unique fixed point $v(t) \in AA(\mathbb{R}, X)$.

In fact, let $v_1(t), v_2(t)$ be in $AA(\mathbb{R}, X)$, and similar to the above proof of the continuity of Λ, one has

$$\|[\Lambda v_1](t) - [\Lambda v_2](t)\| \le \frac{CML (1 + \|B\|_{\mathbb{L}(X)}) |\omega|^{-1/\alpha} \pi}{\alpha \sin (\pi/\alpha)} \|v_1 - v_2\|_{\infty}, \quad (66)$$

which implies

$$\|[\Lambda v_1](t) - [\Lambda v_2](t)\|_{\infty} \le \frac{CML (1 + \|B\|_{\mathbb{L}(X)}) |\omega|^{-1/\alpha} \pi}{\alpha \sin (\pi/\alpha)} \|v_1 - v_2\|_{\infty}. \quad (67)$$

Together with (61), this proves that Λ is a contraction on $AA(\mathbb{R}, X)$. Thus, Banach's fixed point theorem implies that Λ has a unique fixed point $v(t) \in AA(\mathbb{R}, X)$.

Step 2. Set

$$\Omega_r := \{\omega(t) \in C_0(\mathbb{R}, X) : \|\omega(t)\| \le r\}. \quad (68)$$

For the above $v(t)$, define $\Gamma := \Gamma^1 + \Gamma^2$ on $C_0(\mathbb{R}, X)$ as

$$(\Gamma^1 \omega)(t) = \int_{-\infty}^{t} S_{\alpha}(t-s) \cdot [F_1(s, v(s) + \omega(s), B(v(s) + \omega(s))) - F_1(s, v(s), Bv(s))] ds, \quad (69)$$

$$(\Gamma^2 \omega)(t) = \int_{-\infty}^{t} S_{\alpha}(t-s) F_2(s, v(s) + \omega(s), B(v(s) + \omega(s))) ds$$

and prove that Γ maps Ω_{k_0} into itself, where k_0 is a given constant.

Firstly, from (54) it follows that, for all $s \in \mathbb{R}$ and $\omega(s) \in X$,

$$\|F_1(s, v(s) + \omega(s), B(v(s) + \omega(s))) - F_1(s, v(s), Bv(s))\| \le L[\|\omega(s)\| + \|B\omega(s)\|] \le L(1 + \|B\|_{\mathbb{L}(X)}) \|\omega(s)\|, \quad (70)$$

which implies that

$$F_1(\cdot, v(\cdot) + \omega(\cdot), B(v(\cdot) + \omega(\cdot))) - F_1(\cdot, v(\cdot), Bv(\cdot)) \in C_0(\mathbb{R}, X) \quad \text{for every } \omega(\cdot) \in C_0(\mathbb{R}, X). \quad (71)$$

According to (55), one has

$$\begin{aligned}\|F_2(s, v(s) + \omega(s), B(v(s) + \omega(s)))\| &\le \beta(s) \cdot \Phi\left(\|\omega(s) + B\omega(s)\| + \sup_{s\in\mathbb{R}} \|v(s) + Bv(s)\|\right) \\ &\le \beta(s) \Phi\left((1 + \|B\|_{\mathbb{L}(X)}) \|\omega(s)\| + (1 + \|B\|_{\mathbb{L}(X)}) \sup_{s\in\mathbb{R}} \|v(s)\|\right) = \beta(s) \\ &\quad \cdot \Phi\left((1 + \|B\|_{\mathbb{L}(X)}) \left[\|\omega(s)\| + \sup_{s\in\mathbb{R}} \|v(s)\|\right]\right)\end{aligned} \quad (72)$$

for all $s \in \mathbb{R}$ and $\omega(s) \in X$ with $\|\omega(s)\| \le r$; then

$$F_2(\cdot, v(\cdot) + \omega(\cdot), B(v(\cdot) + \omega(\cdot))) \in C_0(\mathbb{R}, X) \quad \text{as } \beta(\cdot) \in C_0(\mathbb{R}, \mathbb{R}^+). \quad (73)$$

Those, together with Lemma 21, yield that Γ is well defined and maps $C_0(\mathbb{R}, X)$ into itself.

On the other hand, in view of (55) and (61) it is not difficult to see that there exists a constant $k_0 > 0$ such that

$$\frac{CML (1 + \|B\|_{\mathbb{L}(X)}) |\omega|^{-1/\alpha} \pi}{\alpha \sin (\pi/\alpha)} k_0 + CM\rho_2 \Phi\left((1 + \|B\|_{\mathbb{L}(X)}) \left(k_0 + \sup_{s\in\mathbb{R}} \|v(s)\|\right)\right) \le k_0. \quad (74)$$

This enables us to conclude that, for any $t \in \mathbb{R}$ and $\omega_1(t), \omega_2(t) \in \Omega_{k_0}$,

$$\begin{aligned}\|(\Gamma^1 \omega_1)(t) + (\Gamma^2 \omega_2)(t)\| &\le \left\|\int_{-\infty}^{t} S_{\alpha}(t-s) \cdot [F_1(s, v(s) + \omega_1(s), B(v(s) + \omega_1(s))) - F_1(s, v(s), Bv(s))] ds\right\| + \left\|\int_{-\infty}^{t} S_{\alpha}(t-s) \cdot F_2(s, v(s) + \omega_2(s), B(v(s) + \omega_2(s))) ds\right\| \\ &\le \int_{-\infty}^{t} \|S_{\alpha}(t-s)\| \|F_1(s, v(s) + \omega_1(s), B(v(s) + \omega_1(s))) - F_1(s, v(s), Bv(s))\| ds + \int_{-\infty}^{t} \|S_{\alpha}(t-s)\| \cdot \|F_2(s, v(s) + \omega_2(s), B(v(s) + \omega_2(s)))\| ds \\ &\le CM \int_{-\infty}^{t} \frac{1}{1 + |\omega| (t-s)^{\alpha}} [\|\omega_1(s)\|\end{aligned}$$

$$
\begin{aligned}
&+ \|B\omega_1(s)\|\big] ds \\
&+ CM \int_{-\infty}^{t} \frac{\beta(s)}{1+|\omega|(t-s)^{\alpha}} \Phi\big(\|\omega_2(s)\| + \|B\omega_2(s)\| \\
&+ \|v(s)\| + \|Bv(s)\|\big) ds \le CML\left(1 + \|B\|_{\mathbb{L}(X)}\right) \\
&\cdot \|\omega_1\|_{\infty} \int_{-\infty}^{t} \frac{1}{1+|\omega|(t-s)^{\alpha}} ds + CM\sigma(t)\Phi\big((1 \\
&+ \|B\|_{\mathbb{L}(X)}\big)\left(\|\omega_2\|_{\infty} + \|v(s)\|_{\infty}\right)\big) \\
&= \frac{CML|\omega|^{-1/\alpha}\pi\left(1+\|B\|_{\mathbb{L}(X)}\right)}{\alpha\sin(\pi/\alpha)} \|\omega\|_{\infty} \\
&+ CM\rho_2\Phi\left(\left(1+\|B\|_{\mathbb{L}(X)}\right)\left(\|\omega_2\|_{\infty} + \|v(s)\|_{\infty}\right)\right) \\
&\le \frac{CML\left(1+\|B\|_{\mathbb{L}(X)}\right)|\omega|^{-1/\alpha}\pi}{\alpha\sin(\pi/\alpha)} k_0 + CM\rho_2\Phi\big((1 \\
&+ \|B\|_{\mathbb{L}(X)}\big)\left(k_0 + \|v(s)\|_{\infty}\right)\big) \le k_0,
\end{aligned}
\tag{75}
$$

which implies that $(\Gamma^1\omega_1)(t) + (\Gamma^2\omega_2)(t) \in \Omega_{k_0}$. Thus Γ maps Ω_{k_0} into itself.

Step 3. Show that Γ^1 is a contraction on Ω_{k_0}.

In fact, for any $\omega_1(t), \omega_2(t) \in \Omega_{k_0}$ and $t \in \mathbb{R}$, from (54) it follows that

$$
\begin{aligned}
&\big\|\big[F_1(s, v(s) + \omega_1(s), B(v(s) + \omega_1(s))) \\
&\quad - F_1(s, v(s), Bv(s))\big] \\
&\quad - \big[F_1(s, v(s) + \omega_2(s), B(v(s) + \omega_2(s))) \\
&\quad - F_1(s, v(s), Bv(s))\big]\big\| \le L\big[\|\omega_1(s) - \omega_2(s)\| \\
&\quad + \|B\omega_1(s) - B\omega_2(s)\|\big] \le L\left(1 + \|B\|_{\mathbb{L}(X)}\right)\|\omega_1(s) \\
&\quad - \omega_2(s)\|.
\end{aligned}
\tag{76}
$$

Thus

$$
\begin{aligned}
&\left\|\left(\Gamma^1\omega_1\right)(t) - \left(\Gamma^1\omega_2\right)(t)\right\| = \bigg\|\int_{-\infty}^{t} S_{\alpha}(t-s)\big[\left(F_1(s, v(s) + \omega_1(s), B(v(s) + \omega_1(s))) - F_1(s, v(s), Bv(s))\right) \\
&- \left(F_1(s, v(s) + \omega_2(s), B(v(s) + \omega_2(s))) - F_1(s, v(s), Bv(s))\right)\big] ds\bigg\| \le L\int_{-\infty}^{t} \|S_{\alpha}(t-s)\|\left(1 + \|B\|_{\mathbb{L}(X)}\right)\|\omega_1(s) \\
&- \omega_2(s)\| ds \le CML\left(1 + \|B\|_{\mathbb{L}(X)}\right)\|\omega_1 - \omega_2\|_{\infty} \int_{-\infty}^{t} \frac{1}{1+|\omega|(t-s)^{\alpha}} ds = \frac{CML\left(1+\|B\|_{\mathbb{L}(X)}\right)|\omega|^{-1/\alpha}\pi}{\alpha\sin(\pi/\alpha)} \|\omega_1 \\
&- \omega_2\|_{\infty},
\end{aligned}
\tag{77}
$$

which implies that

$$
\begin{aligned}
&\left\|\left(\Gamma^1\omega_1\right)(t) - \left(\Gamma^1\omega_2\right)(t)\right\|_{\infty} \\
&\quad \le \frac{CML\left(1+\|B\|_{\mathbb{L}(X)}\right)|\omega|^{-1/\alpha}\pi}{\alpha\sin(\pi/\alpha)} \|\omega_1 - \omega_2\|_{\infty}.
\end{aligned}
\tag{78}
$$

Thus, in view of (61), one obtains the conclusion.

Step 4. Show that Γ^2 is completely continuous on Ω_{k_0}.

Given $\varepsilon > 0$. Let $\{\omega_k\}_{k=1}^{+\infty} \subset \Omega_{k_0}$ with $\omega_k \longrightarrow \omega_0$ in $C_0(\mathbb{R}, X)$ as $k \longrightarrow +\infty$. Since $\sigma(t) \in C_0(\mathbb{R}, \mathbb{R}^+)$, one may choose a $t_1 > 0$ big enough such that, for all $t \ge t_1$,

$$
\Phi\left(\left(1 + \|B\|_{\mathbb{L}(X)}\right)\left(k_0 + \|v\|_{\infty}\right)\right)\sigma(t) < \frac{\varepsilon}{3CM}. \tag{79}
$$

Also, in view of (H_1), we have

$$
\begin{aligned}
&F_2(s, v(s) + \omega_k(s), B(v(s) + \omega_k(s))) \\
&\quad \longrightarrow F_2(s, v(s) + \omega_0(s), B(v(s) + \omega_0(s)))
\end{aligned}
\tag{80}
$$

for all $s \in (-\infty, t_1]$ as $k \longrightarrow +\infty$ and

$$
\begin{aligned}
&\|F_2(\cdot, v(\cdot) + \omega_k(\cdot), B(v(\cdot) + \omega_k(\cdot))) \\
&\quad - F_2(\cdot, v(\cdot) + \omega_0(\cdot), B(v(\cdot) + \omega_0(\cdot)))\| \\
&\quad \le 2\Phi\left(\left(1 + \|B\|_{\mathbb{L}(X)}\right)\left(k_0 + \|v\|_{\infty}\right)\right)\beta(\cdot) \\
&\quad \in L^1(-\infty, t_1].
\end{aligned}
\tag{81}
$$

Hence, by the Lebesgue dominated convergence theorem we deduce that there exists an $N > 0$ such that

$$
\begin{aligned}
&CM\int_{-\infty}^{t_1} \frac{1}{1+|\omega|(t-s)^{\alpha}} \|F_2(s, v(s) \\
&\quad + \omega_k(s), B(v(s) + \omega_k(s))) - F_2(s, v(s) \\
&\quad + \omega_0(s), B(v(s) + \omega_0(s)))\| ds \le \frac{\varepsilon}{3}
\end{aligned}
\tag{82}
$$

whenever $k \geq N$. Thus

$$\begin{aligned}\left\|\left(\Gamma^2\omega_k\right)(t)-\left(\Gamma^2\omega_0\right)(t)\right\| &= \left\|\int_{-\infty}^{t} S_\alpha(t-s)F_2(s,v(s)+\omega_k(s),B(v(s)+\omega_k(s)))\,ds - \int_{-\infty}^{t} S_\alpha(t-s)F_2(s,v(s)+\omega_0(s),B(v(s)+\omega_0(s)))\,ds\right\| \\ &\leq CM\int_{-\infty}^{t_1}\frac{1}{1+|\omega|(t-s)^\alpha}\|F_2(s,v(s)+\omega_k(s),B(v(s)+\omega_k(s)))-F_2(s,v(s)+\omega_0(s),B(v(s)+\omega_0(s)))\|\,ds \\ &\quad + 2CM\Phi\left(\left(1+\|B\|_{\mathbb{L}(X)}\right)\left(k_0+\|v\|_\infty\right)\right)\int_{t_1}^{\max\{t,t_1\}}\frac{\beta(s)}{1+|\omega|(t-s)^\alpha}\,ds \\ &\leq CM\int_{-\infty}^{t_1}\frac{1}{1+|\omega|(t-s)^\alpha}\|F_2(s,v(s)+\omega_k(s),B(v(s)+\omega_k(s)))-F_2(s,v(s)+\omega_0(s),B(v(s)+\omega_0(s)))\|\,ds \\ &\quad + 2CM\Phi\left(\left(1+\|B\|_{\mathbb{L}(X)}\right)\left(k_0+\|v\|_\infty\right)\right)\sigma(t) \leq \frac{\varepsilon}{3}+\frac{2\varepsilon}{3}=\varepsilon\end{aligned} \tag{83}$$

whenever $k \geq N$. Accordingly, Γ^2 is continuous on Ω_{k_0}.

In the sequel, we consider the compactness of Γ^2.

Set $B_r(X)$ for the closed ball with center at 0 and radius r in X, $V = \Gamma^2(\Omega_{k_0})$, and $z(t) = \Gamma^2(u(t))$ for $u(t) \in \Omega_{k_0}$. First, for all $\omega(t) \in \Omega_{k_0}$ and $t \in \mathbb{R}$,

$$\begin{aligned}\left\|\left(\Gamma^2\omega\right)(t)\right\| &= \left\|\int_{-\infty}^{t} S_\alpha(t-s)\cdot F_2(s,v(s)+\omega(s),B(v(s)+\omega(s)))\,ds\right\| \\ &\leq CM\sigma(t)\,\Phi\left(\left(1+\|B\|_{\mathbb{L}(X)}\right)\left(k_0+\|v\|_\infty\right)\right),\end{aligned} \tag{84}$$

and in view of $\sigma(t) \in C_0(\mathbb{R},\mathbb{R}^+)$, which follows from Lemma 23, one concludes that

$$\lim_{|t|\longrightarrow+\infty}\left(\Gamma^2\omega\right)(t)=0 \quad \text{uniformly for } \omega(t)\in\Omega_{k_0}. \tag{85}$$

As

$$\begin{aligned}\left\|\left(\Gamma^2\omega\right)(t)\right\| &= \left\|\int_{-\infty}^{t} S_\alpha(t-s)F_2(s,v(s)+\omega(s),B(v(s)+\omega(s)))\,ds\right\| = \left\|\int_{0}^{+\infty} S_\alpha(\tau)\cdot F_2(t-\tau,v(t-\tau)+\omega(t-\tau),B(v(t-\tau)+\omega(t-\tau)))\,d\tau\right\|.\end{aligned} \tag{86}$$

Hence, given $\varepsilon_0 > 0$, one can choose a $\xi > 0$ such that

$$\left\|\int_{\xi}^{+\infty} S_\alpha(\tau)F_2(t-\tau,v(t-\tau)+\omega(t-\tau),B(v(t-\tau)+\omega(t-\tau)))\,d\tau\right\| < \varepsilon_0. \tag{87}$$

Thus we get

$$z(t)\in\overline{\xi c\left(\left\{S_\alpha(\tau)F_2(\lambda,v(\lambda)+\omega(\lambda),B(v(\lambda)+\omega(\lambda))):0\leq\tau\leq\xi,\ t-\xi\leq\lambda\leq\xi,\ \|\omega\|_\infty\leq r\right\}\right)}+B_{\varepsilon_0}(X), \tag{88}$$

where $c(K)$ denotes the convex hull of K. Using that $S_\alpha(\cdot)$ is strongly continuous, we infer that

$$K=\left\{S_\alpha(\tau)F_2(\lambda,v(\lambda)+\omega(\lambda),B(v(\lambda)+\omega(\lambda))):0\leq\tau\leq\xi,\ t-\xi\leq\lambda\leq\xi,\ \|\omega\|_\infty\leq r\right\} \tag{89}$$

is a relatively compact set and $V \subset \overline{\xi c(K)} + B_{\varepsilon_0}(X)$, which implies that V is a relatively compact subset of X.

Next, we verify the equicontinuity of the set $\{(\Gamma^2\omega)(t) : \omega(t) \in \Omega_{k_0}\}$.

Let $k > 0$ be small enough and $t_1, t_2 \in \mathbb{R}$ and $\omega(t) \in \Omega_{k_0}$. Then by (55) we have

$$\begin{aligned}\left\|\left(\Gamma^2\omega\right)(t_2)-\left(\Gamma^2\omega\right)(t_1)\right\| &\leq \int_{t_1}^{t_2}\|S_\alpha(t_2-s)\cdot F_2(s,v(s)+\omega(s),B(v(s)+\omega(s)))\|\,ds \\ &\quad+\int_{-\infty}^{t_1-k}\|[S_\alpha(t_2-s)-S_\alpha(t_1-s)]\cdot F_2(s,v(s)+\omega(s),B(v(s)+\omega(s)))\|\,ds \\ &\quad+\int_{t_1-k}^{t_1}\|[S_\alpha(t_2-s)-S_\alpha(t_1-s)]\cdot F_2(s,v(s)+\omega(s),B(v(s)+\omega(s)))\|\,ds \\ &\leq CM\Phi\left(\left(1+\|B\|_{\mathbb{L}(X)}\right)\left(k_0+\|v\|_\infty\right)\right)\cdot\int_{t_1}^{t_2}\frac{\beta(s)}{1+|\omega|(t_2-s)^\alpha}\,ds \\ &\quad+\Phi\left(\left(1+\|B\|_{\mathbb{L}(X)}\right)\left(k_0+\|v\|_\infty\right)\right)\sup_{s\in[-\infty,t_1-k]}\|S_\alpha(t_2-s)-S_\alpha(t_1-s)\|\cdot\int_{-\infty}^{t_1-k}\beta(s)\,ds \\ &\quad+CM\Phi\left(\left(1+\|B\|_{\mathbb{L}(X)}\right)\left(k_0\right.\right.\end{aligned}$$

$$+ \|v\|_\infty)) \int_{t_1-k}^{t_1} \left(\frac{\beta(s)}{1+|\omega|(t_2-s)^\alpha} + \frac{\beta(s)}{1+|\omega|(t_1-s)^\alpha} \right) ds \longrightarrow 0$$
$$\text{as } t_2 - t_1 \longrightarrow 0,\ k \longrightarrow 0, \tag{90}$$

which implies the equicontinuity of the set $\{(\Gamma^2\omega)(t) : \omega(t) \in \Omega_{k_0}\}$.

Now an application of Lemma 18 justifies the compactness of Γ^2.

Step 5. Show that (39) has at least one asymptotically almost automorphic mild solution.

Firstly, the complete continuity of Γ^2, together with the results of Steps 2 and 3 as well as Lemma 19, yields that Γ has at least one fixed point $\omega(t) \in \Omega_{k_0}$; furthermore $\omega(t) \in C_0(\mathbb{R}, X)$.

Then, consider the following coupled system of integral equations:

$$\begin{aligned} v(t) &= \int_{-\infty}^{t} S_\alpha(t-s) F_1(s, v(s), Bv(s))\, ds, \quad t \in \mathbb{R}, \\ \omega(t) &= \int_{-\infty}^{t} S_\alpha(t-s) \cdot [F_1(s, v(s)+\omega(s), B(v(s)+\omega(s))) - F_1(s, v(s), Bv(s))]\, ds + \int_{-\infty}^{t} S_\alpha(t-s) \\ &\quad \cdot F_2(s, v(s)+\omega(s), B(v(s)+\omega(s)))\, ds, \quad t \in \mathbb{R}. \end{aligned} \tag{91}$$

From the result of Step 1, together with the above fixed point $\omega(t) \in C_0(\mathbb{R}, X)$, it follows that

$$(v(t), \omega(t)) \in AA(\mathbb{R}, X) \times C_0(\mathbb{R}, X) \tag{92}$$

is a solution to system (91). Thus

$$x(t) := v(t) + \omega(t) \in AAA(\mathbb{R}, X) \tag{93}$$

and it is a solution to the integral equation

$$x(t) = \int_{-\infty}^{t} S_\alpha(t-s) F(s, x(s), Bx(s))\, ds, \quad t \in \mathbb{R}; \tag{94}$$

that is, $x(t)$ is an asymptotically almost automorphic mild solution to (39). □

Taking $A = -\rho^\alpha I$ with $\rho > 0$ in (39), the above theorem gives the following corollary.

Corollary 25. *Let $F : \mathbb{R} \times X \times X \longrightarrow X$ satisfy (H_1) and (H_2). Put $\rho_2 := \sup_{t\in\mathbb{R}} \sigma(t)$. Then (39) admits at least one asymptotically almost automorphic mild solution whenever*

$$\frac{CL(1+\|B\|_{\mathbb{L}(X)})\rho\pi}{\alpha \sin(\pi/\alpha)} + C(1+\|B\|_{\mathbb{L}(X)})\rho_1\rho_2 < 1. \tag{95}$$

Remark 26. It is interesting to note that the function $\alpha \longrightarrow \alpha\sin(\pi/\alpha)/\rho\pi$ is increasing from 0 to $2/\rho\pi$ in the interval $1 < \alpha < 2$. Therefore, with respect to condition (61), the class of admissible terms $F_1(t, x(t), Bx(t))$ is the best in the case $\alpha = 2$ and the worst in the case $\alpha = 1$.

Theorem 24 can be extended to the case of $F_1(t, x, y)$ being locally Lipschitz continuous with respect to x and y, where we have the following result.

(H_1') $F(t, x, y) = F_1(t, x, y) + F_2(t, x, y) \in AAA(\mathbb{R} \times X \times X, X)$ with

$$\begin{aligned} &F_1(t, x, y) \in AA(\mathbb{R} \times X \times X, X), \\ &F_2(t, x, y) \in C_0(\mathbb{R} \times X \times X, X) \end{aligned} \tag{96}$$

and for all $x_1, x_2, y_1, y_2 \in X, t \in \mathbb{R}$,

$$\begin{aligned} &\|F_1(t, x_1, x_2) - F_1(t, y_1, y_2)\| \\ &\quad \le L(t)(\|x_1 - y_1\| + \|x_2 - y_2\|), \end{aligned} \tag{97}$$

where $L(t)$ is a function on $\mathbb{R}$.

Theorem 27. *Assume that A is sectorial of type $\omega < 0$. Let $F : \mathbb{R} \times X \times X \longrightarrow X$ satisfy the hypotheses (H_1') and (H_2) with $L(t) \in BC(\mathbb{R}, \mathbb{R}^+)$. Put $\rho_2 := \sup_{t\in\mathbb{R}} \sigma(t)$. Let $\|L\| = \sup_{t\in\mathbb{R}} \int_t^{t+1} L(s)ds$. Then (39) has at least one asymptotically almost automorphic mild solution provided that*

$$\frac{CML\|L\||\omega|^{-1/\alpha}\pi(1+\|B\|_{\mathbb{L}(X)})}{\alpha\sin(\pi/\alpha)} + CM\rho_1\rho_2(1+\|B\|_{\mathbb{L}(X)}) < 1. \tag{98}$$

Proof. The proof is divided into the following five steps.

Step 1. Define a mapping Λ on $AA(\mathbb{R}, X)$ by (62) and prove that Λ has a unique fixed point $v(t) \in AA(\mathbb{R}, X)$.

Firstly, similar to the proof in Step 1 of Theorem 24, we can prove that $(\Lambda v)(t)$ exists. Moreover from $F_1(t, x, y) \in AA(\mathbb{R} \times X \times X, X)$ satisfying (97), together with Lemma 6 and Remark 8, it follows that

$$F_1(\cdot, v(\cdot), Bv(\cdot)) \in AA(\mathbb{R}, X) \quad \text{for every } v(\cdot) \in AA(\mathbb{R}, X). \tag{99}$$

This, together with Lemma 21, implies that Λ is well defined and maps $AP(\mathbb{R}, X)$ into itself.

In the sequel, we verify that Λ is continuous.

Let $v_n(t), v(t)$ be in $AA(\mathbb{R}, X)$ with $v_n(t) \longrightarrow v(t)$ as $n \longrightarrow \infty$; then one has

$$\begin{aligned} \|[\Lambda v_n](t) - [\Lambda v](t)\| &= \left\| \int_{-\infty}^{t} S_\alpha(t-s) \cdot [F_1(s, v_n(s), Bv_n(s)) - F_1(s, v(s), Bv(s))]\, ds \right\| \le \int_{-\infty}^{t} L(s) \|S_\alpha(t \\ &\quad - s)\| [\|v_n(s) - v(s)\| + \|Bv_n(s) - Bv(s)\|]\, ds \end{aligned}$$

$$\leq CM\int_{-\infty}^{t}\frac{L(s)}{1+|\omega|(t-s)^{\alpha}}\left(1+\|B\|_{\mathbb{L}(X)}\right)\|v_n(s)-v(s)\|\,ds \leq CM\left(1+\|B\|_{\mathbb{L}(X)}\right)\cdot\left(\sum_{m=0}^{+\infty}\int_{t-(m+1)}^{t-m}\frac{L(s)}{1+|\omega|(t-s)^{\alpha}}ds\right)\|v_n-v\|_{\infty}$$
$$\leq CM\left(1+\|B\|_{\mathbb{L}(X)}\right)\cdot\left(\sum_{m=0}^{+\infty}\frac{1}{1+|\omega|m^{\alpha}}\int_{t-(m+1)}^{t-m}L(s)\,ds\right)\|v_n-v\|_{\infty}$$
$$\leq \frac{CM\|L\||\omega|^{-1/\alpha}\pi\left(1+\|B\|_{\mathbb{L}(X)}\right)}{\alpha\sin(\pi/\alpha)}\|v_n-v\|_{\infty}. \quad (100)$$

Therefore, as $n \longrightarrow \infty$ and $\Lambda v_n \longrightarrow \Lambda v$, hence Λ is continuous.

Next, we prove that Λ is a contraction on $AA(\mathbb{R}, X)$ and has a unique fixed point $v(t) \in AA(\mathbb{R}, X)$.

In fact, for $v_1(t), v_2(t)$ in $AA(\mathbb{R}, X)$, similar to the above proof of the continuity of Λ, one has

$$\|(\Lambda v_1)(t)-(\Lambda v_2)(t)\| \leq \frac{CM\|L\||\omega|^{-1/\alpha}\pi\left(1+\|B\|_{\mathbb{L}(X)}\right)}{\alpha\sin(\pi/\alpha)}\|v_1-v_2\|_{\infty}, \quad (101)$$

which implies that

$$\|(\Lambda v_1)(t)-(\Lambda v_2)(t)\|_{\infty} \leq \frac{CM\|L\||\omega|^{-1/\alpha}\pi\left(1+\|B\|_{\mathbb{L}(X)}\right)}{\alpha\sin(\pi/\alpha)}\|v_1-v_2\|_{\infty}. \quad (102)$$

Hence, by (98), together with the contraction principle, Λ has a unique fixed point $v(t) \in AA(\mathbb{R}, X)$.

Step 2. Set

$$\Omega_r := \{\omega(t) \in C_0(\mathbb{R}, X) : \|\omega(t)\| \leq r\}. \quad (103)$$

For the above $v(t)$, define $\Gamma := \Gamma^1 + \Gamma^2$ on $C_0(\mathbb{R}, X)$ as (69) and prove that Γ maps Ω_{k_0} into itself, where k_0 is a given constant.

Firstly, from (97) it follows that, for all $s \in \mathbb{R}$, $\omega(s) \in X$,

$$\|F_1(s, v(s)+\omega(s), B(v(s)+\omega(s))) - F_1(s, v(s), Bv(s))\| \leq L(s)[\|\omega(s)\| + \|B\omega(s)\|] \leq L(s)\left(1+\|B\|_{\mathbb{L}(X)}\right)\|\omega(s)\|, \quad (104)$$

which together with $L(s) \in BC(\mathbb{R}, \mathbb{R}^+)$ implies that

$$F_1(\cdot, v(\cdot)+\omega(\cdot), B(v(\cdot)+\omega(\cdot))) - F_1(\cdot, v(\cdot), Bv(\cdot)) \in C_0(\mathbb{R}, X) \quad \text{for every } \omega(\cdot) \in C_0(\mathbb{R}, X). \quad (105)$$

According to (55), one has

$$\|F_2(s, v(s)+\omega(s), B(v(s)+\omega(s)))\| \leq \beta(s)\cdot\Phi\left(\|\omega(s)+B\omega(s)\| + \sup_{s\in\mathbb{R}}\|v(s)+Bv(s)\|\right) \leq \beta(s)\Phi\left(\left(1+\|B\|_{\mathbb{L}(X)}\right)\|\omega(s)\| + \left(1+\|B\|_{\mathbb{L}(X)}\right)\sup_{s\in\mathbb{R}}\|v(s)\|\right) \leq \beta(s)\cdot\Phi\left(\left(1+\|B\|_{\mathbb{L}(X)}\right)\left[\|\omega(s)\| + \sup_{s\in\mathbb{R}}\|v(s)\|\right]\right) \quad (106)$$

for all $s \in \mathbb{R}$ and $\omega(s) \in X$ with $\|\omega(s)\| \leq r$; then

$$F_2(\cdot, v(\cdot)+\omega(\cdot), B(v(\cdot)+\omega(\cdot))) \in C_0(\mathbb{R}, X) \quad \text{as } \beta(\cdot) \in C_0(\mathbb{R}, \mathbb{R}^+). \quad (107)$$

Those, together with Lemma 21, yield that Γ is well defined and maps $C_0(\mathbb{R}, X)$ into itself.

On the other hand, in view of (55) and (98) it is not difficult to see that there exists a constant $k_0 > 0$ such that

$$\frac{CM\|L\||\omega|^{-1/\alpha}\pi\left(1+\|B\|_{\mathbb{L}(X)}\right)}{\alpha\sin(\pi/\alpha)}k_0 + CM\rho_2\Phi\left(\left(1+\|B\|_{\mathbb{L}(X)}\right)\left(k_0 + \sup_{s\in\mathbb{R}}\|v(s)\|\right)\right) \leq k_0. \quad (108)$$

This enables us to conclude that, for any $t \in \mathbb{R}$ and $\omega_1(t)$, $\omega_2(t) \in \Omega_{k_0}$,

$$\left\|\left(\Gamma^1\omega_1\right)(t) + \left(\Gamma^2\omega_2\right)(t)\right\| \leq \left\|\int_{-\infty}^{t}S_{\alpha}(t-s)\cdot[F_1(s, v(s)+\omega_1(s), B(v(s)+\omega_1(s))) - F_1(s, v(s), Bv(s))]\,ds\right\| + \left\|\int_{-\infty}^{t}S_{\alpha}(t-s)\cdot F_2(s, v(s)+\omega_2(s), B(v(s)+\omega_2(s)))\,ds\right\|$$
$$\leq \int_{-\infty}^{t}L(s)\|S_{\alpha}(t-s)\|[\|\omega_1(s)\| + \|B\omega_1(s)\|]\,ds + CM\int_{-\infty}^{t}\frac{\beta(s)}{1+|\omega|(t-s)^{\alpha}}\Phi\cdot(\|\omega_2(s)\| + \|B\omega_2(s)\| + \|v(s)\| + \|Bv(s)\|)\,ds$$
$$\leq CM\int_{-\infty}^{t}\frac{L(s)}{1+|\omega|(t-s)^{\alpha}}\left(1+\|B\|_{\mathbb{L}(X)}\right)\cdot\|\omega_1(s)\|\,ds + CM\int_{-\infty}^{t}\frac{\beta(s)}{1+|\omega|(t-s)^{\alpha}}\Phi((1 + \|B\|_{\mathbb{L}(X)})(\|\omega_2(s)\| + \|v(s)\|)\,ds \leq CM(1$$

$$+\|B\|_{\mathbb{L}(X)})\|\omega_1\|_\infty\int_{-\infty}^{t}\frac{L(s)}{1+|\omega|(t-s)^\alpha}ds$$
$$+CM\sigma(t)\Phi\Bigg((1+\|B\|_{\mathbb{L}(X)})\Bigg(\|\omega_2\|_\infty$$
$$+\sup_{s\in\mathbb{R}}\|v(s)\|\Bigg)\Bigg)$$
$$\le CM\left(\sum_{m=0}^{+\infty}\int_{t-(m+1)}^{t-m}\frac{L(s)}{1+|\omega|(t-s)^\alpha}ds\right)(1$$
$$+\|B\|_{\mathbb{L}(X)})\|\omega_1\|_\infty+CM\rho_2\Phi\Bigg((1+\|B\|_{\mathbb{L}(X)})$$
$$\cdot\Bigg(\|\omega_2\|_\infty+\sup_{s\in\mathbb{R}}\|v(s)\|\Bigg)\Bigg)$$
$$\le CM\left(\sum_{m=0}^{+\infty}\frac{1}{1+|\omega|m^\alpha}\int_{t-(m+1)}^{t-m}L(s)ds\right)(1$$
$$+\|B\|_{\mathbb{L}(X)})\|\omega_1\|_\infty+CM\rho_2\Phi\Bigg((1+\|B\|_{\mathbb{L}(X)})\Bigg(k_0$$
$$+\sup_{s\in\mathbb{R}}\|v(s)\|\Bigg)\Bigg)\le CM\left(\sum_{m=0}^{+\infty}\frac{1}{1+|\omega|m^\alpha}\right)\|L\|(1$$
$$+\|B\|_{\mathbb{L}(X)})k_0+CM\rho_2\Phi\Bigg((1+\|B\|_{\mathbb{L}(X)})\Bigg(k_0$$
$$+\sup_{s\in\mathbb{R}}\|v(s)\|\Bigg)\Bigg)$$
$$=\frac{CM\|L\||\omega|^{-1/\alpha}\pi(1+\|B\|_{\mathbb{L}(X)})}{\alpha\sin(\pi/\alpha)}k_0$$
$$+CM\rho_2\Phi\left((1+\|B\|_{\mathbb{L}(X)})\left(k_0+\sup_{s\in\mathbb{R}}\|v(s)\|\right)\right)$$
$$\le k_0, \tag{109}$$

which implies that $(\Gamma^1\omega_1)(t)+(\Gamma^2\omega_2)(t)\in\Omega_{k_0}$. Thus Γ maps Ω_{k_0} into itself.

Step 3. Show that Γ^1 is a contraction on Ω_{k_0}.

In fact, for any $\omega_1(t),\omega_2(t)\in\Omega_{k_0}$ and $t\in\mathbb{R}$, from (97) it follows that

$$\begin{aligned}&\|[F_1(s,v(s)+\omega_1(s),B(v(s)+\omega_1(s)))\\&-F_1(s,v(s),Bv(s))]\\&-[F_1(s,v(s)+\omega_2(s),B(v(s)+\omega_2(s)))\\&-F_1(s,v(s),Bv(s))]\|\le L(s)[\|\omega_1(s)-\omega_2(s)\|\\&+\|B\omega_1(s)-B\omega_2(s)\|]\le L(s)(1+\|B\|_{\mathbb{L}(X)})\|\omega_1(s)\\&-\omega_2(s)\|.\end{aligned}\tag{110}$$

Thus

$$\begin{aligned}&\|(\Gamma^1\omega_1)(t)-(\Gamma^1\omega_2)(t)\|=\Bigg\|\int_{-\infty}^{t}S_\alpha(t-s)[(F_1(s,v(s)+\omega_1(s),B(v(s)+\omega_1(s)))-F_1(s,v(s),Bv(s)))\\&-(F_1(s,v(s)+\omega_2(s),B(v(s)+\omega_2(s)))-F_1(s,v(s),Bv(s)))]ds\Bigg\|\le\int_{-\infty}^{t}L(s)\|S_\alpha(t-s)\|(1+\|B\|_{\mathbb{L}(X)})\|\omega_1(s)\\&-\omega_2(s)\|ds\le CM\int_{-\infty}^{t}\frac{L(s)}{1+|\omega|(t-s)^\alpha}(1+\|B\|_{\mathbb{L}(X)})\|\omega_1(s)-\omega_2(s)\|ds\\&\le CM\left(\sum_{m=0}^{+\infty}\int_{t-(m+1)}^{t-m}\frac{L(s)}{1+|\omega|(t-s)^\alpha}ds\right)(1+\|B\|_{\mathbb{L}(X)})\|\omega_1-\omega_2\|_\infty\le CM\left(\sum_{m=0}^{+\infty}\frac{1}{1+|\omega|m^\alpha}\int_{t-(m+1)}^{t-m}L(s)ds\right)(1\\&+\|B\|_{\mathbb{L}(X)})\|\omega_1-\omega_2\|_\infty\le CM\left(\sum_{m=0}^{+\infty}\frac{1}{1+|\omega|m^\alpha}\right)\|L\|(1+\|B\|_{\mathbb{L}(X)})\|\omega_1-\omega_2\|_\infty\\&=\frac{CM\|L\||\omega|^{-1/\alpha}\pi(1+\|B\|_{\mathbb{L}(X)})}{\alpha\sin(\pi/\alpha)}\|\omega_1-\omega_2\|_\infty,\end{aligned}\tag{111}$$

which implies that

$$\begin{aligned}&\|(\Gamma^1\omega_1)(t)-(\Gamma^1\omega_2)(t)\|_\infty\\&\le\frac{CM\|L\||\omega|^{-1/\alpha}\pi(1+\|B\|_{\mathbb{L}(X)})}{\alpha\sin(\pi/\alpha)}\|\omega_1-\omega_2\|_\infty.\end{aligned}\tag{112}$$

Thus, in view of (98), one obtains the conclusion.

Step 4. Show that Γ^2 is completely continuous on Ω_{k_0}.

The proof is similar to the proof in Step 4 of Theorem 24.

Step 5. Show that (39) has at least one asymptotically almost automorphic mild solution.

The proof is similar to the proof in Step 5 of Theorem 24. □

Taking $A = -\rho^\alpha I$ with $\rho > 0$ in (39), Theorem 27 gives the following corollary.

Corollary 28. *Let $F : \mathbb{R} \times X \times X \longrightarrow X$ satisfy (H_1') and (H_2) with $L(t) \in BC(\mathbb{R}, \mathbb{R}^+)$. Put $\rho_2 := \sup_{t\in\mathbb{R}} \sigma(t)$. Let $\|L\| = \sup_{t\in\mathbb{R}} \int_t^{t+1} L(s)ds$. Then (39) admits at least one asymptotically almost automorphic mild solution whenever*

$$\frac{C\|L\|\rho\pi\left(1+\|B\|_{\mathbb{L}(X)}\right)}{\alpha\sin(\pi/\alpha)} + C\rho_1\rho_2\left(1+\|B\|_{\mathbb{L}(X)}\right) < 1. \quad (113)$$

Now we consider a more general case of equations introducing a new class of functions $L(t)$. We have the following result.

(H_2') There exists a function $\beta(t) \in C_0(\mathbb{R}, \mathbb{R}^+)$ such that, for all $t \in \mathbb{R}$ and $x, y \in X$,

$$\|F_2(t,x,y)\| \le \beta(t)\left(\|x\| + \|y\|\right). \quad (114)$$

Theorem 29. *Assume that A is sectorial of type $\omega < 0$. Let $F : \mathbb{R} \times X \times X \longrightarrow X$ satisfy the hypotheses (H_1') and (H_2') with $L(t) \in BC(\mathbb{R}, \mathbb{R}^+)$. Moreover the integral $\int_{-\infty}^t \max\{L(s), \beta(s)\}ds$ exists for all $t \in \mathbb{R}$. Then (39) has at least one asymptotically almost automorphic mild solution.*

Proof. The proof is divided into the following five steps.

Step 1. Define a mapping Λ on $AA(\mathbb{R}, X)$ by (62) and prove that Λ has a unique fixed point $v(t) \in AA(\mathbb{R}, X)$.

Firstly, similar to the proof in Step 1 of Theorem 27, we can prove that Λ is well defined and maps $AP(\mathbb{R}, X)$ into itself; moreover Λ is continuous.

Next, we prove that Λ is a contraction on $AA(\mathbb{R}, X)$ and has a unique fixed point $v(t) \in AA(\mathbb{R}, X)$.

In fact, for $v_1(t), v_2(t)$ is in $AA(\mathbb{R}, X)$ and defines a new norm

$$|||x||| := \sup_{t\in\mathbb{R}} \{\mu(t)\|x(t)\|\}, \quad (115)$$

where $\mu(t) := e^{-k\int_{-\infty}^t \max\{L(s),\beta(s)\}ds}$ and k is a fixed positive constant. Let $C_\alpha := \sup_{t\in\mathbb{R}}\|S_\alpha(t)\|$; then we have

$$\begin{aligned}
&\mu(t)\|\Lambda v_1(t) - \Lambda v_2(t)\| = \mu(t)\left\|\int_{-\infty}^t S_\alpha(t-\sigma)\right.\\
&\cdot [F_1(\sigma, v_1(\sigma), Bv_1(\sigma))\\
&\left. - F_1(\sigma, v_2(\sigma), Bv_2(\sigma))]\, d\sigma\right\|\\
&\le C_\alpha \int_{-\infty}^t \mu(t) L(\sigma)\left[\|v_1(\sigma) - v_2(\sigma)\| + \|Bv_1(\sigma)\right.\\
&\left. - Bv_2(\sigma)\|\right] d\sigma = C_\alpha \int_{-\infty}^t \mu(t)\mu(\sigma) L(\sigma)\\
&\cdot \mu(\sigma)^{-1}\left(1+\|B\|_{\mathbb{L}(X)}\right)\|v_1(\sigma) - v_2(\sigma)\|\, d\sigma\\
&\le C_\alpha\left(1+\|B\|_{\mathbb{L}(X)}\right)|||v_1 - v_2||| \int_{-\infty}^t \mu(t)\mu(\sigma)^{-1}\\
&\cdot L(\sigma)\, d\sigma = \frac{C_\alpha\left(1+\|B\|_{\mathbb{L}(X)}\right)}{k}|||v_1 - v_2|||\\
&\cdot \int_{-\infty}^t k e^{-k\int_\sigma^t \max\{L(\tau),\beta(\tau)\}d\tau} L(\sigma)\, d\sigma\\
&\le \frac{C_\alpha\left(1+\|B\|_{\mathbb{L}(X)}\right)}{k}|||v_1 - v_2|||\\
&\cdot \int_{-\infty}^t k e^{-k\int_\sigma^t L(\tau)d\tau} L(\sigma)\, d\sigma\\
&= \frac{C_\alpha\left(1+\|B\|_{\mathbb{L}(X)}\right)}{k}|||v_1 - v_2|||\\
&\cdot \int_{-\infty}^t \frac{d}{d\sigma}\left(e^{k\int_t^\sigma L(\tau)d\tau}\right) d\sigma\\
&= \frac{C_\alpha\left(1+\|B\|_{\mathbb{L}(X)}\right)}{k}\left(1 - e^{-k\int_{-\infty}^t L(\tau)d\tau}\right)|||v_1 - v_2|||\\
&\le \frac{C_\alpha\left(1+\|B\|_{\mathbb{L}(X)}\right)}{k}|||v_1 - v_2|||,
\end{aligned} \quad (116)$$

which implies that

$$|||\Lambda x(t) - \Lambda y(t)||| \le \frac{C_\alpha\left(1+\|B\|_{\mathbb{L}(X)}\right)}{k}|||x - y|||. \quad (117)$$

Hence Λ has a unique fixed point $x \in AA(\mathbb{R}, X)$ when k is greater than $C_\alpha(1 + \|B\|_{\mathbb{L}(X)})$.

Step 2. Set $\Theta_r := \{\omega(t) \in C_0(\mathbb{R}, X) : |||\omega(t)||| \le r\}$. For the above $v(t)$, define $\Gamma := \Gamma^1 + \Gamma^2$ on $C_0(\mathbb{R}, X)$ as (69) and prove that Γ maps Θ_{k_0} into itself, where k_0 is a given constant.

Firstly, from (97) it follows that, for all $s \in \mathbb{R}$, $\omega(s) \in X$,

$$\begin{aligned}
&\|F_1(s, v(s) + \omega(s), B(v(s) + \omega(s)))\\
&- F_1(s, v(s), Bv(s))\| \le L(s)\left[\|\omega(s)\| + \|B\omega(s)\|\right]\\
&\le L(s)\left(1+\|B\|_{\mathbb{L}(X)}\right)\|\omega(s)\| + \|B\omega(s)\|,
\end{aligned} \quad (118)$$

which together with $L(s) \in BC(\mathbb{R}, \mathbb{R}^+)$ implies that

$$\begin{aligned}
&F_1(\cdot, v(\cdot) + \omega(\cdot), B(v(\cdot) + \omega(\cdot))) - F_1(\cdot, v(\cdot), Bv(\cdot))\\
&\in C_0(\mathbb{R}, X) \quad \text{for every } \omega(\cdot) \in C_0(\mathbb{R}, X).
\end{aligned} \quad (119)$$

According to (114), one has

$$\begin{aligned}
&\|F_2(s, v(s) + \omega(s), B(v(s) + \omega(s)))\| \le \beta(s)\\
&\cdot\left(\|\omega(s) + B\omega(s)\| + \|v(s) + Bv(s)\|\right) \le \beta(s)\\
&\cdot\left(\left(1+\|B\|_{\mathbb{L}(X)}\right)\|\omega(s)\| + \left(1+\|B\|_{\mathbb{L}(X)}\right)\|v(s)\|\right)\\
&\le \beta(s)\left(\left(1+\|B\|_{\mathbb{L}(X)}\right)\left[\|\omega(s)\| + \|v(s)\|\right]\right)
\end{aligned} \quad (120)$$

for all $s \in \mathbb{R}$ and $\omega(s) \in X$ with $\|\omega(s)\| \le r$; then

$$F_2(\cdot, v(\cdot) + \omega(\cdot), B(v(\cdot) + \omega(\cdot))) \in C_0(\mathbb{R}, X) \quad \text{as } \beta(\cdot) \in C_0(\mathbb{R}, \mathbb{R}^+). \tag{121}$$

Those, together with Lemma 21, yield that Γ is well defined and maps $C_0(\mathbb{R}, X)$ into itself.

On the other hand, it is not difficult to see that there exists a constant $k_0 > 0$ such that

$$\frac{2C_\alpha (1 + \|B\|_{\mathbb{L}(X)})}{k} k_0 + \frac{C_\alpha (1 + \|B\|_{\mathbb{L}(X)})}{k} |||v(s)||| \le k_0, \tag{122}$$

when k is large enough. This enables us to conclude that, for any $t \in \mathbb{R}$ and $\omega_1(t), \omega_2(t) \in \Theta_{k_0}$,

$$\begin{aligned}
&\mu(t) \left\|(\Gamma^1 \omega_1)(t) + (\Gamma^2 \omega_2)(t)\right\| \le \mu(t) \left\| \int_{-\infty}^{t} S_\alpha(t-s) \cdot [F_1(s, v(s) + \omega_1(s), B(v(s) + \omega_1(s))) - F_1(s, v(s), Bv(s))] \, ds \right\| + \mu(t) \left\| \int_{-\infty}^{t} S_\alpha(t \\
&- s) F_2(s, v(s) + \omega_2(s), B(v(s) + \omega_2(s))) \, ds \right\| \le C_\alpha \int_{-\infty}^{t} \mu(t) L(s) (\|\omega_1(s)\| + \|B\omega_1(s)\|) \, ds \\
&+ C_\alpha \int_{-\infty}^{t} \mu(t) \beta(s) (\|\omega_2\| + \|v(s)\| + \|B\omega_2\| + \|Bv(s)\|) \, ds = C_\alpha \int_{-\infty}^{t} \mu(t) \mu(s) L(s) \mu(s)^{-1} \\
&\cdot (1 + \|B\|_{\mathbb{L}(X)}) \|\omega_1(s)\| \, ds + C_\alpha \int_{-\infty}^{t} \mu(t) \mu(s) \cdot \beta(s) \mu(s)^{-1} (1 + \|B\|_{\mathbb{L}(X)}) (\|\omega_2\| + \|v(s)\|) \, ds \\
&\le C_\alpha (1 + \|B\|_{\mathbb{L}(X)}) |||\omega_1||| \int_{-\infty}^{t} \mu(t) \mu(s)^{-1} L(s) \, ds + C_\alpha (1 + \|B\|_{\mathbb{L}(X)}) (|||\omega_2||| + |||v(s)|||) \int_{-\infty}^{t} \mu(t) \\
&\cdot \mu(s)^{-1} \beta(s) \, ds = \frac{C_\alpha (1 + \|B\|_{\mathbb{L}(X)})}{k} |||\omega_1||| \cdot \int_{-\infty}^{t} k e^{-k \int_t^s \max\{L(\tau), \beta(\tau)\} d\tau} L(s) \, ds + C_\alpha (1 \\
&+ \|B\|_{\mathbb{L}(X)}) (|||\omega_2||| + |||v(s)|||) \cdot \int_{-\infty}^{t} k e^{-k \int_t^s \max\{L(\tau), \beta(\tau)\} d\tau} \beta(s) \, ds \\
&\le \frac{C_\alpha (1 + \|B\|_{\mathbb{L}(X)})}{k} |||\omega_1||| \cdot \int_{-\infty}^{t} k e^{-k \int_t^s L(\tau) d\tau} L(s) \, ds + C_\alpha (1 + \|B\|_{\mathbb{L}(X)}) \cdot (|||\omega_2||| + |||v(s)|||) \int_{-\infty}^{t} k e^{-k \int_t^s \beta(\tau) d\tau} \beta(s) \, ds \\
&= \frac{C_\alpha (1 + \|B\|_{\mathbb{L}(X)})}{k} |||\omega_1||| \int_{-\infty}^{t} \frac{d}{ds} \left(e^{k \int_t^s L(\tau) d\tau}\right) ds + C_\alpha (1 + \|B\|_{\mathbb{L}(X)}) (|||\omega_2||| + |||v(s)|||) \cdot \int_{-\infty}^{t} \frac{d}{ds} \left(e^{k \int_t^s \beta(\tau) d\tau}\right) ds \\
&= \frac{C_\alpha (1 + \|B\|_{\mathbb{L}(X)})}{k} \left(1 - e^{-k \int_{-\infty}^{t} L(\tau) d\tau}\right) |||\omega_1||| + \frac{C_\alpha (1 + \|B\|_{\mathbb{L}(X)})}{k} \left(1 - e^{-k \int_{-\infty}^{t} \beta(\tau) d\tau}\right) (|||\omega_2||| \\
&+ |||v(s)|||) \le \frac{C_\alpha (1 + \|B\|_{\mathbb{L}(X)})}{k} |||\omega_1||| + \frac{C_\alpha}{k} (1 + \|B\|_{\mathbb{L}(X)}) (|||\omega_2||| + |||v(s)|||) \le k_0,
\end{aligned} \tag{123}$$

which implies that $(\Gamma^1 \omega_1)(t) + (\Gamma^2 \omega_2)(t) \in \Theta_{k_0}$. Thus Γ maps Θ_{k_0} into itself.

Step 3. Show that Γ^1 is a contraction on Θ_{k_0}.

In fact, for any $\omega_1(t), \omega_2(t) \in \Theta_{k_0}$ and $t \in \mathbb{R}$, from (97) it follows that

$$\begin{aligned}
&\|[F_1(s, v(s) + \omega_1(s), B(v(s) + \omega_1(s))) - F_1(s, v(s), Bv(s))] \\
&- [F_1(s, v(s) + \omega_2(s), B(v(s) + \omega_2(s))) - F_1(s, v(s), Bv(s))]\| \le L(s) [\|\omega_1(s) - \omega_2(s)\| \\
&+ \|B\omega_1(s) - B\omega_2(s)\|] \le L(s) (1 + \|B\|_{\mathbb{L}(X)}) \cdot \|\omega_1(s) - \omega_2(s)\|.
\end{aligned} \tag{124}$$

Thus

$$\begin{aligned}
&\mu(t) \left\|(\Gamma^1 \omega_1)(t) - (\Gamma^1 \omega_2)(t)\right\| = \mu(t) \left\| \int_{-\infty}^{t} S_\alpha(t-s) [(F_1(s, v(s) + \omega_1(s), B(v(s) + \omega_1(s))) - F_1(s, v(s), Bv(s))) \right. \\
&\left. - (F_1(s, v(s) + \omega_2(s), B(v(s) + \omega_2(s))) - F_1(s, v(s), Bv(s)))] \, ds \right\| \le C_\alpha \int_{-\infty}^{t} \mu(t) L(\sigma) (1 + \|B\|_{\mathbb{L}(X)}) \|\omega_1(\sigma)
\end{aligned}$$

$$\begin{aligned}
&-\omega_2(\sigma)\| \mathrm{d}\sigma = C_\alpha \int_{-\infty}^{t} \mu(t)\mu(\sigma)L(\sigma)\mu(\sigma)^{-1}\left(1+\|B\|_{\mathbb{L}(X)}\right)\|\omega_1(\sigma)-\omega_2(\sigma)\| \mathrm{d}\sigma \le C_\alpha\left(1+\|B\|_{\mathbb{L}(X)}\right)\||\omega_1 \\
&-\omega_2\|| \int_{-\infty}^{t} \mu(t)\mu(\sigma)^{-1}L(\sigma)\mathrm{d}\sigma = \frac{C_\alpha\left(1+\|B\|_{\mathbb{L}(X)}\right)}{k}\||\omega_1-\omega_2\|| \int_{-\infty}^{t} ke^{-k\int_\sigma^t \max\{L(\tau),\beta(\tau)\}\mathrm{d}\tau}L(\sigma)\mathrm{d}\sigma \\
&\le \frac{C_\alpha\left(1+\|B\|_{\mathbb{L}(X)}\right)}{k}\||\omega_1-\omega_2\|| \int_{-\infty}^{t} ke^{-k\int_\sigma^t L(\tau)\mathrm{d}\tau}L(\sigma)\mathrm{d}\sigma = \frac{C_\alpha\left(1+\|B\|_{\mathbb{L}(X)}\right)}{k}\||\omega_1-\omega_2\|| \int_{-\infty}^{t} \frac{\mathrm{d}}{\mathrm{d}\sigma}\left(e^{k\int_t^\sigma L(\tau)\mathrm{d}\tau}\right)\mathrm{d}\sigma \\
&= \frac{C_\alpha\left(1+\|B\|_{\mathbb{L}(X)}\right)}{k}\left(1-e^{-k\int_{-\infty}^t L(\tau)\mathrm{d}\tau}\right)\||\omega_1-\omega_2\|| \le \frac{C_\alpha\left(1+\|B\|_{\mathbb{L}(X)}\right)}{k}\||\omega_1-\omega_2\||,
\end{aligned} \tag{125}$$

which implies

$$\begin{aligned}
&\left\|\left|\left(\Gamma^1\omega_1\right)(t)-\left(\Gamma^1\omega_2\right)(t)\right\|\right| \\
&\quad \le \frac{C_\alpha\left(1+\|B\|_{\mathbb{L}(X)}\right)}{k}\||\omega_1-\omega_2\||.
\end{aligned} \tag{126}$$

Thus, when k is greater than $C_\alpha(1+\|B\|_{\mathbb{L}(X)})$, one obtains the conclusion.

Step 4. Show that Γ^2 is completely continuous on Θ_{k_0}.

Given $\varepsilon > 0$. Let $\{\omega_n\}_{n=1}^{+\infty} \subset \Theta_{k_0}$ with $\omega_n \longrightarrow \omega_0$ in Θ_{k_0} as $n \longrightarrow +\infty$. Since $\sigma(t) \in C_0(\mathbb{R},\mathbb{R}^+)$, one may choose a $t_1 > 0$ big enough such that, for all $t \ge t_1$,

$$\left(1+\|B\|_{\mathbb{L}(X)}\right)\left(k_0+\||v\||\right)\sigma(t) < \frac{\varepsilon}{3CM}. \tag{127}$$

Also, in view of (H_1'), we have

$$\begin{aligned}
&F_2(s, v(s)+\omega_k(s), B(v(s)+\omega_k(s))) \\
&\quad \longrightarrow F_2(s, v(s)+\omega_0(s), B(v(s)+\omega_0(s)))
\end{aligned} \tag{128}$$

for all $s \in (-\infty, t_1]$ as $k \longrightarrow +\infty$ and

$$\begin{aligned}
&\mu(\cdot)\|F_2(\cdot, v(\cdot)+\omega_n(\cdot), B(v(\cdot)+\omega_n(\cdot))) \\
&\quad - F_2(\cdot, v(\cdot)+\omega_0(\cdot), Bv(\cdot)+\omega_0(\cdot)))\| \le \mu(\cdot) \\
&\quad \cdot \beta(\cdot)\left(\|\omega_n(\cdot)\| + \|v(\cdot)\| + \|B\omega_n(\cdot)\| + \|Bv(\cdot)\|\right. \\
&\quad \left. + \|\omega_0(\cdot)\| + \|v(\cdot)\| + \|B\omega_0(\cdot)\| + \|Bv(\cdot)\|\right) \le \beta(\cdot) \\
&\quad \cdot \left(\||\omega_n\|| + \||v\|| + \||B\omega_n\|| + \||Bv\|| + \||\omega_0\||\right. \\
&\quad \left. + \||v\|| + \||B\omega_0\|| + \||Bv\||\right) \le \beta(\cdot) \\
&\quad \cdot \left(2\left(1+\|B\|_{\mathbb{L}(X)}\right)\left(k_0+\||v\||\right)\right) \in L^1(-\infty, t_1].
\end{aligned} \tag{129}$$

Hence, by the Lebesgue dominated convergence theorem we deduce that there exists an $N > 0$ such that

$$\begin{aligned}
&CM\int_{-\infty}^{t_1} \frac{1}{1+|\omega|(t-s)^\alpha}\mu(t) \\
&\quad \cdot \|F_2(s, v(s)+\omega_k(s), B(v(s)+\omega_k(s))) \\
&\quad - F_2(s, v(s)+\omega_0(s), B(v(s)+\omega_0(s)))\| \mathrm{d}s \\
&\quad \le \frac{\varepsilon}{3}
\end{aligned} \tag{130}$$

whenever $k \ge N$. Thus

$$\begin{aligned}
&\mu(t)\left\|\left(\Gamma^2\omega_k\right)(t)-\left(\Gamma^2\omega_0\right)(t)\right\| = \mu(t)\left\|\int_{-\infty}^{t} S_\alpha(t-s)\right. \\
&\quad \cdot F_2(s, v(s)+\omega_k(s), B(v(s)+\omega_k(s)))\mathrm{d}s \\
&\quad - \int_{-\infty}^{t} S_\alpha(t-s) \\
&\quad \left.\cdot F_2(s, v(s)+\omega_0(s), B(v(s)+\omega_0(s)))\mathrm{d}s\right\| \\
&\quad \le CM\int_{-\infty}^{t_1} \frac{1}{1+|\omega|(t-s)^\alpha}\mu(t) \\
&\quad \cdot \|F_2(s, v(s)+\omega_k(s), B(v(s)+\omega_k(s))) \\
&\quad - F_2(s, v(s)+\omega_0(s), B(v(s)+\omega_0(s)))\| \mathrm{d}s \\
&\quad + CM\left(2\left(1+\|B\|_{\mathbb{L}(X)}\right)\left(k_0+\||v\||\right)\right) \\
&\quad \cdot \int_{t_1}^{\max\{t,t_1\}} \frac{\beta(s)}{1+|\omega|(t-s)^\alpha}\mathrm{d}s \\
&\quad \le CM\int_{-\infty}^{t_1} \frac{1}{1+|\omega|(t-s)^\alpha}\mu(t) \\
&\quad \cdot \|F_2(s, v(s)+\omega_k(s), B(v(s)+\omega_k(s))) \\
&\quad - F_2(s, v(s)+\omega_0(s), B(v(s)+\omega_0(s)))\| \mathrm{d}s \\
&\quad + CM\sigma(t)\left(2\left(1+\|B\|_{\mathbb{L}(X)}\right)\left(k_0+\||v\||\right)\right) \le \frac{\varepsilon}{3} \\
&\quad + \frac{2\varepsilon}{3} = \varepsilon
\end{aligned} \tag{131}$$

whenever $k \ge N$. Accordingly, Γ^2 is continuous on Θ_{k_0}.

In the sequel, we consider the compactness of Γ^2.

Set $B_r(X)$ for the closed ball with center at 0 and radius r in X, $V = \Gamma^2(\Theta_{k_0})$, and $z(t) = \Gamma^2(u(t))$ for $u(t) \in \Theta_{k_0}$. First, for all $\omega(t) \in \Theta_{k_0}$ and $t \in \mathbb{R}$,

$$\mu(t)\left\|\left(\Gamma^2\omega\right)(t)\right\| = \mu(t)\left\|\int_{-\infty}^{t} S_\alpha(t-s) \cdot F_2(s, v(s)+\omega(s), B(v(s)+\omega(s)))\,ds\right\|$$
$$\leq CM\int_{-\infty}^{t}\frac{1}{1+|\omega|(t-s)^\alpha}\mu(t) \cdot \left\|F_2(s, v(s)+\omega(s), B(v(s)+\omega(s)))\right\| ds$$
$$\leq CM\int_{-\infty}^{t}\frac{\beta(s)}{1+|\omega|(t-s)^\alpha}\mu(t)(\|v(s)\|+\|\omega(s)\ ) + \|Bv(s)\|+\|B\omega(s)\ )\|)\,ds$$
$$\leq CM\int_{-\infty}^{t}\frac{\beta(s)}{1+|\omega|(t-s)^\alpha}\mu(t)\left(1+\|B\|_{\mathbb{L}(X)}\right) \cdot (\|v(s)\|+\|\omega(s)\ ))\,ds \leq CM\sigma(t)(1 + \|B\|_{\mathbb{L}(X)})(k_0+|||v(s)|||), \tag{132}$$

in view of $\sigma(t) \in C_0(\mathbb{R}, \mathbb{R}^+)$ which follows from Lemma 23; one concludes that

$$\lim_{|t|\to+\infty}\left(\Gamma^2\omega\right)(t) = 0 \quad \text{uniformly for } \omega(t) \in \Theta_{k_0}. \tag{133}$$

as

$$\left(\Gamma^2\omega\right)(t) = \int_{-\infty}^{t} S_\alpha(t-s) F_2(s, v(s) + \omega(s), B(v(s)+\omega(s)))\,ds = \int_0^{+\infty} S_\alpha(\tau) \cdot F_2(t-\tau, v(t-\tau) + \omega(t-\tau), B(v(t-\tau)+\omega(t-\tau)))\,d\tau. \tag{134}$$

Hence, for given $\varepsilon_0 > 0$, one can choose a $\xi > 0$ such that

$$\left\|\int_{\xi}^{+\infty} S_\alpha(\tau) F_2(t-\tau, v(t-\tau)+\omega(t-\tau), B(v(t-\tau)+\omega(t-\tau)))\,d\tau\right\| < \varepsilon_0. \tag{135}$$

Thus we get

$$z(t) \in \overline{\xi c(\{S_\alpha(\tau) F_2(\lambda, v(\lambda)+\omega(\lambda), B(v(\lambda)+\omega(\lambda))) : 0 \leq \tau \leq \xi,\ t-\xi \leq \lambda \leq \xi,\ |||\omega||| \leq k_0\})} + B_{\varepsilon_0}\left(\Theta_{k_0}\right), \tag{136}$$

where $c(K)$ denotes the convex hull of K. Using the fact that $S_\alpha(\cdot)$ is strongly continuous, we infer that

$$K = \{S_\alpha(\tau) F_2(\lambda, v(\lambda)+\omega(\lambda), B(v(\lambda)+\omega(\lambda))) : 0 \leq \tau \leq \xi,\ t-\xi \leq \lambda \leq \xi,\ |||\omega||| \leq k_0\} \tag{137}$$

is a relatively compact set and $V \subset \overline{\xi c(K)} + B_{\varepsilon_0}(\Theta_{k_0})$, which implies that V is a relatively compact subset of Θ_{k_0}.

Next, we verify the equicontinuity of the set $\{(\Gamma^2\omega)(t) : \omega(t) \in \Theta_{k_0}\}$, given $\varepsilon_1 > 0$. In view of (114), together with the continuity of $\{S_\alpha(t)\}_{t>0}$, there exists an $\eta > 0$ such that, for all $\omega(t) \in \Omega_{k_0}$ and $t_2 \geq t_1$ with $t_2 - t_1 < \eta$,

$$\begin{gathered}\int_{t_1}^{t_2} |||S_\alpha(t_2-s) F_2(s, v(s)+\omega(s), B(v(s)+\omega(s)))|||\,ds < \frac{\varepsilon_1}{4},\\ \int_{t_1-\eta}^{t_1} |||[S_\alpha(t_2-s) - S_\alpha(t_1-s)] F_2(s, v(s)+\omega(s), B(v(s)+\omega(s)))|||\,ds < \frac{\varepsilon_1}{4}.\end{gathered} \tag{138}$$

Also, one can choose a $k > 0$ such that

$$\begin{gathered}\int_{t_1-k}^{t_1-\eta} |||[S_\alpha(t_2-s) - S_\alpha(t_1-s)] F_2(s, v(s)+\omega(s), B(v(s)+\omega(s)))|||\,ds < \frac{\varepsilon_1}{4}\\ (1+\|B\|_{\mathbb{L}(X)})(k_0+|||v|||) \sup_{s\in[-\infty, t_1-k]} \|S_\alpha(t_2-s) - S_\alpha(t_1-s)\| \int_{-\infty}^{t_1-k} \beta(s)\,ds < \frac{\varepsilon_1}{4},\end{gathered} \tag{139}$$

which implies that, for all $\omega(t) \in \Omega_{k_0}$ and $t_2 \geq t_1$,

$$\begin{aligned}&\int_{-\infty}^{t_1-k} \left|\left|\left| \left[S_\alpha(t_2-s) - S_\alpha(t_1-s)\right] F_2(s, v(s)+\omega(s), B(v(s)+\omega(s)))\right|\right|\right| ds \\ &\leq \left(1+\|B\|_{\mathbb{L}(X)}\right)\left(k_0 + |||v|||\right) \sup_{s\in[-\infty, t_1-k]} \left\|S_\alpha(t_2-s) - S_\alpha(t_1-s)\right\| \int_{-\infty}^{t_1-k} \beta(s)\, ds < \frac{\varepsilon_1}{4}.\end{aligned} \tag{140}$$

Then one has

$$\begin{aligned}&\left|\left|\left|\left(\Gamma^2\omega\right)(t_2) - \left(\Gamma^2\omega\right)(t_1)\right|\right|\right| \\ &= \left|\left|\left| \int_{-\infty}^{t_2} S_\alpha(t_2-s) F_2(s, v(s)+\omega(s), B(v(s)+\omega(s)))\, ds - \int_{-\infty}^{t_1} S_\alpha(t_1-s) F_2(s, v(s)+\omega(s), B(v(s)+\omega(s)))\, ds \right|\right|\right| \\ &\leq \int_{t_1}^{t_2} \left|\left|\left| S_\alpha(t_2-s) F_2(s, v(s)+\omega(s), B(v(s)+\omega(s)))\right|\right|\right| ds \\ &+ \int_{t_1-\eta}^{t_1} \left|\left|\left| \left[S_\alpha(t_2-s) - S_\alpha(t_1-s)\right] F_2(s, v(s)+\omega(s), B(v(s)+\omega(s)))\right|\right|\right| ds \\ &+ \int_{-\infty}^{t_1-k} \left|\left|\left| \left[S_\alpha(t_2-s) - S_\alpha(t_1-s)\right] F_2(s, v(s)+\omega(s), B(v(s)+\omega(s)))\right|\right|\right| ds \\ &+ \int_{t_1-k}^{t_1-\eta} \left|\left|\left| \left[S_\alpha(t_2-s) - S_\alpha(t_1-s)\right] F_2(s, v(s)+\omega(s), B(v(s)+\omega(s)))\right|\right|\right| ds < \varepsilon_1,\end{aligned} \tag{141}$$

which implies the equicontinuity of the set $\{(\Gamma^2\omega)(t) : \omega(t) \in \Theta_{k_0}\}$.

Now an application of Lemma 18 justifies the compactness of Γ^2.

Step 5. Show that (39) has at least one asymptotically almost automorphic mild solution.

The proof is similar to the proof in Step 5 of Theorem 24. □

Taking $A = -\rho^\alpha I$ with $\rho > 0$ in (39), Theorem 29 gives the following corollary.

Corollary 30. *Let $F : \mathbb{R} \times X \times X \longrightarrow X$ satisfy (H_1') and (H_2') with $L(t) \in BC(\mathbb{R}, \mathbb{R}^+)$. Moreover the integral $\int_{-\infty}^{t} \max\{L(s), \beta(s)\} ds$ exists for all $t \in \mathbb{R}$. Then (39) has at least one asymptotically almost automorphic mild solution.*

4. Applications

In this section we give an example to illustrate the above results.

Consider the following fractional relaxation-oscillation equation:

$$\begin{aligned}&\partial_t^\alpha u(t,x) = \partial_x^2 u(t,x) - pu(t,x) \\ &+ \partial_t^{\alpha-1}\left[\mu a(t) \sin\left(\frac{1}{2+\cos t + \cos\sqrt{2}t}\right) \times \left[\sin u(t,x) + u(t,x)\right] + \nu e^{-|t|}\left[u(t,x) + \sin u(t,x)\right]\right], \\ &\qquad t \in \mathbb{R},\ x \in [0,\pi], \\ &u(t,0) = u(t,\pi) = 0, \quad t \in \mathbb{R},\end{aligned} \tag{142}$$

where $a(t) \in BC(\mathbb{R}, \mathbb{R}^+)$ is a function and p, μ, and ν are positive constants.

Take $X = L^2([0,\pi])$ and define the operator A by

$$A\varphi := \varphi'' - p\varphi, \quad \varphi \in D(A), \tag{143}$$

where

$$D(A) := \left\{\varphi \in X : \varphi'' \in X,\ \varphi(0) = \varphi(\pi)\right\} \subset X. \tag{144}$$

It is well known that $Bu = u''$ is self-adjoint, with compact resolvent, and is the infinitesimal generator of an analytic semigroup on X. Hence, $pI - B$ is sectorial of type $\omega = -p < 0$. Let

$$\begin{aligned}&F_1(t, x(\xi), y(\xi)) := \mu a(t) \cdot \sin\left(\frac{1}{2+\cos t + \cos\sqrt{2}t}\right)\left[\sin x(\xi) + y(\xi)\right], \end{aligned} \tag{145}$$

$$F_2(t, x(\xi), y(\xi)) := \nu e^{-|t|}\left[x(\xi) + \sin y(\xi)\right].$$

Then it is easy to verify that $F_1, F_2 : \mathbb{R} \times X \times X \longrightarrow X$ are continuous and $F_1(t, x, y) \in AA(\mathbb{R} \times X \times X, X)$ satisfying

$$\begin{aligned} &\left\|F_1(t, x_1, y_1) - F_1(t, x_2, y_2)\right\|_2^2 \\ &\le \int_0^{\pi} \mu^2 \left|a(t) \sin\left(\frac{1}{2 + \cos t + \cos\sqrt{2}t}\right)\right|^2 \\ &\cdot \left|[\sin x_1(s) + y_1(s)] - [\sin x_2(s) + y_2(s)]\right| ds \\ &\le \mu^2 a^2(t) \left|\sin\left(\frac{1}{2 + \cos t + \cos\sqrt{2}t}\right)\right|^2 \\ &\cdot \left(\left\|x_1 - x_2\right\|_2^2 + \left\|y_1 - y_2\right\|_2^2\right), \end{aligned} \tag{146}$$

that is,

$$\begin{aligned} &\left\|F_1(t, x_1, y_1) - F_1(t, x_2, y_2)\right\|_2 \\ &\le \mu a(t) \left(\left\|x_1 - x_2\right\|_2 + \left\|y_1 - y_2\right\|_2\right) \\ &\qquad \forall t \in \mathbb{R},\ x_1, y_1, x_2, y_2 \in X; \end{aligned} \tag{147}$$

furthermore

$$\begin{aligned} &\left\|F_1(t, x_1, y_1) - F_1(t, x_2, y_2)\right\|_2 \\ &\le \mu \left\|a\right\|_{\infty} \left(\left\|x_1 - x_2\right\|_2 + \left\|y_1 - y_2\right\|_2\right) \\ &\qquad \forall t \in \mathbb{R},\ x_1, y_1, x_2, y_2 \in X. \end{aligned} \tag{148}$$

And

$$\begin{aligned} \left\|F_2(t, x, y)\right\|_2^2 &\le \int_0^{\pi} \nu^2 e^{-2|t|} \left|x(s) + \sin y(s)\right| ds \\ &\le \nu^2 e^{-2|t|} \left(\left\|x\right\|_2^2 + \left\|y\right\|_2^2\right), \end{aligned} \tag{149}$$

that is,

$$\begin{aligned} &\left\|F_2(t, x, y)\right\|_2 \le \nu e^{-|t|} \left(\left\|x\right\|_2 + \left\|y\right\|_2\right) \\ &\qquad \forall t \in \mathbb{R},\ x, y \in X, \end{aligned} \tag{150}$$

which implies $F_2(t, x, y) \in C_0(\mathbb{R} \times X \times X, X)$. Furthermore

$$\begin{aligned} F(t, x, y) &= F_1(t, x, y) + F_2(t, x, y) \\ &\in AAA(\mathbb{R} \times X \times X, X). \end{aligned} \tag{151}$$

Thus, (142) can be reformulated as the abstract problem (39) and the assumptions (H_1) and (H_2) hold with

$$\begin{aligned} L &= \mu \left\|a\right\|_{\infty}, \\ \Phi(r) &= r, \\ \beta(t) &= \nu e^{-|t|}, \\ \rho_1 &= 1, \\ \rho_2 &\le \nu, \end{aligned} \tag{152}$$

the assumption (H_1') holds with $L(t) = \mu a(t)$, and the assumption (H_2') holds.

In consequence, the fractional relaxation-oscillation equation (142) has at least one asymptotically almost automorphic mild solutions if either

$$\frac{\mu C M \left\|a\right\|_{\infty} \pi \left|p\right|^{-1/\alpha}}{\alpha \sin(\pi/\alpha)} + C M \nu < \frac{1}{2} \tag{153}$$

(Theorem 24) or

$$\frac{\mu C M \left\|a\right\| \pi \left|p\right|^{-1/\alpha}}{\alpha \sin(\pi/\alpha)} + C M \nu < \frac{1}{2} \tag{154}$$

(Theorem 27), where $\|a\| = \sup_{t\in\mathbb{R}} \int_t^{t+1} a(s) ds$ or the integral

$$\int_{-\infty}^{t} \max\left\{\mu a(s), \nu e^{-|t|}\right\} ds \tag{155}$$

exists for all $t \in \mathbb{R}$ (Theorem 29).

Conflicts of Interest

The authors declare that they have no conflicts of interest.

Acknowledgments

This research was supported by the NNSF of China (no. 11561009) and (no. 41665006), the Guangdong Province Natural Science Foundation (no. 2015A030313896), the Characteristic Innovation Project (Natural Science) of Guangdong Province (no. 2016KTSCX094), the Science and Technology Program Project of Guangzhou (no. 201707010230), and the Guangxi Province Natural Science Foundation (no. 2016GXNSFAA380240).

References

[1] S. Bochner, "Continuous mappings of almost automorphic and almost periodic functions," *Proceedings of the National Acadamy of Sciences of the United States of America*, vol. 52, pp. 907–910, 1964.

[2] S. Bochner, "Uniform convergence of monotone sequences of functions," *Proceedings of the National Acadamy of Sciences of the United States of America*, vol. 47, pp. 582–585, 1961.

[3] S. Bochner, "A new approach to almost periodicity," *Proceedings of the National Acadamy of Sciences of the United States of America*, vol. 48, pp. 2039–2043, 1962.

[4] S. Bochner and J. Von Neumann, "On compact solutions of operational-differential equations," *I. Annals of Mathematics: Second Series*, vol. 36, no. 1, pp. 255–291, 1935.

[5] G. M. N'Guérékata, *Almost Automorphic Functions and Almost Periodic Functions in Abstract Spaces*, Kluwer Academic/Plenum Publishers, New York, London, Moscow, 2001.

[6] G. M. N'Guerekata, *Topics in Almost Automorphy*, Springer, New York, NY, USA, 2005.

[7] W. A. Veech, "Almost automorphic functions," *Proceedings of the National Acadamy of Sciences of the United States of America*, vol. 49, pp. 462–464, 1963.

[8] J. Campos and M. Tarallo, "Almost automorphic linear dynamics by Favard theory," *Journal of Differential Equations*, vol. 256, no. 4, pp. 1350–1367, 2014.

[9] T. Caraballo and D. Cheban, "Almost periodic and almost automorphic solutions of linear differential/difference equations without Favard's separation condition," *I. Journal of Differential Equations*, vol. 246, no. 1, pp. 108–128, 2009.

[10] L. Mahto and S. Abbas, "PC-almost automorphic solution of impulsive fractional differential equations," *Mediterranean Journal of Mathematics*, vol. 12, no. 3, pp. 771–790, 2015.

[11] D. Araya and C. Lizama, "Almost automorphic mild solutions to fractional differential equations," *Nonlinear Analysis. Theory, Methods & Applications. An International Multidisciplinary Journal*, vol. 69, no. 11, pp. 3692–3705, 2008.

[12] G. M. Mophou and G. M. N'Guérékata, "On some classes of almost automorphic functions and applications to fractional differential equations," *Computers & Mathematics with Applications. An International Journal*, vol. 59, no. 3, pp. 1310–1317, 2010.

[13] L. Abadias and C. Lizama, "Almost automorphic mild solutions to fractional partial difference-differential equations," *Applicable Analysis: An International Journal*, vol. 95, no. 6, pp. 1347–1369, 2016.

[14] M. Fu and Z. Liu, "Square-mean almost automorphic solutions for some stochastic differential equations," *Proceedings of the American Mathematical Society*, vol. 138, no. 10, pp. 3689–3701, 2010.

[15] J. Cao, Q. Yang, and Z. Huang, "Existence and exponential stability of almost automorphic mild solutions for stochastic functional differential equations," *Stochastics. An International Journal of Probability and Stochastic Processes*, vol. 83, no. 3, pp. 259–275, 2011.

[16] Z. Liu and K. Sun, "Almost automorphic solutions for stochastic differential equations driven by Lévy noise," *Journal of Functional Analysis*, vol. 266, no. 3, pp. 1115–1149, 2014.

[17] G. M. N'guérékata, "Comments on almost automorphic and almost periodic functions in Banach spaces," *Far East Journal of Mathematical Sciences (FJMS)*, vol. 17, no. 3, pp. 337–344, 2005.

[18] G. M. N'Guérékata, "Sur les solutions presqu automorphes d'équations différentielles abstraites," *Annales des Sciences Mathématiques du Québec*, no. 1, pp. 69–79, 1981.

[19] D. Bugajewski and G. M. N'Guérékata, "On the topological structure of almost automorphic and asymptotically almost automorphic solutions of differential and integral equations in abstract spaces," *Nonlinear Analysis: Theory, Methods & Applications*, vol. 59, no. 8, pp. 1333–1345, 2004.

[20] T. Diagana, E. M. Hernández, and J. dos Santos, "Existence of asymptotically almost automorphic solutions to some abstract partial neutral integro-differential equations," *Nonlinear Analysis. Theory, Methods & Applications. An International Multidisciplinary Journal*, vol. 71, no. 1-2, pp. 248–257, 2009.

[21] H.-S. Ding, T.-J. Xiao, and J. Liang, "Asymptotically almost automorphic solutions for some integrodifferential equations with nonlocal initial conditions," *Journal of Mathematical Analysis and Applications*, vol. 338, no. 1, pp. 141–151, 2008.

[22] J.-Q. Zhao, Y.-K. Chang, and G. M. N'guérékata, "Existence of asymptotically almost automorphic solutions to nonlinear delay integral equations," *Dynamic Systems and Applications*, vol. 21, no. 2-3, pp. 339–349, 2012.

[23] Y.-K. Chang and C. Tang, "Asymptotically almost automorphic solutions to stochastic differential equations driven by a Lévy process," *Stochastics. An International Journal of Probability and Stochastic Processes*, vol. 88, no. 7, pp. 980–1011, 2016.

[24] Z.-H. Zhao, Y.-K. Chang, and J. J. Nieto, "Square-mean asymptotically almost automorphic process and its application to stochastic integro-differential equations," *Dynamic Systems and Applications*, vol. 22, no. 2-3, pp. 269–284, 2013.

[25] G. M. N'Guérékata, *Spectral Theory for Bounded Functions and Applications to Evolution Equations*, Nova Science Publishers, NY, USA, 2017.

[26] A. A. Kilbas, H. M. Srivastava, and J. J. Trujillo, *Theory and Applications of Fractional Differential Equations*, vol. 24 of *North-Holland Mathematics Studies*, Elsevier Science B.V., Amsterdam, 2006.

[27] S. G. Samko, A. A. Kilbas, and O. I. Marichev, *Fractional integral and derivatives: Theory and applications*, Gordon and Breach Science Publishers, Switzerland, 1993.

[28] K. Diethelm, *The Analysis of Fractional Differential Equations*, Lecture Notes in Mathematics, Springer Verlag, Berlin, Heidelberg, Germany, 2010.

[29] R. Hilfer, *Applications of Fractional Calculus in Physics*, World Scientific, Singapore, 2000.

[30] I. Podlubny, *Fractional Differential Equations*, Academic Press, CA, USA, 1999.

[31] K. S. Miller and B. Ross, *An Introduction to the Fractional Calculus and Fractional Differential Equations*, A Wiley Interscience Publication, John Wiley & Sons, NY, USA, 1993.

[32] Y. Zhou, *Basicheory of Fractional Diferential Equations*, World Scientiic, Singapore, 2014.

[33] R. P. Agarwal, M. Belmekki, and M. Benchohra, "A survey on semilinear differential equations and inclusions involving Riemann-Liouville fractional derivative," *Advances in Difference Equations*, vol. 2009, Article ID 981728, 47 pages, 2009.

[34] R. P. Agarwal, V. Lakshmikantham, and J. J. Nieto, "On the concept of solution for fractional differential equations with uncertainty," *Nonlinear Analysis. Theory, Methods & Applications. An International Multidisciplinary Journal*, vol. 72, no. 6, pp. 2859–2862, 2010.

[35] M. Benchohra, J. Henderson, S. K. Ntouyas, and A. Ouahab, "Existence results for fractional order functional differential equations with infinite delay," *Journal of Mathematical Analysis and Applications*, vol. 338, no. 2, pp. 1340–1350, 2008.

[36] R. P. Agarwal, M. Benchohra, and S. Hamani, "A survey on existence results for boundary value problems of nonlinear fractional differential equations and inclusions," *Acta Applicandae Mathematicae*, vol. 109, no. 3, pp. 973–1033, 2010.

[37] M. M. El-Borai, "Some probability densities and fundamental solutions of fractional evolution equations," *Chaos, Solitons & Fractals*, vol. 14, no. 3, pp. 433–440, 2002.

[38] V. Lakshmikantham, "Theory of fractional functional differential equations," *Nonlinear Analysis: Theory, Methods & Applications*, vol. 60, pp. 3337–3343, 2008.

[39] V. Lakshmikantham and A. S. Vatsala, "Basic theory of fractional differential equations," *Nonlinear Analysis. Theory, Methods & Applications. An International Multidisciplinary Journal*, vol. 69, no. 8, pp. 2677–2682, 2008.

[40] V. Lakshmikantham and A. S. Vatsala, "Theory of fractional differential inequalities and applications," *Communications in Applied Analysis*, vol. 11, no. 3-4, pp. 395–402, 2007.

[41] V. Lakshmikantham and J. V. Devi, "Theory of fractional differential equations in a Banach space," *European Journal of Pure and Applied Mathematics*, vol. 1, no. 1, pp. 38–45, 2008.

[42] G. Mophou, O. Nakoulima, and G. M. N'Guérékata, "Existence results for some fractional differential equations with nonlocal conditions," *Nonlinear Studies. The International Journal*, vol. 17, no. 1, pp. 15–21, 2010.

[43] G. M. Mophou and G. M. N'Guérékata, "Existence of the mild solution for some fractional differential equations with nonlocal conditions," *Semigroup Forum*, vol. 79, no. 2, pp. 315–322, 2009.

[44] G. M. Mophou and G. M. N'Guérékata, "On integral solutions of some nonlocal fractional differential equations with nondense domain," *Nonlinear Analysis. Theory, Methods & Applications. An International Multidisciplinary Journal*, vol. 71, no. 10, pp. 4668–4675, 2009.

[45] G. M. Mophou, "Existence and uniqueness of mild solutions to impulsive fractional differential equations," *Nonlinear Analysis. Theory, Methods & Applications. An International Multidisciplinary Journal*, vol. 72, no. 3-4, pp. 1604–1615, 2010.

[46] G. M. N'Guérékata, "A Cauchy problem for some fractional abstract differential equation with non local conditions," *Nonlinear Analysis. Theory, Methods & Applications. An International Multidisciplinary Journal*, vol. 70, no. 5, pp. 1873–1876, 2009.

[47] Y. Zhou and L. Peng, "On the time-fractional Navier-Stokes equations," *Computers & Mathematics with Applications. An International Journal*, vol. 73, no. 6, pp. 874–891, 2017.

[48] Y. Zhou and L. Peng, "Weak solutions of the time-fractional Navier-Stokes equations and optimal control," *Computers & Mathematics with Applications. An International Journal*, vol. 73, no. 6, pp. 1016–1027, 2017.

[49] Y. Zhou and L. Zhang, "Existence and multiplicity results of homoclinic solutions for fractional Hamiltonian systems," *Computers & Mathematics with Applications. An International Journal*, vol. 73, no. 6, pp. 1325–1345, 2017.

[50] Y. Zhou, V. Vijayakumar, and R. Murugesu, "Controllability for fractional evolution inclusions without compactness," *Evolution Equations and Control Theory*, vol. 4, no. 4, pp. 507–524, 2015.

[51] C. Cuevas and C. Lizama, "Almost automorphic solutions to a class of semilinear fractional differential equations," *Applied Mathematics Letters*, vol. 21, no. 12, pp. 1315–1319, 2008.

[52] R. P. Agarwal, B. de Andrade, and C. Cuevas, "On type of periodicity and ergodicity to a class of fractional order differential equations," *Advances in Difference Equations*, Article ID 179750, 2010.

[53] R. P. Agarwal, C. Cuevas, and H. Soto, "Pseudo-almost periodic solutions of a class of semilinear fractional differential equations," *Applied Mathematics and Computation*, vol. 37, no. 1-2, pp. 625–634, 2011.

[54] R. P. Agarwal, B. de Andrade, and C. Cuevas, "Weighted pseudo-almost periodic solutions of a class of semilinear fractional differential equations," *Nonlinear Analysis: Real World Applications*, vol. 11, no. 5, pp. 3532–3554, 2010.

[55] H.-S. Ding, J. Liang, and T.-J. Xiao, "Almost automorphic solutions to abstract fractional differential equations," *Advances in Difference Equations*, Article ID 508374, 2010.

[56] C. Cuevas, A. Sepúlveda, and H. Soto, "Almost periodic and pseudo-almost periodic solutions to fractional differential and integro-differential equations," *Applied Mathematics and Computation*, vol. 218, no. 5, pp. 1735–1745, 2011.

[57] Y.-K. Chang, R. Zhang, and G. M. N'Guérékata, "Weighted pseudo almost automorphic mild solutions to semilinear fractional differential equations," *Computers & Mathematics with Applications. An International Journal*, vol. 64, no. 10, pp. 3160–3170, 2012.

[58] B. He, J. Cao, and B. Yang, "Weighted Stepanov-like pseudo-almost automorphic mild solutions for semilinear fractional differential equations," *Advances in Difference Equations*, vol. 2015, 74 pages, 2015.

[59] J. Cao, Q. Yang, and Z. Huang, "Existence of anti-periodic mild solutions for a class of semilinear fractional differential equations," *Communications in Nonlinear Science and Numerical Simulation*, vol. 17, no. 1, pp. 277–283, 2012.

[60] C. Lizama and F. Poblete, "Regularity of mild solutions for a class of fractional order differential equations," *Applied Mathematics and Computation*, vol. 224, pp. 803–816, 2013.

[61] Z. Xia, M. Fan, and R. P. Agarwal, "Pseudo almost automorphy of semilinear fractional differential equations in Banach spaces," *Fractional Calculus and Applied Analysis*, vol. 19, no. 3, pp. 741–764, 2016.

[62] S. Abbas, V. Kavitha, and R. Murugesu, "Stepanov-like weighted pseudo almost automorphic solutions to fractional order abstract integro-differential equations," *Journal of Fractional Calculus and Applications*, vol. 4, pp. 1–19, 2013.

[63] G. M. Mophou, "Weighted pseudo almost automorphic mild solutions to semilinear fractional differential equations," *Applied Mathematics and Computation*, vol. 217, no. 19, pp. 7579–7587, 2011.

[64] Y.-K. Chang and X.-X. Luo, "Pseudo almost automorphic behavior of solutions to a semi-linear fractional differential equation," *Mathematical Communications*, vol. 20, no. 1, pp. 53–68, 2015.

[65] V. Kavitha, S. Abbas, and R. Murugesu, "(μ1,μ2)-pseudo almost automorphic solutions of fractional order neutral integro-differential equations," *Nonlinear Stud*, vol. 24, pp. 669–685, 2017.

[66] V. Kavitha, S. Abbas, and R. Murugesu, "Asymptotically almost automorphic solutions of fractional order neutral integro-differential equations," *Bulletin of the Malaysian Mathematical Sciences Society*, vol. 39, no. 3, pp. 1075–1088, 2016.

[67] S. Abbas, "Weighted pseudo almost automorphic solutions of fractional functional differential equations," *Cubo (Temuco)*, vol. 16, pp. 21–35, 2014.

[68] E. Bazhlekova, *Fractional Evolution Equations in Banach Spaces*, Eindhoven University of Technology, 2001.

[69] S. D. Eidelman and A. N. Kochubei, "Cauchy problem for fractional diffusion equations," *Journal of Differential Equations*, vol. 199, no. 2, pp. 211–255, 2004.

[70] R. Gorenflo and F. Mainardi, "Fractional calculus: Integral and differential equations of fractional order," in *Fractals and Fractional Calculus in Continuum Mechanics*, A. Carpinteri and F. Mainardi, Eds., pp. 223–276, Springer-Verlag, NY, USA, Vienna, Austria, 1997.

[71] V. V. Anh and R. Mcvinish, "Fractional differential equations driven by Lévy noise," *Journal of Applied Mathematics and Stochastic Analysis*, vol. 16, no. 2, pp. 97–119, 2003.

[72] H.-S. Ding, J. Liang, and T.-J. Xiao, "Some properties of Stepanov-like almost automorphic functions and applications to abstract evolution equations," *Applicable Analysis: An International Journal*, vol. 88, no. 7, pp. 1079–1091, 2009.

[73] J. Liang, J. Zhang, and T.-J. Xiao, "Composition of pseudo almost automorphic and asymptotically almost automorphic functions," *Journal of Mathematical Analysis and Applications*, vol. 340, no. 2, pp. 1493–1499, 2008.

[74] M. Haase, "The functional calculus for sectorial operators," in *Operator Theory: Advances and Applications*, vol. 169, Birkhuser Verlag, Basel, Switzerland, 2006.

[75] E. Cuesta, "Asymptotic behaviour of the solutions of fractional integro-differential equations and some time discretizations," *Discrete and Continuous Dynamical Systems - Series A*, pp. 277–285, 2007.

[76] W. M. Ruess and W. H. Summers, "Compactness in spaces of vector valued continuous functions and asymptotic almost periodicity," *Mathematische Nachrichten*, vol. 135, pp. 7–33, 1988.

[77] D. R. Smart, *Fixed Point Theorems*, Cambridge University Press, London, UK, 1980.

18

Application of Optimal Homotopy Asymptotic Method to Some Well-Known Linear and Nonlinear Two-Point Boundary Value Problems

Muhammad Asim Khan,[1] **Shafiq Ullah,**[2] **and Norhashidah Hj. Mohd Ali**[1]

[1]*School of Mathematical Sciences, Universiti Sains Malaysia, 11800 Penang, Malaysia*
[2]*Department of Mathematics, University of Peshawar, Pakistan*

Correspondence should be addressed to Muhammad Asim Khan; asim.afg@gmail.com

Guest Editor: Dongfang Li

The objective of this paper is to obtain an approximate solution for some well-known linear and nonlinear two-point boundary value problems. For this purpose, a semianalytical method known as optimal homotopy asymptotic method (OHAM) is used. Moreover, optimal homotopy asymptotic method does not involve any discretization, linearization, or small perturbations and that is why it reduces the computations a lot. OHAM results show the effectiveness and reliability of OHAM for application to two-point boundary value problems. The obtained results are compared to the exact solutions and homotopy perturbation method (HPM).

1. Introduction

Two-point boundary value problems (TPBVP) have many applications in the field of science and engineering [1, 2]. These problems arise in many physical situations like modeling of chemical reactions, heat transfer, viscous fluids, diffusions, deflection of beams, the solution of optimal control problems, etc. Due to the wide applications and importance of boundary value problems (BVP) in science and engineering we need solutions to these problems.

There are many techniques available for the solution of-of BVP like Adomian Decomposition Method (ADM) [3–7], Extended Adomian Decomposition Method (EADM)[8], Differential Transformation Method (DTM) [9], Variational Iteration Method (VIM) [10], Perturbation methods(PMs) [1, 11–13], and so on. Perturbation methods are easy to solve but they require small parameters which are sometimes not an easy task. Recently V. Marinca et al. presented optimal homotopy asymptotic method (OHAM) [14] for the solution of BVP, which did not require small parameters. The method can also be applied to solve the stationary solution of some partial differential equations, e.g., gKdv equation, nonlinear parabolic problems, and so on [15–20]. In OHAM, the concept of homotopy is used together with the perturbation techniques. Here, OHAM is applied to TPBVP to check the applicability of OHAM for TPBVP.

2. Basics of OHAM

Let us take the BVP whose general form is the following:

$$\mathscr{L}(w(\xi)) + \mathbb{N}(w(\xi)) + F(\xi) = 0, \quad B\left(w, \frac{dw}{d\xi}\right) \tag{1}$$

where $\mathscr{L}$ is a linear operator, ξ is independent variable, $\mathbb{N}$ is the nonlinear operator, $F(\xi)$ is a known function, and B is a boundary operator.

Homotopy on OHAM can be constructed as

$$\begin{aligned}(1-\text{þ})(\mathscr{L}(\varphi(\xi,\text{þ}) + F(\xi)) &= H(\text{þ})(\mathscr{L}(\varphi(\xi,\text{þ}) + F(\xi) + \mathbb{N}(\varphi(\xi,\text{þ}))), \\ B\left(\varphi(\xi,\text{þ}), \frac{\partial\varphi(\xi,\text{þ})}{\partial\xi}\right)\end{aligned} \tag{2}$$

where $\text{þ} \in [0, 1]$ is an embedding parameter, $\varphi(\xi, \text{þ})$ is an unknown function, $H(\text{þ})$ is a nonzero auxiliary function for $\text{þ} \neq 0$, and $H(\text{þ})$ is of the form

$$H(\text{þ}) = \text{þ}\mathbb{C}_1 + \text{þ}^2\mathbb{C}_2 + \text{þ}^3\mathbb{C}_3 + \cdots \qquad (3)$$

Clearly when $\text{þ} = 0$ then $H(0) = 0$. And obviously, when $\text{þ} = 0$ then $\varphi(\xi, 0) = w_0(\xi)$. When $\text{þ} = 1$ then $\varphi(\xi, 1) = w(\xi)$. So as þ increases from 0 to 1, the solution $\varphi(\xi, \text{þ})$ varies from $w_0(\xi)$ to the exact solution $w(\xi)$, where $w_0(\xi)$ is obtained from (2) for $\text{þ} = 0$

$$\mathscr{L}(w_0(\xi) + F(\xi)) = 0 \qquad (4)$$

The proposed solution of (1) will be of the form

$$\varphi(\xi, \text{þ}, \mathbb{C}_i) = w_0(\xi) + \sum_{k \geq 1} w_k(\xi, \mathbb{C}_i)\text{þ}^k, \quad i = 1, 2, 3, \ldots \qquad (5)$$

Substituting this value of $\varphi(\xi, \text{þ}, \mathbb{C}_i)$ into (1), after some calculations, we can obtain the governing equations of $w_0(\xi)$ by using (4) and $w_k(\xi)$, that is,

$$\mathscr{L}(w_1(\xi)) = \mathbb{C}_1 \mathbb{N}_0(w_0(\xi)),$$

$$B\left(w_1, \frac{dw_1}{d\xi}\right) = 0 \qquad (6)$$

$$\mathscr{L}(w_k(\xi) - w_{k-1}(\xi))$$

$$= \mathbb{C}_k \mathbb{N}_0(w_0(\xi)) + \sum_{i=1}^{k-1} \mathbb{C}_i \mathscr{L}(w_{k-1}(\xi)$$

$$+ \mathbb{N}_{k-1}(w_0(\xi), w_1(\xi), w_2(\xi), \ldots, w_{k-1}(\xi)), \qquad (7)$$

$$k = 2, 3, 4, \ldots,$$

$$B\left(w_k, \frac{dw_k}{d\xi}\right) = 0,$$

where $\mathbb{N}_m(w_0(\xi), w_1(\xi), w_2(\xi), \ldots, w_m(\xi))$ is the coefficient of þ^m in the series expansion of $\mathbb{N}(\varphi(\xi, \text{þ}, \mathbb{C}_i))$ with respect to the embedding parameter þ. And

$$\mathbb{N}(\varphi(\xi, \text{þ}, \mathbb{C}_i)) = \mathbb{N}_0(w_0(\xi))$$

$$+ \sum_{m \geq 1} \mathbb{N}_m(w_0, w_1, w_2, \ldots, w_m)\text{þ}^m, \qquad (8)$$

$$i = 1, 2, 3, \ldots, m$$

where $\varphi(\xi, \text{þ}, \mathbb{C}_i)$ is given by (5). The convergence of series (5) depends on the convergence of the constants $\mathbb{C}_i's$, if these constants are convergent at $\text{þ} = 1$, then the solution becomes

$$w(\xi, \mathbb{C}_i) = w_0(\xi) + \sum_{k \geq 1} (w_k(\xi, \mathbb{C}_i)). \qquad (9)$$

Generally, the mth order solution of the problem can be obtained in the form

$$w^{(m)}(\xi, \mathbb{C}_i) = w_0(\xi) + \sum_{k=1}^{m} (w_k(\xi, \mathbb{C}_i)), \qquad (10)$$

$$i = 1, 2, 3, \ldots, m$$

Putting this solution in (1) we get the following residual:

$$\mathfrak{R}(\xi, \mathbb{C}_i) = \mathscr{L}(w^{(m)}(\xi, \mathbb{C}_i) + F(\xi))$$

$$+ \mathbb{N}(w^{(m)}(\xi, \mathbb{C}_i)), \quad i = 1, 2, 3, \ldots, m \qquad (11)$$

If $\mathfrak{R}(\xi, \mathbb{C}_i) = 0$, then the solution is going to be exact, but generally, such a situation does not arise in nonlinear problems but the functional defined below can be minimized

$$J(\xi, \mathbb{C}_i) = \int_{x_0}^{x_1} \mathfrak{R}^2(\xi, \mathbb{C}_i)\, d\xi, \qquad (12)$$

where x_0 and x_1 are two constants depending on the given problem. The values of $\mathbb{C}_i's$ can be optimally found by the condition

$$\frac{\partial J}{\partial \mathbb{C}_1} = \frac{\partial J}{\partial \mathbb{C}_2} = \cdots = \frac{\partial J}{\partial \mathbb{C}_m} = 0 \qquad (13)$$

After knowing these constants, the solution (10) is well determined.

3. Examples

To check the applicability of OHAM for TPBVP, in this section four examples of TPBVP are presented in which one example is linear and the remaining are nonlinear.

3.1. Example 1. Let us consider the linear problem [1] of second order

$$w''(\xi) = w'(\xi) - e^{(\xi-1)} - 1, \quad 0 < \xi < 1,$$

$$w(0) = 0, \qquad (14)$$

$$w(1) = 1$$

The exact solution of problem (14) is $\xi(1 - e^{\xi-1})$. Now according to OHAM $\mathscr{L}(w_0(\xi)) = w''(\xi) - w'(\xi)$, the nonlinear part $\mathbb{N}(w(\xi)) = 0$ and $F(\xi) = e^{(\xi-1)} + 1$.

The zeroth-order problem is

$$w_0''(\xi) - w_0'(\xi) = 1 + e^{\xi-1},$$

$$w_0(0) = 0, \qquad (15)$$

$$w_0(1) = 1$$

The solution of (15) is

$$w_0(\xi) = \frac{(e - e^{\xi})\xi}{\xi} \qquad (16)$$

The first-order problem is

$$w_1''(\xi) - w_1'(\xi) = -1 - e^{\xi-1} - \mathbb{C}_1 - \mathbb{C}_1 e^{\xi-1} + w_0'(\xi)$$

$$+ \mathbb{C}_1 w_0'(\xi) - w_0''(\xi)$$

$$- \mathbb{C}_1 w_0''(\xi), \qquad (17)$$

$$w_1(0) = 0,$$

$$w_1(1) = 0.$$

TABLE 1: Comparison of the third-order OHAM solution with the exact solution and HPM.

ξ	OHAM Solution ($w^{(2)}$)	Exact	HPM [1]	$\lvert w^{(2)} - Exact\rvert$
0.1	0.059343	0.059343	0.05934820	1.38778×10^{-17}
0.3	0.151024	0.151024	0.15103441	2.77556×10^{-17}
0.5	0.196735	0.196735	0.19673826	2.77556×10^{-17}
0.7	0.181427	0.181427	0.18142196	5.55112×10^{-17}
0.9	0.0856463	0.0856463	0.08564186	5.55112×10^{-17}

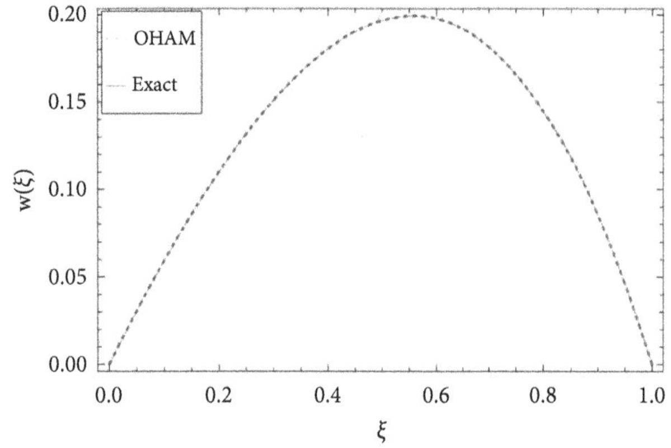

FIGURE 1: Comparison between exact solution (dashed line) and approximate solution (dotted line) for example 1.

The solution of (17) is

$$w_1(\xi) = 0 \tag{18}$$

The second-order problem is

$$\begin{aligned} w_2''(\xi) &= -\mathbb{C}_2 - \exp(\xi - 1)\mathbb{C}_2 + \mathbb{C}_2 w_0'(\xi) + w_1'(\xi) \\ &\quad + \mathbb{C}_1 w_1'(\xi) - w_2'(\xi) - \mathbb{C}_2 w_0''(\xi) - w_1''(\xi) \\ &\quad - \mathbb{C}_1 w_1''(\xi), \end{aligned} \tag{19}$$

$$w_2(0) = 0,$$

$$w_2(1) = 0$$

The solution of (19) is

$$w_2(\xi) = 0 \tag{20}$$

And the third-order approximate solution of the bvp (14) is as follows:

$$w^{(2)}(\xi) = w_0(\xi) + w_1(\xi) + w_2(\xi) \tag{21}$$

$$w^{(2)}(\xi) = \frac{\left(e - e^{\xi}\right)\xi}{\xi} \tag{22}$$

Table 1 shows the comparison between the exact solution and the approximate solution obtained by OHAM. Figure 1 of the solution also shows well agreement with the exact solution.

3.2. *Example 2.* Consider the nonlinear two-point boundary value problem [1] of the type

$$\begin{aligned} w''(\xi) &= w^3(\xi) - w(\xi)\,w'(\xi), \quad \xi \in [1,2], \\ w(1) &= \frac{1}{2}, \\ w(2) &= \frac{1}{3} \end{aligned} \tag{23}$$

According to OHAM $\mathscr{L}(w(x)) = w''(\xi)$ and $\mathbb{N}(w(\xi)) = u(\xi)u'(\xi) - w^3(\xi)$, while $_F(\xi) = 0$. The exact solution of (23) is $1/(\xi + 1)$. Now proceeding with the same lines as above we have the following zeroth-order problem:

$$\begin{aligned} w_0''(\xi) &= 0, \\ w_0(1) &= \frac{1}{2}, \\ w_0(2) &= \frac{1}{3} \end{aligned} \tag{24}$$

The solution of (24) is

$$w_0(\xi) = \frac{4 - \xi}{6} \tag{25}$$

Now the first-order problem is

$$\begin{aligned} w_1''(\xi) &= \mathbb{C}_1 w_0^3(\xi) - \mathbb{C}_1 w_0(\xi)\,w_0'(\xi) - w_0''(\xi) \\ &\quad - \mathbb{C}_1 w_0''(\xi) \end{aligned} \tag{26}$$

$$\begin{aligned} w_1(1) &= 0, \\ w_1(2) &= 0 \end{aligned} \tag{27}$$

The solution of (26) is

$$w_1(\xi) = \frac{-930\mathbb{C}_1 + 1649\xi\mathbb{C}_1 - 880\xi_1^2\mathbb{C} + 180\xi_1^3\mathbb{C} - 20\xi_1^4\mathbb{C} + \xi_{5\mathbb{C}1}}{4320} \tag{28}$$

The second-order problem is

$$\begin{aligned} w_2''(\xi) &= \mathbb{C}_2 w_0^3(\xi) + 3\mathbb{C}_1 w_0^2(\xi)\,w_1(\xi) \\ &\quad - \mathbb{C}_2 w_0(\xi)\,w_0'(\xi) - \mathbb{C}_1 w_1(\xi)\,w_0'(\xi) \\ &\quad - \mathbb{C}_1 w_0(\xi)\,w_1'(\xi) - \mathbb{C}_2 w_0''(\xi) - w_1''(\xi) \\ &\quad - \mathbb{C}_1 w_1''(\xi), \end{aligned} \tag{29}$$

$$w_2(1) = 0,$$

$$w_2(2) = 0.$$

TABLE 2: Comparison of second-order OHAM solution with the exact solution for example 2.

ξ	**OHAM Solution** ($w^{(3)}$)	**Exact**	$\lvert w^{(3)} - Exact \rvert$
1.1	0.47619	0.47619	2.0597×10^{-7}
1.3	0.434783	0.434783	5.38284×10^{-7}
1.5	0.400001	0.4	1.39261×10^{-6}
1.7	0.370371	0.37037	6.7426×10^{-7}
1.9	0.344828	0.344828	8.10196×10^{-8}

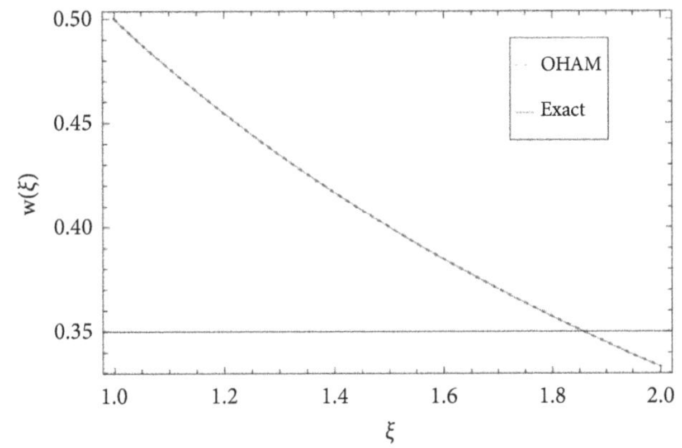

FIGURE 2: Comparison between exact solution (dashed line) and approximate solution (dotted line) for example 2.

The solution of (29) is

$$\begin{aligned} w_2(\xi) = {} & \frac{1}{130636800}(\xi-2)(\xi-1)\{30240\left(-465+\xi(127+(\xi-17)\xi)\mathbb{C}_1\right. \\ & -(8985375+\xi(4253423+\xi(-3664113+\xi(1100519+5\xi(-36795+\xi(3709+7(-33+\xi)\xi))))))\mathbb{C}_1^2 \\ & \left.+30240(-465+\xi(127+(-17+\xi)\xi))\mathbb{C}_2\right) \end{aligned} \tag{30}$$

The third-order problem is

$$\begin{aligned} & \mathbb{C}_3 w_0^3(\xi) + 3\mathbb{C}_2 w_0^2(\xi) w_1(\xi) + 3\mathbb{C}_1 w_0(\xi) w_1^2(\xi) \\ & + 3\mathbb{C}_1 w_0^2(\xi) w_2(\xi) - 3\mathbb{C}_3 w_0(\xi) w_0'(\xi) \\ & - \mathbb{C}_2 w_1(\xi) w_0'(\xi) - \mathbb{C}_1 w_2(\xi) w_0'(\xi) \\ & - \mathbb{C}_2 w_0(\xi) w_1'(\xi) - \mathbb{C}_1 w_1(\xi) w_1'(\xi) \\ & - \mathbb{C}_1 w_0(\xi) w_2'(\xi) - \mathbb{C}_3 w_0''(\xi) - \mathbb{C}_2 w_1''(\xi) \\ & - w_2''(\xi) - \mathbb{C}_1 w_2''(\xi) + w_3''(\xi) = 0 \end{aligned} \tag{31}$$

The solution of the third-order problem results a large output, therefore not included here.

Now the third-order approximate solution is

$$w^{(3)}(\xi) = w_0(\xi) + w_1(\xi) + w_2(\xi) + w_3(\xi) \tag{32}$$

$\mathbb{C}_i's$ has the following values and then substituting in the above solution we will get the approximate solution. $w^{(3)}(\xi)$ is given in Appendix (A.1).

$$\begin{aligned} \mathbb{C}_1 &= -0.9637924142971654, \\ \mathbb{C}_2 &= -0.0002296939939480446, \\ \mathbb{C}_3 &= -0.000014314891134337846, \end{aligned} \tag{33}$$

The solution at the points given in Table 2 and the graph of the solution is shown in Figure 2. Here it is third-order OHAM solution while the HPM [1] gives the accuracy up to 9 decimal places in 7th order.

3.3. Example 3. Now we consider higher order TPBVP of order four. The problem is

$$\frac{d^4 w(\xi)}{d\xi^4} = w^2(\xi) + F(\xi), \quad 0 \le \xi \le 1 \tag{34}$$

with the boundary conditions $w(0) = 0$, $w'(0) = 0$, $w(1) = 1$, and $w'(1) = 1$.

Table 3: Comparison of second-order OHAM solution with the exact solution for example 3.

ξ	OHAM Solution $(w^{(2)})$	Exact	$\lvert w^{(2)} - Exact\rvert$
0.2	0.077119	0.0771200	6.152306×10^{-11}
0.4	0.279039	0.2790400	1.314346×10^{-10}
0.6	0.538559	0.5385600	1.001054×10^{-10}
0.8	0.788479	0.7884800	2.415356×10^{-11}

Where $\mathscr{L}(w(\xi)) = d^4w(\xi)/dx^4$, $\mathbb{N}(w(\xi)) = w^2(\xi)$, and $F(\xi) = -\xi^{10} + 4\xi^9 - 4\xi^8 - 4\xi^7 + 8\xi^6 - 4\xi^4 + 120\xi - 48$, the exact solution of problem (34) is $w_{exact} = \xi^5 - 2\xi^4 + 2\xi^2$. After solving this by the method described in Section 2, we have the following zeroth-order problem:

$$48 - 120\xi + 4\xi^4 - 8\xi^6 + 4\xi^7 + 4\xi^8 - 4\xi^9 + \xi^{10} + \frac{d^4 w_0(\xi)}{d\xi^4} = 0 \quad (35)$$

$$w_0(0) = 0, \quad w_0'(0) = 0, \quad w_0(1) = 1, \quad w_0'(1) = 1 \quad (36)$$

The solution to (35) is

$$w_0(\xi) = \frac{1}{1081080}\left(2155683\xi^2 + 8038\xi^3 - 2162160\xi^4 + 1081080\xi^5 - 2574\xi^8 + 1716\xi^{10} - 546\xi^{11} - 364\xi^{12} + 252\xi^{13} - 45\xi^{14}\right) \quad (37)$$

The first-order problem is

$$-48 + 120\xi - 4\xi^4 + 8\xi^6 - 4\xi^7 - 4\xi^8 + 4\xi^9 - \xi^{10} - 48\mathbb{C}_1 + 120\xi\mathbb{C}_1 - 4\xi_1^4\mathbb{C} + 8\xi^6\mathbb{C}_1 - 4\xi^7\mathbb{C}_1 - 4\xi^8\mathbb{C}_1 + 4\xi^9\mathbb{C}_1 - \xi^{10}\mathbb{C}_1 + \mathbb{C}_1 w_0^2(\xi) - (1 + \mathbb{C}_1)\frac{d^4 w_0(\xi)}{d\xi^4} + \frac{d^4 w_1(\xi)}{d\xi^4} = 0 \quad (38)$$

$$w_1(0) = 0, \quad w_1'(0) = 0, \quad w_1(1) = 0, \quad w_1'(1) = 0. \quad (39)$$

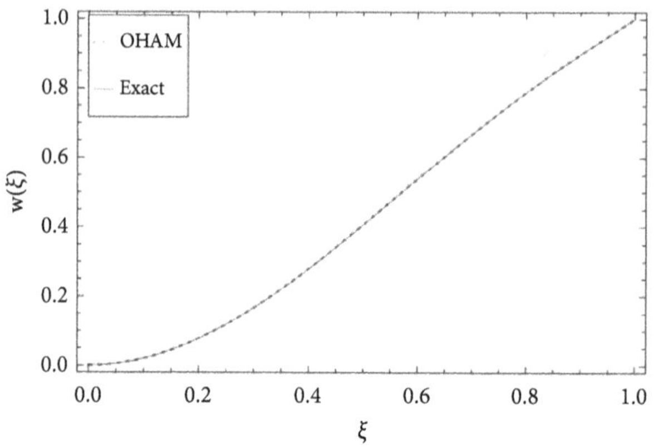

Figure 3: Comparison between exact solution (dashed line) and approximate solution (dotted line) for example 3.

The second-order problem is

$$-48\mathbb{C}_2 + 120\xi\mathbb{C}_2 - 4\xi^4\mathbb{C}_2 + 8\xi^6\mathbb{C}_2 - 4\xi^7\mathbb{C}_2 - 4\xi^8\mathbb{C}_2 + 4\xi^9\mathbb{C}_2 - \xi^{10}\mathbb{C}_2 + \mathbb{C}_2 w_0^2(\xi) + 2\mathbb{C}_1 w_0(\xi) w_1(\xi) - \mathbb{C}_2 w_0^{(4)}(\xi) - w_1^{(4)}(\xi) - \mathbb{C}_1 w_1^{(4)}(\xi) + w_2^{(4)}(\xi) = 0 \quad (40)$$

$$w_2(0) = 0, \quad w_2'(0) = 0, \quad w_2(1) = 0, \quad w_2'(1) = 0. \quad (41)$$

The solutions of problem (38) and (40) are very large; therefore we did not write it here. The constants $\mathbb{C}_1$ and $\mathbb{C}_2$ have the values -1.0011320722175725 and $-1.079468959963785\times 10^{-6}$, respectively. Table 3 and Figure 3 show a good agreement with the exact values. The approximate solution $w_2(\xi)$ is given in Appendix (A.2).

3.4. *Example 4.* At last, consider the second-order nonlinear TPBVP[1]

$$w''(\xi) = w^2(\xi) + 2\pi^2\cos(2\pi\xi) - \sin^2(2\pi\xi), \quad 0 \le \xi \le 1 \quad (42)$$

$$w(0) = 0, \quad w(1) = 0. \quad (43)$$

TABLE 4: Comparison of third-order OHAM solution with the exact solution.

ξ	OHAM Solution ($w^{(3)}$)	Exact	$\lvert w^{(3)} - Exact\rvert$
0.1	0.0954915	0.0954915	5.59262×10^{-9}
0.3	0.654508	0.654508	1.23779×10^{-8}
0.5	0.999999	1.	1.51565×10^{-8}
0.7	0.654508	0.654508	1.23779×10^{-8}
0.9	0.0954915	0.0954915	5.59262×10^{-9}

The exact solution of (42) is $\sin^2(\pi\xi)$. Solving (42) by the method depicted in Section 2, we have the following zeroth order problem:

$$-2\pi^2\cos(2\pi\xi) + \sin^4(\pi\xi) + w_0''(\xi) = 0, \quad w_0(0) = 0, \quad w_0(1) = 0 \tag{44}$$

The solution of (44) is given by

$$w_0(\xi) = \frac{1}{128\pi^2}\left(15 - 64\pi^2 + 24\pi^2\xi - 24\pi^2\xi^2 - 16\cos(2\pi\xi) - 64\pi^2\cos(2\pi\xi) + \cos(4\pi\xi)\right). \tag{45}$$

The first-, second-, and third-order problems are given in (46), (47), and (48) respectively.

$$\begin{aligned} &2\pi^2\cos(2\pi\xi)(1 + C_1) - \sin^4(\pi\xi)(1 + C_1) \\ &\quad + C_1 w_0^2(\xi) - w_0''(\xi) - C_1 w_0''(\xi) + w_1''(\xi) = 0, \\ &w_1(0) = 0, \\ &w_1(1) = 0 \end{aligned} \tag{46}$$

$$\begin{aligned} &\left\{2\pi^2\cos(2\pi\xi) - \sin^4(\pi\xi)\right\} C_2 + C_2 w_0^2(\xi) \\ &\quad + 2C_1 w_0(\xi) w_1(\xi) - C_2 w_0''(\xi) - w_1''(\xi) \\ &\quad - C_1 w_1''(\xi) + w_2''(\xi) = 0, \\ &w_2(0) = 0, \\ &w_2(1) = 0 \end{aligned} \tag{47}$$

$$\begin{aligned} &\left\{2\pi^2\cos(2\pi\xi) - \sin^4(\pi\xi)\right\} C_3 + C_3 w_0^2(\xi) \\ &\quad + 2C_2 w_0(\xi) w_1(\xi) + C_1 w_1^2(\xi) + 2C_1 w_0(\xi) w_2(\xi) \\ &\quad + C_3 w_0''(\xi) - C_2 w_1''(\xi) - w_2''(\xi) - C_1 w_2''(\xi) \\ &\quad + w_3''(\xi) = 0 \end{aligned} \tag{48}$$

$$w_3(0) = 0, \quad w_3(1) = 0 \tag{49}$$

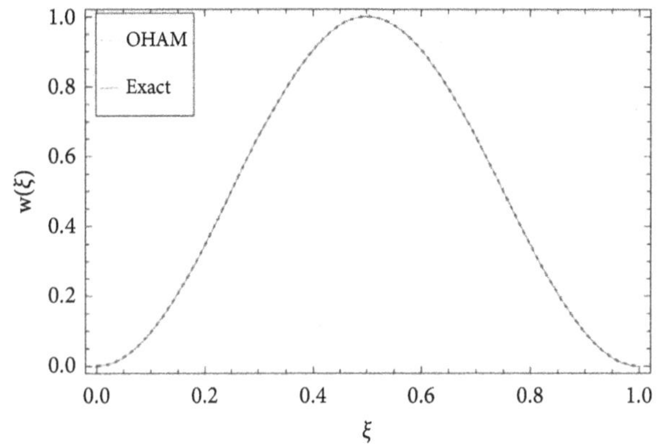

FIGURE 4: Comparison between exact solution (dashed line) and approximate solution (dotted line) for example 4.

The solutions of problem (46), (47), and (48) are very large and therefore cannot be written here but the table of values and the graph are shown in Table 4 and Figure 4, respectively. The approximate solution $w^{(3)}(\xi)$ is written in Appendix (A.3). The values of the constants $C_i's$ can be found by (13) which are given as follows:

$$\begin{aligned} C_1 &= -0.9030981665320986, \\ C_2 &= -0.00569345107292796, \\ C_3 &= 0.00021560218552318884 \end{aligned} \tag{50}$$

4. Conclusion

This paper reveals that OHAM is a very strong method for solving TPBVP and gives us a more accurate solution as compared to other methods. In these examples only second- and third-order solution gives us the accuracy up to 8 or 10 decimal places; therefore it is concluded that this method converges very fast to the exact solution and in some problems like example 1 it gives us the exact solution. The plots and tables show well agreement with the exact solution.

Appendix

$$
\begin{aligned}
\overset{(3)}{w}(\xi) = {} & \frac{4-\xi}{6} + \frac{1}{4320}\left(-930\mathbb{C}_1 + 1649\xi\mathbb{C}_1 - 880\xi^2\mathbb{C}_1 + 180\xi^3\mathbb{C}_1 - 20\xi^4\mathbb{C}_1 + \xi^5\mathbb{C}_1\right) + \frac{1}{130636800}\left(-28123200\mathbb{C}_1\right. \\
& + 49865760\xi\mathbb{C}_1 - 26611200\xi^2\mathbb{C}_1 + 5443200\xi^3\mathbb{C}_1 - 604800\xi^4\mathbb{C}_1 + 30240\xi^5\mathbb{C}_1 - 17970750\mathbb{C}_1^2 + 18449279\xi\mathbb{C}_1^2 \\
& + 11103120\xi^2\mathbb{C}_1^2 - 17446800\xi^3\mathbb{C}_1^2 + 7333620\xi^4\mathbb{C}_1^2 - 1689534\xi^5\mathbb{C}_1^2 + 241920\xi^6\mathbb{C}_1^2 - 22080\xi^7\mathbb{C}_1^2 + 1260\xi^8\mathbb{C}_1^2 \\
& \left. - 35\xi^8\mathbb{C}_1^2 - 28123200\mathbb{C}_2 + 49865760\xi\mathbb{C}_2 - 26611200\xi^2\mathbb{C}_2 + 5443200\xi^3\mathbb{C}_2 - 604800\xi^4\mathbb{C}_2 + 30240\xi^5\mathbb{C}_2\right) \\
& + \frac{1}{13450364928000}\left(-2895564672000\mathbb{C}_1 + 5134178649600\xi\mathbb{C}_1 - 2739889152000\xi^2\mathbb{C}_1 + 560431872000\xi^3\mathbb{C}_1\right. \\
& - 62270208000\xi^4\mathbb{C}_1 + 3113510400\xi^5\mathbb{C}_1 - 3700536840000\mathbb{C}_1^2 + 3799075531680\xi\mathbb{C}_1^2 + 2286354470400\xi^2\mathbb{C}_1^2 \\
& - 3592645056000\xi^3\mathbb{C}_1^2 + 1510139030400\xi^4\mathbb{C}_1^2 - 347908841280\xi^5\mathbb{C}_1^2 + 49816166400\xi^6\mathbb{C}_1^2 - 4546713600\xi^7\mathbb{C}_1^2 \\
& + 259459200\xi^8\mathbb{C}_1^2 - 7207200\xi^9\mathbb{C}_1^2 - 1161287826270\mathbb{C}_1^3 + 270053629823\xi\mathbb{C}_1^3 + 1988069792280\xi^2\mathbb{C}_1^3 \\
& - 963479454080\xi^3\mathbb{C}_1^3 - 507564676190\xi^4\mathbb{C}_1^3 + 549609764847\xi^5\mathbb{C}_1^3 - 221549328000\xi^6\mathbb{C}_1^3 + 54219285120\xi^7\mathbb{C}_1^3 \\
& - 9069733530\xi^8\mathbb{C}_1^3 + 1086825025\xi^9\mathbb{C}_1^3 - 93716480\xi^{10}\mathbb{C}_1^3 + 5653440\xi^{11}\mathbb{C}_1^3 - 220220\xi^{12}\mathbb{C}_1^3 + 4235\xi^{13}\mathbb{C}_1^3 \\
& - 2895564672000\mathbb{C}_2 + 5134178649600\xi\mathbb{C}_2 - 2739889152000\xi^2\mathbb{C}_2 + 560431872000\xi^3\mathbb{C}_2 - 62270208000\xi^4\mathbb{C}_2 \\
& + 3113510400\xi^5\mathbb{C}_2 - 3700536840000\mathbb{C}_1\mathbb{C}_2 + 3799075531680\xi\mathbb{C}_1\mathbb{C}_2 + 2286354470400\xi^2\mathbb{C}_1\mathbb{C}_2 \\
& - 3592645056000\xi^3\mathbb{C}_1\mathbb{C}_2 + 1510139030400\xi^4\mathbb{C}_1\mathbb{C}_2 - 347908841280\xi^5\mathbb{C}_1\mathbb{C}_2 + 49816166400\xi^6\mathbb{C}_1\mathbb{C}_2 \\
& - 4546713600\xi^7\mathbb{C}_1\mathbb{C}_2 + 259459200\xi^8\mathbb{C}_1\mathbb{C}_2 - 7207200\xi^9\mathbb{C}_1\mathbb{C}_2 - 2895564672000\mathbb{C}_3 + 5134178649600\xi\mathbb{C}_3 \\
& \left. - 2739889152000\xi^2\mathbb{C}_3 + 560431872000\xi^3\mathbb{C}_3 - 62270208000\xi^4\mathbb{C}_3 + 3113510400\xi^5\mathbb{C}_3\right).
\end{aligned}
\tag{A.1}
$$

$$
\begin{aligned}
w^{(2)}(\xi) = {} & \left(2155683\xi^2 + 8038\xi^3 - 2162160\xi^4 + 1081080\xi^5 - 2574\xi^8 + 1716\xi^{10} - 546\xi^{11} - 364\xi^{12} + 252\xi^{13} - 45\xi^{14}\right)/ \\
& 1081080 + 1/2360410309588661890560000\left(-1411382850381345391135 9\xi^2\mathbb{C}_1\right. \\
& + 175129157667046669623 22\xi^3\mathbb{C}_1 - 5586404113887189501420\xi^8\mathbb{C}_1 - 2314477400095222 6800\xi^9\mathbb{C}_1 \\
& + 3735433519034813810160\xi^{10}\mathbb{C}_1 - 1179691619002973294400\xi^{11}\mathbb{C}_1 - 797705495192034374400\xi^{12}\mathbb{C}_1 \\
& + 550212193377310464000\xi^{13}\mathbb{C}_1 - 97319245468388853600\xi^{14}\mathbb{C}_1 + 2551023644304960\xi^{15}\mathbb{C}_1 \\
& - 8567292999988720 80\xi^{16}\mathbb{C}_1 + 279036341094724200\xi^{17}\mathbb{C}_1 + 247466505045614400\xi^{18}\mathbb{C}_1 \\
& - 155275084044250800\xi^{19}\mathbb{C}_1 - 36505192535330 40\xi^{2}0\mathbb{C}_1 + 26398020790188000\xi^{2}1\mathbb{C}_1 \\
& - 8405056667479680\xi^{2}2\mathbb{C}_1 + 897927547083840\xi^{2}3\mathbb{C}_1 - 381597192288 00\xi^{2}4\mathbb{C}_1 + 21095475553152\xi^{2}5\mathbb{C}_1 \\
& + 4049797421376\xi^{2}6\mathbb{C}_1 - 6052906241984\xi^{2}7\mathbb{C}_1 + 1221181635008\xi^{2}8\mathbb{C}_1 + 475889853600\xi^{2}9\mathbb{C}_1 \\
& \left. - 295593372480\xi^{3}0\mathbb{C}_1 + 60656299200\xi^{3}1\mathbb{C}_1 - 4738773375\xi^{3}2\mathbb{C}_1\right) \\
& + \left(\left(-3266968253516603817780056271740345177478720 0\xi^2\mathbb{C}_1\right.\right. \\
& + 405376470465654982768725806942188930704576 00\xi^3\mathbb{C}_1 \\
& - 129310093901543591689242666969205624671360 00\xi^8\mathbb{C}_1
\end{aligned}
$$

$$+ 405376470465654982768725806942188930704576000\xi^3\mathbb{C}_1$$

$$- 1293100939015435916892426669692056246713600\xi^8\mathbb{C}_1$$

$$- 5357387038924076963068897715678742144000\xi^9\mathbb{C}_1$$

$$+ 86465148109963491224184203077213392161280\xi^{10}\mathbb{C}_1$$

$$- 273066593318814352267252241234258775552000\xi^{11}\mathbb{C}_1$$

$$- 184647172646597956123620297067672061952000\xi^{12}\mathbb{C}_1$$

$$+ 12735919017123762373862880265978355712000\xi^{13}\mathbb{C}_1$$

$$- 2252676411049710692001622000328823628800\xi^{14}\mathbb{C}_1 + 59049273963210763661687274856519680000\xi^{15}\mathbb{C}_1$$

$$- 198309581570838358908091345502728166400\xi^{16}\mathbb{C}_1 + 6458933999995720474035815391174393600\xi^{17}\mathbb{C}_1$$

$$+ 572817797505678403219571120181611520000\xi^{18}\mathbb{C}_1$$

$$+ 862046688684628002096517676757223157085151\xi^{10}\mathbb{C}_1^2$$

$$- 27019424542416038674372835931239409782685\xi^{11}\mathbb{C}_1^2$$

$$- 1853296246170788567680120913554535247983040\xi^{12}\mathbb{C}_1^2$$

$$+ 12735919017123762373862880265978355712000\xi^{13}\mathbb{C}_1^2$$

$$- 22312755109850374780609930388825424427188\xi^{14}\mathbb{C}_1^2 + 241887531112728162536201595588156497768\xi^{15}\mathbb{C}_1^2$$

$$- 395429499911928902670749978002715174240\xi^{16}\mathbb{C}_1^2 + 1280476731768992936554600505857043216034\xi^{17}\mathbb{C}_1^2$$

$$+ 1146299848262837493979094855639474837488\xi^{18}\mathbb{C}_1^2 - 716583570253989121237950089315154553020\xi^{19}\mathbb{C}_1^2$$

$$- 18153410829102066767079156365109506436\xi^2 0\mathbb{C}_1^2 + 12232506247599725530975551889772833788\xi^2 1\mathbb{C}_1^2$$

$$- 38583478912638559610674295035606805760\xi^2 2\mathbb{C}_1^2 + 40647085107228344914023460357811779 20\xi^2 3\mathbb{C}_1^2$$

$$- 3049444827689904471550031859215952 00\xi^2 4\mathbb{C}_1^2 + 17213182431610251871836854208282240 0\xi^2 5\mathbb{C}_1^2$$

$$+ 3122598976922878410874222728210650 4\xi^2 6\mathbb{C}_1^2 - 486960649472624354043524692778784 96\xi^2 7\mathbb{C}_1^2$$

$$+ 995595411043267481549629461389241 6\xi^2 8\mathbb{C}_1^2 + 37680517224202509563809356490608 00\xi^2 9\mathbb{C}_1^2$$

$$- 234122003038524567782675624738688 0\xi^3 0\mathbb{C}_1^2 + 4745262945911464867597339852327 68\xi^3 1\mathbb{C}_1^2$$

$$- 40917013297075779490615187391192\xi^3 2\mathbb{C}_1^2 + 3867571077895433794192598216016\xi^3 3\mathbb{C}_1^2$$

$$- 52161594559300577150208932796 0\xi^3 4\mathbb{C}_1^2 - 57865189063228775007105204637 2\xi^3 5\mathbb{C}_1^2$$

$$+ 265019543551924982873921756160\xi^3 6\mathbb{C}_1^2 - 7980251045802826679782141248\xi^3 7\mathbb{C}_1^2$$

$$- 252689233858232871709767137 13452\xi^3 8\mathbb{C}_1^2 + 8700984701390689923286015560\xi^3 9\mathbb{C}_1^2$$

$$- 13403098910307828206072427 84\xi^4 0\mathbb{C}_1^2 + 13504516660880363746270843 2\xi^4 1\mathbb{C}_1^2$$

$$- 1868346413550687790020787 2\xi^4 2\mathbb{C}_1^2 - 3279443942818588507522560\xi^4 3\mathbb{C}_1^2$$

$$+ 3820823675617325912753400\xi^4 4\mathbb{C}_1^2 - 8051944202434111906967 04\xi^4 5\mathbb{C}_1^2 - 1435825488225501086968 80\xi^4 6\mathbb{C}_1^2$$

$$+ 11543632097509125029052 0\xi^4 7\mathbb{C}_1^2 - 27795688158510545773500\xi^4 8\mathbb{C}_1^2 + 3304269870772182097500\xi^4 9\mathbb{C}_1^2$$

$$- 1652134935386091048 75\xi^5 0\mathbb{C}_1^2 - 3266968253516603817780056271740345177478 7200\xi^2\mathbb{C}_2$$

$$+ 405376470465654982768725806942188930704576 00\xi^3\mathbb{C}_2$$

$$- 129310093901543591689242666969205624671360 00\xi^8\mathbb{C}_2$$

$$- 535738703892407696306889771567874214400 00\xi^9\mathbb{C}_2$$

$$+ 864651481099634912241842030772133921612 8000\xi^{10}\mathbb{C}_2$$

$$- 273066593318814352267252241234258775552 0000\xi^{11}\mathbb{C}_2$$

$$- 184647172646597956123620297067672061952 0000\xi^{12}\mathbb{C}_2$$

$$+ 127359190171237623738628802659783557120 0000\xi^{13}\mathbb{C}_2$$

$$- 225267641104971069200162200032882362880000\xi^{14}\mathbb{C}_2 + 590492739632107636616872748565196 8000\xi^{15}\mathbb{C}_2$$

$$- 198309581570838358908091345502728166 4000\xi^{16}\mathbb{C}_2 + 645893399999572047403581539117439360000\xi^{17}\mathbb{C}_2$$

$$+ 572817797505678403219571120181611520000\xi^{18}\mathbb{C}_2 - 359419678365531276056662133636240640000\xi^{19}\mathbb{C}_2$$

$$- 844996133183935042193575050001843 2000\xi^2 0\mathbb{C}_2 + 611042537847995408644654707045504 00000\xi^2 1\mathbb{C}_2$$

$$- 194554250777844861581489650876477440 00\xi^2 2\mathbb{C}_2 + 207845857662811349123372092569907 2000\xi^2 3\mathbb{C}_2$$

$$- 883293935801421228703444656230400 00\xi^2 4\mathbb{C}_2 + 488303006560990812585801712521216 00\xi^2 5\mathbb{C}_2$$

$$+ 937418192748622056129586038190080 0\xi^2 6\mathbb{C}_2 - 140108352094056200088829490767872 00\xi^2 7\mathbb{C}_2$$

$$+ 282670405997259066205625206640640 0\xi^2 8\mathbb{C}_2 + 110155585599030838957118316288000 0\xi^2 9\mathbb{C}_2$$

$$- 684218434127313503766165801984000\xi^3 0\mathbb{C}_2 + 140402870708442139347389199360000\xi^3 1\mathbb{C}_2$$

$$- 109689742740970421365147812000 00\xi^3 2\mathbb{C}_2)) / 546370925834609405838717563410471460044 8000000$$

$$+ ((-638124094436097170281835052952590015088100643695530983999965982720 0000\xi^2\mathbb{C}_1)) /$$

$$106720489830021596828753345862376975380339080587078787885747404800 0000000$$

(A.2)

$$w^{(3)}(\xi) = \left(15 + 64\pi^2 + 24\pi^2\xi - 24\pi^2\xi^2 - 16\cos[2\pi\xi] - 64\pi^2\cos[2\pi\xi] + \cos[4\pi\xi]\right) / \left(128\pi^2\right) + 1/\left(94371840\pi^6\right)$$

$$\cdot (2312275\mathbb{C}_1 + 12037120\pi^2\mathbb{C}_1 + 11059200\pi^4\mathbb{C}_1 + 1018080\pi^2\xi\mathbb{C}_1 + 8824320\pi^4\xi\mathbb{C}_1 + 19224576\pi^6\xi\mathbb{C}_1$$

$$- 1018080\pi^2\xi^2\mathbb{C}_1 - 8478720\pi^4\xi^2\mathbb{C}_1 - 17694720\pi^6\xi^2\mathbb{C}_1 - 691200\pi^4\xi^3\mathbb{C}_1 - 2949120\pi^6\xi^3\mathbb{C}_1 + 345600\pi^4\xi^4\mathbb{C}_1$$

$$+ 1198080\pi^6\xi^4\mathbb{C}_1 + 331776\pi^6\xi^5\mathbb{C}_1 - 110592\pi^6\xi^6\mathbb{C}_1 - 2373120\cos[2\pi\xi]\mathbb{C}_1 - 12441600\pi^2\cos[2\pi\xi]\mathbb{C}_1$$

$$- 11796480\pi^4\cos[2\pi\xi]\mathbb{C}_1 - 1105920\pi^2\xi\cos[2\pi\xi]\mathbb{C}_1 - 4423680\pi^4\xi\cos[2\pi\xi]\mathbb{C}_1 + 1105920\pi^2\xi^2\cos[2\pi\xi]\mathbb{C}_1$$

$$+ 4423680\pi^4\xi^2\cos[2\pi\xi]\mathbb{C}_1 + 63360\cos[4\pi\xi]\mathbb{C}_1 + 414720\pi^2\cos[4\pi\xi]\mathbb{C}_1 + 737280\pi^4\cos[4\pi\xi]\mathbb{C}_1$$

$$+ 17280\pi^2\xi\cos[4\pi\xi]\mathbb{C}_1 - 17280\pi^2\xi^2\cos[4\pi\xi]\mathbb{C}_1 - 2560\cos[6\pi\xi]\mathbb{C}_1 - 10240\pi^2\cos[6\pi\xi]\mathbb{C}_1 + 45\cos[8\pi\xi]\mathbb{C}_1$$

$$+ 1105920\pi\sin[2\pi\xi]\mathbb{C}_1 + 4423680\pi^3\sin[2\pi\xi]\mathbb{C}_1 - 2211840\pi\xi\sin[2\pi\xi]\mathbb{C}_1 - 8847360\pi^3\xi\sin[2\pi\xi]\mathbb{C}_1$$

$$- 8640\pi\sin[4\pi\xi]\mathbb{C}_1 + 17280\pi\xi\sin[4\pi\xi]\mathbb{C}_1) + 1/\left(487049291366400\pi^{10}\right)\left(11933558784000\pi^4\mathbb{C}_1\right) + 1/$$

$$\left(356598009166823424 00\pi^{12}\right)\xi\left(384696643800268800\pi^8\mathbb{C}_1 + 3334400329855795200\pi^{10}\mathbb{C}_1\right.$$

$$+ 7264291475800719360\pi^{12}\mathbb{C}_1 + 317056740153753600\pi^4\mathbb{C}_1^2 + 3055348438990848000\pi^6\mathbb{C}_1^2$$

$$
\begin{aligned}
&+ 976132441066438656 0\pi^{8}\mathbb{C}_1^2 + 14367361192521891840\pi^{10}\mathbb{C}_1^2 + 15851242925787709440\pi^{12}\mathbb{C}_1^2 \\
&+ 2221247340746984 55\mathbb{C}_1^3 + 21714041899000811 36\pi^{2}\mathbb{C}_1^3 + 7100075693557310464\pi^{4}\mathbb{C}_1^3 \\
&+ 10842199196917637120\pi^{6}\mathbb{C}_1^3 + 14250836120871567360\pi^{8}\mathbb{C}_1^3 + 11649501141852487680\pi^{10}\mathbb{C}_1^3 \\
&+ 8684066594856370176\pi^{12}\mathbb{C}_1^3 + 3846966438002688 00\pi^{8}\mathbb{C}_2 + 3334400329855795200\pi^{10}\mathbb{C}_2 \\
&+ 7264291475800719360\pi^{12}\mathbb{C}_2 + 317056740153753600\pi^{4}\mathbb{C}_1\mathbb{C}_2 + 3055348438990848000\pi^{6}\mathbb{C}_1\mathbb{C}_2 \\
&+ 9761324410664386560\pi^{8}\mathbb{C}_1\mathbb{C}_2 + 14367361192521891840\pi^{10}\mathbb{C}_1\mathbb{C}_2 + 15851242925787709440\pi^{12}\mathbb{C}_1\mathbb{C}_2 \\
&+ 384696643800268800\pi^{8}\mathbb{C}_3 + 3334400329855795200\pi^{10}\mathbb{C}_3 + 7264291475800719360\pi^{12}\mathbb{C}_3\big) + 1/ \\
&\big(3770462866155503616000000\pi^{14}\big)\big(923829293128089600000000\pi^{8}\mathbb{C}_1 + 4809222112810106880000000\pi^{10}\mathbb{C}_1 \\
&+ 4418511171275980800000000\pi^{12}\mathbb{C}_1 + 241156678171145748480000\pi^{4}\mathbb{C}_1^2 + 1273402703047211089920000\pi^{6}\mathbb{C}_1^2 \\
&+ 1659580957335748608000000\pi^{8}\mathbb{C}_1^2 + 1923688845124042752000000\pi^{10}\mathbb{C}_1^2 \\
&+ 8837022342551961600000000\pi^{12}\mathbb{C}_1^2 + 5940855552826154184 20429\mathbb{C}_1^3 + 28624364400527844294 31616\pi^{2}\mathbb{C}_1^3 \\
&+ 2597952006766286551449600\pi^{4}\mathbb{C}_1^3 + 3066666484153535692800000\pi^{6}\mathbb{C}_1^3 + 2139343103453036544000000\pi^{8}\mathbb{C}_1^3 \\
&+ 1442766633843032064000000\pi^{10}\mathbb{C}_1^3 + 4418511171275980800000000\pi^{12}\mathbb{C}_1^3 + 923829293128089600000000\pi^{8}\mathbb{C}_2 \\
&+ 4809222112810106880000000\pi^{10}\mathbb{C}_2 + 4418511171275980800000000\pi^{12}\mathbb{C}_2 \\
&+ 241156678171145748480000\pi^{4}\mathbb{C}_1\mathbb{C}_2 + 1273402703047211089920000\pi^{6}\mathbb{C}_1\mathbb{C}_2 \\
&+ 1659580957335748608000000\pi^{8}\mathbb{C}_1\mathbb{C}_2 + 1923688845124042752000000\pi^{10}\mathbb{C}_1\mathbb{C}_2 \\
&+ 8837022342551961600000000\pi^{12}\mathbb{C}_1\mathbb{C}_2 + 923829293128089600000000\pi^{8}\mathbb{C}_3 + 4809222112810106880000000\pi^{10}\mathbb{C}_3 \\
&+ 4418511171275980800000000\pi^{12}\mathbb{C}_3\big) + 1/\big(124684618589798400\pi^{12}\big)\big(-1345093160140800\pi^{8}\xi^{2}\mathbb{C}_1\big)
\end{aligned}
\tag{A.3}
$$

Conflicts of Interest

The authors declare that they have no conflicts of interest.

References

[1] C. Chun and R. Sakthivel, "Homotopy perturbation technique for solving two-point boundary value problems—comparison with other methods," *Computer Physics Communications*, vol. 181, no. 6, pp. 1021–1024, 2010.

[2] H. B. Keller, *Numerical Methods for Two-Point Boundary-Value Problems*, Courier Dover Publications, 2018.

[3] G. Adomian, "A review of the decomposition method in applied mathematics," *Journal of Mathematical Analysis and Applications*, vol. 135, no. 2, pp. 501–544, 1988.

[4] M. Tatari and M. Dehghan, "The use of the Adomian decomposition method for solving multipoint boundary value problems," *Physica Scripta*, vol. 73, no. 6, pp. 672–676, 2006.

[5] M. Inc and D. J. Evans, "The decomposition method for solving of a class of singular two-point boundary value problems," *International Journal of Computer Mathematics*, vol. 80, no. 7, pp. 869–882, 2003.

[6] B. Jang, "Exact solutions to one dimensional non-homogeneous parabolic problems by the homogeneous Adomian decomposition method," *Applied Mathematics and Computation*, vol. 186, no. 2, pp. 969–979, 2007.

[7] A. M. Wazwaz, "A reliable algorithm for obtaining positive solutions for nonlinear boundary value problems," *Computers & Mathematics with Applications*, vol. 41, pp. 1237–1244, 2001.

[8] B. Jang, "Two-point boundary value problems by the extended Adomian decomposition method," *Journal of Computational and Applied Mathematics*, vol. 219, no. 1, pp. 253–262, 2008.

[9] H. Yaghoobi and M. Torabi, "The application of differential transformation method to nonlinear equations arising in heat transfer," *International Communications in Heat and Mass Transfer*, vol. 38, no. 6, pp. 815–820, 2011.

[10] D. D. Ganji, G. A. Afrouzi, and R. A. Talarposhti, "Application of variational iteration method and homotopy-perturbation method for nonlinear heat diffusion and heat transfer equations," *Physics Letters Section A: General, Atomic and Solid State Physics*, vol. 368, no. 6, pp. 450–457, 2007.

[11] A. H. Nayfeh, *Introduction to Perturbation Techniques*, John Wiley & Sons, New York, NY, USA, 1981.

[12] A. H. Nayfeh, *Perturbation Methods*, John Wiley & Sons, New York, NY, USA, 2000.

[13] A. Rafiq, M. Y. Malik, and T. Abbasi, "Solution of nonlinear pull-in behavior in electrostatic micro-actuators by using He's homotopy perturbation method," *Computers & Mathematics with Applications. An International Journal*, vol. 59, no. 8, pp. 2723–2733, 2010.

[14] V. Marinca and N. Herişanu, "Application of optimal homotopy asymptotic method for solving nonlinear equations arising in heat transfer," *International Communications in Heat and Mass Transfer*, vol. 35, no. 6, pp. 710–715, 2008.

[15] M. Javidi and A. Golbabai, "A new domain decomposition algorithm for generalized Burger's-Huxley equation based on Chebyshev polynomials and preconditioning," *Chaos, Solitons & Fractals*, vol. 39, no. 2, pp. 849–857, 2009.

[16] M. Wang, D. Li, C. Zhang, and Y. Tang, "Long time behavior of solutions of gKdV equations," *Journal of Mathematical Analysis and Applications*, vol. 390, no. 1, pp. 136–150, 2012.

[17] A. Acosta, P. García, H. Leiva, and A. Merlitti, "Finite Time Synchronization of Extended Nonlinear Dynamical Systems Using Local Coupling," *International Journal of Differential Equations*, vol. 2017, Article ID 1946304, 7 pages, 2017.

[18] D. Li and J. Zhang, "Efficient implementation to numerically solve the nonlinear time fractional parabolic problems on unbounded spatial domain," *Journal of Computational Physics*, vol. 322, pp. 415–428, 2016.

[19] D. Li, J. Zhang, and Z. Zhang, "Unconditionally optimal error estimates of a linearized galerkin method for nonlinear time fractional reaction–subdiffusion equations," *Journal of Scientific Computing*, pp. 1–19, 2018.

[20] A. J. Khattak, "A computational meshless method for the generalized Burger's-Huxley equation," *Applied Mathematical Modelling*, vol. 33, no. 9, pp. 3718–3729, 2009.

19

On a Singular Second-Order Multipoint Boundary Value Problem at Resonance

S. A. Iyase and O. F. Imaga

Department of Mathematics, Covenant University, PMB 1023, Ota, Ogun State, Nigeria

Correspondence should be addressed to O. F. Imaga; imaga.ogbu@covenantuniversity.edu.ng

Academic Editor: Qingkai Kong

The aim of this paper is to derive existence results for a second-order singular multipoint boundary value problem at resonance using coincidence degree arguments.

1. Introduction

In this paper we derive existence results for the second-order singular multipoint boundary value problem of the form

$$\begin{aligned} x''(t) &= f\left(t, x(t), x'(t)\right) + g(t), \quad 0 < t < 1, \\ x'(0) &= 0, \\ x(1) &= \sum_{i=1}^{m-2} a_i x(\xi_i), \end{aligned} \tag{1}$$

where $f : [0,1] \times \mathbb{R}^2 \rightarrow \mathbb{R}$ is Caratheodory's function (i.e., for each $(x, y) \in \mathbb{R}^2$ the function $f(\cdot, x, y)$ is measurable on $[0,1]$; for a.e. $t \in [0,1]$, the function $f(t, \cdot, \cdot)$ is continuous on $\mathbb{R}^2$). Let $\xi_i \in (0,1)$, $i = 1, 2, \ldots, m-2$, $0 < \xi_1 < \xi_2 < \cdots < \xi_{m-2} < 1$, $a_i \in (0,1)$ $i = 1, 2, \ldots, m-2$, and $\sum_{i=1}^{m-2} a_i = 1$, where f and g have singularity at $t = 1$.

In [1] Gupta et al. studied the above equation when f and g have no singularity and $\sum_{i=1}^{m-2} a_i \neq 1$. They obtained existence of a $C^1[0,1]$ solution by utilising the Leray-Schauder continuation principle. In [2] Ma and O'Regan derived existence results for the same equation when f and g have a singularity at $t = 1$ and $\sum_{i=1}^{m-2} a_i \neq 1$. They also utilised the Leray-Schauder continuation method. These results correspond to the nonresonance case. The purpose of this article is therefore to derive existence results for (1) when $\sum_{i=1}^{m-2} a_i = 1$ (the resonance case) and when f and g have a singularity at $t = 1$. We shall employ coincidence degree arguments in obtaining our results. In this case, the methods used in [1, 2] are not valid.

Research on singular differential equations is important because singular differential equations are useful in the modeling of many problems in the physical and engineering sciences; see [3].

In general singular boundary value problems can be difficult to solve because they may blow up near the singularity. The existence and multiplicity of solutions for second-order nonsingular boundary value problems have been extensively studied by many researchers. However to the best of our knowledge the corresponding problem for second-order differential equations at resonance and with a singularity had not received much attention in the literature. For recent results in these directions see [1, 2, 4–9] and references therein.

The rest of this paper is organised as follows. In Section 2, we present some definitions, lemmas, and theorems necessary for obtaining our main results. In Section 3, we derive some lemmas and the main theorem. In what follows we shall utilise the following assumptions:

(A0) For $\xi_i \in (0,1)$, $i = 1, 2, \ldots, m-2$, $0 < \xi_1 < \xi_2 < \cdots < \xi_{m-2} < 1$ and $\sum_{i=1}^{m-2} a_i = 1$.

(A1) There exist $a(t), c(t) \in L^1[0,1]$ with $(1-t)a(t), (1-t)c(t), b(t) \in L^1[0,1]$ and $|f(t,x,y)| \leq a(t)|x| + b(t)|y| + c(t)$, a.e., $t \in [0,1]$, $(x, y) \in \mathbb{R}^2$.

(A2) $g : [0,1] \to \mathbb{R}$ is such that $\int_0^1 (1-t)|g(t)| < \infty$.

2. Preliminaries

In this section we state some definitions, theorems, and lemmas that will be used in the subsequent section.

Definition 1. Let X and Z be real Banach spaces. One says that the linear operator $L : \mathrm{dom}\, L \subset X \to Z$ is a Fredholm mapping of index zero if $\mathrm{Ker}\, L$ and $Z/\mathrm{Im}\, L$ are of finite dimension, where $\mathrm{Im}\, L$ denotes the image of L.

As a result of Definition 1, we will require the continuous projections $P : X \to X, Q : Z \to Z$ such that $\mathrm{Im}\, P = \mathrm{Ker}\, L$, $\mathrm{Ker}\, Q = \mathrm{Im}\, L$, $X = \mathrm{Ker}\, L \oplus \mathrm{Ker}\, P$, $Z = \mathrm{Im}\, L \oplus \mathrm{Im}\, Q$, and $L|_{\mathrm{dom}\, L \cap \mathrm{Ker}\, P} : \mathrm{dom}\, L \cap \mathrm{Ker}\, P \to \mathrm{Im}\, L$ is an isomorphism.

Definition 2. Let L be a Fredholm mapping of index zero and Ω a bounded open subset of X such that $\mathrm{dom}\, L \cap \Omega \neq \phi$. The map $N : X \to Z$ is called L-compact on $\overline{\Omega}$, if the map $QN(\overline{\Omega})$ is bounded and $K_p(I-Q)$ is compact, where one denotes by $K_p : \mathrm{Im}\, L \to \mathrm{dom}\, L \cap \mathrm{Ker}\, P$ the generalised inverse of L. In addition N is L-completely continuous if it is L-compact on every bounded $\Omega \subset X$.

Theorem 3 (see [10]). *Let L be a Fredholm operator of index zero and let N be L-compact on $\overline{\Omega}$. Assume that the following conditions are satisfied:*

(i) *$Lx \neq \lambda Nx$ for every $(x,\lambda) \in [(\mathrm{dom}\, L \setminus \mathrm{Ker}\, L) \cap \partial\Omega] \times (0,1)$.*

(ii) *$Nx \notin \mathrm{Im}\, L$, for every $x \in \mathrm{Ker}\, L \cap \partial\Omega$.*

(iii) *$\deg(QN|_{\mathrm{Ker}\, L \cap \partial\Omega}, \Omega \cap \mathrm{Ker}\, L, 0) \neq 0$,*

with $Q : Z \to Z$ being a continuous projection such that $\mathrm{Ker}\, Q = \mathrm{Im}\, L$. Then the equation $Lx = Nx$ has at least one solution in $\mathrm{dom}\, L \cap \overline{\Omega}$.

In what follows, we shall make use of the following classical spaces, $C[0,1]$, $C^1[0,1]$, $L^1[0,1]$, and $L^\infty[0,1]$. Let $AC[0,1]$ denote the space of all absolute continuous functions on $[0,1]$, $AC^1[0,1] = \{x \in C^1[0,1] : x'(t) \in AC[0,1]\}$, $L^1_{\mathrm{loc}}[0,1] = \{x : x|_{[0,d]} \in L^1[0,d]$ for every compact interval $[0,d] \subseteq [0,1)\}$.

$AC_{\mathrm{loc}}[0,1) = \{x : x|_{[0,d]} \in AC[0,d]\}$.

Let Z be the Banach space defined by

$$Z = \{y \in L^1_{\mathrm{loc}}[0,1) : (1-t)\, y(t) \in L^1[0,1]\}, \quad (2)$$

with the norm

$$\|y\|_Z = \int_0^1 (1-t)\,|y(t)|\, dt. \quad (3)$$

Let X be the Banach space

$$X = \Big\{x \in C^1[0,1) : x \in C[0,1],\ \lim_{t\to 1^-} (1-t)\, x'(t) \text{ exists}\Big\}, \quad (4)$$

with the norm

$$\|x\|_X = \max\left\{\|x\|_\infty, \left\|(1-t)\, x'(t)\right\|_\infty\right\} \quad \text{where } \|x\|_\infty = \sup_{t\in[0,1]} |x(t)|. \quad (5)$$

We denote the norm in $L^1[0,1]$ by $\|\cdot\|_1$. We define the linear operator $L : \mathrm{dom}\, L \subset X \to Z$ by

$$Lx = x''(t), \quad (6)$$

where $\mathrm{dom}\, L = \{x \in X : x'(0) = 0,\ x(1) = \sum_{i=1}^{m-2} a_i x(\xi_i)\}$ and $N : X \to Z$ is defined by

$$Nx = f(t, x(t), x'(t)) + g(t). \quad (7)$$

Then boundary value problem (1) can be written as

$$Lx = Nx. \quad (8)$$

Lemma 4 (see [2]). *Let $y \in Z$. Then*

(i) $\int_0^t y(s)ds \in L^1[0,1]$.

(ii) $\lim_{t\to 1^-} (1-t) \int_0^t y(s)ds = 0$.

Lemma 5. *If $\sum_{i=1}^{m-2} a_i = 1$ then*

(i) $\mathrm{Ker}\, L = \{x \in \mathrm{dom}\, L : x(t) = c,\ c \in \mathbb{R},\ t \in [0,1]\}$*;*

(ii) $\mathrm{Im}\, L = \{y \in Z : \sum_{i=1}^{m-2} a_i \int_{\xi_i}^1 \int_0^s y(\tau)d\tau\, ds = 0\}$*;*

(iii) *$L : \mathrm{dom}\, L \subset X \to Z$ is a Fredholm operator of index zero and the continuous operator $Q : Z \to Z$ can be defined by*

$$Qy = \frac{e^t}{h} \sum_{i=1}^{m-2} a_i \int_{\xi_i}^1 \int_0^s y(\tau)\, d\tau\, ds, \quad (9)$$

where $h = \sum_{i=1}^{m-2} a_i[e + \xi_i - e^{\xi_i} - 1] \neq 0$.

(iv) *The linear operator $K_p : \mathrm{Im}\, L :\to \mathrm{dom}\, L \cap \mathrm{Ker}\, P$ can be defined as*

$$K_p y = \int_0^t \int_0^s y(\tau)\, d\tau\, ds. \quad (10)$$

(v) *$\|K_p y\|_X \leq \|y\|_Z$ for all $y \in Z$.*

Proof. (i) It is obvious that

$$\mathrm{Ker}\, L = \{x \in \mathrm{dom}\, L : x(t) = c,\ c \in \mathbb{R}\}. \quad (11)$$

(ii) We show that

$$\mathrm{Im}\, L = \left\{y \in Z : \sum_{i=1}^{m-2} a_i \int_{\xi_i}^1 \int_0^s y(\tau)\, d\tau\, ds = 0\right\}. \quad (12)$$

To do this, we consider the problem

$$x''(t) = y(t), \quad (13)$$

and we show that (13) has a solution $x(t)$ satisfying $x'(0) = 0$, $x(1) = \sum_{i=1}^{m-2} a_i x(\xi_i)$ if and only if

$$\sum_{i=1}^{m-2} a_i \int_{\xi_i}^{1} \int_0^s y(\tau)\, d\tau\, ds = 0. \quad (14)$$

Suppose (13) has a solution $x(t)$ satisfying $x'(0) = 0$, $x(1) = \sum_{i=1}^{m-2} a_i x(\xi_i)$; then we obtain from (13) that

$$x(t) = x(0) + \int_0^t \int_0^s y(\tau)\, d\tau\, ds, \quad (15)$$

and applying the boundary conditions we get

$$\sum_{i=1}^{m-2} a_i \int_0^{\xi_i} \int_0^s y(\tau)\, d\tau\, ds = \int_0^1 \int_0^s y(\tau)\, d\tau\, ds, \quad (16)$$

since $\sum_{i=1}^{m-2} a_i = 1$, and using (i) of Lemma 4 we get

$$\sum_{i=1}^{m-2} a_i \int_{\xi_i}^{1} \int_0^s y(\tau)\, d\tau\, ds = 0. \quad (17)$$

On the other hand if (14) holds, let $x_0 \in \mathbb{R}$; then $x(t) = x_0 + \int_0^t \int_0^s y(\tau) d\tau\, ds$, where $y \in Z$ and $x'(t) \in AC_{\text{loc}}[0, 1)$. Then from Lemma 4 $\int_0^t y(\tau) d\tau \in L^1[0, 1]$ and $\lim_{t\to 1^-}(1-t)x'(t) = \lim_{t\to 1^-}(1-t)\int_0^t y(\tau) d\tau = 0$. Hence

$$x''(t) = y(t). \quad (18)$$

(iii) For $y \in Z$, we define the projection Qy as

$$Qy = \frac{e^t}{h} \sum_{i=1}^{m-2} a_i \int_{\xi_i}^{1} \int_0^s y(\tau)\, d\tau\, ds, \quad t \in [0, 1], \quad (19)$$

where $h = \sum_{i=1}^{m-2} a_i [e + \xi_i - e^{\xi_i} - 1] \neq 0$.

We show that $Q : Z \to Z$ is well defined and bounded.

$$\begin{aligned} |Qy(t)| &\le \frac{|e^t|}{|h|} \sum_{i=1}^{m-2} |a_i| \int_0^1 (1-s)\, |y(s)|\, ds \\ &= \frac{1}{|h|} \sum_{i=1}^{m-2} |a_i|\, \|y\|_Z \left|e^t\right|, \\ \|Qy\|_Z &\le \int_0^1 (1-t)\, |Qy(t)|\, dt \\ &\le \frac{1}{|h|} \sum_{i=1}^{m-2} |a_i|\, \|y\|_Z \int_0^1 (1-t) \left|e^t\right| dt \\ &= \frac{1}{|h|} \sum_{i=1}^{m-2} |a_i|\, \|y\|_Z \left\|e^t\right\|_Z. \end{aligned} \quad (20)$$

In addition it is easily verified that

$$Q^2 y = Qy, \quad y \in Z. \quad (21)$$

We therefore conclude that $Q : Z \to Z$ is a projection. If $y \in \operatorname{Im} L$, then from (14) $Qy(t) = 0$. Hence $\operatorname{Im} L \subseteq \operatorname{Ker} Q$. Let $y_1 = y - Qy$; that is, $y_1 \in \operatorname{Ker} Q$. Then

$$\begin{aligned} &\sum_{i=1}^{m-2} a_i \int_{\xi_i}^{1} \int_0^s y_1(\tau)\, d\tau\, ds \\ &= \sum_{i=1}^{m-2} a_i \int_{\xi_i}^{1} \int_0^s y(\tau)\, d\tau\, ds - \frac{1}{h} \sum_{i=1}^{m-2} \int_{\xi_i}^{1} \int_0^1 y(\tau)\, d\tau\, ds \\ &\cdot h = 0. \end{aligned} \quad (22)$$

Thus, $y_1 \in \operatorname{Im} L$ and therefore $\operatorname{Ker} Q \subseteq \operatorname{Im} L$ and hence $Z = \operatorname{Im} L + \operatorname{Im} Q = \operatorname{Im} L + \mathbb{R}$. It follows that since $\operatorname{Im} L \cap \mathbb{R} = \{0\}$, then $Z = \operatorname{Im} L \oplus \operatorname{Im} Q$. Therefore

$$\dim \operatorname{Ker} L = \dim \operatorname{Im} Q = \dim \mathbb{R} = \operatorname{codim} \operatorname{Im} L = 1. \quad (23)$$

This implies that L is Fredholm mapping of index zero.

(iv) We define $P : X \to X$ by

$$Px = x(0), \quad (24)$$

and clearly P is continuous and linear and $P^2 x = P(Px) = Px(0) = x(0) = Px$ and $\operatorname{Ker} P = \{x \in X : x(0) = 0\}$. We now show that the generalised inverse $K_p : \operatorname{Im} L \to \operatorname{dom} L \cap \operatorname{Ker} P$ of L is given by

$$K_p y = \int_0^t \int_0^s y(\tau)\, d\tau\, ds. \quad (25)$$

For $y \in \operatorname{Im} L$ we have

$$\left(LK_p\right) y(t) = \left[\left(K_p y\right)(t)\right]'' = y(t) \quad (26)$$

and for $x \in \operatorname{dom} L \cap \operatorname{Ker} P$ we know that

$$\begin{aligned} \left(K_p L\right) x(t) &= \int_0^t \int_0^s x''(\tau)\, d\tau\, ds \\ &= \int_0^t (t-s)\, x''(s)\, ds \\ &= x(t) - x'(0)\, t - x(0) = x(t) \end{aligned} \quad (27)$$

since $x \in \operatorname{dom} L \cap \operatorname{Ker} P$, $x(0) = 0$, and $Px = 0$.

This shows that $K_p = (L|_{\operatorname{dom} L \cap \operatorname{Ker} P})^{-1}$.

(v)

$$\begin{aligned} \left\|K_p y\right\|_\infty &\le \max_{t\in[0,1]} \int_0^t (t-s)\, |y(s)|\, ds \\ &\le \int_0^1 (1-s)\, |y(s)|\, ds \le \|y\|_Z, \\ \left\|(1-t)\left(K_p y\right)'\right\|_\infty &\le \max_{t\in[0,1]} \int_0^t (1-s)\, |y(s)|\, ds \\ &\le \|y\|_Z. \end{aligned} \quad (28)$$

We conclude that

$$\|K_P y\|_X \le \|y\|_Z. \tag{29}$$

□

Lemma 6. *The operator $N : X \to Z$ defined by*

$$Nx(t) = f(t, x(t), x'(t)) + g(t), \quad t \in (0,1) \tag{30}$$

is L-completely continuous.

Proof. Suppose Ω is an open bounded subset of X. Let $R_1 = \sup\{\|x\|_X : x \in \Omega\}$. From condition (A1) and each $x_n \in \Omega$ we have

$$\begin{aligned} |Nx_n(t)| &\le |f(t), x_n(t), x_n'(t)| + |g(t)| \\ &\le a(t) R_1 + b(t)\frac{R_1}{1-t} + r(t) + |g(t)| \\ &= \varphi(t). \end{aligned} \tag{31}$$

We can deduce from (A1) and (A2) that $\varphi(t) \in Z$:

$$\begin{aligned} |QNx_n(t)| &= \left| \frac{e^t}{h} \sum_{i=1}^{m-2} a_i \int_{\xi_i}^{1} \int_0^s Nx_n(\tau)\, d\tau\, ds \right| \\ &\le \frac{1}{|h|} \sum_{i=1}^{m-2} |a_i| \int_0^1 (1-s) |Nx_n(x)|\, ds \left|e^t\right| \\ &\le \frac{1}{|h|} \sum_{i=1}^{m-2} |a_i| \int_0^1 (1-s) |\varphi(s)|\, ds \left|e^t\right| \\ &\le \frac{1}{|h|} \sum_{i=1}^{m-2} |a_i| \|\varphi\|_Z \left|e^t\right|, \quad t \in (0,1), \\ \|QNx_n\|_Z &\le \frac{1}{|h|} \sum_{i=1}^{m-2} |a_i| \|\varphi\|_Z \int_0^1 (1-t) \left|e^t\right| dt \\ &= \frac{1}{|h|} \sum_{i=1}^{m-2} |a_i| \|\varphi\|_Z \left\|e^t\right\|_Z. \end{aligned} \tag{32}$$

This shows that $QN(\overline{\Omega})$ is bounded in Z and QN is continuous by using the Lebesgue Dominated Convergence Theorem. Next we show that $K_{P,Q}N(\overline{\Omega}) = K_P(I-Q)N(\overline{\Omega})$ is compact.

By using (31) we derive

$$\begin{aligned} |K_P N_n(t)| &\le \int_0^t (t-s) |Nx_n(s)|\, ds \\ &\le \int_0^1 (1-s) |\varphi(s)|\, ds = \|\varphi\|_Z, \end{aligned}$$

$$\begin{aligned} |Nx_n(t) - QNx_n(t)| &\le |Nx_n(t)| + |QNx_n(t)| \\ &\le |\varphi(t)| + \frac{1}{|h|} \sum_{i=1}^{m-2} |a_i| \|\varphi\|_Z \left|e^t\right| \\ &= \alpha_r(t), \\ |K_{P,Q}Nx_n(t)| &= |K_p(Nx_n - QNx_n)(t)| \\ &\le K_P \alpha_r(t) \le \|\alpha_r\|_Z. \end{aligned} \tag{33}$$

This indicates that the sequence $\{K_{P,Q}Nx_n(\overline{\Omega})\}$ is uniformly bounded in $C[0,1]$. Also for $t \in [0,1)$

$$\begin{aligned} &\left|(1-t)(K_{P,Q}Nx_n)'(t)\right| \\ &= \left|(1-t)\int_0^t (Nx_n(s) - QNx_n(s))\, ds\right| \\ &\le \int_0^1 (1-s)\alpha_r(s)\, ds \le \|\alpha_r\|_Z. \end{aligned} \tag{34}$$

Hence the sequence $K_{P,Q}Nx_n(t)$ is bounded in $C[0,1]$ and $\lim_{t\to 1^-}(1-t)(K_{P,Q}Nx_n)'(t) = 0$. Thus $K_{P,Q}Nx_n(t)$ is bounded in X.

Next we show that the sequence $\{K_{P,Q}Nx_n(t)\}$ is equicontinuous. Let $t_1, t_2 \in [0,1]$, $t_1 < t_2$; then

$$\begin{aligned} |K_{P,Q}x_n(t_2) - K_{P,Q}x_n(t_1)| &= \left|\int_{t_1}^{t_2} (K_{P,Q}Nx_n)'(s)\, ds\right| \\ &\le \int_{t_1}^{t_2} \int_0^s |(Nx_n(\tau) - QNx_n(\tau))|\, d\tau\, ds \\ &\le \int_{t_1}^{t_2} \int_0^s \alpha_r(\tau)\, d\tau\, ds, \end{aligned} \tag{35}$$

for every $t_1, t_2 \in [0,1]$. By (i) of Lemma 4 $\int_0^s \alpha_r(\tau) d\tau \in L^1[0,1]$. Thus the sequence $\{K_{P,Q}Nx_n(t)\}$ is equicontinuous on $[0,1]$ and by Arzela-Ascoli Theorem is convergent. Next we prove that the sequence $\{(1-t)(K_{P,Q}Nx_n)'\}$ is also equicontinuous on $[0,1]$. We have for $t \in [0,1]$

$$\begin{aligned} &\left|\left[(1-t)(K_{P,Q}Nx_n)'(t)\right]'\right| \\ &= \left|-(K_{P,Q}Nx_n)'(t) + (1-t)(K_{P,Q}Nx_n)''(t)\right| \\ &\le \left|\int_0^t (Nx_n(s) - QNx_n(s))\, ds\right| + |(1-t)Nx_n(t)| \\ &\le \int_0^t \alpha_r(s)\, ds + (1-t)\varphi(t) = \psi(t). \end{aligned} \tag{36}$$

Using (i) of Lemma 4 and the fact that $\alpha_r(t)$ and $\varphi(t)$ are in Z we conclude that $\psi(t) \in L^1[0,1]$. Therefore

$$\begin{aligned} &\left|(1-t_2)(K_{P,Q}Nx_n)'(t_2) - (1-t_1)(K_{P,Q}Nx_n)'(t_1)\right| \\ &= \left|\int_{t_1}^{t_2} \left[(1-s)(K_{P,Q}Nx_n)'(s)\right]' ds\right| \\ &\le \int_{t_1}^{t_2} \left|\left[(1-s)(K_{P,Q}Nx_n)'(s)\right]'\right| ds \le \int_{t_1}^{t_2} \psi(s)\,ds. \end{aligned} \tag{37}$$

The sequence $\{(1-t)(K_{P,Q}Nx_n)'(t)\}$ is therefore equicontinuous on $[0,1)$ and therefore converges to some $(1-t)(K_{P,Q}Nx_0)'(t) \in C[0,1]$ with $\lim_{t\to 1^-}(1-t)[(1-t)(K_{P,Q}Nx_n)'(t)] = 0, t \in [0,1)$.

We then conclude that $K_{P,Q}$ is relatively compact and since $QN(\overline{\Omega})$ is bounded we conclude from Definition 2 that N is L-compact on every bounded subset Ω of X and hence N is L-completely continuous. □

3. Main Result

In this section we will state and prove the main existence results for problem (1).

Theorem 7. *Assume that the following conditions are satisfied:*

(H1) *There exists a positive constant B_1 such that, for each $x \in \operatorname{dom} L$, if $|x(t)| > B_1$ for all $t \in [0,1]$ then*

$$\begin{aligned} &QNx(t) \\ &= \frac{e^t}{h}\sum_{i=1}^{m=2} a_i \int_{\xi_i}^1 \int_0^s \left[f\left(\tau, x(\tau), x'(\tau)\right) + g(\tau)\right] d\tau\, ds \\ &\neq 0. \end{aligned} \tag{38}$$

(H2) *There exists a positive constant B_2 such that for $c \in \mathbb{R}$ and $|c| > B_2$ either (i) $QN(c) \ge 0$ or (ii) $QN(c) \le 0$.*

Then (1) has at least one solution in X provided

$$\|a\|_Z + \|b\|_1 < \frac{1}{2}. \tag{39}$$

To prove Theorem 7, we first establish some lemmas.

Lemma 8. *Let $\Omega_1 = \{x \in \operatorname{dom} L \setminus \operatorname{Ker} L : Lx = \lambda Nx, \lambda \in (0,1)\}$ then Ω_1 is bounded in X.*

Proof. Let $x \in \Omega_1$. We let $Lx = \lambda Nx$, $0 < \lambda < 1$. Since $\lambda \neq 0$ it is clear that $Nx \in \operatorname{Im} L = \operatorname{Ker} Q$; hence $QNx = 0$ for all $t \in [0,1]$. Therefore by assumption (H1) there exist $t_0 \in [0,1]$ such that $|x(t_0)| < B_1$. Now

$$\begin{aligned} \int_0^t \int_0^s x''(\tau)\,d\tau\,ds &= \int_0^{t_0} (t_0 - s)\,x''(s)\,ds \\ &= x(t_0) - x(0) \end{aligned} \tag{40}$$

$$\begin{aligned} \|Px\|_X = |x(0)| &= \left|x(t_0) + \int_0^{t_0} (t_0 - s)\,x''(s)\,ds\right| \\ &\le |x(t_0)| + \int_0^1 (1-s)\left|x''(s)\right| ds \\ &\le B_1 + \|Lx\|_Z \le B_1 + \|Nx\|_Z. \end{aligned} \tag{41}$$

We note that $(I-P)x \in \operatorname{dom} L \cap \operatorname{Ker} P$:

$$\begin{aligned} \|(I-P)x\|_X &= \|K_P L(I-P)x\|_X \le \|K_P Lx\|_X \\ &\le \|Lx\|_Z < \|Nx\|_Z. \end{aligned} \tag{42}$$

From (41) and (42) we get

$$\begin{aligned} \|x\|_X &= \|Px + (I-P)x\|_X \le \|Px\|_X + \|(I-P)x\|_X \\ &< B_1 + 2\|Nx\|_Z. \end{aligned} \tag{43}$$

From the definition of N we obtain

$$\begin{aligned} \|Nx\|_Z &= \|(1-t)(Nx)(t)\|_1 = \int_0^1 (1-t) \\ &\cdot \left[f\left(t, x(t), x'(t)\right) + g(t)\right] dt \\ &\le \int_0^1 \left[(1-t)\,a(t)\,|x(t)| + |b(t)|\,(1-t)\,x'(t)\right. \\ &\left.+ (1-t)\,|r(t)| + (1-t)\left|g(t)\right|\right] dt \le \|a\|_Z \\ &\cdot \|x\|_\infty + \|b\|_1 \left\|(1-t)\,x'\right\|_\infty + \|r\|_Z + \|g\|_Z. \end{aligned} \tag{44}$$

From (43) and (44) we get

$$\begin{aligned} \|x\|_X &< B_1 \\ &+ 2\left[\|a\|_Z\|x\|_X + \|b\|_1\|x\|_X + \|r\|_Z + \|g\|_Z\right]. \end{aligned} \tag{45}$$

Since $1 - 2[\|a\|_Z + \|b\|_1] > 0$ we obtain that

$$\|x\|_X < \frac{B_1 + 2\|r\|_Z}{1 - 2\left[\|a\|_Z + \|b\|_1\right]} + \frac{2\|g\|_Z}{1 - 2\left[\|a\|_Z + \|b\|_1\right]}. \tag{46}$$

Therefore Ω_1 is bounded in X. □

Lemma 9. *The set $\Omega_2 = \{x \in \operatorname{Ker} L : Nx \in \operatorname{Im} L\}$ is a bounded subset of X.*

Proof. Let $x \in \Omega_2$ with $x(t) = c$, $c \in \mathbb{R}$. Then $QN(c) = 0$ implies $N(c) \in \operatorname{Im} L = \operatorname{Ker} Q$. We therefore derive from (H2) that

$$\|x\|_X = |c| = \max\{|c|, \|(1-t)\,0\|\} = |c| < B_2. \tag{47}$$

□

Lemma 10. *The sets* $\Omega_3^+ = \{x \in \operatorname{Ker} L : \lambda x + (1-\lambda)QNx = 0,\ \lambda \in [0,1]\}$ *and* $\Omega_3^- = \{x \in \operatorname{Ker} L : -\lambda x + (1-\lambda)QNx = 0,\ \lambda \in [0,1]\}$ *are bounded in* X *provided (H2)(i) and (H2)(ii) are satisfied simultaneously.*

Proof. If $QN(c) \geq 0$ then, for $x \in \Omega_3^+$ with $x(t) = c$, $c \in \mathbb{R}$, we have

$$\lambda c = -(1-\lambda) QN(c). \tag{48}$$

If $\lambda = 0$, it follows from (48) that $N(c) \in \operatorname{Ker} Q = \operatorname{Im} L$; that is, $N(c) \in \Omega_2$, and therefore by Lemma 9. we have $\|x\|_X \leq B_2$. However if $\lambda \in (0,1)$ and $\|c\| > B_2$ then using assumption (H2)(i) we obtain the contradiction

$$\lambda c^2 = -(1-\lambda) cQN(c) \leq 0. \tag{49}$$

Thus $\|x\|_X = |c| < B_2$. Hence Ω_3^+ is bounded in X. We can use the same argument to prove that Ω_3^- is also bounded in X. □

Proof of Theorem 7. We show that the conditions of Theorem 3 are satisfied where Ω is an open and bounded set such that $\bigcup_{i=1}^3 \Omega_i \subset \Omega$. It is easily seen that conditions (i) and (ii) of Theorem 3 are satisfied by using Lemmas 8 and 9. To verify the third condition we set $H(x,\lambda) = \pm\lambda x + (1-\lambda)QNx$. We choose the isomorphism $J : \operatorname{Im} Q \to \operatorname{Ker} L$ defined by $J(c) = c$, $c \in \mathbb{R}$. By Lemma 10, we derive that $H(x,\lambda) \neq 0$ for all $(x,\lambda) \in (\operatorname{Ker} L \cap \partial\Omega) \times [0,1]$. Hence

$$\begin{aligned} &\deg\left(QN|_{\operatorname{Ker} L}, \Omega \cap \operatorname{Ker} L, 0\right) \\ &\quad = \deg\left(\pm J, \Omega \cap \operatorname{Ker} L, 0\right) \neq 0. \end{aligned} \tag{50}$$

Therefore problem (1) has at least one solution in X. □

Conflicts of Interest

The authors declare that there are no conflicts of interest regarding the publication of this paper.

Acknowledgments

This work is supported by the Covenant University Centre for Research, Innovation and Discovery (CUCRID).

References

[1] C. P. Gupta, S. K. Ntouyas, and P. C. Tsamatos, "Solvability of an m-point boundary value problem for second order ordinary differential equations," *Journal of Mathematical Analysis and Applications*, vol. 189, no. 2, pp. 575–584, 1995.

[2] R. Ma and D. O'Regan, "Solvability of singular second order m-point boundary value problems," *Journal of Mathematical Analysis and Applications*, vol. 301, no. 1, pp. 124–134, 2005.

[3] R. P. Agarwal and D. O'Regan, *Singular Differential and Integral Equations with Applications*, Kluwer Academic Publishers, London, UK, 2003.

[4] H. Asakawa, "Nonresonant singular two-point boundary value problems," *Nonlinear Analysis: Theory, Methods & Applications*, vol. 44, no. 6, pp. 791–809, 2001.

[5] G. Infante and M. a. Zima, "Positive solutions of multi-point boundary value problems at resonance," *Nonlinear Analysis: Theory, Methods & Applications*, vol. 69, no. 8, pp. 2458–2465, 2008.

[6] N. Kosmatov, "A singular non-local problem at resonance," *Journal of Mathematical Analysis and Applications*, vol. 394, no. 1, pp. 425–431, 2012.

[7] R. Ma, "Existence of positive solutions for superlinear semipositone m-point boundary-value problems," *Proceedings of the Edinburg Mathematics Society*, vol. 46, no. 2, pp. 279–292, 2003.

[8] D. O'Regan, *Theory of Singular Boundary Value Problems*, World Scientific, River Edge, NJ, Usa, 1994.

[9] Z. Zhang and J. Wang, "The upper and lower solution method for a class of singular nonlinear second order three-point boundary value problems," *Journal of Computational and Applied Mathematics*, vol. 147, no. 1, pp. 41–52, 2002.

[10] J. Mawhin, *Topological Degree Methods in Nonlinear Boundary Value Problems*, vol. 40 of *NSFCBMS Regional Conference Series in Mathematics*, American Mathematical Society, Providence, RI, USA, 1979.

20

Computational Optimization of Residual Power Series Algorithm for Certain Classes of Fuzzy Fractional Differential Equations

Mohammad Alaroud,[1] Mohammed Al-Smadi,[2] Rokiah Rozita Ahmad,[1] and Ummul Khair Salma Din[1]

[1]*School of Mathematical Sciences, Faculty of Science and Technology, Universiti Kebangsaan Malaysia, 43600 Bangi, Selangor, Malaysia*
[2]*Department of Applied Science, Ajloun College, Al-Balqa Applied University, Ajloun 26816, Jordan*

Correspondence should be addressed to Rokiah Rozita Ahmad; rozy@ukm.edu.my

Academic Editor: Carla Pinto

This paper aims to present a novel optimization technique, the residual power series (RPS), for handling certain classes of fuzzy fractional differential equations of order $1 < \gamma \leq 2$ under strongly generalized differentiability. The proposed technique relies on generalized Taylor formula under Caputo sense aiming at extracting a supportive analytical solution in convergent series form. The RPS algorithm is significant and straightforward tool for creating a fractional power series solution without linearization, limitation on the problem's nature, sort of classification, or perturbation. Some illustrative examples are provided to demonstrate the feasibility of the RPS scheme. The results obtained show that the scheme is simple and reliable and there is good agreement with exact solution.

1. Introduction

Fuzzy fractional differential equation is hot and important branch of mathematics. It has attracted much attention recently due to potential applications in artificial intelligence, industrial engineering, physics, chemistry, and other fields of science. Parameters and variables in many of the nature studies and technological processes that were designed utilizing the fractional differential equation (FDE) are specific and completely defined. Indeed, such information may be vague and uncertain because of experimentation and measurement errors that then lead to uncertain models, which cannot handle these studies. The process of analyzing the relative influence of uncertainty in inputs information to outputs led us to study solutions to the qualitative behavior of equations. Therefore, it is necessary to obtain some mathematical tools to understand the complex structure of uncertainty models [1–5]. On the other hand, the theory of fractional calculus, which is a generalization of classical calculus, deals with the discussion of the integrals and derivatives of noninteger order, has a long history, and dates back to the seventeenth century [6–10]. Different forms of fractional operators are introduced to study FDEs such as Riemann–Liouville, Grunwald-Letnikov, and Caputo. Out of these forms, the Caputo concept is an appropriate tool for modeling practical situations due to its countless benefits as it allows the process to be performed based on initial and boundary conditions as is traditional and its derivative is zero for constant [11–17].

The residual power series (RPS) method developed in [18] is considered as an effective optimization technique to determine and define the power series solution's values of coefficients of first- and second-order fuzzy differential equations [19–22]. Furthermore, the RPS is characterized as an applicable and easy technique to create power series solutions for strongly linear and nonlinear equations without being linearized, discretized, or exposed to perturbation [23–27]. Unlike the classical power series method, the RPS neither requires comparing the corresponding coefficients nor is a recursion relation needed as well. Besides that, it calculates the power series coefficients through chain of equations of one or more variables and offers convergence of a series solution whose terms approach quickly, especially when the exact solution is polynomial.

The remainder of this paper is organized as follows. In Section 2, essential facts and results related to the fuzzy fractional calculus will be shown. In Section 3, the concept of Caputo's H-differentiability will be presented together with some closely related results. In Section 4, basic idea of the RPS method will be presented to solve the fuzzy FDEs of order $1 < \gamma \leq 2$. In Section 5, numerical application will be performed to show capability, potentiality, and simplicity of the method. Conclusions will be given in Section 6.

2. Preliminaries

In this section, necessary definitions and results relating to fuzzy fractional calculus are presented. For the fuzzy derivative concept, the strongly generalized differentiability will be adopted, which is considered H-differentiability modification.

A fuzzy set v in a nonempty set U is described by its membership function $v : U \to [0,1]$. So, for each $\eta \in U$ the degree of membership of η in v is defined by $v(\eta)$.

Definition 1 ([28]). Suppose that v is a fuzzy subset of $\mathbb{R}$. Then, v is called a fuzzy number such that v is upper semicontinuous membership function of bounded support, normal, and convex.

If v is a fuzzy number, then $[v]^{\sigma} = [v_1(\sigma), v_2(\sigma)]$, where $v_1(\sigma) = \min\{\eta \mid \eta \in [v]^{\sigma}\}$ and $v_2(\sigma) = \max\{\eta \mid \eta \in [v]^{\sigma}\}$ for each $\sigma \in [0,1]$. The symbol $[v]^{\sigma}$ is called the σ-level representation or the parametric form of a fuzzy number v.

Theorem 2 ([29]). *Suppose that $v_1, v_2 : [0,1] \to \mathbb{R}$ satisfy the following conditions:*

(1) *v_1 is a bounded nondecreasing function.*

(2) *v_2 is a bounded nonincreasing function.*

(3) $v_1(1) \leq v_2(1)$.

(4) *for each $k \in (0,1]$, $\lim_{\sigma \to k^-} v_1(\sigma) = v_1(k)$ and $\lim_{\sigma \to k^-} v_2(\sigma) = v_2(k)$.*

(5) $\lim_{\sigma \to 0^+} v_1(\sigma) = v_1(0)$ *and* $\lim_{\sigma \to 0^+} v_2(\sigma) = v_2(0)$.

Then $v : \mathbb{R} \to [0,1]$ given by $v(x) = \sup\{\sigma \mid v_1(\sigma) \leq x \leq v(\sigma)\}$ is a fuzzy number with parameterization $[v_1(\sigma), v_2(\sigma)]$.

Definition 3 ([29]). Let $v, w \in \mathbb{R}_{\mathcal{F}}$. If there exists an element $\mathcal{P} \in \mathbb{R}_{\mathcal{F}}$ such that $v = w + \mathcal{P}$, then we say that $\mathcal{P}$ is the Hukuhara difference (H-difference) of v and w, denoted by $v \ominus w$.

The sign $\ominus$ stands always for Hukuhara difference. Thus, it should be noted that $v \ominus w \neq v + (-1)w$. Normally, $v + (-1)w$ is denoted by $v - w$. If the H-difference $v \ominus w$ exists, then $[v \ominus w]^{\sigma} = [v_1(\sigma) - w_1(\sigma), v_2(\sigma) - w_2(\sigma)]$.

Definition 4 ([30]). The complete metric structure on $\mathbb{R}_{\mathcal{F}}$ is given by the Hausdorff distance mapping $D_H : \mathbb{R}_{\mathcal{F}} \times \mathbb{R}_{\mathcal{F}} \to \mathbb{R}^+ \cup \{0\}$ such that

$$D_H(v, w) = \sup_{0 \leq \sigma \leq 1} \max\left\{\left|v_{1\sigma} - w_{1\sigma}\right|, \left|v_{2\sigma} - w_{2\sigma}\right|\right\}, \quad (1)$$

for arbitrary fuzzy numbers $v = (v_1, v_2)$ and $w = (w_1, w_2)$.

Definition 5 ([30]). Let $\varphi : [a,b] \to \mathbb{R}_{\mathcal{F}}$. Then the function φ is continuous at $x_0 \in [a,b]$ if for every $\epsilon > 0$, $\exists \delta = \delta(x_0, \epsilon) > 0$ such that $D_H(\varphi(x), \varphi(x_0)) < \epsilon$, for each $x \in [a,b]$, whenever $|x - x_0| < \delta$.

Remark 6. If the function $\varphi(x)$ is continuous for each $x \in [a,b]$, where the continuity is one-sided at endpoints of $[a,b]$, then $\varphi(x)$ is continuous function on $[a,b]$. This means that $\varphi(x)$ is continuous on $[a,b]$ if and only if $\varphi_{1\sigma}$ and $\varphi_{2\sigma}$ are continuous on $[a,b]$.

Definition 7 ([28]). For fixed $x_0 \in [a,b]$ and $\varphi : [a,b] \to \mathbb{R}_{\mathcal{F}}$, the function φ is called a strongly generalized differentiable at x_0, if there is an element $\varphi'(x_0) \in \mathbb{R}_{\mathcal{F}}$ such that either

(i) the H-differences $\varphi(x_0 + \xi) \ominus \varphi(x_0), \varphi(x_0) \ominus \varphi(x_0 - \xi)$ exist, for each $\xi>0$ sufficiently tends to 0 and $\lim_{\xi \to 0^+}((\varphi(x_0 + \xi) \ominus \varphi(x_0))/\xi) = \varphi'(x_0) = \lim_{\xi \to 0^+}((\varphi(x_0) \ominus \varphi(x_0 - \xi))/\xi)$, or

(ii) the H-differences $\varphi(x_0) \ominus \varphi(x_0 + \xi), \varphi(x_0 - \xi) \ominus \varphi(x_0)$ exist, for each $\xi>0$ sufficiently tends to 0 and $\lim_{\xi \to 0^+}((\varphi(x_0) \ominus \varphi(x_0 + \xi))/-\xi) = \varphi'(x_0) = \lim_{\xi \to 0^+}((\varphi(x_0 - \xi) \ominus \varphi(x_0))/-\xi)$,

where the limit here is taken in the complete metric space $(\mathbb{R}_{\mathcal{F}}, D_H)$.

Theorem 8 ([31]). *Suppose that $\varphi : [a,b] \to \mathbb{R}_{\mathcal{F}}$, where $[\varphi(x)]^{\sigma} = [\varphi_{1\sigma}(x), \varphi_{2\sigma}(x)]$, $\forall \sigma \in [0,1]$, then*

(1) *the functions $\varphi_{1\sigma}$ and $\varphi_{2\sigma}$ are two differentiable functions and $[D_1^1\varphi(x)]^{\sigma} = [\varphi'_{1\sigma}(x), \varphi'_{2\sigma}(x)]$, when φ is (1)-differentiable;*

(2) *the functions $\varphi_{1\sigma}$ and $\varphi_{2\sigma}$ are two differentiable functions and $[D_2^1\varphi(x)]^{\sigma} = [\varphi'_{2\sigma}(x), \varphi'_{1\sigma}(x)]$, when φ is (2)-differentiable.*

Definition 9 ([31]). Suppose that $\varphi : [a,b] \to \mathbb{R}_{\mathcal{F}}$. One can say that φ is (n,m)-differentiable at $x_0 \in (a,b)$, if $D_n^1\varphi$ exists on a neighborhood of x_0 as a fuzzy function and it is (m)-differentiable at x_0. The second-order derivatives of φ at x are indicated by $\varphi''(x) = D_{n,m}^2\varphi(x)$ for $n, m = \{1,2\}$.

Theorem 10 ([32]). *Let $D_1^1\varphi : [a,b] \to \mathbb{R}_{\mathcal{F}}$ and $D_2^1\varphi : [a,b] \to \mathbb{R}_{\mathcal{F}}$, where $[\varphi(x)]^{\sigma} = [\varphi_{1\sigma}(x), \varphi_{2\sigma}(x)]$ for each $\sigma \in [0,1]$:*

(1) *If $D_1^1\varphi$ is (1)-differentiable, then $\varphi'_{1\sigma}$ and $\varphi'_{2\sigma}$ are differentiable functions and $[D_{1,1}^2\varphi(x)]^{\sigma} = [\varphi''_{1\sigma}(x), \varphi''_{2\sigma}(x)]$,*

(2) *If $D_1^1\varphi$ is (2)-differentiable, then $\varphi'_{1\sigma}$ and $\varphi'_{2\sigma}$ are differentiable functions and $[D_{1,2}^2\varphi(x)]^{\sigma} = [\varphi''_{2\sigma}(x), \varphi''_{1\sigma}(x)]$,*

(3) *If* $D_2^1\varphi$ *is (1)-differentiable, then* $\varphi_{1\sigma}'$ *and* $\varphi_{2\sigma}'$ *are differentiable functions and* $[D_{2,1}^2\varphi(x)]^\sigma = [\varphi_{2\sigma}''(x), \varphi_{1\sigma}''(x)]$,

(4) *If* $D_2^1\varphi$ *is (2)-differentiable, then* $\varphi_{1\sigma}'$ *and* $\varphi_{2\sigma}'$ *are differentiable functions and* $[D_{2,2}^2\varphi(x)]^\sigma = [\varphi_{1\sigma}''(x), \varphi_{2\sigma}''(x)]$.

Definition 11 ([32]). Let $\varphi : [a,b] \to \mathbb{R}_{\mathcal{F}}$ and $\varphi \in C^{\mathcal{F}}[a,b] \cap L^{\mathcal{F}}[a,b]$. One can say that φ is Caputo fuzzy H-differentiable at x when $({}^CD_{a+}^\gamma\varphi)(x) = (1/\Gamma(1-\gamma))\int_a^x \varphi'(\tau)/(x-\tau)^\gamma d\tau$ exists, where $0 < \gamma \le 1$. Also, we say that φ is Caputo $[(1)-\gamma]$-differentiable if φ is (1)-differentiable and φ is Caputo $[(2)-\gamma]$ differentiable if φ is (2)-differentiable, where $C^{\mathcal{F}}[a,b]$ and $L^{\mathcal{F}}[a,b]$ stand for the space of all continuous and Lebesque integrable fuzzy-valued functions on $[a,b]$, respectively.

Theorem 12 ([33]). *Let* $0 < \gamma \le 1$ *and* $\varphi \in C^{\mathcal{F}}[a,b]$. *Then, for each* $\sigma \in [0,1]$, *the Caputo fuzzy fractional derivative exists on* (a,b) *such that*

$$\left[\left({}^CD_{a+}^\gamma\varphi\right)(x)\right]^\sigma = \left[\frac{1}{\Gamma(1-\gamma)} \cdot \int_a^x \frac{\varphi_{1\sigma}'(\tau)}{(x-\tau)^\gamma}d\tau, \frac{1}{\Gamma(1-\gamma)}\int_a^x \frac{\varphi_{2\sigma}'(\tau)}{(x-\tau)^\gamma}d\tau\right] \quad (2)$$

for (1)-differentiable and

$$\left[\left({}^CD_{a+}^\gamma\varphi\right)(x)\right]^\sigma = \left[\frac{1}{\Gamma(1-\gamma)} \cdot \int_a^x \frac{\varphi_{2\sigma}'(\tau)}{(x-\tau)^\gamma}d\tau, \frac{1}{\Gamma(1-\gamma)}\int_a^x \frac{\varphi_{1\sigma}'(\tau)}{(x-\tau)^\gamma}d\tau\right] \quad (3)$$

for (2)-differentiable.

The next characterization theorem shows a way to convert the FFDEs into a system of ordinary fractional differential equations (OFDEs), ignoring the fuzzy setting approach.

Theorem 13 ([34]). *Consider the below fuzzy fractional IVPs*

$$\left({}^CD_{t_0^+}^\gamma\varphi\right)(t) = f(t,\varphi(t)), \quad t > t_0, \quad (4)$$

subject to

$$\varphi(t_0) = \varphi_0, \quad (5)$$

where $f : [a,b] \times \mathbb{R}_{\mathcal{F}} \to \mathbb{R}_{\mathcal{F}}$ *such that*

(i) $[f(t,\varphi(t))]^\sigma = [f_{1\sigma}(t,\varphi_{1\sigma}(t),\varphi_{2\sigma}(t)), f_{2\sigma}(t,\varphi_{1\sigma}(t),\varphi_{2\sigma}(t))]$.

(ii) for any $\epsilon > 0$ *there exist* $\delta > 0$ *such that* $|f_{1\sigma}(t,s,u) - f_{1\sigma}(t_1,s_1,u_1)| < \epsilon$ *and* $|f_{2\sigma}(t,s,u) - f_{2\sigma}(t_1,s_1,u_1)| < \epsilon$, $\forall\sigma \in [0,1]$, *whenever* (t,s,u) *and* $(t_1,s_1,u_1) \in [a,b]\times\mathbb{R}^2$, $\|(t,s,u)-(t_1,s_1,u_1)\|_{\mathbb{R}^3} < \delta$ *and* $f_{1\sigma}, f_{2\sigma}$ *are uniformly bounded on any bounded set.*

(iii) there is a constant (say) $\ell > 0$ *such that*

$$|f_{1\sigma}(t_2,s_2,u_2) - f_{1\sigma}(t_1,s_1,u_1)| \le \ell.\max\{|s_2-s_1|,|u_2-u_1|\}, \quad \forall\sigma\in[0,1] \quad (6)$$

and

$$|f_{2\sigma}(t_2,s_2,u_2) - f_{2\sigma}(t_1,s_1,u_1)| \le \ell.\max\{|s_2-s_1|,|u_2-u_1|\}, \quad \forall\sigma\in[0,1]. \quad (7)$$

Therefore, there are two systems of OFDEs that are equivalent to FFDEs (4) and (5) as follows:

Case 1. When $\varphi(t)$ *is Caputo [(1)-*γ*]-differentiable*

$$\begin{aligned}\left({}^CD_{t_0^+}^\gamma\varphi_{1\sigma}\right)(t) &= f_{1\sigma}(t,\varphi_{1\sigma}(t),\varphi_{2\sigma}(t)),\\ \left({}^CD_{t_0^+}^\gamma\varphi_{2\sigma}\right)(t) &= f_{2\sigma}(t,\varphi_{1\sigma}(t),\varphi_{2\sigma}(t)),\end{aligned} \quad (8)$$

with $\varphi_{1\sigma}(t_0) = \varphi_{01\sigma}$, $\varphi_{2\sigma}(t_0) = \varphi_{02\sigma}$.

Case 2. When $\varphi(t)$ *is Caputo [(2)-*γ*]-differentiable*

$$\begin{aligned}\left({}^CD_{t_0^+}^\gamma\varphi_{1\sigma}\right)(t) &= f_{2\sigma}(t,\varphi_{1\sigma}(t),\varphi_{2\sigma}(t)),\\ \left({}^CD_{t_0^+}^\gamma\varphi_{2\sigma}\right)(t) &= f_{1\sigma}(t,\varphi_{1\sigma}(t),\varphi_{2\sigma}(t)),\end{aligned} \quad (9)$$

with $\varphi_{1\sigma}(t_0) = \varphi_{01\sigma}$, $\varphi_{2\sigma}(t_0) = \varphi_{02\sigma}$.

3. Formulation of Fuzzy Fractional IVPs of Order $1 < \gamma \le 2$

Consider the below fuzzy fractional differential equation

$$\left({}^CD_{a^+}^\gamma\varphi\right)(t) = g(t)\varphi'(t) + f(t,\varphi(t)), \quad a \le t \le b, \ 1 < \gamma \le 2, \quad (10)$$

subject to fuzzy initial conditions

$$\begin{aligned}\varphi(a) &= \alpha,\\ \varphi'(a) &= \beta.\end{aligned} \quad (11)$$

where $\alpha, \beta \in \mathbb{R}_{\mathcal{F}}$, $f : [a,b]\times\mathbb{R}_{\mathcal{F}} \to \mathbb{R}_{\mathcal{F}}$ is a linear or nonlinear continuous fuzzy-valued function, $g(t)$ is a continuous real valued function with nonnegative values on $[a,b]$, and $\varphi(t)$ is unknown analytical fuzzy function to be determined. We assume that the fuzzy fractional IVPs (10) and (11) have unique smooth solution on the domain of interest.

Next, some theorems and definitions which are used later in this paper are presented.

Definition 14. Let $\varphi : [a,b] \to \mathbb{R}_{\mathcal{F}}$ be fuzzy function such that $\varphi, \varphi' \in C^{\mathcal{F}}[a,b] \cap L^{\mathcal{F}}[a,b]$. Then, for $1 < \gamma \le 2$, Caputo's H-derivative of φ at $x \in (a,b)$ is defined as

$$\left({}^CD_{a+}^\gamma\varphi\right)(x) = \frac{1}{\Gamma(2-\gamma)}\int_a^x \varphi''(\tau)(x-\tau)^{1-\gamma}d\tau. \quad (12)$$

Also, we say that φ is Caputo $[(n,m)-\gamma]$-differentiable for $n, m \in \{1,2\}$, when $({}^CD_{a+}^\gamma\varphi)(x)$ exists, and φ is (n,m)-differentiable.

Theorem 15. *Let $\varphi, \varphi' \in C^{\mathcal{F}}[a,b]$, such that $[\varphi(x)]^{\sigma} = [\varphi_{1\sigma}(x), \varphi_{2\sigma}(x)]$, $\forall \sigma \in [0,1]$. Caputo's H-derivative of order $1 < \gamma \leq 2$ exists on (a,b) such that*

(i) If φ is (1,1)-differentiable, then $[({}^{C}D^{\gamma}_{a+}\varphi)(x)]^{\sigma} = [M \int_a^x \varphi''_{1\sigma}(\tau)(x-\tau)^{1-\gamma}d\tau, M \int_a^x \varphi''_{2\sigma}(\tau)(x-\tau)^{1-\gamma}d\tau] = [({}^{C}D^{\gamma}_{a+}\varphi_{1\sigma})(x), ({}^{C}D^{\gamma}_{a+}\varphi_{2\sigma})(x)]$.

(ii) If φ is (1,2)-differentiable, then $[({}^{C}D^{\gamma}_{a+}\varphi)(x)]^{\sigma} = [M \int_a^x \varphi''_{2\sigma}(\tau)(x-\tau)^{1-\gamma}d\tau, M \int_a^x \varphi''_{1\sigma}(\tau)(x-\tau)^{1-\gamma}d\tau] = [({}^{C}D^{\gamma}_{a+}\varphi_{2\sigma})(x), ({}^{C}D^{\gamma}_{a+}\varphi_{1\sigma})(x)]$.

(iii) If φ is (2,1)-differentiable, then $[({}^{C}D^{\gamma}_{a+}\varphi)(x)]^{\sigma} = [M \int_a^x \varphi''_{2\sigma}(\tau)(x-\tau)^{1-\gamma}d\tau, M \int_a^x \varphi''_{1\sigma}(\tau)(x-\tau)^{1-\gamma}d\tau] = [({}^{C}D^{\gamma}_{a+}\varphi_{2\sigma})(x), ({}^{C}D^{\gamma}_{a+}\varphi_{1\sigma})(x)]$.

(iv) If φ is (2,2)-differentiable, then $[({}^{C}D^{\gamma}_{a+}\varphi)(x)]^{\sigma} = [M \int_a^x \varphi''_{1\sigma}(\tau)(x-\tau)^{1-\gamma}d\tau, M \int_a^x \varphi''_{2\sigma}(\tau)(x-\tau)^{1-\gamma}d\tau] = [({}^{C}D^{\gamma}_{a+}\varphi_{1\sigma})(x), {}^{C}D^{\gamma}_{a+}\varphi_{2\sigma})(x)]$, where $M = 1/\Gamma(2-\gamma)$.

The (n,m)-solution of fuzzy fractional IVPs (10) and (11) is a function $\varphi : [a,b] \to \mathbb{R}_{\mathcal{F}}$ that has Caputo $[(n,m)-\gamma]$-differentiable and satisfies the FFIVPs (10) and (11). To compute it, we firstly convert the fuzzy problem into equivalent system of second OFDEs, called correspondence (n,m)-system, based upon the type of derivative chosen. Then, by utilizing the σ-cut representation of $\varphi(t)$, $f(t,\varphi(t))$, and the initial data in (11) such that $[\varphi(t)]^{\sigma} = [\varphi_{1\sigma}(t), \varphi_{2\sigma}(t)]$, $[f(t,\varphi(t))]^{\sigma} = [f_{1\sigma}(t,\varphi_{1\sigma}(t),\varphi_{2\sigma}(t)), f_{2\sigma}(t,\varphi_{1\sigma}(t),\varphi_{2\sigma}(t))]$, $[\varphi(a)]^{\sigma} = [\varphi_{1\sigma}(a), \varphi_{2\sigma}(a)] = [\alpha_{1\sigma}, \alpha_{2\sigma}]$, and $[\varphi'(a)]^{\sigma} = [\varphi'_{1\sigma}(a), \varphi'_{2\sigma}(a)] = [\beta_{1\sigma}, \beta_{2\sigma}]$, the following corresponding (n,m)-systems will be hold:

(i) (1,1)-system such that

$$\begin{aligned} \left({}^{C}D^{\gamma}_{a^+}\varphi_{1\sigma}\right)(t) &= g(t)\varphi'_{1\sigma}(t) + f_{1\sigma}(t,\varphi_{1\sigma}(t),\varphi_{2\sigma}(t)), \\ \left({}^{C}D^{\gamma}_{a^+}\varphi_{2\sigma}\right)(t) &= g(t)\varphi'_{2\sigma}(t) + f_{2\sigma}(t,\varphi_{1\sigma}(t),\varphi_{2\sigma}(t)), \end{aligned} \tag{13}$$

(ii) the (1,2)-system such that

$$\begin{aligned} \left({}^{C}D^{\gamma}_{a^+}\varphi_{2\sigma}\right)(t) &= g(t)\varphi'_{1\sigma}(t) + f_{1\sigma}(t,\varphi_{1\sigma}(t),\varphi_{2\sigma}(t)), \\ \left({}^{C}D^{\gamma}_{a^+}\varphi_{1\sigma}\right)(t) &= g(t)\varphi'_{2\sigma}(t) + f_{2\sigma}(t,\varphi_{1\sigma}(t),\varphi_{2\sigma}(t)), \end{aligned} \tag{14}$$

(iii) the (2,1)-system such that

$$\begin{aligned} \left({}^{C}D^{\gamma}_{a^+}\varphi_{2\sigma}\right)(t) &= g(t)\varphi'_{2\sigma}(t) + f_{1\sigma}(t,\varphi_{1\sigma}(t),\varphi_{2\sigma}(t)), \\ \left({}^{C}D^{\gamma}_{a^+}\varphi_{1\sigma}\right)(t) &= g(t)\varphi'_{1\sigma}(t) + f_{2\sigma}(t,\varphi_{1\sigma}(t),\varphi_{2\sigma}(t)), \end{aligned} \tag{15}$$

(iv) the (2,2)-system such that

$$\begin{aligned} \left({}^{C}D^{\gamma}_{a^+}\varphi_{1\sigma}\right)(t) &= g(t)\varphi'_{2\sigma}(t) + f_{1\sigma}(t,\varphi_{1\sigma}(t),\varphi_{2\sigma}(t)), \\ \left({}^{C}D^{\gamma}_{a^+}\varphi_{2\sigma}\right)(t) &= g(t)\varphi'_{1\sigma}(t) + f_{2\sigma}(t,\varphi_{1\sigma}(t),\varphi_{2\sigma}(t)), \end{aligned} \tag{16}$$

subject to initial conditions

$$\begin{aligned} \varphi_{1\sigma}(a) &= \alpha_{1\sigma}, \\ \varphi'_{1\sigma}(a) &= \beta_{1\sigma}, \\ \varphi_{2\sigma}(a) &= \alpha_{2\sigma}, \\ \varphi'_{2\sigma}(a) &= \beta_{2\sigma}. \end{aligned} \tag{17}$$

Theorem 16 ([33]). *Let $n,m \in \{1,2\}$ and let $[\varphi(t)]^{\sigma} = [\varphi_{1\sigma}(t), \varphi_{2\sigma}(t)]$ be an (n,m)-solution of FFIVPs (10) and (11) on $[a,b]$. Then, $\varphi_{1\sigma}(t)$ and $\varphi_{2\sigma}(t)$ will be a solution to the associated (n,m)-system.*

Theorem 17 ([33]). *Let $n,m \in \{1,2\}$ and let $\varphi_{1\sigma}(t)$ and $\varphi_{2\sigma}(t)$ be the solution of (n,m)-system for each $\sigma \in [0,1]$. If $[\varphi(t)]^{\sigma} = [\varphi_{1\sigma}(t), \varphi_{2\sigma}(t)]$ has valid level sets and $\varphi(t)$ is Caputo $[(n,m)-\gamma]$-differentiable, then $\varphi(t)$ is an (n,m)-solution of FFIVPs (10) and (11) on $[a,b]$.*

The aim of the next algorithm is to perform a strategy to solve the FFIVPs (10) and (11) in terms of its σ-cut representation form. Indeed, there are four cases that depend on type of differentiability.

Algorithm 18. To determine the solutions of FFIVPs (10) and (11), do the following:

Case (I). If $\varphi(t)$ is Caputo $[(1,1)\text{-}\gamma]$-differentiable and the FFIVPs (10) and (11) will be converted to crisp system described in (13) and (17), then do the following steps:

Step 1: Solve the required system.

Step 2: Ensure that $[\varphi_{1\sigma}(t), \varphi_{2\sigma}(t)]$, $[\varphi'_{1\sigma}(t), \varphi'_{2\sigma}(t)]$ and $[\varphi''_{1\sigma}(t), \varphi''_{2\sigma}(t)]$ are valid level sets for each $\sigma \in [0,1]$.

Step 3: Construct (1,1)-solution $\varphi(t)$ whose σ-cut representation is $[\varphi_{1\sigma}(t), \varphi_{2\sigma}(t)]$.

Case (II). If $\varphi(t)$ is Caputo $[(1,2)\text{-}\gamma]$-differentiable and the FFIVPs (10) and (11) will be converted to crisp system described in (14) and (17), then do the following steps:

Step 1: Solve the required system.

Step 2: Ensure that $[\varphi_{1\sigma}(t), \varphi_{2\sigma}(t)]$, $[\varphi'_{1\sigma}(t), \varphi'_{2\sigma}(t)]$ and $[\varphi''_{2\sigma}(t), \varphi''_{1\sigma}(t)]$ are valid level sets for each $\sigma \in [0,1]$.

Step 3: Construct (1,2)-solution $\varphi(t)$ whose σ-cut representation is $[\varphi_{1\sigma}(t), \varphi_{2\sigma}(t)]$.

Case (III). If $\varphi(t)$ is Caputo $[(2,1)\text{-}\gamma]$-differentiable and the FFIVPs (10) and (11) will be converted to crisp system described in (15) and (17), then do the following steps:

Step 1: Solve the required system.

Step 2: Ensure that $[\varphi_{1\sigma}(t), \varphi_{2\sigma}(t)]$, $[\varphi'_{2\sigma}(t), \varphi'_{1\sigma}(t)]$ and $[\varphi''_{2\sigma}(t), \varphi''_{1\sigma}(t)]$ are valid level sets for each $\sigma \in [0,1]$.

Step 3: Construct (2,1)-solution $\varphi(t)$ whose σ-cut representation is $[\varphi_{1\sigma}(t), \varphi_{2\sigma}(t)]$.

Case (IV). If $\varphi(t)$ is Caputo $[(2,2)\text{-}\gamma]$-differentiable and the FFIVPs (10) and (11) will be converted to crisp system described in (16) and (17), then do the following steps:

Step 1: Solve the required system.

Step 2: Ensure that $[\varphi_{1\sigma}(t), \varphi_{2\sigma}(t)]$, $[\varphi'_{2\sigma}(t), \varphi'_{1\sigma}(t)]$ and $[\varphi''_{1\sigma}(t), \varphi''_{2\sigma}(t)]$ are valid level sets for each $\sigma \in [0,1]$.

Step 3: Construct (2,2)-solution $\varphi(t)$ whose σ-cut representation is $[\varphi_{1\sigma}(t), \varphi_{2\sigma}(t)]$.

4. Description of Fractional RPS Method

In this section, the RPS scheme is presented for constructing an analytical solution of FFIVPs (10) and (11) through substituting the expansion of fractional power series (FPS) among the truncated residual functions. In view of that, the resultant equation helps us to derive a recursion formula for the coefficients' computation, where the coefficients can be computed recursively through the recurrent fractional differentiating of the truncated residual function.

Definition 19 ([35]). A fractional power series (FPS) representation at t_0 has the following form:

$$\sum_{n=0}^{\infty} c_n (t-t_0)^{n\gamma} = c_0 + c_1 (t-t_0)^{\gamma} + c_2 (t-t_0)^{2\gamma} + \ldots, \quad (18)$$

where $0 \le m-1 < \gamma \le m$, $t \ge t_0$, and c_n's are the coefficients of the series.

Theorem 20 ([35]). *Suppose that f has the following FPS representation at t_0:*

$$f(t) = \sum_{n=0}^{\infty} c_n (t-t_0)^{n\gamma}, \quad (19)$$

where $f(t) \in C[t_0, t_0+R)$ and ${}^{C}D^{n\gamma} f(t) \in C(t_0, t_0+R)$ for $n = 0,1,2,\ldots$; then the coefficients c_n will be in the form $c_n = {}^{C}D^{n\gamma} f(t_0)/\Gamma(1+n\gamma)$ such that ${}^{C}D^{n\gamma} = {}^{C}D^{\gamma} \cdot {}^{C}D^{\gamma} \cdot \ldots \cdot {}^{C}D^{\gamma}$ (n-times).

Conveniently, for obtaining (n,m)-solution of FFIVPs (10) and (11) utilizing the solution of the corresponding (n,m)-system, we will explain the fashion to determine $(1,1)$-solution equivalent to the solution for the system of OFDEs (13) and (17). Further, same manner can be applied to construct other type of (n,m)-solutions. To achieve our goal, assume that the solution of OFDEs (13) and (17) at $t_0 = 0$ has the following form:

$$\varphi_{1\sigma}(t) = \sum_{n=0}^{\infty} c_n \frac{t^{n\gamma}}{\Gamma(1+n\gamma)},$$
$$\varphi_{2\sigma}(t) = \sum_{n=0}^{\infty} d_n \frac{t^{n\gamma}}{\Gamma(1+n\gamma)}. \quad (20)$$

Since $\varphi_{1\sigma}(t)$ and $\varphi_{2\sigma}(t)$ satisfy the initial conditions in (17), then the following polynomials $\varphi_{1\sigma}(t) = \alpha_{1\sigma} + \beta_{1\sigma} t$ and $\varphi_{2\sigma}(t) = \alpha_{2\sigma} + \beta_{2\sigma} t$ will be the initial guesses for the system and the solutions can also be represented by

$$\varphi_{1\sigma}(t) = \alpha_{1\sigma} + \beta_{1\sigma} t + \sum_{n=1}^{\infty} c_n \frac{t^{n\gamma}}{\Gamma(1+n\gamma)},$$
$$\varphi_{2\sigma}(t) = \alpha_{2\sigma} + \beta_{2\sigma} t + \sum_{n=1}^{\infty} d_n \frac{t^{n\gamma}}{\Gamma(1+n\gamma)}. \quad (21)$$

Consequently, the kth-truncated series solutions can be given by

$$\varphi_{k,1\sigma}(\mathrm{t}) = \alpha_{1\sigma} + \beta_{1\sigma} t + \sum_{n=1}^{k} c_n \frac{t^{n\gamma}}{\Gamma(1+n\gamma)},$$
$$\varphi_{k,2\sigma}(\mathrm{t}) = \alpha_{2\sigma} + \beta_{2\sigma} t + \sum_{n=1}^{k} d_n \frac{t^{n\gamma}}{\Gamma(1+n\gamma)}. \quad (22)$$

The residual functions $Res_{1\sigma}(t)$ and $Res_{2\sigma}(t)$ are defined as follows:

$$Res_{1\sigma}(t) = \left({}^{C}D_{0^+}^{\gamma} \varphi_{1\sigma}\right)(t) - g(t)\varphi'_{1\sigma}(t) - f_{1\sigma}(t, \varphi_{1\sigma}(t), \varphi_{2\sigma}(t)),$$
$$Res_{2\sigma}(t) = \left({}^{C}D_{0^+}^{\gamma} \varphi_{2\sigma}\right)(t) - g(t)\varphi'_{2\sigma}(t) - f_{2\sigma}(t, \varphi_{1\sigma}(t), \varphi_{2\sigma}(t)), \quad (23)$$

and the kth-residual functions $Res_{k,1\sigma}(t)$ and $Res_{k,2\sigma}(t)$ for $k = 1,2,3,\ldots n$ are defined as follows:

$$Res_{k,1\sigma}(t) = \left({}^{C}D_{0^+}^{\gamma} \varphi_{k,1\sigma}\right)(t) - g(t)\varphi'_{k,1\sigma}(t) - f_{1\sigma}(t, \varphi_{k,1\sigma}(t), \varphi_{k,2\sigma}(t)),$$
$$Res_{k,2\sigma}(t) = \left({}^{C}D_{0^+}^{\gamma} \varphi_{k,2\sigma}\right)(t) - g(t)\varphi'_{k,2\sigma}(t) - f_{2\sigma}(t, \varphi_{k,1\sigma}(t), \varphi_{k,2\sigma}(t)). \quad (24)$$

From (23), we have $Res_{n\sigma}(t) = 0$ and $\lim_{k\to\infty} Res_{k,n\sigma}(t) = Res_{n\sigma} \equiv 0$ for $n = 1,2$ and each $t \ge 0$, which leads to ${}^{C}D_t^{m\gamma} Res_{n\sigma}(t) = 0$. Also, the fractional derivatives ${}^{C}D_t^{m\gamma} Res_{n\sigma}(t)$ and ${}^{C}D_t^{m\gamma} Res_{k,n\sigma}(t)$ are equivalent at $t = 0$ for each $m = 0,1,2,\ldots,k$, that is, ${}^{C}D_t^{m\gamma} Res_{n\sigma}(0) = {}^{C}D_t^{m\gamma} Res_{k,n\sigma}(0) = 0$. However, ${}^{C}D_t^{(k-1)\gamma} Res_{k,n\sigma}(0) = 0$ holds for $n = 1,2$.

Regarding employing the RPS algorithm to obtain the 1st unknown coefficients, c_1 and d_1, substitute the 1st approximations $\varphi_{1,1\sigma}(t) = \alpha_{1\sigma} + \beta_{1\sigma}t + c_1(t^{\gamma}/\Gamma(1+\gamma))$ and $\varphi_{1,2\sigma}(t) = \alpha_{2\sigma} + \beta_{2\sigma}t + d_1(t^{\gamma}/\Gamma(1+\gamma))$ into the 1st residual functions $Res_{1,1\sigma}(t)$ and $Res_{1,2\sigma}(t)$ of (24) such that

$$\begin{aligned} Res_{1,1\sigma}(t) &= \left({}^{C}D_{0^+}^{\gamma}\varphi_{1,1\sigma}\right)(t) - g(t)\varphi_{1,1\sigma}'(t) \\ &\quad - f_{1\sigma}(t,\varphi_{1,1\sigma}(t),\varphi_{1,2\sigma}(t)), \\ Res_{1,2\sigma}(t) &= \left({}^{C}D_{0^+}^{\gamma}\varphi_{1,2\sigma}\right)(t) - g(t)\varphi_{1,2\sigma}'(t) \\ &\quad - f_{2\sigma}(t,\varphi_{1,1\sigma}(t),\varphi_{1,2\sigma}(t)), \end{aligned} \tag{25}$$

and based upon the facts $Res_{1,1\sigma}(0) = Res_{1,2\sigma}(0) = 0$, we have $c_1 = g(0)\varphi_{1,1\sigma}'(0) - f_{1\sigma}(0,\alpha_{1\sigma},\alpha_{2\sigma})$ and $d_1 = g(0)\varphi_{1,2\sigma}'(0) - f_{2\sigma}(0,\alpha_{1\sigma},\alpha_{2\sigma})$. Therefore, the 1st RPS approximate solutions can be written as

$$\begin{aligned} &\varphi_{1,1\sigma}(t) \\ &= \alpha_{1\sigma} + \beta_{1\sigma}t \\ &\quad + \left(g(0)\varphi_{1,1\sigma}'(0) - f_{1\sigma}(0,\alpha_{1\sigma},\alpha_{2\sigma})\right)\frac{t^{\gamma}}{\Gamma(1+\gamma)}, \\ &\varphi_{1,2\sigma}(t) \\ &= \alpha_{2\sigma} + \beta_{2\sigma}t \\ &\quad + \left(g(0)\varphi_{1,2\sigma}'(0) - f_{2\sigma}(0,\alpha_{1\sigma},\alpha_{2\sigma})\right)\frac{t^{\gamma}}{\Gamma(1+\gamma)}. \end{aligned} \tag{26}$$

Currently, for the 2nd unknown coefficients, c_2 and d_2 substitute $\varphi_{2,1\sigma}(t) = \alpha_{1\sigma} + \beta_{1\sigma}t + \sum_{n=1}^{2} c_n(t^{n\gamma}/\Gamma(1+n\gamma))$ and $\varphi_{2,2\sigma}(t) = \alpha_{2\sigma} + \beta_{2\sigma}t + \sum_{n=1}^{2} d_n(t^{n\gamma}/\Gamma(1+n\gamma))$ into the 2nd residual functions, $Res_{2,1\sigma}(t)$ and $Res_{2,2\sigma}(t)$ of (24) such that

$$\begin{aligned} Res_{2,1\sigma}(t) &= \left({}^{C}D_{0^+}^{\gamma}\varphi_{2,1\sigma}\right)(t) - g(t)\varphi_{2,1\sigma}'(t) \\ &\quad - f_{1\sigma}(t,\varphi_{2,1\sigma}(t),\varphi_{2,2\sigma}(t)), \\ Res_{2,2\sigma}(t) &= \left({}^{C}D_{0^+}^{\gamma}\varphi_{2,2\sigma}\right)(t) - g(t)\varphi_{2,2\sigma}'(t) \\ &\quad - f_{2\sigma}(t,\varphi_{2,1\sigma}(t),\varphi_{2,2\sigma}(t)). \end{aligned} \tag{27}$$

Then, by applying the fractional derivative ${}^{C}D_t^{\gamma}$ on both sides of $Res_{2,1\sigma}(t)$ and $Res_{2,2\sigma}(t)$, using the facts ${}^{C}D_t^{\gamma}Res_{2,1\sigma}(0) = {}^{C}D_t^{\gamma}Res_{2,2\sigma}(0) = 0$ as well, the values of c_2 and d_2 will be given by

$$\begin{aligned} c_2 &= \frac{\Gamma(2\gamma)}{\Gamma(\gamma)}\left(\frac{\Gamma(\gamma)\beta_{1\sigma}\left({}^{C}D_t^{\gamma}(g(t))\right)\big|_{t=0} + c_1\left({}^{C}D_t^{\gamma}(g(t).t^{\gamma-1})\right)\big|_{t=0} + f_{1\sigma}(0,c_1,d_1)}{\Gamma(2\gamma) - \left({}^{C}D_t^{\gamma}(g(t).t^{2\gamma-1})\right)\big|_{t=0}}\right), \\ d_2 &= \frac{\Gamma(2\gamma)}{\Gamma(\gamma)}\left(\frac{\Gamma(\gamma)\beta_{2\sigma}\left({}^{C}D_t^{\gamma}(g(t))\right)\big|_{t=0} + d_1\left({}^{C}D_t^{\gamma}(g(t).t^{\gamma-1})\right)\big|_{t=0} + f_{2\sigma}(0,c_1,d_1)}{\Gamma(2\gamma) - \left({}^{C}D_t^{\gamma}(g(t).t^{2\gamma-1})\right)\big|_{t=0}}\right). \end{aligned} \tag{28}$$

For the 3rd unknown coefficients, c_3 and d_3 substitute $\varphi_{3,1\sigma}(t) = \alpha_{1\sigma} + \beta_{1\sigma}t + \sum_{n=1}^{3} c_n(t^{n\gamma}/\Gamma(1+n\gamma))$ and $\varphi_{3,2\sigma}(t) = \alpha_{2\sigma} + \beta_{2\sigma}t + \sum_{n=1}^{3} d_n(t^{n\gamma}/\Gamma(1+n\gamma))$ into the 3rd residual functions, $Res_{3,1\sigma}(t)$ and $Res_{3,2\sigma}(t)$ of (24), and then by computing ${}^{C}D_t^{2\gamma}Res_{3,1\sigma}(t)$ and ${}^{C}D_t^{2\gamma}Res_{3,2\sigma}(t)$ and using the facts ${}^{C}D_t^{2\gamma}Res_{3,1\sigma}(0) = {}^{C}D_t^{2\gamma}Res_{3,2\sigma}(0) = 0$, the coefficients, c_3 and d_3, will be given such that

$$\begin{aligned} &c_3 \\ &= \frac{\Gamma(3\gamma)}{\Gamma(\gamma)\Gamma(2\gamma)}\left(\frac{\Gamma(\gamma)\Gamma(2\gamma)\beta_{1\sigma}\left({}^{C}D_t^{2\gamma}(g(t))\right)\big|_{t=0} + c_1\left({}^{C}D_t^{2\gamma}(g(t).t^{\gamma-1})\right)\big|_{t=0} + c_2\left({}^{C}D_t^{2\gamma}(g(t).t^{2\gamma-1})\right)\big|_{t=0} + f_{1\sigma}(0,c_2,d_2)}{\Gamma(3\gamma) - \left({}^{C}D_t^{2\gamma}(g(t).t^{3\gamma-1})\right)\big|_{t=0}}\right), \\ &d_3 \\ &= \frac{\Gamma(3\gamma)}{\Gamma(\gamma)\Gamma(2\gamma)}\left(\frac{\Gamma(\gamma)\Gamma(2\gamma)\beta_{2\sigma}\left({}^{C}D_t^{2\gamma}(g(t))\right)\big|_{t=0} + d_1\left({}^{C}D_t^{2\gamma}(g(t).t^{\gamma-1})\right)\big|_{t=0} + d_2\left({}^{C}D_t^{2\gamma}(g(t).t^{2\gamma-1})\right)\big|_{t=0} + f_{2\sigma}(0,c_2,d_2)}{\Gamma(3\gamma) - \left({}^{C}D_t^{2\gamma}(g(t).t^{3\gamma-1})\right)\big|_{t=0}}\right). \end{aligned} \tag{29}$$

Using similar argument, the 4th unknown coefficients, c_4 and d_4, will be given utilizing the facts ${}^{C}D_t^{3\gamma}Res_{4,1\sigma}(0) = {}^{C}D_t^{3\gamma}Res_{3,1\sigma}(0) = 0$. The same manner can be repeated until we obtain on the coefficients' arbitrary order of the FPS solution for the OFDE (13).

5. Numerical Simulation and Discussion

This section aims to verify the efficiency and applicability of the proposed algorithm by applying the RPS method to a numerical example. Here, all necessary calculations and analysis are done using Mathematica 10.

For this purpose, let us consider the fuzzy fractional differential equation

$$\left({}^{C}D_{0^+}^{\gamma}\varphi\right)(t) = \mu, \quad 0 \le t \le 1, \tag{30}$$

with the fuzzy initial conditions

$$\varphi(0) = \alpha, \qquad \varphi'(0) = \beta, \tag{31}$$

where $\gamma \in (1, 2]$ and μ, α, β are the fuzzy numbers whose σ-cut representation is $[\sigma - 1, 1 - \sigma]$.

Based on the type of differentiability, the FFIVPs (30) and (31) can be converted into one of the following systems.

Case 1. If $\varphi(t)$ is (1,1)-solution, then the corresponding (1,1)-system will be

$$\begin{aligned}\left({}^{C}D_{0^+}^{\gamma}\varphi_{1\sigma}\right)(t) &= \sigma - 1,\\ \left({}^{C}D_{0^+}^{\gamma}\varphi_{2\sigma}\right)(t) &= 1 - \sigma,\\ \varphi_{1\sigma}(0) = \varphi'_{1\sigma}(0) &= \sigma - 1,\\ \varphi_{2\sigma}(0) = \varphi'_{2\sigma}(0) &= 1 - \sigma.\end{aligned} \tag{32}$$

If $\gamma = 2$, then the exact solution of (32) is $[\varphi(t)]^{\sigma} = [\sigma - 1, 1 - \sigma](1 + t + t^{\gamma}/\Gamma(\gamma + 1))$, $t \in [0, 1]$. In finding the fuzzy (1,1)-solution of FFDEs (30), let $\varphi(t)$ be Caputo $[(1,1)\text{-}\gamma]$-differentiable. Sequentially, after selecting the initial guesses as $\varphi_{0,1\sigma}(t) = (\sigma - 1) + (\sigma - 1)t$ and $\varphi_{0,2\sigma}(t) = (1 - \sigma) + (1 - \sigma)t$, the FPS expansion of solutions for OFDEs (32) can be represented as follows:

$$\begin{aligned}\varphi_{1\sigma}(t) &= (\sigma - 1) + (\sigma - 1)t + \sum_{n=1}^{\infty} c_n \frac{t^{n\gamma}}{\Gamma(1 + n\gamma)},\\ \varphi_{2\sigma}(t) &= (1 - \sigma) + (1 - \sigma)t + \sum_{n=1}^{\infty} d_n \frac{t^{n\gamma}}{\Gamma(1 + n\gamma)}.\end{aligned} \tag{33}$$

To determine the 1st RPS approximate solution for OFDEs (32), substitute the 1st-truncated series $\varphi_{1,1\sigma}(t) = (\sigma - 1) + (\sigma - 1)t + c_1(t^{\gamma}/\Gamma(1 + \gamma))$ and $\varphi_{1,2\sigma}(t) = (1 - \sigma) + (1 - \sigma)t + d_1(t^{\gamma}/\Gamma(1 + \gamma))$ into the 1st-residual functions $Res_{1,1\sigma}(t)$ and $Res_{1,2\sigma}(t)$ such that $Res_{1,1\sigma}(t) = 1 - \sigma + c_1$ and $Res_{1,2\sigma}(t) = \sigma - 1 + d_1$. Thus, based upon the facts $Res_{1,1\sigma}(0) = 0$ and $Res_{1,2\sigma}(0) = 0$, we have $c_1 = \sigma - 1$ and $d_1 = 1 - \sigma$. Hence, the 1st RPS approximate solution for OFDEs (32) can be written in the form of

$$\begin{aligned}\varphi_{1,1\sigma}(t) &= (\sigma - 1) + (\sigma - 1)t + (\sigma - 1)\frac{t^{\gamma}}{\Gamma(1 + \gamma)},\\ \varphi_{1,2\sigma}(t) &= (1 - \sigma) + (1 - \sigma)t + (1 - \sigma)\frac{t^{\gamma}}{\Gamma(1 + \gamma)}.\end{aligned} \tag{34}$$

Similarly, to find out the 2nd RPS approximate solution for OFDEs (32), substitute the 2nd truncated series $\varphi_{2,1\sigma}(t) = (\sigma - 1) + (\sigma - 1)t + c_1(t^{\gamma}/\Gamma(1 + \gamma)) + c_2(t^{2\gamma}/\Gamma(1 + 2\gamma))$ and $\varphi_{2,2\sigma}(t) = (1 - \sigma) + (1 - \sigma)t + d_1(t^{\gamma}/\Gamma(1 + \gamma)) + d_2(t^{2\gamma}/\Gamma(1 + 2\gamma))$ into the 2nd residual functions $Res_{2,1\sigma}(t)$ and $Res_{2,2\sigma}(t)$ such that $Res_{2,1\sigma}(t) = ({}^{C}D_{0^+}^{\gamma}\varphi_{2,1\sigma})(t) - (\sigma - 1) = 1 - \sigma + c_1 + c_2(t^{\gamma}/\Gamma(1 + \gamma))$ and $Res_{2,2\sigma}(t) = ({}^{C}D_{0^+}^{\gamma}\varphi_{2,2\sigma})(t) - (1 - \sigma) = -1 + \sigma + d_1 + d_2(t^{\gamma}/\Gamma(1 + \gamma))$. Now, applying the fractional derivative ${}^{C}D_t^{\gamma}$ on both sides of $Res_{2,1\sigma}(t)$ and $Res_{2,2\sigma}(t)$ yields the following: ${}^{C}D_t^{\gamma}Res_{2,1\sigma}(t) = c_2$ and ${}^{C}D_t^{\gamma}Res_{2,2\sigma}(t) = d_2$. So, the 2nd unknown coefficients are $c_2 = 0$ and $d_2 = 0$ through using the facts ${}^{C}D_t^{\gamma}Res_{2,1\sigma}(0) = {}^{C}D_t^{\gamma}Res_{2,2\sigma}(0) = 0$. Therefore, the 2nd RPS approximate solution for OFDEs (32) is given by

$$\begin{aligned}\varphi_{2,1\sigma}(t) &= (\sigma - 1) + (\sigma - 1)t + (\sigma - 1)\frac{t^{\gamma}}{\Gamma(1 + \gamma)},\\ \varphi_{2,2\sigma}(t) &= (1 - \sigma) + (1 - \sigma)t + (1 - \sigma)\frac{t^{\gamma}}{\Gamma(1 + \gamma)}.\end{aligned} \tag{35}$$

Accordingly, the unknown coefficients c_n and d_n will be vanished for $n \ge 3$ by continuing in the similar approach, that is, $\sum_{n=3}^{\infty} c_n(t^{n\gamma}/\Gamma(1 + n\gamma)) = 0$ and $\sum_{n=3}^{\infty} d_n(t^{n\gamma}/\Gamma(1 + n\gamma)) = 0$. Hence, the RPS approximate solutions corresponding to (1,1)-system are coinciding well with the exact solutions $\varphi_{1\sigma}(t) = (1 + t + (t^{\gamma}/\Gamma(1 + \gamma)))(\sigma - 1)$ and $\varphi_{2\sigma}(t) = (1 + t + t^{\gamma}/\Gamma(1 + \gamma))(1 - \sigma)$. Here, $[\varphi_{1\sigma}(t), \varphi_{2\sigma}(t)]$, $[\varphi'_{1\sigma}(t), \varphi'_{2\sigma}(t)]$, and $[\varphi''_{1\sigma}(t), \varphi''_{2\sigma}(t)]$ are valid level sets for $\sigma \in [0, 1]$ and $t \in [0, 1]$. Moreover, $\varphi(t) = \mu(1 + t + t^{\gamma}/\Gamma(1 + \gamma))$ is a (1,1)-solution for FFIVPs (30) and (31) on $[0, 1]$.

Case 2. If $\varphi(t)$ is (1,2)-solution, then the corresponding (1,2)-system will be

$$\begin{aligned}\left({}^{C}D_{0^+}^{\gamma}\varphi_{1\sigma}\right)(t) &= 1 - \sigma,\\ \left({}^{C}D_{0^+}^{\gamma}\varphi_{2\sigma}\right)(t) &= \sigma - 1,\\ \varphi_{1\sigma}(0) = \varphi'_{1\sigma}(0) &= \sigma - 1,\\ \varphi_{2\sigma}(0) = \varphi'_{2\sigma}(0) &= 1 - \sigma,\end{aligned} \tag{36}$$

If $\gamma = 2$, then the exact solution of (36) is $[\varphi(t)]^{\sigma} = [\sigma - 1, 1 - \sigma](1 + t - t^{\gamma}/\Gamma(\gamma + 1))$, $t \in [0, 1]$. In finding the fuzzy (1,2)-solution of FFDEs (30), let $\varphi(t)$ be Caputo $[(1,2)\text{-}\gamma]$-differentiable. Sequentially, after selecting the initial guesses as in case 1, the FPS expansion of solutions for OFDEs (36) can be represented by

$$\begin{aligned}\varphi_{1\sigma}(t) &= (\sigma - 1) + (\sigma - 1)t + \sum_{n=1}^{\infty} c_n \frac{t^{n\gamma}}{\Gamma(1 + n\gamma)},\\ \varphi_{2\sigma}(t) &= (1 - \sigma) + (1 - \sigma)t + \sum_{n=1}^{\infty} d_n \frac{t^{n\gamma}}{\Gamma(1 + n\gamma)}.\end{aligned} \tag{37}$$

To determine the 1st RPS approximate solution for OFDEs (36), substitute the 1st truncated series $\varphi_{1,1\sigma}(t) =$

$(\sigma - 1) + (\sigma - 1)t + c_1(t^\gamma/\Gamma(1+\gamma))$ and $\varphi_{1,2\sigma}(t) = (1 - \sigma) + (1-\sigma)t + d_1(t^\gamma/\Gamma(1+\gamma))$ into the 1st residual functions $Res_{1,1\sigma}(t)$ and $Res_{1,2\sigma}(t)$ such that $Res_{1,1\sigma}(t) = \sigma - 1 + c_1$ and $Res_{1,2\sigma}(t) = 1 - \sigma + d_1$. Thus, based upon the facts $Res_{1,1\sigma}(0) = Res_{1,2\sigma}(0) = 0$, we have $c_1 = 1-\sigma$ and $d_1 = \sigma - 1$. Hence, the 1st RPS approximate solution for OFDEs (36) can be written in the form of

$$\begin{aligned}\varphi_{1,1\sigma}(t) &= (\sigma-1) + (\sigma-1)t + (1-\sigma)\frac{t^\gamma}{\Gamma(1+\gamma)},\\ \varphi_{1,2\sigma}(t) &= (1-\sigma) + (1-\sigma)t + (\sigma-1)\frac{t^\gamma}{\Gamma(1+\gamma)}.\end{aligned} \tag{38}$$

Similarly, to find out the 2nd RPS approximate solution for OFDEs (36), substitute the 2nd truncated series $\varphi_{2,1\sigma}(t) = (\sigma - 1) + (\sigma - 1)t + c_1(t^\gamma/\Gamma(1+\gamma)) + c_2(t^{2\gamma}/\Gamma(1+2\gamma))$ and $\varphi_{2,2\sigma}(t) = (1-\sigma)+(1-\sigma)t+d_1(t^\gamma/\Gamma(1+\gamma))+d_2(t^{2\gamma}/\Gamma(1+2\gamma))$ into the 2nd residual functions $Res_{2,1\sigma}(t)$ and $Res_{2,2\sigma}(t)$ such that $Res_{2,1\sigma}(t) = 1 - \sigma + c_1 + c_2(t^\gamma/\Gamma(1+\gamma))$ and $Res_{2,2\sigma}(t) = -1 + \sigma + d_1 + d_2(t^\gamma/\Gamma(1+\gamma))$. Then, applying the fractional derivative ${}^CD_t^\gamma$ on both sides of $Res_{2,1\sigma}(t)$ and $Res_{2,2\sigma}(t)$yields the following: ${}^CD_t^\gamma Res_{2,1\sigma}(t)=c_2$ and ${}^CD_t^\gamma Res_{2,2\sigma}(t)=d_2$. So, the 2nd unknown coefficients are $c_2 = 0$ and $d_2 = 0$ through using the facts ${}^CD_t^\gamma Res_{2,1\sigma}(0) = {}^CD_t^\gamma Res_{2,2\sigma}(0) = 0$. Therefore, the 2nd RPS approximate solution for OFDEs (36) is given by

$$\begin{aligned}\varphi_{2,1\sigma}(t) &= (\sigma-1) + (\sigma-1)t + (1-\sigma)\frac{t^\gamma}{\Gamma(1+\gamma)},\\ \varphi_{2,2\sigma}(t) &= (1-\sigma) + (1-\sigma)t + (\sigma-1)\frac{t^\gamma}{\Gamma(1+\gamma)}.\end{aligned} \tag{39}$$

By continuing in the similar manner, the unknown coefficients c_n and d_n will be vanished for $n \geq 3$, that is, $\sum_{n=3}^{\infty} c_n(t^{n\gamma}/\Gamma(1+n\gamma)) = 0$ and $\sum_{n=3}^{\infty} d_n(t^{n\gamma}/\Gamma(1+n\gamma)) = 0$. Hence, the RPS approximate solutions corresponding to (1,2)-system are coinciding well with the exact solutions $\varphi_{1\sigma}(t) = (1+t-t^\gamma/\Gamma(1+\gamma))(\sigma-1)$ and $\varphi_{2\sigma}(t) = (1+t-t^\gamma/\Gamma(1+\gamma))(1-\sigma)$. Here, $[\varphi_{1\sigma}(t), \varphi_{2\sigma}(t)]$, $[\varphi'_{1\sigma}(t), \varphi'_{2\sigma}(t)]$, and $[\varphi''_{1\sigma}(t), \varphi''_{2\sigma}(t)]$ are valid level sets for $\sigma \in [0,1]$ and $t \in [0,1]$. On the other hand, $\varphi(t) = \mu(1+t-t^\gamma/\Gamma(1+\gamma))$ is a (1,2)-solution for FFIVPs (30) and (31) on $[0,1]$.

Case 3. If $\varphi(t)$ is (2,1)-solution, then the corresponding (2,1)-system will be

$$\begin{aligned}\left({}^CD_{0^+}^\gamma \varphi_{1\sigma}\right)(t) &= 1-\sigma,\\ \left({}^CD_{0^+}^\gamma \varphi_{2\sigma}\right)(t) &= \sigma-1,\\ \varphi_{1\sigma}(0) = \varphi'_{2\sigma}(0) &= \sigma-1,\\ \varphi_{2\sigma}(0) = \varphi'_{1\sigma}(0) &= 1-\sigma,\end{aligned} \tag{40}$$

If $\gamma = 2$, then the exact solution of (40) is $[\varphi(t)]^\sigma = [\sigma - 1, 1-\sigma](1 - t - t^\gamma/\Gamma(1+\gamma))$, $t \in (0, \sqrt{3}-1)$. To obtain the fuzzy (2,1)-solution of FFDEs (30), let $\varphi(t)$ is Caputo $[(2,1)\text{-}\gamma]$-differentiable. By using the same manner in previous cases, the solutions for (2,1)-system can be obtained such as $\varphi_{1\sigma}(t) = (1-t-t^\gamma/\Gamma(1+\gamma))(\sigma-1)$ and $\varphi_{2\sigma}(t) = (1-t-t^\gamma/\Gamma(1+\gamma))(1-\sigma)$. It is easy to check that $[\varphi_{1\sigma}(t), \varphi_{2\sigma}(t)]$, $[\varphi'_{2\sigma}(t), \varphi'_{1\sigma}(t)]$ and $[\varphi''_{2\sigma}(t), \varphi''_{1\sigma}(t)]$ are also valid level sets for $\sigma \in [0,1]$ and $t \in [0, \sqrt{3}-1]$. Thus, $\varphi(t) = \mu(1-t-t^\gamma/\Gamma(1+\gamma))$ is a (2,1)-solution for FFIVPs (30) and (31) on $(0, \sqrt{3}-1]$.

Case 4. If $\varphi(t)$ is (2,2)-solution, then the corresponding (2,2)-system will be

$$\begin{aligned}\left({}^CD_{0^+}^\gamma \varphi_{1\sigma}\right)(t) &= \sigma-1,\\ \left({}^CD_{0^+}^\gamma \varphi_{2\sigma}\right)(t) &= 1-\sigma,\\ \varphi_{1\sigma}(0) = \varphi'_{2\sigma}(0) &= \sigma-1,\\ \varphi_{2\sigma}(0) = \varphi'_{1\sigma}(0) &= 1-\sigma,\end{aligned} \tag{41}$$

If $\gamma = 2$, then the exact solution of OFDEs (41) is $[\varphi(t)]^\sigma = [\sigma - 1, 1-\sigma](1 - t + t^\gamma/\Gamma(\gamma+1))$, $t \in [0,1]$. Finally, to determine the fuzzy (2,2)-solution of FFDEs (30), let $\varphi(t)$ be Caputo $[(2,2)\text{-}\gamma]$-differentiable. By using the same manner in previous cases, the solutions for (2,2)-system can be obtained such as $\varphi_{1\sigma}(t) = (\sigma-1)(1-t+t^\gamma/\Gamma(1+\gamma))$ and $\varphi_{2\sigma}(t) = (1-\sigma)(1-t+t^\gamma/\Gamma(1+\gamma))$. Here, $[\varphi_{1\sigma}(t), \varphi_{2\sigma}(t)]$, $[\varphi'_{2\sigma}(t), \varphi'_{1\sigma}(t)]$ and $[\varphi''_{1\sigma}(t), \varphi''_{2\sigma}(t)]$ are also valid level sets for $\sigma \in [0,1]$ and $t \in [0,1]$. However, $\varphi(t) = \mu(1-t+t^\gamma/\Gamma(\gamma+1))$ defines as a (2,2)-solution for FFIVPs (30) and (31) on $(0,1]$.

To demonstrate the agreement between the exact and approximate solution, Table 1 shows the absolute error of the 10th PRS approximate solution for FFIVPs (30) and (31) obtained for different values of σ-cut representations and nodes with fractional order $\gamma = 1.9$. Some graphical results are also presented in Figures 1 and 2. The numerical results obtained indicate that the RPS approximate solutions are in good agreement with each other and with the exact solutions for all cases of differentiability.

6. Conclusion

In this paper, the RPS algorithm is successfully developed, investigated, and applied to solve the fuzzy differential equation of fractional order $1 < \gamma \leq 2$ with fuzzy initial constraints under the fuzzy concept of Caputo H-differentiability. The fuzziness is represented using upper semicontinuous membership function of bounded support, convex, and normalized fuzzy numbers based on its single parametric form. The behavior of approximate solution for different values of fractional order γ is discussed quantitatively as well as graphically. The numerical results in this paper demonstrate the efficiency of the algorithm. We conclude that the proposed scheme is highly accurate in solving widely array of fuzzy fractional issues.

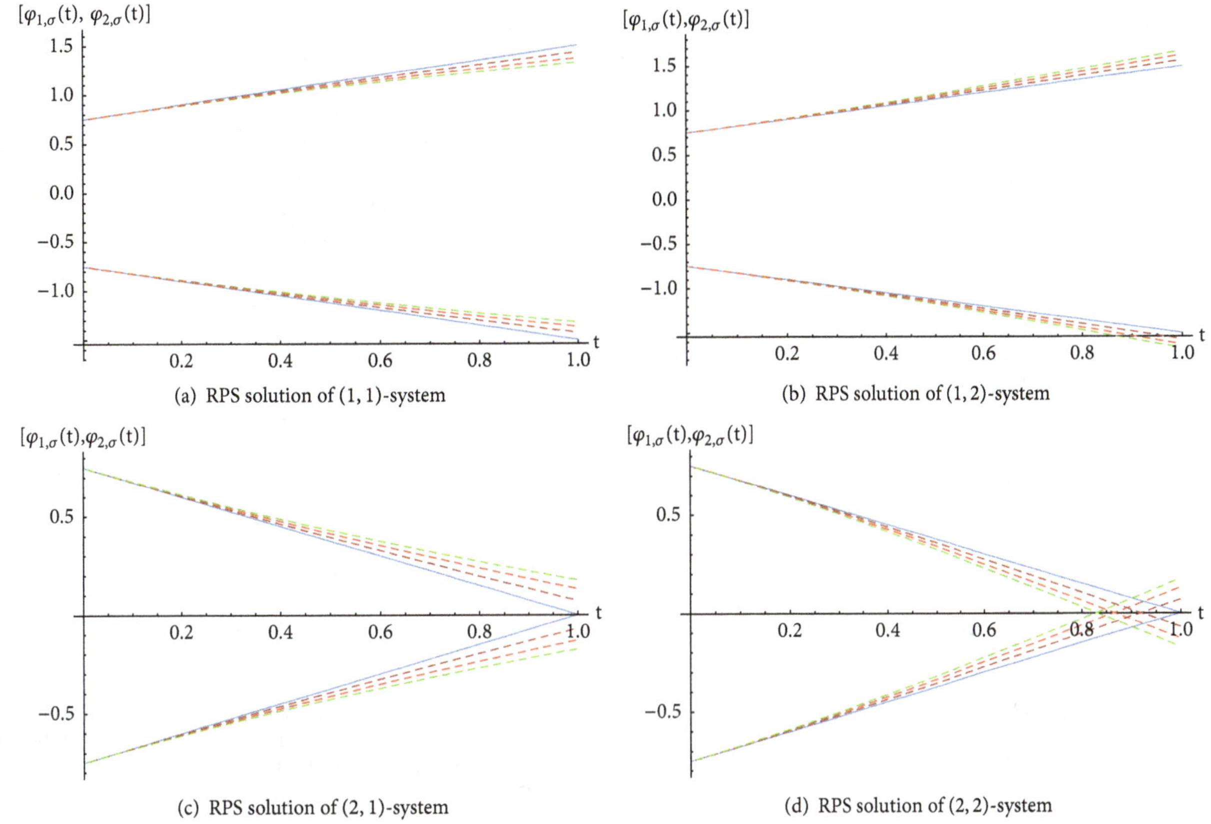

FIGURE 1: Plots of α-cut representations of $\varphi_{k,1\sigma}(t), \varphi_{k,2\sigma}(t)$ with $k = 10$, $\sigma = 0.25$, and different values of $\gamma \in \{2, 1.9, 1.8, 1.7\}$ (--- Exact, ··· RPS-approximation).

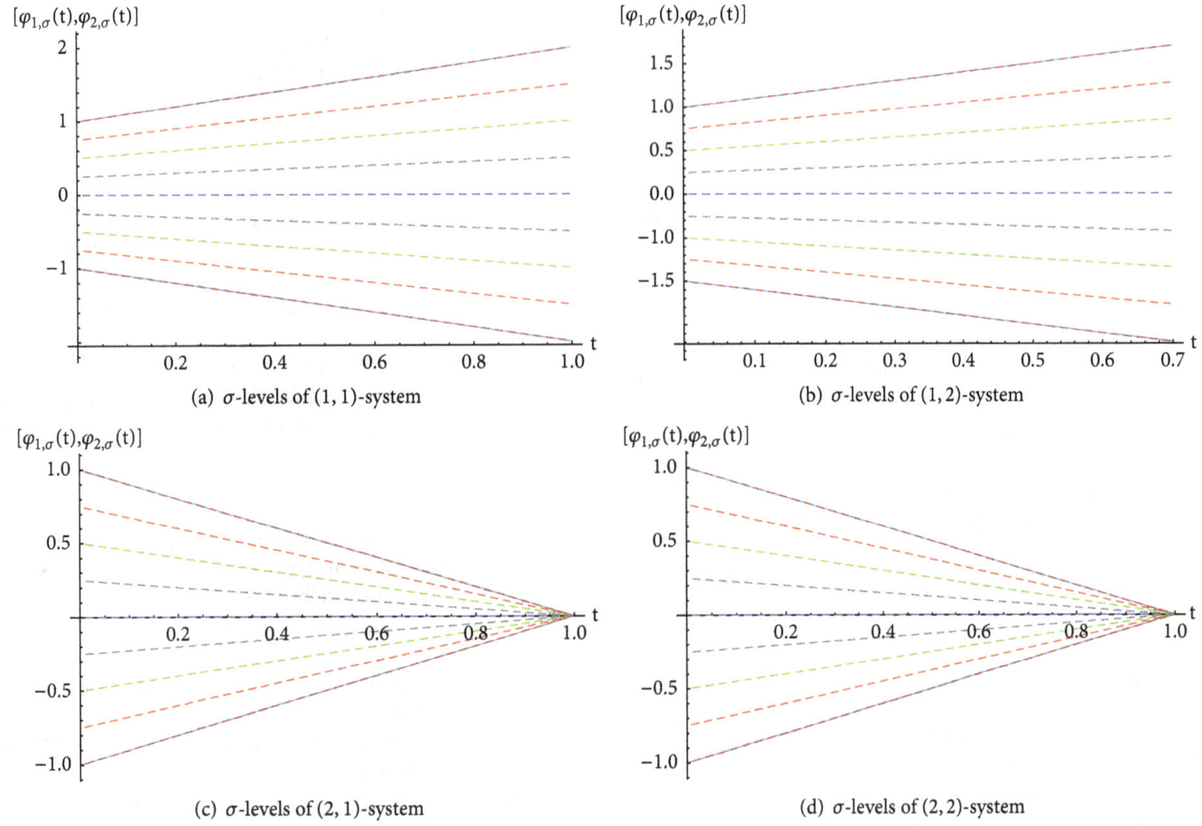

FIGURE 2: Plots of exact and RPS-approximation at $\gamma = 2$ with different values of σ-levels, $\sigma \in \{0, 0.25, 0.5, 0.75, 1\}$ (--- Exact, ··· RPS-approximation).

TABLE 1: The absolute error of 10th approximation of FFIVPs (30) and (31).

(n, m)-solution	t	$\sigma = 0$	$\sigma = 0.25$	$\sigma = 0.5$	$\sigma = 0.75$
$(1, 1)$-system	0.2	2.0578567×10^{-4}	1.5433927×10^{-4}	1.0289283×10^{-4}	5.1446419×10^{-5}
	0.4	3.8465419×10^{-4}	2.8849064×10^{-4}	1.9232709×10^{-4}	9.6163547×10^{-5}
	0.6	5.5428820×10^{-4}	4.1571615×10^{-4}	2.7714410×10^{-4}	1.3857205×10^{-4}
	0.8	7.1821803×10^{-4}	5.3866352×10^{-4}	3.5910901×10^{-4}	1.7955450×10^{-4}
$(1, 2)$ -system	0.2	2.9676224×10^{-4}	2.2257168×10^{-4}	1.4838112×10^{-4}	5.1446419×10^{-5}
	0.4	4.7033136×10^{-4}	3.5274852×10^{-4}	2.3516568×10^{-4}	9.6163547×10^{-5}
	0.6	6.3684312×10^{-4}	4.7763234×10^{-4}	2.7714410×10^{-4}	1.3857205×10^{-4}
	0.8	7.9857642×10^{-4}	5.3866352×10^{-4}	3.5910901×10^{-4}	1.7955450×10^{-4}
$(2, 1)$ -system	0.2	2.0578567×10^{-4}	1.5433925×10^{-4}	1.0289283×10^{-4}	5.1446419×10^{-5}
	0.4	3.8465419×10^{-4}	2.8849064×10^{-4}	1.9232709×10^{-4}	9.6163547×10^{-5}
	0.6	5.5428820×10^{-4}	4.1571615×10^{-4}	2.7714410×10^{-4}	1.3857205×10^{-4}
	0.8	6.3684312×10^{-4}	5.3746353×10^{-4}	3.5910901×10^{-4}	1.7955450×10^{-4}
$(2, 2)$ -system	0.2	2.0578567×10^{-4}	1.5433925×10^{-4}	1.0289283×10^{-4}	5.1446419×10^{-5}
	0.4	3.8465419×10^{-4}	2.8849064×10^{-4}	1.9232709×10^{-4}	9.6163547×10^{-5}
	0.6	5.5428820×10^{-4}	4.1571615×10^{-4}	2.7714410×10^{-4}	1.3857205×10^{-4}
	0.8	7.1821803×10^{-4}	5.3866352×10^{-4}	3.5910901×10^{-4}	1.7955450×10^{-4}

Conflicts of Interest

The authors declare that they have no conflicts of interest.

Acknowledgments

This research was financially supported by the UKM (Grant no. GP-K007788 and GP-K006926).

References

[1] A. A. Kilbas, H. M. Srivastava, and J. J. Trujillo, *Theory and Applications of Fractional Differential Equations*, Elsevier Science B.V., Amsterdam, 2006.

[2] I. Podlubny, *Fractional Differential Equations*, vol. 198 of *Mathematics in Science and Engineering*, Academic Press, San Diego, Calif, USA, 1999.

[3] S. Momani, O. Abu Arqub, A. Freihat, and M. Al-Smadi, "Analytical approximations for Fokker-Planck equations of fractional order in multistep schemes," *Applied and Computational Mathematics*, vol. 15, no. 3, pp. 319–330, 2016.

[4] H. Khalil, R. A. Khan, M. H. Al-Smadi, and A. A. Freihat, "Approximation of solution of time fractional order three-dimensional heat conduction problems with Jacobi polynomials," *The Punjab University. Journal of Mathematics*, vol. 47, no. 1, pp. 35–56, 2015.

[5] S. Arshad and V. Lupulescu, "Fractional differential equation with the fuzzy initial condition," *Electronic Journal of Differential Equations*, vol. 34, pp. 1–8, 2011.

[6] M. Al-Smadi, A. Freihat, H. Khalil, S. Momani, and R. A. Khan, "Numerical multistep approach for solving fractional partial differential equations," *International Journal of Computational Methods*, vol. 14, no. 3, Article ID 1750029, pp. 1–15, 2017.

[7] V. Lakshmikantham and R. N. Mohapatra, *Theory of Fuzzy Differential Equations and Inclusions*, vol. 6, Taylor & Francis, Ltd, London, UK, 2003.

[8] M. AL-Smadi, A. Freihat, M. A. Hammad, S. Momani, and O. A. Arqub, "Analytical approximations of partial differential equations of fractional order with multistep approach," *Journal of Computational and Theoretical Nanoscience*, vol. 13, no. 11, pp. 7793–7801, 2016.

[9] G. Gumah, K. Moaddy, M. Al-Smadi, and I. Hashim, "Solutions to uncertain Volterra integral equations by fitted reproducing kernel Hilbert space method," *Journal of Function Spaces*, Art. ID 2920463, 11 pages, 2016.

[10] O. Abu Arqub and M. Al-Smadi, "Numerical algorithm for solving two-point, second-order periodic boundary value problems for mixed integro-differential equations," *Applied Mathematics and Computation*, vol. 243, pp. 911–922, 2014.

[11] M. Al-Smadi, O. Abu Arqub, and S. Momani, "A computational method for two-point boundary value problems of fourth-order mixed integrodifferential equations," *Mathematical Problems in Engineering*, vol. 2013, Article ID 832074, pp. 1–10, 2013.

[12] K. Moaddy, A. Freihat, M. Al-Smadi, E. Abuteen, and I. Hashim, "Numerical investigation for handling fractional-order Rabinovich–Fabrikant model using the multistep approach," *Soft Computing*, vol. 22, no. 3, pp. 773–782, 2018.

[13] M. Al-Smadi, O. A. Arqub, N. Shawagfeh, and S. Momani, "Numerical investigations for systems of second-order periodic boundary value problems using reproducing kernel method," *Applied Mathematics and Computation*, vol. 291, pp. 137–148, 2016.

[14] O. Abu Arqub and M. Al-Smadi, "Numerical algorithm for solving time-fractional partial integrodifferential equations subject to initial and Dirichlet boundary conditions," *Numerical Methods for Partial Differential Equations*, pp. 1–21, 2017.

[15] H. Khalil, R. A. Khan, M. A. Smadi, and A. Freihat, "A generalized algorithm based on Legendre polynomials for numerical solutions of coupled system of fractional order differential equations," *Journal of Fractional Calculus and Applications*, vol. 6, no. 2, pp. 123–143, 2015.

[16] H. Khalil, R. A. Khan, M. H. Al-Smadi, A. A. Freihat, and N. Shawagfeh, "New operational matrix of shifted Legendre polynomials and fractional differential equations with variable

coefficients," *The Punjab University. Journal of Mathematics*, vol. 47, no. 1, pp. 81–103, 2015.

[17] H. Khalil, M. Al-Smadi, K. Moaddy, R. A. Khan, and I. Hashim, "Toward the approximate solution for fractional order nonlinear mixed derivative and nonlocal boundary value problems," *Discrete Dynamics in Nature and Society*, Article ID 5601821, pp. 1–12, 2016.

[18] O. Abu Arqub, "Series solution of fuzzy differential equations under strongly generalized differentiability," *Journal of Advanced Research in Applied Mathematics*, vol. 5, no. 1, pp. 31–52, 2013.

[19] I. Komashynska, M. Al-Smadi, A. Ateiwi, and S. Al-Obaidy, "Approximate analytical solution by residual power series method for system of Fredholm integral equations," *Applied Mathematics & Information Sciences*, vol. 10, no. 3, pp. 975–985, 2016.

[20] I. Komashynska, M. Al-Smadi, O. A. Arqub, and S. Momani, "An efficient analytical method for solving singular initial value problems of nonlinear systems," *Applied Mathematics & Information Sciences*, vol. 10, no. 2, pp. 647–656, 2016.

[21] O. Abu Arqub, M. Al-Smadi, S. Momani, and T. Hayat, "Application of reproducing kernel algorithm for solving second-order, two-point fuzzy boundary value problems," *Soft Computing*, vol. 21, no. 23, pp. 7191–7206, 2017.

[22] B. S. Keerthi and B. Raja, "Coefficient inequality for certain new subclasses of analytic bi-univalent functions," *Theoretical Mathematics & Applications*, vol. 3, no. 1, pp. 1–10, 2013.

[23] K. Moaddy, M. AL-Smadi, and I. Hashim, "A novel representation of the exact solution for differential algebraic equations system using residual power-series method," *Discrete Dynamics in Nature and Society*, Article ID 205207, pp. 1–12, 2015.

[24] A. El-Ajou, O. Abu Arqub, and M. Al-Smadi, "A general form of the generalized Taylor's formula with some applications," *Applied Mathematics and Computation*, vol. 256, pp. 851–859, 2015.

[25] A. El-Ajou, O. Abu Arqub, and S. Momani, "Approximate analytical solution of the nonlinear fractional KdV-Burgers equation: a new iterative algorithm," *Journal of Computational Physics*, vol. 293, pp. 81–95, 2015.

[26] R. Abu-Gdairi, M. Al-Smadi, and G. Gumah, "An expansion iterative technique for handling fractional differential equations using fractional power series scheme," *Journal of Mathematics and Statistics*, vol. 11, no. 2, pp. 29–38, 2015.

[27] R. Saadeh, M. Al-Smadi, G. Gumah, H. Khalil, and R. A. Khan, "Numerical investigation for solving two-point fuzzy boundary value problems by reproducing kernel approach," *Applied Mathematics & Information Sciences*, vol. 10, no. 6, pp. 1–13, 2016.

[28] O. Kaleva, "Fuzzy differential equations," *Fuzzy Sets and Systems*, vol. 24, no. 3, pp. 301–317, 1987.

[29] J. Goetschel and W. Voxman, "Elementary fuzzy calculus," *Fuzzy Sets and Systems*, vol. 18, no. 1, pp. 31–43, 1986.

[30] M. L. Puri and D. A. Ralescu, "Fuzzy random variables," *Journal of Mathematical Analysis and Applications*, vol. 114, no. 2, pp. 409–422, 1986.

[31] H. T. Nguyen, "A note on the extension principle for fuzzy sets," *Journal of Mathematical Analysis and Applications*, vol. 64, no. 2, pp. 369–380, 1978.

[32] B. Bede and S. G. Gal, "Generalizations of the differentiability of fuzzy-number-valued functions with applications to fuzzy differential equations," *Fuzzy Sets and Systems*, vol. 151, no. 3, pp. 581–599, 2005.

[33] A. Khastan, F. Bahrami, and K. Ivaz, "New Results on multiple solutions for Nth-order fuzzy differential equations under generalized differentiability, Boundary Value Problems," *Boundary Value Problems*, vol. 2009, Article ID 395714, 2009.

[34] S. Salahshour, T. Allahviranloo, S. Abbasbandy, and D. Baleanu, "Existence and uniqueness results for fractional differential equations with uncertainty," *Advances in Difference Equations*, vol. 112, pp. 1–12, 2012.

[35] A. El-Ajou, O. Abu Arqub, Z. Al Zhour, and S. Momani, "New results on fractional power series: theories and applications," *Entropy. An International and Interdisciplinary Journal of Entropy and Information Studies*, vol. 15, no. 12, pp. 5305–5323, 2013.

Permissions

 Every chapter published in this book has been scrutinized by our experts. Their significance has been extensively debated. The topics covered herein carry significant findings which will fuel the growth of the discipline. They may even be implemented as practical applications or may be referred to as a beginning point for another development.

The contributors of this book come from diverse backgrounds, making this book a truly international effort. This book will bring forth new frontiers with its revolutionizing research information and detailed analysis of the nascent developments around the world.

We would like to thank all the contributing authors for lending their expertise to make the book truly unique. They have played a crucial role in the development of this book. Without their invaluable contributions this book wouldn't have been possible. They have made vital efforts to compile up to date information on the varied aspects of this subject to make this book a valuable addition to the collection of many professionals and students.

This book was conceptualized with the vision of imparting up-to-date information and advanced data in this field. To ensure the same, a matchless editorial board was set up. Every individual on the board went through rigorous rounds of assessment to prove their worth. After which they invested a large part of their time researching and compiling the most relevant data for our readers.

The editorial board has been involved in producing this book since its inception. They have spent rigorous hours researching and exploring the diverse topics which have resulted in the successful publishing of this book. They have passed on their knowledge of decades through this book. To expedite this challenging task, the publisher supported the team at every step. A small team of assistant editors was also appointed to further simplify the editing procedure and attain best results for the readers.

Apart from the editorial board, the designing team has also invested a significant amount of their time in understanding the subject and creating the most relevant covers. They scrutinized every image to scout for the most suitable representation of the subject and create an appropriate cover for the book.

The publishing team has been an ardent support to the editorial, designing and production team. Their endless efforts to recruit the best for this project, has resulted in the accomplishment of this book. They are a veteran in the field of academics and their pool of knowledge is as vast as their experience in printing. Their expertise and guidance has proved useful at every step. Their uncompromising quality standards have made this book an exceptional effort. Their encouragement from time to time has been an inspiration for everyone.

The publisher and the editorial board hope that this book will prove to be a valuable piece of knowledge for researchers, students, practitioners and scholars across the globe.

List of Contributors

Adnane Boukhouima and Noura Yousfi
Laboratory of Analysis, Modeling and Simulation (LAMS), Faculty of Sciences Ben M'sik, Hassan II University, Sidi Othman, Casablanca, Morocco

Khalid Hattaf
Laboratory of Analysis, Modeling and Simulation (LAMS), Faculty of Sciences Ben M'sik, Hassan II University, Sidi Othman, Casablanca, Morocco
Centre Régional des Métiers de l'Education et de la Formation (CRMEF), 20340 Derb Ghalef, Casablanca, Morocco

Süleyman Cengizci
Institute of Applied Mathematics, Middle East Technical University, 06800 Ankara, Turkey

T. A. Biala
Department of Mathematics and Computer Science, Sule Lamido University, PMB 048, Kafin Hausa, Nigeria

S. N. Jator
Department of Mathematics and Statistics, Austin Peay State University, Clarksville, TN 37044, USA

Abdeluaab Lidouh and Rachid Messaoudi
Department of Mathematics and Computer Science, Laboratory LACSA, Faculty of Sciences, Mohammed 1*st* University, BV Mohammed VI, Oujda, Morocco

Adil Misir and Banu Mermerkaya
Department of Mathematics, Faculty of Science, Gazi University, Teknikokullar, 06500 Ankara, Turkey

Joseph Páez Chávez
Center for Applied Dynamical Systems and Computational Methods (CADSCOM), Faculty of Natural Sciences and Mathematics, Escuela Superior Politécnica del Litoral, Guayaquil, Ecuador
Center for Dynamics, Department of Mathematics, TU Dresden, 01062 Dresden, Germany

Stefan Siegmund
Center for Dynamics, Department of Mathematics, TU Dresden, 01062 Dresden, Germany

Dirk Jungmann
Institute of Hydrobiology, Faculty of Environmental Sciences, TU Dresden, 01062 Dresden, Germany

Anas Arafa
Department of Mathematics and Computer Science, Faculty of Science, Port Said University, Port Said, Egypt

Ghada Elmahdy
Department of Basic science, Canal High Institute of Engineering and Technology, Suez, Egypt

Anael Verdugo
Department of Mathematics, California State University, Fullerton, CA, USA
Center for Computational and Applied Mathematics, California State University, Fullerton, CA, USA

Taras Lukashiv and Igor Malyk
Department of the System Analysis and Insurance and Financial Mathematics, Yuriy Fedkovych Chernivtsi National University, 28 Unversitetska St., Chernivtsi 58012, Ukraine

Haixia Wang and Lingdi Zhao
School of Economics, Ocean University of China, Qingdao, Shandong 266100, China

Mingzhao Hu
University of California Santa Barbara, Santa Barbara, CA 93106-3110, USA

Haide Gou and Baolin Li
College of Mathematics and Statistics, Northwest Normal University, Lanzhou 730070, China

Hongyu Qin
Wenhua College, Wuhan 430074, China

Zhiyong Wang
School of Mathematical Sciences, University of Electronic Science and Technology of China, Sichuan 611731, China

Fumin Zhu
College of Economics, Shenzhen University, Shenzhen 518060, China

Jinming Wen
Department of Electrical and Computer Engineering, University of Toronto, Toronto, Canada M5S3G4

T. Suebcharoen
Center of Excellence in Mathematics and Applied Mathematics, Department of Mathematics, Faculty of Science, Chiang Mai University, Chiang Mai 50200, Thailand

Junfei Zhang
School of Statistics and Mathematics, Central University of Finance and Economics, Beijing 100081, China

U. Al Khawaja
Physics Department, United Arab Emirates University, Al Ain, UAE

Qasem M. Al-Mdallal
Department of Mathematical Sciences, United Arab Emirates University, Al Ain, UAE

Abdulnasir Isah
Department of Mathematics, Ahmadu Bello University, Zaria, Nigeria
Department of Mathematics and Statistics, Faculty of Science, Technology and Human Development, Universiti Tun Hussein OnnMalaysia, Batu Pahat, Malaysia

Chang Phang
Department of Mathematics and Statistics, Faculty of Science, Technology and Human Development, Universiti Tun Hussein OnnMalaysia, Batu Pahat, Malaysia

Piau Phang
Faculty of Computer Science and Information Technology, Universiti Malaysia Sarawak, Sarawak, Malaysia

Junfei Cao
Department of Mathematics, Guangdong University of Education, Guangzhou 510303, China

Zaitang Huang
School of Mathematical Sciences, Guangxi Teachers Education University, Nanning 530023, China

Gaston M. N'Guérékata
Department of Mathematics, Morgan State University, Baltimore, MD 21251, USA

Muhammad Asim Khan and Norhashidah Hj. Mohd Ali
School of Mathematical Sciences, Universiti Sains Malaysia,11800 Penang, Malaysia

Shafiq Ullah
Department of Mathematics, University of Peshawar, Pakistan

S. A. Iyase and O. F. Imaga
Department of Mathematics, Covenant University, PMB 1023, Ota, Ogun State, Nigeria

Mohammad Alaroud, Rokiah Rozita Ahmad and Ummul Khair Salma Din
School of Mathematical Sciences, Faculty of Science and Technology, Universiti Kebangsaan Malaysia, 43600 Bangi, Selangor, Malaysia

Mohammed Al-Smadi
Department of Applied Science, Ajloun College, Al-Balqa Applied University, Ajloun 26816, Jordan

Index

www.ingramcontent.com/pod-product-compliance
Lightning Source LLC
Chambersburg PA
CBHW080255230326
41458CB00097B/4981